Inventing the AIDS Virus

INVENTING THE AIDS VIRUS

Dr. Peter Duesberg

REGNERY PUBLISHING, INC.
Washington, D.C.

Copyright © 1996 by Peter Duesberg and Bryan J. Ellison

The original manuscript for *Inventing the AIDS Virus* was co-authored by Peter H. Duesberg and Bryan J. Ellison. Mr. Ellison did not participate in the final editing of the book or the preparation of the appendices.

All rights reserved. No part of this publication may be reproduced or transmitted in any form or by any means electronic or mechanical, including photocopy, recording, or any information storage and retrieval system now known or to be invented, without permission in writing from the publisher, except by a reviewer who wishes to quote brief passages in connection with a review written for inclusion in a magazine, newspaper, or broadcast.

Library of Congress Cataloging-in-Publication Data

Duesberg, Peter.
    Inventing the AIDS virus/ Peter H. Duesberg.
    p. cm.
    Includes bibliographical references and index.
    ISBN 0-89526-470-6
    1. AIDS (Disease)—Etiology. 2. HIV infections. 3. HIV (Viruses)
I. Title.
RC607.A26D84   1995
616.97'92071—dc20   95-25754
                                                              CIP

Published in the United States by   Distributed to the trade by
Regnery Publishing, Inc.   National Book Network
An Eagle Publishing Company   4720-A Boston Way
422 First Street, SE, Suite 300   Lanham, MD 20706
Washington, DC 20003

Printed on acid-free paper.
Manufactured in the United States of America

10 9 8 7 6 5 4 3 2 1

Text design by Dori Miller

Books are available in quantity for promotional or premium use. Write to Director of Special Sales, Regnery Publishing, Inc., 422 First Street, SE, Suite 300, Washington, DC 20003, for information on discounts and terms or call (202) 546-5005.

Contents

	Publisher's Preface	vii
	Acknowledgments	ix
	Foreword by Kary B. Mullis	xi
ONE	Losing the War on AIDS	3
TWO	The Great Bacteria Hunt	31
THREE	Virus Hunting Takes Over	61
FOUR	Virologists in the War on Cancer	89
FIVE	AIDS: The Virus Hunters Converge	131
SIX	A Fabricated Epidemic	169
SEVEN	Dissension in the Ranks	219
EIGHT	So What Is AIDS?	255
NINE	With Therapies Like This, Who Needs Disease?	299
TEN	Marching Off to War	361
ELEVEN	Proving the Drug-AIDS Hypothesis, the Solution to AIDS	409
TWELVE	The AIDS Debate Breaks the Wall of Silence	435
APPENDIX A	Foreign-Protein–Mediated Immunodeficiency in Hemophiliacs With and Without HIV	465
APPENDIX B	AIDS Acquired by Drug Consumption and Other Noncontagious Risk Factors	505

APPENDIX C	*The HIV Gap in National AIDS Statistics*	643
APPENDIX D	*"The Duesberg Phenomenon": Duesberg and Other Voices*	651
	Notes	655
	Index	705

Publisher's Preface

As one reviewer said, "At last! This is the book every AIDS-watcher has been awaiting, in which the most prominent and persistent critic of HIV as the cause of AIDS presents his case most exhaustively and popularly."

The book you are about to read has been a long time in coming. Why? It is at once enormously controversial and impeccably documented. It comes from a scientist and writer of great ability and courage. It will cause, we believe, a firestorm of yet undetermined proportions in both the scientific and lay communities. And it is, I think I am safe in saying, about the most difficult book that the Regnery Company has published in nearly 50 years in the business.

If Duesberg is right in what he says about AIDS, and we think he is, he documents one of the great science scandals of the century. AIDS is the first political disease, the disease that consumes more government research money, more press time, and indeed probably more heartache—much of it unnecessary—than any other. Duesberg tells us why.

Regnery is the third publisher to have contracted to publish *Inventing the AIDS Virus*. Addison Wesley initially announced the book in 1993. St. Martin's signed it in January 1994 and subsequently assigned its contract to us in January 1995. We announced it, initially, in the fall of 1995 and finally published it in February 1996.

Bryan Ellison, Duesberg's former research assistant and original co-author, became disenchanted with Duesberg's and his publisher's insistence on careful documentation and self-published his own version under the title *Why We Will Never Win the War on AIDS* in 1994. We sued Ellison for breach of contract and copyright violation and, after a two-week federal court jury trial, were awarded a six-figure verdict and an injunction against Ellison's edition.

Inventing the AIDS Virus has been edited by at least five editors, has been agonized over by the publishers of three major publishing firms, and concurrently praised and damned by countless critics.

We anticipate that the prepublication controversy may be just a precursor of what is to follow. In our tradition of presenting to the public provocative books, we are proud to be Peter Duesberg's publisher.

Acknowledgments

I AM GRATEFUL TO all the dissidents against the HIV-AIDS hypothesis—whether scientists, journalists, or public-spirited citizens—who have decided that the truth is more important than the comfort of compromise. Many people around the world, too numerous to mention, have kept me abreast of the latest developments in the AIDS epidemic or have helped inform the public at risk of their own careers and social status. This debate over AIDS has only flourished because of their courage and integrity.

For providing me with all manner of information for the background and content of this book, I especially thank Harvey Bialy, science editor of *Bio/Technology* in New York; Fred Cline of San Francisco; Michael Ellner of HEAL in New York; Hector Gildemeister of Meditel Productions, Ltd., in London, England; Harry Haverkos of the National Institute on Drug Abuse in Rockville, Maryland; Phillip Johnson of the University of California at Berkeley (for information as well as critical advice); Abraham Karpas of the University of Cambridge; Serge Lang of Yale University; John Lauritsen of New York; Ruhong Li of the University of California at Berkeley; Ilse Lass of Berlin; Charles Ortleb of the *New York Native*; Ingrid Radke, librarian at the University of California at Berkeley; Harry Rubin of the University of California at Berkeley; David Schryer of Hampton, Virginia; Joan Shenton of Meditel Productions, Ltd., in London, England; Richard Strohman of the University of California at Berkeley; Etsuro Totsuka of London; Michael Verney-Elliott of Meditel Productions, Ltd., in London, England; and Bernhard Witkop of the National Institutes of Health.

I am also indebted to those people who not only provided critical information, but who also consented to be interviewed for, or quoted in, this book.

This book would not have succeeded without the well-timed advice and experience of my literary agent, Linda Chester, and of Laurie Fox, who patiently worked through the minefield of publishing negotiations.

I am particularly grateful to Patrick Miller for his final editing of the book.

I thank Judith Lopez, Rosy Paterson, and Russell Schoch for their thorough review and comments, which contributed most to strengthening the final manuscript.

I thank the Council for Tobacco Research, New York; the Foundation for the Advancement in Cancer Therapy, New York; a foundation from New York that prefers to remain anonymous; and numerous private donors for support of the research that led to this book.

Finally, I extend my gratitude to my most critical opponents in the AIDS debate, who have unwittingly provided me the great volume of evidence by which I have disproved the virus-AIDS hypothesis and exposed the political maneuverings behind the war on AIDS.

Foreword

In 1988 I was working as a consultant at Specialty Labs in Santa Monica, setting up analytic routines for the Human Immunodeficiency Virus (HIV). I knew a lot about setting up analytic routines for anything with nucleic acids in it because I had invented the Polymerase Chain Reaction. That's why they had hired me.

Acquired Immune Deficiency Syndrome (AIDS), on the other hand, was something I did not know a lot about. Thus, when I found myself writing a report on our progress and goals for the project, sponsored by the National Institutes of Health, I recognized that I did not know the scientific reference to support a statement I had just written: "HIV is the probable cause of AIDS."

So I turned to the virologist at the next desk, a reliable and competent fellow, and asked him for the reference. He said I didn't need one. I disagreed. While it's true that certain scientific discoveries or techniques are so well established that their sources are no longer referenced in the contemporary literature, that didn't seem to be the case with the HIV/AIDS connection. It was totally remarkable to me that the individual who had discovered the cause of a deadly and as-yet-uncured disease would not be continually referenced in the scientific papers until that disease was cured and forgotten. But as I would soon learn, the name of that individual—who would surely be Nobel material—was on the tip of no one's tongue.

Of course, this simple reference had to be out there *somewhere*. Otherwise, tens of thousands of public servants and esteemed scientists of many callings, trying to solve the tragic deaths of a large number of homosexual and/or intravenous (IV) drug-using men between the ages of twenty-five and forty, would not have allowed their research to settle into one narrow channel of investigation. Everyone wouldn't fish in the same pond unless it was well established that all the other ponds were empty. There had to be a published paper, or perhaps several of them, which taken together indicated that HIV was the probable cause of AIDS. There just had to be.

I did computer searches, but came up with nothing. Of course, you can miss something important in computer searches by not putting in just the right key words. To be certain about a scientific issue, it's best to ask other scientists directly. That's one thing that scientific conferences in faraway places with nice beaches are *for*.

I was going to a lot of meetings and conferences as part of my job. I got in the habit of approaching anyone who gave a talk about AIDS and asking him or her what reference I should quote for that increasingly problematic statement, "HIV is the probable cause of AIDS."

After ten or fifteen meetings over a couple years, I was getting pretty upset when *no one* could cite the reference. I didn't like the ugly conclusion that was forming in my mind: The entire campaign against a disease increasingly regarded as a twentieth-century Black Plague was based on a hypothesis whose origins no one could recall. That defied both scientific and common sense.

Finally, I had an opportunity to question one of the giants in HIV and AIDS research, Dr. Luc Montagnier of the Pasteur Institute, when he gave a talk in San Diego. It would be the last time I would be able to ask my little question without showing anger, and I figured Montagnier would know the answer. So I asked him.

With a look of condescending puzzlement, Montagnier said, "Why don't you quote the report from the Centers for Disease Control?"

I replied, "It doesn't really address the issue of whether or not HIV is the probable cause of AIDS, does it?"

"No," he admitted, no doubt wondering when I would just go away. He looked for support to the little circle of people around

Foreword ▪ xiii

him, but they were all awaiting a more definitive response, like I was.

"Why don't you quote the work on SIV [Simian Immunodeficiency Virus]?" the good doctor offered.

"I read that too, Dr. Montagnier," I responded. "What happened to those monkeys didn't remind me of AIDS. Besides, that paper was just published only a couple of months ago. I'm looking for the *original* paper where somebody showed that HIV caused AIDS."

This time, Dr. Montagnier's response was to walk quickly away to greet an acquaintance across the room.

Cut to the scene inside my car just a few years ago. I was driving from Mendocino to San Diego. Like everyone else by now, I knew a lot more about AIDS than I wanted to. But I still didn't know who had determined that it was caused by HIV. Getting sleepy as I came over the San Bernardino Mountains, I switched on the radio and tuned in a guy who was talking about AIDS. His name was Peter Duesberg, and he was a prominent virologist at Berkeley. I'd heard of him, but had never read his papers or heard him speak. But I listened, now wide awake, while he explained exactly why I was having so much trouble finding the references that linked HIV to AIDS. *There weren't any.* No one had ever proved that HIV causes AIDS. When I got home, I invited Duesberg down to San Diego to present his ideas to a meeting of the American Association for Chemistry. Mostly skeptical at first, the audience stayed for the lecture, and then an hour of questions, and then stayed talking to each other until requested to clear the room. Everyone left with more questions than they had brought.

I like and respect Peter Duesberg. I don't think he knows necessarily what causes AIDS; we have disagreements about that. But we're both certain about what *doesn't* cause AIDS.

We have not been able to discover any good reasons why most of the people on earth believe that AIDS is a disease caused by a virus called *HIV*. There is simply no scientific evidence demonstrating that this is true.

We have also not been able to discover why doctors prescribe a *toxic* drug called *AZT* (Zidovudine) to people who have no other complaint than the presence of antibodies to HIV in their blood. In fact, we cannot understand why humans would take that drug for any reason.

We cannot understand how all this madness came about, and having both lived in Berkeley, we've seen some strange things indeed. We know that to err is human, but the HIV/AIDS hypothesis is one hell of a mistake.

I say this rather strongly as a warning. Duesberg has been saying it for a long time. Read this book.

Kary B. Mullis
Nobel Prize in Chemistry, 1993

Inventing the AIDS Virus

CHAPTER ONE

■

Losing the War on AIDS

BY ANY MEASURE, the war on AIDS has been a colossal failure. In the twelve years since the Human Immunodeficiency Virus (HIV) was announced to be the cause of AIDS (Acquired Immune Deficiency Syndrome), our leading scientists and policymakers cannot demonstrate that their efforts have saved a single life. This dismal picture applies as much to the United States as to Europe and Africa.

This war has been fought in the name of the virus-AIDS hypothesis, which holds that HIV, the AIDS virus, is a *new* cause of thirty old diseases, including Kaposi's sarcoma, tuberculosis, dementia, pneumonia, weight loss, diarrhea, leukemia, and twenty-three others (see chapter 6). If any of these previously known diseases now occurs in a patient who has antibodies against HIV (but rarely ever any HIV), then his or her disease is diagnosed as AIDS and is blamed on HIV. If the same disease occurs in a patient without HIV-antibodies, his or her disease is diagnosed by its old name and blamed on conventional chemical or microbial causes. The following examples illustrate this point:

   1. Kaposi's sarcoma + HIV-antibody = AIDS
      Kaposi's sarcoma − HIV-antibody = Kaposi's sarcoma

2. Tuberculosis + HIV-antibody = AIDS
   Tuberculosis − HIV-antibody = Tuberculosis

3. Dementia + HIV-antibody = AIDS
   Dementia − HIV-antibody = Dementia

No scientist or doctor has stepped forward to claim credit for discovering a vaccine to prevent AIDS nor is any vaccine expected for several more years, at a minimum. In contrast, the post–World War II polio epidemic was declared ended in little more than a decade once the vaccines of Jonas Salk and Albert Sabin became widely available. Nor have any useful drugs to treat AIDS been produced. AIDS patients can only choose Zidovudine (AZT) or, in certain cases, dideoxyinosine (ddI) or dideoxycytidine (ddC). All these drugs were originally developed for chemotherapy to kill human cancer cells, and they bring with them all the usual effects: hair loss, muscle degeneration, anemia, nausea, and vomiting—a severe price for questionable benefits. Indeed, these drugs appear to cause AIDS-like symptoms on their own. Physicians can do little more than comfort the dying patient, monitor his condition, and hope for the best.

Public health officials still cannot show that their efforts have curbed the epidemic or that they have stopped anyone from contracting AIDS. Despite various preventive educational programs in schools and in the community at large, as well as various official and unofficial efforts to distribute condoms or sterile hypodermic needles in Europe and the United States, no actual decrease in the number of new AIDS cases can be seen anywhere. On the contrary, each year brings a greater number of new AIDS patients. Perhaps more astoundingly, even the screening of the nation's blood supply has not led to any noticeable reduction in AIDS-defining diseases (including pneumonia, candidiasis, and lymphoma) nor in death rates among blood transfusion recipients, including hemophiliacs.[1]

Worse yet, the experts have found their estimates and projections of the epidemic to be embarrassingly inaccurate. The so-called latency period—the time between when a person is infected with HIV and develops clinical AIDS—was originally calculated in 1984 to be ten months.[2] Almost every year since, this incubation period has been revised upward. Now it is placed at ten years or longer. Even at the clinical level, doctors find the prognosis of any single infected patient frustratingly unpredictable. They cannot anticipate when a healthy HIV-infected person will become sick and which disease will affect him—a yeast infection, a pneumonia, a cancer of the blood, dementia—or perhaps no sickness at all.

Estimating the spread of the virus has meanwhile led to another problem: Officials have continually predicted the explosion of AIDS into the general population through sexual transmission of HIV, striking males and females equally, as well as homosexuals and heterosexuals, to be followed by a corresponding increase in the rate of death. However, despite the extensive use of the test for HIV antibodies—commonly known as the *AIDS test*—which first led officials to announce that 1 million Americans were already infected with the virus as of 1985, the number of HIV-positive Americans now is the same as that in 1985—1 million.[3] In short, the alleged viral disease does not seem to be spreading from the 1 million HIV-positive Americans to the remaining 250 million. AIDS itself has not yet affected larger numbers of women nor has it entered the heterosexual population outside of drug addicts: Nine of every ten AIDS patients is still male, and more than 95 percent still fall into the same risk categories—homosexuals, heroin addicts, or, in a few cases, hemophiliacs.[4] In Africa, the six million to eight million people who were said to be infected for more than a decade have translated into a mere 250,000 AIDS victims, some 3 percent to 4 percent of the HIV-positive people. The Caribbean nation of Haiti, where 6 percent of the population was known to be infected with HIV by 1985, has meanwhile remained relatively untouched by the AIDS epidemic.[5]

Something is very wrong with this picture. How could the largest and most sophisticated scientific establishment in history have failed so miserably in saving lives and even in forecasting the epidemic's toll? Certainly not for lack of resources. With an annual federal AIDS budget now more than $7 billion, AIDS has become the best-funded epidemic of all time. Not only are tens of thousands of scientists employed in a permanent, round-the-clock race to unravel the syndrome's mysteries, but the researchers have access to the most sensitive medical technology in history. With these techniques, researchers now have achieved the ability to detect and manipulate individual molecules, an ability unimaginable to the scientists who fought smallpox, tuberculosis, and polio just years earlier. Nor have AIDS researchers suffered any lack of scientific data. With more than one hundred thousand papers having already been published on this one syndrome, literature on AIDS has been surpassed only by the combined literature on all cancers generated throughout this century.

The ultimate test of any medical hypothesis lies in the public health benefits it generates; but the virus-AIDS hypothesis has produced none. Faced with this medical debacle, scientists should re-open a simple but most essential question: What causes AIDS?

The answers to the epidemic do not lie in increased funding or efforts to make science more productive. The answers will instead be found by reinterpreting existing information. Science's most important task, much more than unearthing new data, is to make sense of the data already in hand. Without going back to check its underlying assumptions, the AIDS establishment will never make sense of its mountains of raw data. The colossal failure of the war on AIDS is a predictable consequence if scientists are operating from a fundamentally flawed assumption upon which they have built a huge artifice of mistaken ideas. The single flaw that determined the destiny of AIDS research since 1984 was the assumption that AIDS is infectious. After taking this wrong turn scientists had to make many more bad assumptions upon which they have built a huge artifice of mistaken ideas.

The only solution is to rethink the basic assumption that AIDS is infectious and is caused by HIV. But the federal and industrial downpour of funding has created an army of HIV-AIDS experts that includes scientists, journalists, and activists who cannot afford to question the direction of their crusade. Thousands compete for a bigger slice of AIDS funding and AIDS publicity by producing ever more of the same science than the competition. In that climate, rethinking the basics could be fatal to the livelihood and prosperity of thousands.

Before becoming an HIV-AIDS advocate, John Maddox, the editor of *Nature*, the world's oldest scientific journal, described the dilemma:

> Is there a danger, in molecular biology, that the accumulation of data will get so far ahead of its assimilation into a conceptual framework that the data will eventually prove an encumbrance? Part of the trouble is that excitement of the chase leaves little room for reflection. And there are grants for producing data, but hardly any for standing back in contemplation.[6]

INFECTIOUS AIDS—DID WE MAKE THE RIGHT CHOICE?

Any new disease or epidemic forces medical experts to search for the new cause, which they hope to bring under control. From the start, however, they have a responsibility to consider both possible causes for an epidemic: (1) a contagious, infectious agent such as a microbe or a virus or (2) some noninfectious cause such as poor diet or some toxic substance present in the environment or a toxin consumed in an unusually large quantity. Lives depend on the right answer to this primary question. A contagious disease must be handled very differently from a noncontagious one. Unnecessary public hysteria, inappropriate prevention measures, and toxic therapies are the price for misidentifying a noncontagious disease for one that is contagious.

The period of research into the cause of AIDS in which both infectious and noninfectious agents were considered lasted only three years. It started with the identification of AIDS in 1981 and officially ended in April 1984 with the announcement of the "AIDS virus" at an international press conference conducted by the secretary of Health and Human Services and the federal AIDS researcher Robert Gallo in Washington, D.C.[7]

This announcement was made prior to the publication of any scientific evidence confirming the virus theory. With this unprecedented maneuver, Gallo's discovery bypassed review by the scientific community. Science by press conference was substituted for the conventional process of scientific validation, which is based on publications in the professional literature. The "AIDS virus" became instant national dogma, and the tremendous weight of federal resources was diverted into just one race—the race to study the AIDS virus. For the National Institutes of Health (NIH), the Centers for Disease Control (CDC), the Food and Drug Administration (FDA), the National Institute on Drug Abuse (NIDA), and all other divisions of the federal Department of Health and Human Services and for all researchers who received federal grants and contracts, the search for the cause of AIDS was over. The only questions to be studied from 1984 on were how HIV causes AIDS and what could be done about it. The scientists directing this search, including Robert Gallo, David Baltimore, and Anthony Fauci, had previously risen to the top of the biomedical research establishment as experts on viruses or contagious disease. Naturally the virologists chose to employ their familiar logic and tools, rather than dropping their old habits to meet new challenges, when AIDS appeared in 1981.

But serious doubts are now surfacing about HIV, the so-called AIDS virus. Dozens of prominent scientists have been questioning the HIV hypothesis openly during the past eight years, and the controversy gains momentum with each passing week. The consensus on the virus hypothesis of AIDS is falling apart, with its advocates digging in their heels even as its opponents grow in number.

As with most diseases today in the industrial world, AIDS appears not to be a contagious syndrome. The evidence for this exists in the scientific literature, but this evidence is widely neglected by researchers intent on viewing the data through the single lens of virology. If biomedical science has erred, if AIDS is not caused by a virus, then the entire medical and public health approach to the syndrome is misdirected. People are not being warned about the true risks for developing AIDS, doctors are using ineffective or dangerous treatments, and public fear is being exploited.

In view of the omnipotence of modern science, an error in identifying the cause of AIDS may seem inconceivable. How could a whole new generation of more than one hundred thousand AIDS experts, including medical doctors, virologists, immunologists, cancer researchers, pharmacologists, and epidemiologists—including more than half a dozen Nobel Laureates—be wrong? How could a scientific world that so freely exchanges all information from every corner of this planet have missed an alternative explanation of AIDS?

Faith in the infallibility of modern science has deep and solid roots. Rightfully, medical science is admired for its knowledge about infectious diseases and its virtuosity in dealing with them. The elimination of infectious diseases with vaccines and antibiotics has, in fact, been the most complete success story in the history of medicine. Today all infectious diseases combined cannot claim 1 percent of the lives of modern Americans and Europeans anymore.[8] Since the late nineteenth and early twentieth centuries, when Robert Koch found the tuberculosis bacillus and Walter Reed found the yellow fever virus, ever more victories have been won against infectious diseases.

These pioneers established models that every scientist confronted with an unexplained disease wants to imitate: Pick an unexplained disease, discover a causative virus or microbe and invent a curative drug or vaccine, and become a medical legend just like Koch, Pasteur, Semmelweis, and Reed. The Koch-Pasteur model set off a medical gold rush of microbe and virus hunters that came to a happy end when all major infectious diseases were apparently eliminated from the Western world, the last being polio in the 1950s.

Only noninfectious diseases like cancer, emphysema, multiple sclerosis, Alzheimer's, and osteoporosis have not yielded to medical control. On the contrary, these diseases have increased their shares as causes of death and illness, having taken the place that infectious diseases once held.

It was on the basis of this impressive record of triumphs over infectious diseases that the secretary of Health and Human Services and the virus researcher Robert Gallo promised so confidently at their international press conference in 1984 to stop the AIDS epidemic in just two years with a vaccine against the "AIDS virus."[9] Is it possible that this promise could not be kept because the hypothesis was simply wrong and that AIDS might not even be caused by a virus? Could a medical science that had broken the secrets of infectious diseases long ago have prematurely misdiagnosed AIDS as an infectious disease?

Because of their inherent potential to spread beyond control, infectious diseases are the first concern of public health officials, politicians, and taxpayers. Given the human tendency to fear the worst, the public is readily inclined to believe in infectious causes of disease. Among scientists, the infectious disease experts are the primary beneficiaries of the fear of contagion. With the argument of caution on their side, the infectious disease experts claim the privilege to convict suspect microbes without trial—while putting the burden of proof on all alternative hypotheses.

But the premature assumption of contagiousness has many times in the past obstructed free investigation for the treatment and prevention of noninfectious disease—sometimes for years, at the cost of many thousands of lives. Even when nontransmissible causes would have provided much better explanations and much easier prevention than hypothetical microbes, the microbes were pursued because antibiotics and antiviral vaccines promised proven therapies and prevention as well as professional and commercial gratification. As the research establishment becomes more centralized, bureaucratized, and fraught with commercial conflicts of interest, each episode achieves more monstrous proportions. The U.S. Department of Health and Human Services' premature endorsement of the hypothesis that AIDS is a sexually transmitted, infectious epidemic caused by the newly discovered "AIDS virus" could be the most costly and most harmful of these fatal errors in the history of medicine if AIDS proves to be not infectious.

THE SMON FIASCO

Indeed, blaming noninfectious diseases on infectious microbes has occurred many times before. Hidden in foreign-language materials and the footnotes of obscure sources lies the story of SMON, a frightening disease epidemic that struck Japan while the war on polio was accelerating in the 1950s. In many ways, SMON anticipated the later AIDS epidemic. For fifteen years the syndrome was mismanaged by the Japanese science establishment, where virtually all research efforts were controlled by virus hunters. Ignoring strong evidence to the contrary, researchers continued to assume the syndrome was contagious and searched for one virus after another. Year after year the epidemic grew, despite public health measures to prevent the spread of an infectious agent. And in the end, medical doctors were forced to admit that their treatment had actually caused SMON in the first place.[10]

Once the truth about SMON could no longer be ignored, the episode dissolved into lawsuits for the thousands of remaining victims. This story has remained untold outside of Japan, ignored as being too embarrassing for the virus hunters. It deserves to be told in full here.

The patient was middle aged, suffering from a mysterious nerve disorder that had already paralyzed both her legs. Reisaku Kono was there to observe the victim because of his work studying poliovirus, which in a few infected individuals would break into the central nervous system, causing progressive paralysis and sometimes a slow, miserable, death. While the condition he examined that day in 1959 was not polio, it bore a certain resemblance to it. And the suspicion was growing that this, too, could be the result of some undiscovered virus, perhaps one similar to poliovirus.

Kono was visiting the patient at the hospital affiliated with Mie University's medical school. Hiroshi Takasaki, a professor of medicine at the university, told Kono about a number of these cases he had recently seen at the hospital. They now realized they were facing an outbreak of something new, not just a minor mystery that doctors would catalog and forget. Just the previous year, medical Professor Kenzo Kusui had published a report of another such case in central Japan: The patient had suffered a similarly strange combination of intestinal problems, manifesting as internal bleeding and diarrhea, with symptoms of nerve degeneration. This illness, stomach pains or diarrhea followed by nerve damage, had been noticed in a few isolated cases as early as 1955, but was now turning into a local epidemic.

More published reports began accumulating after Kono's visit to the hospital. The next five years saw seven major regional epidemics of the new polio-like syndrome, with the annual number of new cases increasing from several dozen in 1959 to 161 victims by 1964—an alarming rate for those small areas. Scientists jumped to conclusions, believing they had every reason to assume the disease was infectious. Just its sudden appearance was enough evidence to convince them. The disease also broke out in clusters around specific towns or cities, and clusters were seen within families. The first person to develop the condition in each of these families was followed by a relative within several weeks. Many outbreaks were centered around hospitals, places notorious for spreading disease. The annual peak of new patients occurred in late summer, hinting at possible spread of the disease through insects. Those scientists who first thought the disease might be related to some noncontagious occupational hazard were quickly dissuaded once the data showed that the disease lacked the expected preferences. Farmers, for example, who would be more easily exposed to pesticides, had a lower-than-average incidence. Medical workers, on the other hand, had a rather high rate of this condition—further suggesting it was contagious.

However, the scientists investigating the epidemic did notice some important contradictions. For instance, the disease had an odd, amazingly consistent bias for striking middle-aged women, but was less common among men and could hardly be found among children, who normally transmit virtually any infectious disease. Careful medical inspection showed that the symptoms did not coincide with those typically expected for an infection. Blood and other bodily fluids, which usually circulate a virus throughout the body, showed no abnormalities, nor did the patients manifest any fevers, rashes, or other signs of fighting off some invading germ. These important pieces of evidence should have raised doubts about the viral hypothesis.

The virus hunt pressed onward. Scientists were expecting to find a virus that primarily induced diarrhea, as was the case in polio. Looking back on this period, Kono has since become admirably frank about his early biases, shared at the time by his fellow virologists: "I was at that time engaged in poliovirus research, so I suspected such a virus to be the cause."[11] Despite years spent searching for the elusive virus, he never could isolate a single one from any patient. Kono patiently reported his null results as he plodded forward.

Meanwhile the epidemic was growing and the 1964 Olympic Games were approaching. Ninety-six new cases had been diagnosed the previous year, and the increased number of cases was being accompanied by new symptoms. Some victims, for example, were now suffering debilitating blindness. Preparing to host tourists from around the world for the 1964 Olympics, Japan could ill afford to have an uncontrolled plague. To make matters worse, forty-six new patients suddenly appeared around the city of Toda, one of the locations for Olympic events. Embarrassingly dubbed the "Toda disease," this outbreak directly threatened Japan's reputation and tourist industry while focusing public fear on the epidemic. Etsuro Totsuka, later to become a lawyer for victims of the disease, summarized the public mood at the time: "Even I was quite worried at the time, as a university student studying physics. The general public, including me, was extremely worried; we didn't know how to prevent it, and there was no cure."[12]

In May of 1964, at the 61st General Meeting of the Japanese Society of Internal Medicine, the disease was raised as a formal topic. Kenzo Kusui, one of the first doctors to report patients stricken with this condition, chaired that session. The participating researchers gave the disease a formal name, *Subacute Myelo-Optico-Neuropathy* (SMON), and they agreed on a standardized clinical diagnosis. The Japanese Ministry of Health and Welfare quickly provided a research grant and launched a formal commission to investigate the epidemic under the leadership of Magojiro Maekawa, a medical professor at Kyoto University. Kono was one of several virologists named to the commission, thereby establishing its mandate as a formal search for a virus.

The same year brought the first sign of a possible breakthrough. Masahisa Shingu, a virologist at Kurume University and a fellow member of the commission, announced his discovery of a virus in excretions from SMON patients. The virus was classified as an *echovirus*—an acronym for enteric cytopathogenic human orphan virus. The viruses were called *orphans* because they had been discovered accidentally during polio research but caused no disease. Echoviruses were known for infecting the stomach or intestines, and Shingu found evidence of infection in various SMON sufferers. He excitedly drew the conclusion that this orphan virus had finally been matched with a disease. Perhaps, he speculated, this virus could also occasionally break into the nervous system, much like poliovirus. He published the finding in 1965, unabashedly boasting he had isolated the syndrome's cause.

But Kono, knowing the potentially disastrous results of blaming the wrong microbe for the disease, took a more cautious attitude. In 1967, after three years of research trying to confirm Shingu's claims, Kono could only report to a SMON symposium that he had not isolated the virus from patients, nor could he find even indirect evidence that the patients had previously been infected. Kono's better judgment saved Japanese science from stampeding in the wrong direction. He was fully vindicated four years later when other researchers announced the same lack of evidence to suggest any danger from Shingu's virus.

In the midst of this fruitless investigation, the Maekawa team made a surprising observation that was tragically brushed aside. According to surveys of hospitals, about half the SMON patients had previously been prescribed a diarrhea-fighting drug known by the brand name Entero-vioform, and the other half had received a compound marketed under the name Emaform. Both drugs were prescribed for problems of the digestive tract—the early symptom of SMON. The suspicion naturally arose that these drugs might play some role in the syndrome, but the commission, intent on the viral hypothesis, bowed to the consensus view of SMON as contagious and quickly dismissed this, noting that two different drugs should not cause the same new disease. Had the commission researchers checked further, however, they would have discovered that the two drugs were merely different brand names applied to the same drug, a fact that did not surface for several years.

The SMON commission dissolved in 1967, a failure. The cumulative total of reported SMON cases had meanwhile reached nearly two thousand by the end of 1966, a significant but not terrifying number. If not for the quiet growth of the disease epidemic, the floundering virus hunt might have killed public interest in SMON research altogether.

Almost immediately after the official commission was dissolved, two rural areas in the Okayama province began reeling from a new explosive outbreak of the syndrome. Dozens of elderly women, and some men in their thirties, began filling the nearby hospitals, totaling almost 3 percent of the local population by 1971. Scientific attention was again focused on SMON, with the specter of a resurgent epidemic recharging the virus hunt.

Two researchers issued reports in 1968 describing a new virus found in tissues of SMON patients, stirring a wave of excitement. The agent fell under the classification of "Coxsackie" viruses, a type of passenger virus known to infect the digestive tract and originally discovered as a by-product of polio research. It was another false alarm: The virus proved to be an accidental laboratory contamination.

In 1969 the Japanese Ministry of Health and Welfare, anxious about the expanding epidemic, again decided to form an official investigating body. With more than ten times the funding of the old 1964 commission, the SMON Research Commission became the largest Japanese research program ever devoted to a single disease. Its first meeting was held in the heavily affected Okayama province in early September. The consensus view among Japanese scientists had completely focused on some unknown virus as the probable cause of the disease. The naming of Kono, Japan's most respected virologist, as chairman symbolically established the new commission's priorities.

So far, after more than a decade of persistent research, the virologists had come up painfully empty-handed. Kono, though himself a virologist, now saw the need to explore alternative hypotheses. Kono divided the commission's work into four sections, each led by top Japanese medical officials. An epidemiologist was put in charge of a group conducting nationwide surveys on the extent, distribution, and associated risk factors of the disease. Kono himself headed the virology group. A pathologist headed a group focused on analyzing autopsy results, and a neurologist led a group classifying neurological and intestinal SMON symptoms. Altogether, forty top scientists participated in the commission during 1969.

Although Kono had opened the door for alternative research directions, the virus hunt accelerated—for just at this time, some key scientific claims by English and American virologists were beginning to have a profound impact on virus research worldwide, and particularly on SMON research in Japan. The first came in the early 1960s from virologist Carleton Gajdusek of the American National Institutes of Health, who reported finding evidence of the first "slow virus" in humans. (A slow virus is a virus alleged to produce a disease long after the original infection, that is, after a long "latent period." See chapter 3.) He believed it to be the cause of kuru disease among New Guinea natives. Kuru was a slowly progressing neurological disease that led to the debilitation of motor skills. The patients presented with symptoms of tremor and paralysis similar to Parkinson's disease. Gajdusek claimed to have found the kuru virus, but his methods were highly unusual by any scientific standards. He had never actually isolated a virus but instead had ground up the diseased brains of dead kuru victims and injected these unpurified mixtures into the brains of living monkeys. When some of the monkeys showed deficits in motor skills, Gajdusek published his findings in the world's oldest scientific journal, *Nature*, and was lauded by his fellow virologists. The second alleged discovery came from London's Middlesex Hospital in 1964, directly inspired by Gajdusek's claims. Two researchers found a virus that was believed to cause the childhood cancer, Burkitt's lymphoma. It was the first virus ever claimed to cause human cancer and the first known human virus thought to have an incubation time between infection and disease measured in years, rather than days or weeks.

These claims were made by very large and respected research establishments; therefore, Kono could not afford to ignore them. Other medical experts on the SMON commission warned him that the SMON symptoms did not resemble those of standard virus infections, suggesting the condition was not contagious. Kono, however, brushed aside this advice, arguing that if scientists were unwilling to consider the possible existence of nonclassic viruses then "Dr. Gajdusek could not have established a slow virus etiology for kuru."[13] Imitating Gajdusek's methods, he injected unpurified fluids from SMON patients into the brains of experimental mice and monkeys, hoping to cause the disease and isolate the guilty virus. Frustrated, but not willing to give up, he decided the American researchers were better equipped to find such a virus. He mailed the same fluid samples directly to Gajdusek, who repeated the inoculations into the brains of his own chimpanzees; after three years, they, too, remained perfectly normal. With that, Kono finally abandoned the search for a "slow virus."

With their virus research faltering, a few of the investigators began looking for bacteria. One lab found that SMON patients had imbalanced levels of the beneficial bacteria normally growing in everyone's intestines, but it could not isolate any new invading microbe. Kono's own lab, as well as two other researchers, did notice unusually large amounts of a mycoplasma, one type of bacterial parasite, in disease victims. However, since mycoplasma are found in a large percentage of human populations and are usually known for being either relatively harmless or causing some pneumonias, Kono and his fellow researchers decided against pursuing this further.

By 1970, one fact stood out more agonizingly than any other:

Twelve years of microbe research into the SMON epidemic had yielded nothing but dead ends. Yet the pressure continued to mount as the death toll rose. The year 1969 alone claimed almost two thousand new SMON victims, the worst toll ever. Kono and his commission were running out of options.

Fortunately for the Japanese people, several researchers on the commission were not virus hunters, and these scientists actually rediscovered the evidence for a toxin-SMON hypothesis.

The Drug Connection

As the race to find a SMON virus was capturing all the attention, other scientists were turning up some important clues to the mysterious syndrome. One pharmacologist, Dr. H. Beppu, visited the hard-hit Okayama province in 1969 to investigate the increasing outbreak and independently discovered the same coincidence the Maekawa group had years earlier—that SMON victims had taken certain drugs to treat diarrhea. Unlike the Maekawa group, Beppu investigated and found that Entero-vioform and Emaform—the diarrhea-fighting drugs found present in an earlier SMON study—turned out to be different brand names for a substance known as *clioquinol,* a freely available medical drug used against some types of diarrhea and dysentery. Beppu fed the chemical to experimental mice, hoping to see nerve damage like that in SMON, but was disappointed when the mice merely died. He missed the significance of his own results. Clioquinol was sold because it was believed not to be absorbed into the body, instead remaining in the intestines to kill invading germs. The death of Beppu's animals, however, proved that the drug not only entered the body, but could kill many essential tissues in the animal. His experiment led the SMON commission to rediscover this clioquinol connection the following year. "He later confessed to feeling stupid, because he gave up the experiment when the animals died," Totsuka explained of Beppu. "He wanted to prove a neurological disorder, but only proved the drug's severe toxicity."[14]

Meanwhile the SMON commission's first priority lay in

conducting a nationwide survey of SMON cases reported since 1967, gathered by sending questionnaires to doctors and hospitals throughout Japan. In the fall of 1969, shortly after the commission began analyzing survey data, the head of the clinical symptoms section came across several SMON patients with a strange green coating on their tongues, a symptom unnoticed before nationwide data were gathered. At first other researchers on the commission suggested that this new symptom might be caused by *Pseudomonas* bacteria, which can release colorful blue and green pigments. One of the investigators did isolate such a bacterium from some patients but not from others, and the inexplicable symptom merely became a part of the revised SMON definition. The green tongue observation achieved new importance in May of 1970, when one group of doctors encountered two SMON patients with greenish urine. Enough of the pigment could be extracted to perform chemical tests. Within a short time the substance was determined to be an altered form of clioquinol, the same drug previously found by the Maekawa commission and by Beppu.

This raised two very troubling questions. Clioquinol had been marketed for years on the assumptions that it only killed amoeba in the intestinal tract and could not be absorbed into the body; its appearance on the tongue and in the urine now proved this belief wrong. Could the medicine therefore have unexpected side effects? And why would SMON patients manifest the drug by-products so much more obviously than the rest of the population? This latter question particularly bothered one neurology professor at Niigata University, Tadao Tsubaki. Making an educated guess, he openly formulated the hypothesis abandoned by earlier investigators—that SMON might be the result of clioquinol consumption, not of a virus.

As expected, the interpretation of SMON as a noncontagious syndrome did not become popular among the virus hunters. And the suggestion that clioquinol might be guilty met even stronger resistance, for the drug was being used to treat the very abdominal symptoms found in SMON. Doctors, naturally, were reluctant to believe they were exacerbating these abdominal pains and thus adding the severe insult of nerve damage to the injury. Totsuka recalled that "doctors and scientists wanted to believe in a virus, because they prescribed clioquinol. One of the drug's main side effects was constipation and abdominal pain. Now because the drug caused pain, doctors again prescribed the drug."[15] Doctors, ignorant of clioquinol's side effects, assumed the stomach pains resulted from the primary sickness and kept increasing the dose in a vicious cycle.

Tsubaki knew he had to gather strong evidence before they could shoot down the virus-SMON hypothesis. Pulling together several associates, Tsubaki arranged for a small study of SMON patients at seven hospitals. By July of 1970 he had already compiled enough data to draw several important conclusions: 96 percent of SMON victims had definitely taken clioquinol before the disease appeared, and those with the most severe symptoms had taken the highest doses of the medication. The number of SMON cases throughout Japan, moreover, had risen and fallen with the sales of clioquinol.

This clioquinol hypothesis explained all the strangest features of the SMON syndrome, such as its preference for striking middle-aged women, its absence in children (who received fewer and smaller doses of the drug), and its symptomatic differences from typical viral infections. It also shed new light on the supposed evidence that SMON was infectious: its tendency to appear in hospital patients, to cluster in families, to afflict medical workers, and to break out more heavily in the summer—all of these reflected the patterns of clioquinol use. The epidemic itself had begun shortly after approval for pharmaceutical companies to begin manufacturing the drug in Japan.

In 1970 there were thirty-seven SMON cases in January and nearly sixty more cases during the month of July. The Japanese Ministry of Health and Welfare decided not to wait any longer, and promptly released the information about clioquinol to the press. The news of Tsubaki's research reached the public in early August, and the number of new SMON cases for that month dropped to under fifty, presumably because some doctors stopped prescribing clioquinol to their patients. On September 8 the Japanese government banned all sales of the drug, and the total new caseload for that month sank below twenty. The following year, 1971, saw only thirty-six cases. Three more cases were reported in 1972, and one in 1973. The epidemic was over.

For the next few years, the commission's research focused on confirming the role of clioquinol. In 1975 it released a comprehensive report. Systematic epidemiological surveys matched use of the drug with outbreaks of the syndrome, and experiments were performed on animals ranging from mice to chimpanzees. As it turned out, the drug induced SMON-like symptoms most perfectly in dogs and cats. Meanwhile, the investigators began uncovering individual case reports of SMON symptoms from around the world, wherever clioquinol had been marketed. Totaling roughly one hundred cases, the published reports ranged from Argentina in the 1930s to Great Britain, Sweden, and Australia in more recent times, often with the doctor specifically pointing out the association with the use of clioquinol or similar compounds. Ciba-Geigy, the international producer of the drug, had received warnings of these incidents years before the Japanese epidemic, a fact that later became the basis of a successful lawsuit against the pharmaceutical company.

Clioquinol, often marketed under the brand name Enterovioform, has been available for decades throughout many countries in the world. But while doctors outside Japan have published a few reports of SMON-like conditions, no real epidemic of the disease has ever broken out in Europe, India, or other countries with widespread use of the drug. Much of the difference lies in the heavier consumption of clioquinol in Japan, where the stomach, rather than the heart, is considered the seat of the emotions. The general over-prescription of drugs in that country further worsens the problem, such that many SMON victims had histories of using not only clioquinol but also multiple other medications, often at the same time. Government health insurance policies have encouraged this over-medication, paying doctors for every drug prescribed to the patient. As a result, the proportion of the Japanese health insurance budget spent on pharmaceutical drugs grew from 26 percent in 1961 to 40 percent in 1971, a level many times higher than in other nations. By the time the Japanese government decided to ban clioquinol, many of the hardest-hit SMON patients had each consumed hundreds of grams over the course of several months. And whereas the outside world mostly used clioquinol to prevent diarrhea when traveling abroad, the Japanese usually received the drug as hospital patients, having an already weakened condition.

Years later, at a 1979 conference, Reisaku Kono asked, "Why had research on the etiology of SMON not hit upon clioquinol until 1970?" The question has two answers; both pointed out by Kono himself:

> There were at least two occasions when physicians suspected that clioquinol might have something to do with SMON. I know of a certain professor rebuking one of his staff physicians for connecting clioquinol with SMON. In 1967 the study group of the National Hospitals on SMON reported as follows: Entero-vioform (clioquinol's brand name), mesaphylin, Emaform (home producer of clioquinol), chloromycetin and Ilosone were often prescribed to SMON patients, but no link was found between Entero-vioform and SMON. This report referred to Entero-vioform in particular so that clioquinol must have been suspected by someone in the study group. Dr. Tsugane, who was responsible for the survey, said that the survey was not thorough enough to unearth clioquinol as a causative agent. One of the reasons could have been that clioquinol had been used as a drug for the intestinal disorders of SMON, and it was hard to believe that clioquinol was toxic rather than a remedy.[16]

Referring here to the tentative fingering of clioquinol by the Maekawa group, Kono observed that too many medical doctors refused to recognize the possibility of an iatrogenic disease (one caused by the doctor's treatment). They understandably disliked the idea that a drug might cause some of the very symptoms for which it was prescribed in the first place.

Another, more fundamental, reason for overlooking clioquinol lay in the prevailing attitude of the virologists. As expressed by Kono, "We were still within grasp of the ghosts of Pasteur and Koch!"[17] SMON, a vaguely polio-like syndrome, had first appeared in the midst of a war against polio. The polio virologists, Kono included, were naturally inclined to search for a new virus as the cause of the new disease. The Japanese government, having funded poliovirus research, simply kept up the momentum by funding the same virologists to study SMON. Thus, the virus hunters received the lion's share of research moneys and attention, and with that the power to direct the SMON research program. Had it not been for Kono's foresight in also appointing nonvirologists to the commission, the epidemic might have lasted much longer.

At least the epidemic had ended, with the truth universally recognized. The virologists had learned their lesson, and the search for SMON viruses was over.

Or was it? Incredibly, against all evidence, the SMON virus hunt suddenly came back to life within weeks of the epidemic's end. The fight over the cause of the syndrome was to drag on for several more years, with the virus hunters simply ignoring the fact that SMON itself had disappeared after the ban on clioquinol.

The Virus Hunt Revived

In February of 1970, while the SMON Research Commission was still scrambling to find the cause of the epidemic and a few researchers were just beginning to notice the greenish pigments in some patients, Assistant Professor Shigeyuki Inoue at Kyoto University's Institute for Virus Research claimed discovery of a virus in the spinal fluid and excretions of SMON patients. He added the extracts to laboratory culture dishes of hamster tumor cells and found that the new agent killed the cells. With more experimentation, Inoue classified the microbe as a new herpes virus. He was able to isolate this particular virus from nearly all SMON patients he tested, more than forty in all, and found antibodies against the virus in other victims.

Reisaku Kono moved promptly to test these new observations. He used Inoue's own virus isolate and cell cultures, and within three months of Inoue's first report found that the virus could kill some cells. These particular cells, however, were extremely sensitive, prone to spontaneous death even in the uninfected cultures. Kono began to suspect the virus was harmless. He also could not isolate the virus from any SMON patients, unlike Inoue's lab. Perhaps, he openly wondered, the alleged virus might not exist at all.

A number of scientists sided with Kono, insisting they could neither find the virus in SMON victims nor cause cell death in culture dishes by adding virus samples from Inoue's lab. Nor could Inoue's extracts induce symptoms when injected into mice. Indeed, Kono and some of these other investigators could never even find the virus at all, reinforcing the growing question of whether it truly existed. The virus could not even be detected in the samples sent them from Inoue. An occasional mouse injected with Inoue's supposed virus would become sick, but the symptoms did not resemble those of SMON. Kono won allies among his peers when many of them could not reproduce Inoue's observations, a troubling problem for any scientific claim.

Nevertheless, Inoue had meanwhile rapidly achieved celebrity status for his "SMON virus" during 1970, before the clioquinol announcement that August. The Japanese news media had prematurely publicized his results, creating the widespread impression that the cause of SMON had been determined. Hysteria over the contagious plague swept through much of the country, causing frightened family members of SMON patients to avoid contact with their "infected" relatives, and leading many of the victims to commit suicide. "Patients were isolated, many committed suicide, and there was national panic," reflected Totsuka on the horror he witnessed. "I met families who lost relatives. I heard from most or all of my 900 clients; most of the patients said they very much feared and dreaded the disease. Everybody told me about that, about those sufferings. Once they found out about the drug, they were somewhat relieved, because it was not infectious."[18]

The new virus-SMON hypothesis had indeed achieved a life of its own, causing a few scientists to jump on the Inoue bandwagon; months *after* clioquinol had been banned and the epidemic had virtually disappeared, several labs excitedly issued reports claiming they could reproduce Inoue's findings. Inoue himself further insisted he had caused SMON-like symptoms in mice—including weight loss, paralysis, and nerve damage—either by injecting the virus into their brains or feeding the virus to other immune-suppressed mice unable to fight off the infection. Inoue and a collaborating scientist also both claimed to have photographed the virus directly with electron microscopes, although Inoue's colleague eventually retracted his own report as having been mistaken.

A meeting of the SMON Research Commission was finally held in July of 1972 to resolve the controversy. Until that time, Inoue's results had received attention and concern equal to the clioquinol research. But based on the inability of many scientists to produce the same results, which must be done for any scientific hypothesis to be accepted, the members at the meeting decided not to focus any more research efforts on the Inoue virus. Samples were frozen for future study, and the group thereafter devoted its resources to studying clioquinol.

Despite the absence of confirming evidence, and despite the disappearance of SMON following the ban on clioquinol, Inoue and his supporting colleagues continued to publish reports of evidence for the virus hypothesis. This publicity carried the Inoue hypothesis overseas, leading the 1974 edition of the *Review of Medical Microbiology*, an American textbook, to incorporate the Inoue virus hypothesis of SMON.

Shocked and angered by the favorable publicity surrounding Inoue's hypothesis, Kono wrote a letter to the British medical journal *Lancet*; the letter was published in August of 1975. The international popularity of virus research had whetted scientists' appetite for Inoue's hypothesis, but Kono also knew he was battling a nearly complete ignorance of the SMON episode outside Japan:

> Inoue et al. published several papers on SMON virus, and a standard textbook adopted Inoue's virus theory as

confirmed. However, research in the laboratories of the SMON Research Commission in Japan failed to confirm Inoue's results. Unfortunately, this negative information has not been published in English.[19]

The epidemic's toll had officially ended in 1973 with 11,007 victims, including thousands of fatalities. Angered upon learning of Ciba-Geigy's disregard of previously reported clioquinol toxicity, many of these patients filed a lawsuit in May of 1971 against the Japanese government, Ciba-Geigy of Japan, fifteen other distributors of the drug, and twenty-three doctors and hospitals. The ranks of the plaintiffs soon swelled to some forty-five hundred, with legal action initiated in twenty-three Japanese district courts. The largest group of SMON victims sued jointly in the Tokyo District Court. When frustrations mounted over the slow and indecisive actions of their lawyers, nine hundred of the plaintiffs broke away to form a second group. The aggressive investigations conducted by this new legal team reinvigorated the case, bolstering the positions of the plaintiffs in parallel lawsuits. Etsuro Totsuka, one of the thirty members of this legal team, has described the fight:

> We were the only team gathering information outside Japan, inviting foreign experts to testify in Japanese courts, discovering the United States FDA had restricted clioquinol ten years before Japan, and waging an international campaign against Ciba-Geigy...
>
> We found many foreign doctors who had reported clioquinol side effects before. They were contacted by Ciba-Geigy, and except in one or two instances were persuaded not to help us. By the time I saw the doctors, they had already been contacted by the other side. They had been invited on trips, some to Ciba-Geigy's headquarters... We felt they were already compensated, under the condition not to tell us anything.[20]

The two sides slugged it out for several years, but the testimony by members of Kono's SMON Research Commission proved devastating, and a string of legal victories followed in the courts.

Today most scientists and laymen outside Japan have never heard of the virus-SMON controversy, even in the face of the lawsuit against the distributors of clioquinol, television documentaries in Germany and England on clioquinol, and two conferences during the 1970s on iatrogenic (medically caused) disease. The story that SMON research had ignored the evidence of a toxic cause for fifteen years and had sacrificed thousands of human lives to a flawed virus hypothesis is too embarrassing to the virus-hunting establishment to record.

AIDS: AN ENCORE OF THE SMON DISASTER?

When Michael Gottlieb, at the medical center of the University of California, Los Angeles, observed five patients dying from bizarre diseases during the early months of 1981, he already suspected he was opening the curtain on a new epidemic. AIDS, like SMON, did grow dramatically over the next decade, although not explosively as other new, infectious epidemics, like a seasonal flu or cholera epidemic did before the days of antibiotics. AIDS appeared with unnerving suddenness in major cities of the United States and Europe—as well as in Africa and the Caribbean, where mystique-ridden stereotypes of these countries lent credibility to stories of widespread devastation.

Again following the pattern of SMON, AIDS circumstantially appeared to be contagious, with cases turning up among hemophiliacs and other recipients of blood transfusions and with outbreaks of the syndrome found among mutual sex partners in the homosexual community. In other words, potential transmission routes for some unknown virus could be identified. But other evidence actually indicated both syndromes to be noninfectious: Whereas SMON struck middle-aged women more than any other group, AIDS showed an extreme bias for young men in their twenties to their forties, mostly heroin addicts and homosexuals.

SMON, as it turned out, resulted from the use of a prescription drug for the early symptoms of SMON itself, a fact so horrifying to doctors that the possibility was repeatedly cast aside whenever the evidence would emerge. AIDS may also be partly the product of a prescription medicine—AZT, the very one provided as a therapy for AIDS. Once again, that horrifying possibility is cast aside by scientists and doctors.

AIDS, too, became a centrally managed epidemic, with the U.S. National Institutes of Health directing most research and preventive education in this country. Special commissions were also set up by prestigious scientists and government officials, beginning in 1986, to focus all resources and efforts into a concerted war on AIDS.

And from literally the first week after Gottlieb reported his AIDS cases, the virus hunters began the search for an AIDS virus, dominating the research effort just as their Japanese counterparts had done with SMON. Once again, several viruses in turn were blamed, from the herpes-type cytomegalovirus to the retrovirus HTLV-I, until a consensus formed around another retrovirus, the Human Immunodeficiency Virus (HIV).

The SMON epidemic finally ended because Reisaku Kono and other Japanese scientists possessed the wisdom to direct some resources into nonvirological research and listen when those other investigators found answers. But the officials and scientists driving our war on AIDS have had little tolerance for alternatives. Ignoring the lessons of SMON and other diseases, today's biomedical research establishment blocks virtually all research and questions that disagree with the consensus view of infectious AIDS.

If the war on SMON was a molehill of misdirected science, AIDS has become an unmovable mountain. The difference lies in the respective sizes of the scientific establishments involved. Not only is the funding for AIDS research much greater than the amount spent on SMON, but the preexisting structure—measured in number of scientists, size of departments, and sheer volume of published data—now far exceeds the combined size of all scientific endeavors in human history. Thus, errors necessarily become magnified beyond any individual's control, and adjustments to AIDS theory become ever more difficult to change.

SMON and AIDS are even more intimately connected. Both have been episodes in a long series of miscalculations emanating from a single ongoing, self-propagating scientific program—microbe hunting. Microbiology certainly achieved many notable scientific discoveries, especially early in this century. Polio marked the end of the infectious disease epidemics that once ravaged the industrial world. Microbe research has mostly outlived its usefulness, leaving virus and bacteria hunters with little to accomplish, yet they still dominate the increasingly well-funded science establishment. As a result, they have for three decades been misleading science and the public about medical conditions ranging from cervical cancer to leukemia, from Alzheimer's disease to hepatitis C, and many more. All these smaller programs are failing in their public health goals as they prescribe the wrong treatments and preventive measures, while generating unnecessary fear among the lay public.

SMON did not mark the first time microbe hunters falsely blamed viruses or other microbes for noninfectious diseases. "Pellagra is a classic example," Reisaku Kono emphasized in retrospect. "It was once believed to be a communicable disease and, as is well known, Goldberger swallowed fecal extracts of the patients to destroy this notion."[21]

Pellagra, the quintessential human tragedy representing the era of the bacteria hunters, has been too widely forgotten. Chapter 2 tells the story of Goldberger and other scientists who fought the excesses of the first microbe-hunting establishment.

CHAPTER TWO

■

The Great Bacteria Hunt

THE LEADING KILLERS in the industrial world today are the slow-developing conditions of older age, including heart disease, osteoporosis, Alzheimer's disease, and cancer. As our health and life expectancies increase, the more lives these diseases will claim.

But people throughout the Third World and in our own past have faced death at much younger ages, and from a different cause: contagious disease. Pre-industrial societies are marked by frequent and deadly epidemics of every conceivable infectious illness, from flus and pneumonias to tuberculosis and smallpox. Although infectious disease was commonplace in earlier times, people were mystified by these strange conditions that could be passed from one individual to another. Thus, during the many centuries in which infections dominated human mortality, myths ranging from possession by evil spirits to inhalation of miasmal airs were offered as explanations.

Not until the seventeenth century did the first person use lens-making technology to discover the existence of microbes. Antony van Leeuwenhoek, a Dutch janitor with a penchant for constructing microscopes in his spare time, found immense numbers of the tiny one-celled organisms now known as *bacteria* in saliva. The tiny creatures existed not only in bodies of humans and animals but even in the water of rivers and lakes. Leeuwenhoek's discovery did attract the attention of established scientists at the time, but he never supposed that these bacteria might cause disease and considered them mere curiosities. Nor had he any reason to blame them for disease, because no logical rules yet existed for proving such an idea.

Two centuries later, Leeuwenhoek's discovery did give birth to the germ theory of disease. A French chemistry professor named Louis Pasteur was asked by local brewers to determine why some vats fermented and others did not. He learned through his experiments that yeast, a microbial type of fungus, was the organism making the alcohol and that bacteria could prevent the fermentation as well as cause contaminated food to decompose. Physicians and scientists throughout Europe soon made the logical connection with disease, and the hypothesis that such germs might cause sickness became a widespread topic of discussion. Joseph Lister, for example, gained prominence as the doctor who popularized antiseptic surgical techniques in the wake of Pasteur's growing fame. And Ignaz Semmelweis from the University of Vienna correctly deduced that washed hands and germ-free clothes eliminated child bed fever.

Still, no one had actually proven that a particular infectious disease was caused by a corresponding bacterium. Many leading doctors, in fact, refused to believe that disease could result from transmissible microbes at all. Although they ultimately turned out to be mistaken, their healthy skepticism nonetheless played a critical scientific role, forcing the early microbe hunters to formulate objective standards for blaming any disease on a germ. The importance of such proof cannot be underestimated: Many diseases are not infectious, yet a number have been falsely blamed on harmless passenger microbes throughout the nineteenth and twentieth centuries. Such mistakes can be easily avoided only when scientists carefully apply logical standards.

By 1840, Jakob Henle, a professor at Germany's Goettingen University, publicly suggested that infectious disease would be found to be the result of some invisible living organism that could be transmitted from person to person. The problem, as Henle observed, was that to prove this "contagion" caused a disease, it would have to be isolated and grown outside the human body. At the University of Prague, another German professor named Edwin Klebs carried this reasoning one step further during the 1870s. Not only should the microbe be cultured from the diseased body, but it should be able to cause the same disease when injected into another animal. To many European doctors, this proposal certainly made logical sense. But without any examples proven by such experiments, most doctors preferred to suspend judgment on the germ theory.

At this point a German medical doctor named Robert Koch entered the fray. He founded his research on the results of Casimir Joseph Davaine of France, who had demonstrated that blood from cows with anthrax could transfer the disease to newly injected cows. Studying the strain of bacterium found most easily in cattle with anthrax, Koch wanted to prove his suggestion that the microbes could spread disease. He was therefore forced to find some way to grow them under his microscope. He developed a method of growing the bacteria in the eye fluid from slaughtered cattle and quickly proved his point. Koch inoculated mice with these bacterial cultures and discovered that they, too, became sick as their bodies filled with the deadly bacteria. Having initially planned to study bacteria merely for their own sake, he instead published a paper in 1876 boldly announcing he had proved this bacillus to be the cause of anthrax.

Koch thus became the first person to meet the criteria of Edwin Klebs. However, the anthrax bacteria were large and easy to isolate, and they usually caused disease in animals rather than humans. So he next followed his growing interest in the subject of human disease and started his work with the study of open wound infections. Observing samples from various animals and people, he reported that bacteria could hardly be found in healthy organisms, while they were abundant in the blood of the diseased animals. Koch's results led him to add now a third and key condition to the others proposed by Klebs:

> In order to prove that bacteria are the cause of traumatic

infective diseases, it would be absolutely necessary to show *that bacteria are present without exception and that their number and distribution are such that the symptoms of the disease are fully explained* [italics in original].[1]

In other words, a microbe cannot scientifically be proved guilty of causing a disease unless every diseased individual has large amounts of the germ growing in the damaged tissues of the body. A single exception would be enough to pronounce the microbe innocent of creating that disease.

One major problem with meeting such standards of proof lay in the difficulty of culturing pure preparations of any given bacterial species. Koch's 1878 book on his wound infection experiments described his attempts to purify the cultures so that contaminating bacteria could not be blamed for causing the disease, but only in 1881 did he finally publish a paper describing a new technique for pure culture of bacteria. The method used a dish, later improved and named the *Petri dish* after Koch's assistant, that allowed scientists easily to separate or "clone" individual bacteria by growing them apart from one another. Finally, the microbe-hunting tools, both experimental and logical, were in place.

However, the appeal to find even individual microbes in a patient with Koch's new method turned out to be a mixed blessing. Many of Koch's followers triumphantly claimed bacterial causes of nonbacterial, even noninfectious, diseases—without ever checking the titer, or number of bacteria in these diseases. Many of these putative microbial pathogens later proved to be harmless passenger microbes, normal parasites of healthy and ill persons, when subjected to Koch's postulates for criteria to distinguish harmless from pathogenic microbes. The problem of confounding harmless with pathogenic microbes has reached epidemic proportions in recent history as hypersensitive molecular techniques have been invented that allow the detection of dormant, dead, and even defective viruses or microbes. (See chapter 6.)

Koch next focused his attention on tuberculosis, the leading infectious killer of humans at that time. Within months, he found, isolated, and cultured a bacterium from the patients. According to Koch:

> In all tissues in which the tuberculosis process has recently developed and is progressing most rapidly, these bacilli can be found in large numbers... As soon as the peak of the tubercle eruption has passed, the bacilli become rarer.[2]

Having met the first two conditions of proof, he went on to show that guinea pigs injected with the purified bacteria would now become sick with tuberculosis. The proof complete, Koch published his landmark 1882 paper describing the experiments.

He wrote another key paper on tuberculosis in 1884, in which he spelled out the three criteria for proving a microbe guilty of causing a disease:

- First, the germ must be found growing abundantly in every patient and every diseased tissue.
- Second, the germ must be isolated and grown in the laboratory.
- Third, the purified germ must cause the disease again in another host.

Together, these rules have become known as *Koch's postulates*.

Fame quickly followed Koch's work, and scientists and doctors alike jumped on the bandwagon. During the next two decades, bacteria were found and proven guilty of inducing more than a dozen major diseases, including diphtheria, tetanus, food poisoning, some types of pneumonia, and syphilis. But in the rush and popularity of the new microbe hunting, a scientific sloppiness led many researchers to blame newly discovered bacteria prematurely, without having satisfied the universally accepted postulates of Koch. Even Koch himself was partly guilty, for he too maintained an overly enthusiastic ambition to find bacteria in almost every disease. In his study of cholera, for example, he isolated the correct bacterium, but could not find an animal that would become sick when injected with the microbe. He nevertheless declared it the cause of cholera using statistical correlations, rather than testing other animal species to meet the third postulate.

Unable to distinguish an animal that was vaccinated by natural infection from one that was susceptible, Koch may have tested his cholera bacteria in immune animals. At this time, microbe hunters were just beginning to understand how vaccination works. Since immunology was in its infancy, Koch never used artificial vaccination as a reverse means of conducting the test (e.g., rendering an animal resistant to a microbe by vaccination). As it turns out, scientists have since produced cholera in rabbits, dogs, and guinea pigs, though in unimmunized animals. While Koch was lucky on that score, he and others soon made numerous mistakes in identifying disease-causing bacteria.

But the successes did lead to a variety of developments in medical technologies, including the discovery of antibiotics for killing bacteria, the development of new vaccines against various microbes, and an increased emphasis on hygiene. Governments began enforcing public sanitation and vaccination measures—mostly after Koch's appointment to the Imperial Health Office of Germany—policies that soon spread throughout the industrializing world. Nutrition, and standards of living, also improved among industrial nations during the same time period. While controversy exists over the importance of each condition in stopping particular epidemics, the epidemics as such have largely disappeared, and medical intervention against the microbes is widely credited for this.[3] Indeed, no other medical discovery has ever achieved as much acclaim.

Naturally, then, scientists have since kept an ambitious eye out for new microbes, hoping to find the causes of unexplained diseases—often the ticket to fame and fortune. But when scientific standards such as Koch's postulates have been pushed aside in the race for recognition, medical disasters have usually struck. Humans and animals, whether healthy or sick, are host to many hundreds of microbes, the great majority of which cause no harm whatsoever. Some can even be beneficial, such as the *E. coli* bacteria that populate the intestines and aid digestion. Without the rigor of the scientific method, researchers can easily isolate one of these harmless microbes and blame it for a disease, even if the illness is noninfectious.

PLAGUES OF MALNUTRITION

As we know today, scurvy is a disease caused by a lack of vitamin C in the diet. It begins with such characteristic symptoms as bleeding gums, progresses to swollen legs and brain-destroying dementia, and ultimately leads to death. Long before vitamin C was chemically identified and isolated in the 1930s, various observant individuals had noticed that scurvy could be cured, even in its latest stages, by some "antiscorbutic factor" found in such foods as citrus fruits, potatoes, milk, and fresh meat. But the historic preoccupation with contagious disease often obscured this discovery, each time delaying public knowledge of the health benefits of such foods for many more years.

Fear of contagion predates Robert Koch's discovery of disease-causing bacteria. During the mid-sixteenth century, roughly one hundred years before Antony van Leeuwenhoek first saw microbes in his primitive microscope, scurvy was given its first description by physicians that included Ronsseus, an advocate of a dietary hypothesis of the disease. His contemporary, Echthius, on the other hand, watched outbreaks of scurvy among monks in a single monastery and concluded the disease was infectious.

This latter opinion proved influential for centuries, despite an early proof of diet as the cause. Sir Richard Hawkins, a British admiral, confronted scurvy among his sailors on a long voyage in 1593. Upon reaching Brazil, he discovered that eating oranges and lemons would cure the condition. Nevertheless, even he felt obligated partly to blame unsanitary shipboard conditions, and following his death the British navy completely lost all memory of the citrus fruit cure.

While Hawkins still lived, a Frenchman named François Pyrard described an expedition to the East Indies. Unaware of Hawkins's findings, he ascribed scurvy to a "want of cleanliness" and insisted that "it is very contagious even by approaching or breathing another's breath."[4] Yet Pyrard ironically had also discovered the curative power of citrus fruits. His independent dietary discovery was forgotten, as was Hawkins's, and the infectious view continued to prevail.

An outbreak of scurvy occurred on a 1734 voyage of a British ship, affecting one sailor especially severely. Anxious to prevent spread of what he believed to be a contagious disease, the captain marooned the hardest-hit sailor on the nearest island. Fortunately for the sailor, he ate grass, snails, and later shellfish, from which he received enough vitamin C to recover. A passing ship found him, and upon reaching England he astonished many by the very fact that he lived. This was one of the events that stimulated James Lind, British naval surgeon, to begin his experimentation in curing scurvy.[5] After several years of research, he concluded that the key to the cure and prevention of scurvy was some factor found in citrus fruits but missing from sailors' diets. He published this proof as a book in 1753, but he was roundly rejected by the British medical establishment for some forty years.[6]

Only in 1795 was lemon juice finally provided to naval sailors (at that time often called "lime" juice, thus originating the nickname "limeys" for sailors). During this period the English captain James Cook also discovered, on his 1769 voyage, that fresh vegetables and citrus fruits worked, despite no apparent knowledge of Lind's work. But he too insisted on hygienic practices and fresh air, which he believed to be as important as diet in preventing scurvy, thereby helping to confuse the significance of his results.

By the turn of the nineteenth century, the point seemingly should have been settled. However, the role of diet had never been fully accepted outside of England, and even British doctors gradually reduced their emphasis on it as the century progressed. This negligence, combined with the rise of bacteria hunting in the later 1800s, led too many scientists to forget or ignore the earlier discoveries. One could more easily isolate a new bacterium than a new vitamin.

Jean-Antoine Villemin provided one prominent example. A member of the Paris Academy of Medicine, he was the first to demonstrate that tuberculosis was an infectious disease; Robert Koch had based his search for the tuberculosis bacillus on this work. Villemin became a passionate advocate of the germ theory for disease in general and in 1874 began debating the still widely accepted view that bad diet was somehow responsible for scurvy. In one paraphrased version he states:

> Scurvy is a contagious miasm, comparable to typhus, which occurs in epidemic form when people are closely congregated in large groups as in prisons, naval vessels and sieges... We have many examples of well-fed sailors and soldiers going down with scurvy, while others less well fed do not. Also, we have positive evidence of the spread of the disease by contagion—for example, the introduction of scurvy into French military hospitals by veterans returning from the Crimea, and the rapid spread of scurvy from one sailor to another in naval vessels.7

Villemin, of course, was using a poor argument that nonetheless is still repeated today by top scientists for other diseases. Outbreaks of a disease do not really argue for an infectious cause, merely for a factor common to the group in which the disease appears. Another member of the Academy of Medicine responded to Villemin by arguing that some diet common to the afflicted was indeed the reason for those scurvy epidemics. Further, he pointed out the danger of falsely blaming a disease on infection: Medical authorities would justifiably see the need to quarantine patients to protect the public.

The growing popularity of the germ theory, and its clear successes, soon gripped medicine so tightly that it began redirecting research on scurvy. In 1899, British explorer Frederick Jackson teamed up with a professor at London University to perform experiments on the disease in animals. Jackson decided that fresh meat did not contain a vitamin but rather that older meat was contaminated with bacteria that spoiled it and produced "ptomaines," poisons that would cause scurvy. Joseph Lister, the surgeon inspired by Pasteur's discoveries enough to popularize antiseptic surgery to avoid infections, had by this time become president of the Royal Society of London and was only too happy to provide funding for Jackson's research. As he himself put it, he wished to see new research on scurvy in light of the recent microbe discoveries. The two researchers chose monkeys for their experiment, feeding them various diets to see whether diet itself or food contamination would induce scurvy. But since vitamin C had not yet been isolated, the diets were not well controlled for the postulated food factor, and the results showed that monkeys fed tainted meat became sick more often. The president of the Royal Society endorsed and promoted the experimental report, and microbe hunters—believing scurvy to result from digestive tract infection and intent on finding a guilty bacterium—seized on the report in an attempt to silence diet-minded critics.[8]

The obsession with microbe hunting not only distracted scientists from finding vitamin C, but actually helped cause epidemics of scurvy. For instance, Louis Pasteur's technique of sterilizing milk by heating it had spread throughout Europe and America, becoming popular because the microbe hunters had convinced the public of hygiene's primary importance. The pasteurization process unfortunately also tended to destroy the vitamin C in milk, which led to hundreds of new scurvy cases among young children each year. Unwilling to admit their mistake or to read the available history of the disease, the American Pediatric Association issued a report on childhood scurvy in 1898, concluding that bacteria-produced ptomaine poisoning, not the heating of milk, was the real cause of the epidemic.

Researchers simply would not let go of the germ theory in their scurvy research. A popular textbook, Osler's *Modern Medicine*, while recognizing some dietary role in the disease, insisted in 1907 that an unidentified microbe contaminating the food must infect the unsuspecting victim and cause the sickness. Another contemporary view held the disease to be a type of inherited syphilis, itself a genuine bacterial disease. One French scientist actually found a new strain of bacteria in a scorbutic baby and proposed it to be the cause, although other scientists examining the blood of other patients could not find the bacterium. During World War I, another group of scientists isolated a different bacterium from scorbutic guinea pigs and still another from an adult human. The bacilli found in the animals was then injected into healthy guinea pigs, some of which developed symptoms vaguely resembling scurvy. But the bacteria could never be found in the blood of these newly infected animals, and blood from a sick animal would not make another animal sick when injected. Still, the researchers argued they had the scurvy-causing germ. Another report at that time proposed that scurvy could be transmitted through lice. Many or most doctors in Russia meanwhile believed bacteria to be the cause, as did various surgeons in other European armies. And at least one German doctor, sent in 1916 to examine Russian soldiers suffering from scurvy, largely blamed their unsanitary conditions. Of course, all of the germs blamed for scurvy failed to meet Koch's postulates, standards that could have prevented much of the wasted effort, but scientists were busier trying to emulate Koch's success rather than his rigorous logic.[9]

Fortunately, the microbe-hunting craze did not permanently derail the search for vitamin C, which was finally purified by the 1930s. C. P. Stewart, professor of clinical chemistry at the University of Edinburgh, in 1953 summarized the chronic scurvy disaster:

> One factor which undoubtedly held up the development of the concept of deficiency diseases was the discovery of bacteria in the nineteenth century and the consequent preoccupation of scientists and doctors with positive infective agents in disease. So strong was the impetus provided by bacteriology that many diseases which we now know to be due to nutritional or endocrine deficiencies were, as late as 1910, thought to be "toxemias"; in default of any evidence of an active infecting microorganism they were ascribed to the remote effects of imaginary toxins elaborated by bacteria.[10]

Beriberi is a fatal condition brought on by a dietary lack of vitamin B_I (thiamin). The nervous system degenerates, creating paralysis, swelling, and often heart attacks. Though it has primarily plagued Asia throughout history, it appeared with a vengeance in the West after the French Revolution, when the French population rejected the dark bread of peasantry in favor of the royal milled white bread from which the thiamin had been unknowingly removed. Bread processing soon swept throughout Europe and the United States, and beriberi followed closely.[11]

The first person to discover the basic cause of the condition was Kanehiro Takaki, a medical doctor and later surgeon general for the Japanese navy. Concerned about the beriberi epidemic rampant in the Japanese military and in the cities, he carefully studied its characteristics and during the 1880s performed an experiment. By experimenting with the diets of sailors in different ships, he found he could cure and even prevent the disease. The military, responding decisively, altered the official diet for sailors and thereby ended the epidemic in 1885. Takaki then published his persuasive results in the British medical journal *Lancet* in 1887. Instead of acknowledging poor nutrition as the cause of beriberi, the scientific community wantonly disregarded it. The report had arrived during the height of the bacteria-hunting craze, five years after Robert Koch had found the tuberculosis bacillus, and microbe hunters were eager to find new germs. Even in Japan, microbe hunters strongly influenced by Koch and his contemporaries sniped at Takaki, insisting that beriberi was truly infectious and had been cured by better sanitation, not by better diet.

Christiaan Eijkman, a Dutch army doctor, had meanwhile observed firsthand an epidemic of beriberi among the Dutch soldiers in Java. Although the disease mysteriously left the natives alone while ravaging the conscientiously hygienic Dutch, Eijkman's infection-biased medical training led him to assume some germ must cause the disease. He therefore decided to advance his skills in finding bacteria and spent a few months (1885–1886) working in Robert Koch's laboratory in Berlin. Having become desperate, the Dutch Colonial Administration in the meantime formed a team of scientists under Dr. Pekelharing to study the disease. Pekelharing also assumed the condition was infectious, and, after consulting Koch, recruited Eijkman onto the team.

In Java, Pekelharing isolated a bacterium that he promptly blamed for beriberi. He left shortly thereafter, turning over his work to the enthusiastic microbe-hunter Eijkman. But Eijkman, unable to find the microbe in all the sick patients, tried at least to transmit the disease to chickens through blood from patients. At first nothing happened, then all the chickens developed a sickness like beriberi—including those not having received any blood. Confused, he performed several other experiments until he discovered that the sickness was caused by eating polished rice, which had temporarily been fed to the chickens instead of their usual unprocessed rice. This explained the human disease: the Dutch all ate polished rice, while the Javan natives did not. Eijkman convinced the Dutch prison warden in Java to test the idea by feeding unrefined rice to the prisoners. Their beriberi soon disappeared.

Upon presenting his results to his supervisor, Eijkman received only rejection. His superior even went so far as to publish an attack on the chicken and prison studies, and when Eijkman published his own paper in 1890, colleagues criticized him. The Dutch commission to which Eijkman belonged officially concluded that, although blame could not be fixed on the Pekelharing bacillus, the epidemic must be caused by an undiscovered germ. Eijkman himself was so under the hypnotic spell of the germ theory that he continued for at least eight more years to refer to beriberi as a contagious disease caused by microbes, despite his own results.

The peer pressure of scientific consensus must also have intimidated him. At least two dozen of his colleagues continued to find and blame the sickness on a dizzying variety of microbes ranging from bacteria to worms. Scientists isolated bacteria from the digestive system, blood, and urine of beriberi patients. One group found three types of bacteria and blamed them all; another investigative team discovered four types simultaneously. Three groups blamed protozoa, organisms similar to the one causing malaria, and at least two scientists decided fungi growing on moldy rice were the culprits. Even a virus was reported found and falsely convicted in 1900.

No single microbe remained popular for long, however, largely because a fair number of scientists failed in trying to find each germ in all beriberi patients, and they were willing to publish their negative results. Robert Koch himself ironically held high hopes of finding the beriberi bacillus but was unsuccessful during his research on a trip to New Guinea. Koch's careful commitment to logical scientific standards overrode his enthusiasm, and he openly published his lack of results in 1900. Nevertheless, reports of beriberi-causing microbes actually continued after 1910, and the predominant infectious view of the disease led doctors to "treat" it with such compounds as quinine, arsenic, and strychnine. The question of beriberi's cause was finally settled only when vitamin B_I was isolated in 1911 and again in 1926. The vitamin is now added back to white bread, and beriberi has become a rare disease.

Robert Williams, one of the scientists who pioneered the discovery of vitamin B_I, later commented on the dangerous influence of the microbe hunters in emulating Pasteur and Koch too carelessly:

> Because of [the work of Pasteur and Koch] and other dramatic successes bacteriology had advanced, within twenty years after its birth, to become the chief cornerstone of medical education. All young physicians were so imbued with the idea of infection as the cause of disease that it presently came to be accepted as almost axiomatic that disease could have no other cause.
>
> This preoccupation of physicians with infection as a cause of disease was doubtless responsible for many digressions from attention to food as the causal factor of beriberi.[12]

THE PELLAGRA PLAGUE

In terms of the number of people affected, pellagra has probably been the most devastating vitamin deficiency epidemic of all. It manifests itself most visibly by rough and peeling skin with splotches of reddish pigmentation, followed by nerve disorders and dementia, wasting syndrome and diarrhea, and finally death.[13] First described in the eighteenth century, the disease soon grew into an epidemic in Italy and spread throughout the Mediterranean area during the nineteenth century. The name *pellagra* derives from the Italian for "rough skin." As was discovered earlier in this century, niacin deficiency is the cause. Because corn lacks niacin and various populations have turned to corn as a nearly complete substitute for other vegetables, pellagra has usually appeared wherever corn has become a dietary mainstay.

Doctors who wrote the early descriptions of the disease clearly noticed the association with corn diets and poverty. Beginning in the early 1800s, a series of physicians formulated several closely related hypotheses about this connection, speculating either that corn itself caused pellagra or that the fungus on moldy corn produced some sort of poison. Some prescient observers even correctly guessed that corn was not nutritious enough as a complete diet. But most European doctors originally agreed that the syndrome could not be contagious, since it never seriously spread out of the impoverished corn-eating subpopulations. Already in the 1700s several physicians blamed the disease on miasms, or bad airs. And as early as the 1790s, a doctor on occasion would observe that pellagrins (pellagra patients) could be cured with more balanced eating habits.

Despite the clear inability of the disease to spread beyond the risk groups, some doctors unfamiliar with the disease still proposed it to be contagious. The German doctor Titius, himself far removed from epidemic areas, in 1791 simply called it infectious. Prominent French doctor Jean-Marie Hameau, in his 1853 doctoral thesis, decided that since pellagra strikes people living near sheep and sheep have an infectious disease with some symptoms resembling pellagra, the disease was transmitted from sheep. Barring this unlikely possibility Hameau conceded the infection might come from contaminated corn. Although most doctors did not agree with Hameau's view, a typical approach to treating pellagra was nevertheless based on fighting infection, which included artificial bleeding, quinine, and arsenic. The latter, in particular, was the treatment pioneered by one of the early Italian microbe hunters, who believed the common fungus on moldy corn caused the disease.

As the successes of Pasteur and Koch became widely popular, scientists and doctors began flocking into bacteriology. No longer required to invent hypothetical microbes, they could use Koch's simple tools for isolating real bacteria and blame them for the disease. Many wanted to have Koch's success, but few were willing to apply the acid test of Koch's postulates nor even to ask whether the disease in question was truly contagious, as evidenced by spreading out of its initial risk groups.

Thus, bacteria hunting turned to pellagra with a vengeance. In 1881, the Italian doctor Majocchi was first to isolate a bacterium from both spoiled corn and the blood of patients. Several more scientists discovered that this microbe was the same as a previously identified bacterium found in potatoes, and that the rotten mass of corn contaminated by this germ could cause diarrhea in dogs, though not in other animals. However, unlike Majocchi, they could not find the bacterium in the blood of pellagrins, instead finding it growing in the intestines of all humans, including those without disease. So ended Majocchi's bacterium. Another bacterium reported in 1896 by Carraroli was also soon abandoned.

Then, for several years after the turn of the century, an Italian researcher named Ceni generated a remarkable number of scientific papers claiming that a corn fungus excreted by chickens—regardless of whether the fowl had eaten fresh or spoiled corn—also caused the disease in humans. Ceni and his coworkers found these fungal spores in most, but not all, people who had died of pellagra, and tested a variety of animals to show that large amounts of this fungus would make the animals sick, especially when injected into the blood. Ceni soon expanded his list to two, and then four, separate fungi that he thought would all cause pellagra. Even though these fungi could not grow in the body, Ceni insisted they could still release poisons. During these years Carraroli, who had previously isolated a bacterium from pellagrins, now jumped on the fungus bandwagon, alleging that one of Ceni's fungi could be isolated from the fecal matter, urine, blood, saliva, and affected skin of patients. By injecting the fungus into experimental animals, he even produced symptoms he thought resembled pellagra. In fact, Carraroli was so caught up in admiring this microbe that he simultaneously accused it of causing syphilis.

Another researcher inspired by Ceni reported in 1904 two new candidate bacteria for causing pellagra, based on their presence in corn and resistance to the heat of cooking. One of these could cause intestinal sickness when injected into animals. The other was similar to the intestinal bacterium present in all humans that helps digest food, so he decided it released poisons that could act as "cofactors," or enhancers, in helping Ceni's fungi cause pellagra.

The sheer volume of Ceni's ongoing research forced a number of scientists to spend a great deal of effort refuting his results. The fungal spores, as it turned out, neither caused pellagra nor any other disease in animals, nor could they be found in patients having died of the disease. And the full-grown fungi were often simply natural parasites of humans.

Yet the microbe hunt continued. Tizzoni, a prominent Italian researcher and doctor, began reporting from 1906 onward for several years his experiments on two strains of bacteria, both blamed by him for pellagra. Having found the germs in pellagrins, he and other scientists were able to cause some sort of sickness in monkeys and guinea pigs injected with the bacteria. Thus, he brazenly declared, "It would seem to be settled that pellagra is a bacterial disease."[14] However, a number of scientists never could isolate these bacteria from people with pellagra, leaving Tizzoni's work with little impact among European doctors stymied in trying to cure the disease.

The chances of finding a cure, as well as the opportunities for microbe hunters, multiplied dramatically once the pellagra epidemic appeared in the United States. A few cases had passed unnoticed before the twentieth century, but the first recognized instance appeared in Georgia, when a single farmer was diagnosed by his doctor with the disease in 1902. Four years passed without the medical establishment paying any attention. Then an outbreak suddenly appeared at a hospital for the insane in Alabama. Eighty-eight patients became severely ill, most of whom died. Soon dozens of cases began appearing in hospitals throughout southern states and even in Illinois. Facing a now-unnerving epidemic, the head of a hospital for the insane in South Carolina visited Italy in 1908 and decisively concluded that the American epidemic was indeed pellagra.

By mid-1909 hundreds of cases had occurred in more than a dozen states. The Public Health Service, a branch of the federal government that still exists today, established a small laboratory for pellagra research in South Carolina. Their man in charge, Claude Lavinder, pursued three lines of activity: experiments, therapy, and public relations. Having no other serious model to follow, he searched for a microbial cause of the disease by injecting various types of animals with bodily fluids from pellagrins, though to no avail; none of the animals became sick. Lavinder's treatments fared no better, for he used the widely popular arsenic as well as mercury. But his propaganda efforts proved more effective, for the media soon mobilized to convince Americans they were facing a disease that could spread out of control and that would affect everyone, rich and poor alike.

The growing epidemic activated the concern of many medical doctors, who in 1909 held a National Conference on Pellagra in South Carolina. As in Europe, the evidence of pellagra's association with corn-based diets was clearly recognized at the meeting, as was the fact that it struck exclusively poor communities (soon thereafter blacks were recognized as the major risk group), both facts indicating a noncontagious epidemic. But the age of microbe hunting was still in full swing, and although many scientists began investigating the corn connection, the conference also set in motion a revived hunt for a pellagra microbe.

The following year Lavinder was replaced at the pellagra lab by John D. Long, who believed the disease was brought on through a lack of hygiene. He discovered an amoebal microbe in the intestines of most of his pellagra patients and fingered this germ as the cause in his 1910 report. Long, as it turned out, had followed the lead of Louis Sambon, a well-known British doctor who in 1905 had announced after a brief visit to Italy that he believed pellagra to be an infectious disease. Building on his own work on malaria, Sambon declared to the press in 1910 that the disease was transmitted by insects, either flies or buffalo gnats. He failed to notice that, unlike malaria, pellagra did not spread out of its risk groups; even in epidemic areas, only very poor farmers were affected. Sambon did realize that an infectious disease should spread at least somewhat and therefore argued erroneously that children were primary targets of the disease.

Because of his own reputation, and the fact that he had assembled an official commission of top British doctors, Sambon's hypothesis caught on and quickly spread to the United States. One scientist, convinced of the Sambon hypothesis, published evidence in 1912 that airborne insects crowded the areas near water during the seasons pellagra was most prevalent, implying a malaria-like spread of pellagra. Another research team created a complex hypothesis of insect transmission in Kentucky, reasoning that insects picked up the deadly microbe from horses, transferred it to blackbirds that flew to other areas, where more insects now carried the germ to unsuspecting humans. Meanwhile, at least two other prominent doctors actually isolated protozoa from many, but not all, pellagrins, and published these as either causes or cofactors. Even the Department of Agriculture sent a special team of entomologists to study insects in South Carolina in 1912. Potential transmission routes ranging from contaminated drinking water to mosquitoes, even houseflies and bedbugs, were suspected as vectors carrying pellagra, and newspaper articles served to fan public fears as the epidemic grew—not unlike our modern response to the AIDS epidemic:

> So great was the horror of the disease that a diagnosis of pellagra was synonymous with a sentence of ostracism. A severe case of eczema was enough to start a stampede in a community, and pellagrins sometimes covered their hands with gloves or salve, hoping to conceal their condition.

Many hospitals refused admission to pellagra patients. One in Atlanta did so on the grounds that it was an incurable disease. At another hospital in the same city student nurses went on strike when they were required to nurse pellagrins. Physicians and nurses at Johns Hopkins Hospital in Baltimore were forbidden even to discuss pellagra cases which might be there. Fear of the disease spread to schools and hotels, too...

Tennessee began to isolate all its pellagra patients. The state board of health declared pellagra to be a transmissible disease and required physicians to report all cases...

Exhibits on pellagra were prepared for the public, creating fear of the disease along with interest in it...

There was pressure for a quarantine in Kentucky, and pellagra patients at the Western Kentucky Asylum for the Insane were isolated...

Isolation did not prevent spread of pellagra but instead heightened panic over it.[15]

A second National Conference on Pellagra was organized in South Carolina in 1912, and this time the momentum of scientific and medical thought had turned in favor of finding pellagra germs. Earlier that same year, an official federal government commission, the Thompson-McFadden Commission, was created and began studies in the South. One of its three leaders was an Army Medical Corps man who had previously served on Louis Sambon's pellagra commission in England. Not surprisingly, the commission showed a complete bias for infectious causes. Quickly and casually dismissing dietary connections, the commission turned its attention to studies of sewage, insect transmission, bacteria, fungi, and even the suggestion that Italian immigrants had brought the disease with them. Ultimately, the stable fly was officially blamed by the commission for spreading the deadly contagion.

The prestige of this federal commission spurred the Public Health Service to renew its own effort at finding the pellagra microbe in 1913. Lavinder was reassigned to head a group that once again tried in vain to give monkeys the disease from injections of human blood. Yet even then Lavinder could not completely let go of the infection hypothesis, and eventually he gave up pellagra research altogether.

Finally, as the epidemic reached the two-hundred-thousand–victim mark during 1914, and while the Thompson-McFadden Commission continued to issue its reports, the Public Health Service replaced Lavinder with an obscure officer named Dr. Joseph Goldberger as head of their team. This was the turning point in the epidemic.

Within weeks of arriving in the South, Goldberger saw something the entire medical establishment and its experts, obsessed with microbes, had failed to notice: Venturing both into rural areas and insane asylums to see the victims firsthand, he was astonished to find that even where many patients were concentrated, their doctors and nurses did not catch pellagra. He also observed the different diets of the two groups, the doctors eating meat and vegetables and the farmers their customary corn diets. Goldberger drew the inescapable conclusion. Some nutritional deficiency was the cause. After publicly stating his hypothesis in 1914, he was attacked by doctors who insisted the disease was contagious.

Gathering the proof for his notion through a series of experiments in which he completely cured pellagra by changing diets in orphanages, hospitals, and prisons, Goldberger announced his findings in 1915. The *New York Times* carried the story, although on its inside pages. At a medical conference, where the leaders of the Thompson-McFadden Commission presented further findings on infection, Goldberger stirred up intense anger and controversy by critiquing the commission's latest study. When he then presented his own results, the effect was electrifying. Two leading advocates of the contagion view backed down, one of them a leader of the Thompson-McFadden Commission, the other withdrawing his own paper from submission.

But when the news media began giving Goldberger's results favorable publicity, pellagra microbe hunters reacted with alarm and anger. Prominent doctors joined in a growing chorus of protest against the supposedly dangerous nutrition hypothesis, arguing that the public was now being misled. One such doctor at a medical conference "drew applause when he described as 'pernicious' the newspaper publicity that told people there was no danger of pellagra except from poor food and cooking."[16]

The Thompson-McFadden Commission struck back especially hard in 1916 in the pages of medical journals as well as in the *New York Times*. They reiterated their conclusions, including the dangers of insects. Goldberger patiently confronted his critics and answered their objections, but finally reached a point of exasperation. He decided to perform a new experiment to prove the disease noninfectious. He, his wife, and fourteen coworkers injected themselves with samples of blood, feces, mucus, and other bodily fluids from pellagra patients. As he expected, none contracted pellagra. Even this experiment had little effect on medical opinion. Opponents continued to attack or ignore him for several more years, their ranks only gradually thinning with time. Part of the problem lay in pellagra's increasing human toll until the early 1930s, when diets finally began changing to include greater variety. Goldberger continued studying the disease until his death in 1929. Niacin, the vitamin missing in pellagrin diets, was isolated in the mid-1930s.

THE LAST STAND OF THE BACTERIA HUNTERS

By the 1930s, the era of bacterial hunting was rapidly drawing to a close. Improved nutrition had improved everybody's immunity, and improved immunity in turn had reduced disease from microbial infection. Today infectious disease constitutes only about 1 percent of all causes of death in the industrial world. Public fear of contagion evaporated along with the epidemics, and the microbe hunters were forced into relative obscurity for a time.

But today the bacteria hunt is enjoying a modest revival, largely in the wake of the virus-hunting era that currently dominates biomedical research. Syphilis is one example. This is a genuinely infectious venereal disease, first causing genital sores called *chancres* and often spreading from there throughout the body in secondary stages, thereby causing a limited variety of symptoms in different patients. From this ability the disease acquired the unjustified name, "the Great Masquerader." A bacterium was isolated for syphilis in 1905 that fully meets Koch's postulates for causing the disease.

Along with the earlier, well-defined symptoms of syphilis, scientists identified an additional stage, known as *neurosyphilis*, in which the bacterium would supposedly invade the central nervous system, including the brain, years after the original infection and disease. This later manifestation of the disease results in dementia and insanity. However, if dated from the time of infection, this dementia stage develops only after long incubation periods, and syphilis bacteria cannot be isolated in large numbers from the central nervous system even once these symptoms appear. And infected monkeys have never shown neurosyphilis. Neurosyphilis has also suddenly died out in humans once treatment was switched from arsenic compounds in the 1950s to penicillin. The bacterium therefore does not seem to meet Koch's postulates for this particular disease stage.

A better explanation of neurosyphilis may lie, ironically, in the treatment itself. Throughout the nineteenth century, the therapy of choice was mercury, the poisonous heavy metal known to cause nerve and brain damage, especially over long time periods. After the discovery of the syphilis bacterium, doctors began switching their treatment to arsenic-derived compounds developed by Paul Ehrlich and dubbed "magic bullets." Arsenic treatments, however, were not without complications either. Only after the introduction of penicillin, rather than mercury and arsenic, to treat syphilis in the 1950s—and with it the decline of neurosyphilis—did it become apparent that doctors had been mistakenly confusing the poisonous effects of these chemicals with syphilis itself.

Since the introduction of penicillin, mercury and arsenic treatments are no longer used and neurosyphilis has become medical history. But this long-standing belief in the ability of the syphilis bacterium to cause dementia years after infection continues to fascinate scientists. Some researchers who raise questions about the true cause of AIDS, for example, have offered the notion that AIDS might be a disguised form of syphilis or at least that this might explain AIDS dementia. But AIDS may not be infectious at all.

LEGIONNAIRES' DISEASE

Undoubtedly, the most spectacular modern bacterial "epidemic" in America has been Legionnaires' disease, which received an inordinate share of media and official medical attention despite serious questions about the disease itself. The original incident occurred about two weeks after the nation's 1976 bicentennial celebration at the Pennsylvania convention of the American Legion. The convention was headquartered in the Bellevue Stratford Hotel in Philadelphia. Within days after the four thousand plus conventioneers had disbanded and returned home, many of them began showing up in hospitals throughout the state with severe, sometimes lethal, pneumonias. The entire epidemic ended within a few more days, leaving 182 casualties, including 29 deaths.

A special team of investigators from the federal Centers for Disease Control (CDC) spent the next five months trying to isolate the germ responsible. None of more than fifty known viruses, bacteria, fungi, or protozoa could be found in all the victims, but that December one CDC lab researcher discovered a previously unknown bacterium in tissue samples from some of the patients. The CDC immediately declared the microbe guilty of causing Legionnaires' disease, taxonomically designating it *Legionella pneumophila*. According to their hypothesis, the bacteria had infected the legionnaires through the air-conditioning system in the Bellevue Stratford Hotel, where it had quietly been growing. Since that date, CDC officials have retroactively blamed previous mysterious epidemics all around the country on *Legionella* and continue to pin many periodic but small epidemics of flu-like pneumonias on the germ.

But simply finding another germ in such victims cannot tell a scientist whether that microbe actually causes the disease or whether it may simply be one of the many harmless microorganisms found in humans and animals. One microbiologist has stated the point that such germs can always be "secondary invaders," opportunists that take advantage of a weak person's decreased resistance rather than causing the weakness in the first place.[17] The opportunistic microbe defines the diagnostic disease but did not cause the immunodeficiency that allowed it to take over its victim. As discussed throughout this chapter, the only logical standards of proof for causation are Koch's postulates.

Legionella fails the test. The first postulate states the germ must be found in all cases of the disease and must be multiplying actively enough in the appropriate tissues to explain the symptoms. But even among the legionnaires struck in the 1976 outbreak, 10 percent of the victims were never infected by the bacterium. In other pneumonia epidemics, the percentage of sick people positive for the germ has ranged from 1 percent to this example of 90 percent. Even these figures may be high, since other bacteria can mimic *Legionella* in the laboratory tests. Since CDC scientists often do not think to exclude other bacteria, "limited testing for other bacteria may have inflated the frequency of *Legionella* infections."[18]

Koch's second postulate proved to be difficult to meet in those victims who have been infected by the original germ. The microbe appears to be so inactive in the body that it cannot be found in the saliva or mucus. It is, indeed, hard to culture at all, even from the lung tissue it infects.

Koch's third postulate requires the germ to duplicate the sickness in a newly infected host, usually an animal. *Legionella* will cause some symptoms, or even death if injected in large amounts, but only in guinea pigs. While the germ also successfully infects and grows in hamsters and rats, it does not cause serious disease in them. The microbe even seems to have a hard time making the guinea pigs ill, since many cultures of the bacteria fail this experiment.

CDC experts admit the symptoms of Legionnaires' disease are easily confused with other types of pneumonia, suggesting that perhaps other germs are actually causing the symptoms. This possibility now stands confirmed, since many antibiotics that kill *Legionella* in the lab culture dish do not cure the disease in humans, while many that work in humans cannot kill the bacteria in culture. These latter antibiotics must be killing other microbes in the body.

The evidence indicates *Legionella* is actually quite harmless. Since 1976, CDC and public health investigators have found the bacteria all over the country, in water cooling towers, condensers, shower heads, faucets, humidifiers, whirlpools, swimming pools, and even hot-water tanks, assorted plumbing, mud, and lakes. The bacterium is so universal that between 20 percent and 30 percent of the American population has already been infected, yet virtually no one ever develops Legionnaires' disease symptoms. Even laboratories testing for this bacterium find problems, because *Legionella* frequently contaminates the experiments from the surrounding air.

Thus, the CDC should have dropped *Legionella* and searched for other causes long ago. Pneumonias are often caused by microbes already living in the body, rather than new ones infecting from the environment. The body contains many potentially harmful germs that rarely, if ever, cause illness until the immune system becomes weak for some other reason. Legionnaires' disease was probably one example of pneumonia caused by standard germs that take advantage of people whose resistance had been lowered.

So what made the legionnaires susceptible? The CDC has presumed *Legionella* did all the work, but the first question to ask should be whether the original cause was even contagious. One month before the CDC isolated the bacterium, a U.S. House of Representatives Investigative Committee held hearings excoriating the CDC for not having looked for toxic chemicals as a possible cause of the 1976 epidemic.[19] Chairman John Murphy of New York sharply attacked the investigation because "The CDC, for example, did not have a toxicologist present in their initial team of investigators sent to deal with the... epidemic. No apparent precautions were taken to deal with the possibility, however remote at the time, that something else might have been the cause."[20]

The outbreak certainly did not fit the pattern of infectious epidemics. The CDC itself has openly admitted that none of the afflicted legionnaires transmitted the disease to anyone else nor can human-to-human transmission be documented in any other supposed *Legionella* epidemic. The hotel staff in 1976 experienced none of the disease nor have any doctors or nurses caring for such patients ever contracted the illness. Conversely, some of the legionnaires with the disease stayed only in nearby hotels and never spent any time in the Bellevue Stratford. Thus, the disease was not distributed randomly among people exposed to *Legionella*, as contagion should.

The victims, as it turned out, were textbook examples of people at risk for pneumonia. Not just the average legionnaire, the affected people were heavier smokers, had prior heart and lung conditions, were older, and included several who had received kidney transplants and the accompanying immune-suppressive drugs to prevent organ rejection. Because the convention had taken place during the nation's bicentennial, these highly susceptible people also engaged in unusually heavy drinking. The "epidemic," such as it was, resulted from the classical risks for pneumonia. Certainly, it presented no public health threat.

Representative John Murphy delivered the important lesson: "The early investigators of legionnaires' disease focused so intensely on a biological cause—upon a virus, fungus, or bacteria—that chemicals and poisons were apparently largely overlooked."[21] Yet the CDC and the sensational media coverage of the small and short-lived outbreak terrified the American public at large, and they continue to do so in various small epidemics every year. Despite what Congressman Murphy called a "fiasco," the CDC has recovered politically and continues to hold the official view of *Legionella* as a public health threat. The first international conference of scientists studying *Legionella* was held at CDC headquarters in 1978, and a growing number of researchers have earned their salaries producing thousands of papers since that time, creating an entire field of science for studying this modest germ. This deluge of misdirected information has drowned out any public criticism of their flawed hypothesis of infection.

The bacteria hunters of the turn of the century failed to grasp the point that vast numbers of harmless microbes exist in the world and that even potentially pathogenic bacteria only cause life-threatening disease in those whose immune systems are temporarily or chronically impaired. But a scientist who assumes an epidemic to be infectious can always find a harmless, ubiquitous microbe that, whether through occupational exposure or by sheer coincidence, will correlate with the disease. Microbes lived on this planet long before humans. We coexist with a sea of microbes and benefit from many, including those that naturally reside in the human body. Simply finding a microbe is not enough to convict it of causing disease; Koch's postulates must be met. Otherwise, reckless science can obstruct genuine discoveries leading to effective prevention and cure. Ironically, public anxiety about catching a contagious disease actually propels microbe hunting, for desperate people will gratefully provide extraordinary money and power to researchers and public health officials to protect them from microbial epidemics. Scientists with alternative views are pushed aside, *for too many noninfectious diseases would put microbe hunters out of business.*

"Better safe than sorry" is the ultimate argument of those who warn that any unidentified pathogen is infectious unless proven otherwise. But since the establishment of the germ theory by Koch and Pasteur, the medical establishment has never erred on the side of noninfectious causation of disease. Instead, thousands of lives have been lost by misdirected prevention and treatment of noninfectious diseases with microbial measures and "therapies."

Bacteria hunting did actually fade for a time, mostly following the disappearance of serious contagious epidemics from the industrial world. But today microbe hunting has returned in force, searching for viruses as well as bacteria—even though infectious plagues have not returned. The reasons lie in the deep-seated bias for microbial causes of disease and in the explosive growth in funding for biological research, which has built a powerful array of government and private institutions with large vested interests in laboratory medicine and biotechnology. The scientific bureaucracy has become immensely larger, and the techniques for finding microbes incredibly sensitive, allowing even the most minute quantities of inactive germs to be isolated from any diseased patient. The discovery of microbial diseases has become a weekly routine in the scientific press releases—but the rest of the story, that the same microbes are later also found in healthy people, remains hidden in the professional literature. Now follows the story of modern virus hunting and the political infrastructure built around it.

CHAPTER THREE

■

Virus Hunting Takes Over

TRADITIONALLY, THE SCIENTIST HAS been the creative individual who searched for simple explanations of seemingly complex phenomena. Copernicus and Galileo, for instance, reinterpreted the motions of planets in the sky, inferring that the earth and other planets revolve around the sun, not the sun around the earth. Newton puzzled out why apples should fall down and not in other directions and discovered the law of gravity. Koch found a method of proving when a germ causes a disease. Einstein seized on seemingly paradoxical behavior of light and proposed his theory of relativity as an explanation—without having performed any experiments on the subject. Watson and Crick, who never experimented with DNA, took a second look at existing physical and chemical data and deduced the structure of the genetic molecule.

Many pivotal contributions to science throughout history have consisted less of new observations than of new explanations for old data. Classical scientists did not view their occupation in terms of gathering data, but rather in terms of discovering the logical mistakes and simplifying the complexities of the prevailing explanations. Such work tended to wound egos and invited the anger of colleagues whose pet hypotheses had been sunk, but the scientific enterprise in any case achieved its well-deserved reputation for brilliant innovation.

Because experimentation played such a limited role compared to thinking in classical science, the process was relatively inexpensive. Scientists labored nearly in obscurity, driven not by high-stakes politics or finance but by their own curiosity. Nuclear physicist Ralph E. Lapp, a prominent scientist who served as a researcher and advisor on the Manhattan Project, the Atomic Energy Committee, the Office of Naval Research, and other institutions, experienced science before and after the postwar transition. His early training had predated this change, allowing him to describe the classical situation:

> One has to have experienced these lean years in science to remember how frugally money was hoarded for research in physics. In those days no scientist ventured to ask the federal government for funds. He gathered together what money he needed from private sources or earned extra pay as a consultant to pay for his own research. But mostly he acted as a Jack-of-all-trades and built his own equipment. Graduate students were required to take machine-shop practice and learn glass blowing. If he needed Geiger counters he made them himself, and he wired his own electronic circuits. The physicist was the original "do-it-yourself" man on campus...
> When scientists found, as they did after the great crash on Wall Street, that new ideas demanded financial support for their exploitation, they did not think of asking the government to help. Funds to build cyclotrons and other expensive machinery of science were secured from private sources, generally from foundations, and the cost of operations was assumed by colleges, universities, and a few institutes.[1]

All other scientific fields, and indeed academic pursuit in general, faced the same conditions. The little federal money available went mostly into applied biology through the Department of Agriculture.

But then came the Second World War, its immediate aftermath, and the Cold War. The detonation of two nuclear bombs over

Japan, products of a program known as the Manhattan Project, violently brought science into public awareness. A team of scientists, equipped with $2 billion, had invented the new weapon in an around-the-clock engineering effort. This success was soon coupled with the onset of the Cold War, symbolized in the launch of *Sputnik*, the first artificial satellite. This Soviet propaganda coup terrified Americans, creating strong public support for crash science and engineering research efforts to catch up with the Soviets.

The federal government moved to take advantage of this opportunity. The Atomic Energy Commission, formed in 1947, picked up the remains of the Manhattan Project and continued nuclear research. The National Science Foundation was established in 1950 and began disbursing grants for basic scientific research. In the years that followed, a bewildering array of federal science departments and agencies materialized to fund and monitor research of all kinds in government facilities, universities, and independent research labs.

This new science establishment was modeled after the Manhattan Project's team-based investigation. Priorities therefore focused on the practical results of science, an appropriate goal for the engineering- and technology-oriented research that first dominated the new federal spending programs. But recognizing that technology is the applied form of fundamental science, the government soon began throwing money at basic research as well and thus transformed it into a bureaucracy. Creative geniuses were swept aside to make way for skilled administrators who led large teams of specialized technicians, whose only strength was gathering ever-larger quantities of raw data. Where nonconformist individuals once competed with only a handful of peers, they now faced opposition from tens of thousands of irritated colleagues, a crowd that could more easily drown out minority viewpoints. Experiments replaced contemplative thought and analysis, while research became dazzlingly high-tech—and incredibly expensive.

Just before World War II, total research and development funding in the United States, public and private together, amounted to approximately $250 million per year. By the mid-1950s, the federal share alone had grown to more than $2 billion, reaching $63 billion in 1989, and in 1993 becoming half of all research and development spending in the United States at $76 billion.[2] Even adjusting for inflation, this federal spending figure has greatly outpaced the growth in our national economy, becoming 1.25 percent of the entire gross national product by 1989. Federal research money has turned into the major funding source for universities and other institutions, expanding and reshaping departments in its wake.

President Eisenhower summarized the emerging problem well in his 1961 farewell address:

> Today, the solitary inventor, tinkering in his shop, has been overshadowed by task forces of scientists in laboratories and testing fields. In the same fashion, the free university, historically the fountainhead of free ideas and scientific discovery, has experienced a revolution in the conduct of research. Partly because of the huge costs involved, a government contract becomes virtually a substitute for intellectual curiosity. For every old blackboard there are now hundreds of electronics [sic] computers... The prospect of domination of the nation's scholars by federal employment, project allocations, and the power of money is ever present—and is gravely to be regarded.[3]

Ironically, Eisenhower had previously declared in 1957 that "shortages of trained manpower exist in virtually every field" and had pushed for rapid production of more scientists.[4] This supposed Ph.D. shortage defined the basis of an important part of the explosive federal spending, a portion of which was devoted exclusively to the subsidy of graduate students and postdoctoral fellows to work in scientific fields. Universities, especially their science departments, became little more than factories turning out new doctorates as quickly as possible.

The results have been predictable. When the American Association for the Advancement of Science was established in 1848, it had 461 scientists as members. It then reached 36,000 members during World War II and already passed 100,000 during the 1960s.[5] Today it boasts some 135,000 members and is only one of many growing science associations. The National Academy of Sciences, in which membership even today is a unique honor reserved for a few scientists, started in 1863 with 50 members. Those ranks swelled past 600 by the mid-1960s and now stand at 1,650.[6] The total number of science doctorates awarded each year has increased from under 6,000 in 1960 to nearly 17,000 in 1979.[7] By the mid-1980s, the ranks of Ph.D.s and M.D.s working in science or engineering had swelled to 400,000, a figure that for decades has grown much faster than national employment.[8] As a result, "Of every eight scientists who ever lived [in the history of the world], seven are alive today [in 1969]";[9] similar statistics would hold today. Nor has the pressure for further expansion abated until very recently, as evidenced in a 1990 policy statement of the Association of American Universities referring to an "impending Ph.D. shortage."[10] Only in October 1995 did *Science* for the first time begin to worry about the imminent American Ph.D. glut.[11]

Yet we cannot find among them the eight modern Galileos, Plancks, Einsteins, Kochs, Pasteurs, or Mendels that these statistics predict. Increasing numbers of scientists means many more papers being published in scientific journals, with the publish-or-perish stakes rising constantly. According to one summary, "The first scientific journal... began publication in 1665. By 1800 there were 100 journals; by 1900, 10,000 journals; today [1969], over 100,000."[12] By 1986, an unreadable total of nearly 140,000 papers were being published each year just by U.S. scientists, about one-third of the world total.[13]

Such overgrowth in scientific ranks produces regression to the mean. Competition among large numbers of scientists for one or a few central sources of funding restricts freedom of thought and action to a mean that appeals to the majority. The scientist who is very productive, most able to sell research, and well liked for not offending his peers with new hypotheses and ideas is selected by his peers for funding. The eccentric, "absent-minded professor" with "crazy" ideas has been replaced by a new breed of scientist, more like a "yuppie" executive than the quirky genius of old academia. These peers cannot afford a nonconformist, or unpredictable, thinker because every new, alternative hypothesis is a potential threat to their own line of research. Albert Einstein would not get funded for his work by the peer review system, and Linus Pauling did not (for his work on vitamin C and cancer even though he received two Nobel Prizes). The only benefit of the numerous cascades of competitive tests and reviews set up by peer review is the elimination of unsophisticated charlatans and real incompetence. In sum, the review of too many by too many achieves but one result with certainty: regression to the mean. It guarantees first-rate mediocrity. As these armies of new scientists flood the peer review system, they even act to suppress any remaining dissension by the few remaining thoughtful researchers. Peer review, after all, can never check the accuracy of experimental data; it can only censor unacceptable interpretations. A scientist's grants, publications, positions, awards, and even invitations to conferences are entirely controlled by his competitors. As in any other profession, no scientist welcomes being out-competed or having his pet idea disproved by a colleague. Former *Science* editor Dr. Philip Abelson presciently described the pressures against dissenters who raise questions publicly:

> The witness in questioning the wisdom of the establishment pays a price and incurs hazards. He is diverted from his professional activities. He stirs the enmity of powerful foes. He fears that reprisals may extend beyond him to his institution. Perhaps he fears shadows, but in a day when almost all research institutions are highly dependent on federal funds, prudence seems to dictate silence.[14]

Few scientists are any longer willing to question, even privately, the consensus views in any field whatsoever. The successful researcher—the one who receives the biggest grants, the best career positions, the most prestigious prizes, the greatest number of published papers—is the one who generates the most data and the least controversy. The transition from small to big to megascience has created an establishment of skilled technicians but mediocre scientists, who have abandoned real scientific interpretation and who even equate their experiments with science itself. They pride themselves on molding data to fit popular scientific belief, or perhaps in adding nonthreatening discoveries. But when someone strays outside accepted boundaries to ask questions of a more fundamental nature, the majority of researchers close ranks to protect their consensus beliefs.

Biology now constitutes about a third of the total basic science in this country and about half of all academic research—far larger than physics, engineering, mathematics, social science, or any other field. Biology's dominance of research has resulted, naturally, from a massive infusion of federal funds, mostly through the National Institutes of Health (NIH). Formerly a backwater agency buried inside the Public Health Service bureaucracy, the NIH has since the 1950s developed a voracious appetite for money. Its 1955 budget hovered somewhere around $100 million; today it spends closer to $10 billion. NIH research grants not only fund some in-house labs, but they now provide the basic source of funding for universities and other institutes, including research conducted in other nations. Half the total federal research spending on universities and colleges—for all subjects combined—is now provided by the NIH. So while academic institutions formerly provided their own limited monies for research, NIH grants have now become a major source of *income* for the larger and increasingly dependent universities. According to a 1990 article in the *Journal of NIH Research*, "When NIH sneezes, it is the academic community that catches cold."[15]

As both funding and conformism increase, one would expect the potential for disastrous mistakes to increase as well. The new money in biology was grafted onto an establishment long dominated by microbe hunters. Despite the disappearance of infectious plagues, therefore, both bioscience and popular culture have entered a new, revived era of microbiology, now in the form of virus hunting. Because biology is also the foundation underlying medicine, a mistaken hypothesis must inevitably lead to human tragedy. This happened when the successful war on polio, the last triumph of the germ theory, was extended to the misdirected War on Cancer and then climaxed in the failed war on AIDS. Because virus hunting won the war against polio, victorious virus hunters continued to march against cancer and AIDS with the same concepts—but not with the same success.

FROM EARLY VIROLOGY TO POLIO

Unlike bacteria, protozoa, or fungi, viruses are not living microorganisms. Whereas bacteria are single-celled creatures, viruses are much smaller and cannot grow on their own. Composed typically of protein and either DNA or RNA (the genetic molecules), virus particles must infect living cells, tricking their new hosts into producing large numbers of viral molecules, which are then assembled into new viruses like cars on an assembly line. Only by this means do viruses "survive" and go on to infect new hosts. While countless different viruses exist in the world, each can infect only a limited range of living hosts, and then only specific cell types within the host's body. Every category of living organism, whether animal, plant, or bacterium, is susceptible to infection by some of nature's viruses.

The early microbe hunters began accidentally finding viruses while searching for bacteria. During the eighteenth century, Edward Jenner gained fame for his discovery that humans could be immunized against smallpox by injecting material from cowpox. Jenner could not know that he had used a virus, much less what a virus was, and he lived decades before anyone even proposed bacteria to be disease-causing. When Louis Pasteur turned to rabies research in the early 1880s, he correctly discovered that the disease could be transmitted from one animal to another through its saliva but was astonished that he could never find any guilty bacterium. Pasteur guessed the cause to be a bacterium too tiny to see even in the microscope; in fact, this was also a virus. Pasteur then became the second person to invent an immunization, this time for rabies.

Not until 1892 did anyone perform the first actual isolation of a virus. Russian bacteria hunter Dmitri Iwanowski gathered fluid from tobacco plants suffering the mosaic disease. He passed this liquid through a filter so fine that the pores allowed no bacteria through, yet to Iwanowski's surprise the bacteria-free filtered liquid easily made new plants sick with the disease. This observation was repeated independently by the Dutch botanist Martinus Willem Beijerinck in 1898, who recognized that the invisible cause was indeed some altogether different kind of infectious agent. He coined the term that led to the microbe's name—"tobacco mosaic virus."

In the same year as Beijerinck's report, two German scientists purified a liquid containing "filterable viruses" that caused foot-and-mouth disease in cattle. Walter Reed followed in 1901 with a filtrate responsible for yellow fever, and soon dozens of other disease-causing viruses were being found.

The next logical step was to determine what viruses really were. American chemist Wendell M. Stanley accomplished exactly this in 1935 when he created pure crystals of tobacco mosaic virus from an infectious liquid solution. Having these crystals allowed him to study their structure, and he discovered that these crystallized germs could still infect plants with no trouble. In other words, the virus was not a living organism, since it could be crystallized like salt and yet remain infectious. Soon he and other scientists began routinely crystallizing many different viruses. In 1946, Stanley received the first Nobel Prize ever awarded to a virologist, and two years afterward established the Virus Lab at the Berkeley campus of the University of California, where he later supervised the training of Harry Rubin, Peter Duesberg, and other scientists in virus research.

While viruses were joining the ranks of sought-after microbes, the political institutions that would revive microbe hunting after the second World War were developing. Congress had in 1798 formed the Marine Hospital Service for the medical treatment of sailors, an agency that was renamed and expanded in 1912, right in the middle of the microbe-hunting era. This new Public Health Service received a mandate to investigate and cure human disease, inevitably focusing on contemporary contagious or suspected contagious diseases like pellagra. This bias for infectious disease had been reflected in the name of a small subdivision created in 1887, the Hygienic Laboratory, which itself was expanded in 1930 and renamed the National Institutes of Health (NIH). True to form, the medical experts trained by and hired into these structures could think of no other way to fight disease, and they avidly pursued their one skill right on through both world wars. Even Joseph Goldberger, who discovered that a vitamin deficiency caused pellagra, had spent his previous fourteen years with the Public Health Service chasing microbes.

But as infectious plagues gradually disappeared, microbe hunting not only interfered with the genuine scientific challenges of noninfectious diseases, it also determined unsuccessful, if not disastrous, strategies against diseases that proved to be noninfectious. Indeed the victorious war on polio, the last of the serious contagious epidemics of the industrialized world, became the very model for the failed wars on cancer and AIDS.

Polio had always been, and is still throughout much of the Third World today, an awesome illness. Though often fatal, the disease was best known for causing paralysis, and it tended to strike children most commonly. President Franklin Roosevelt, perhaps the most celebrated polio case of all time, in 1938 set up the private National Foundation for Infantile Paralysis (NFIP) to conquer the dread disease. The impetus provided by the Foundation led many key scientists to research poliovirus, as did the sudden, frightening polio epidemic that exploded in the Western nations, brought home by troops returning from the Pacific theater in 1945.

The virus was first isolated as a filtered liquid in 1908 but, as with all viruses, no one could make these nonliving entities grow outside the body. To produce an effective vaccine, the virus had to be produced in a laboratory. Dr. John Enders and two coworkers stumbled on a means of doing so in 1948 by growing the virus in cells cultured from human placentas cast away at birth. A Nobel Prize was awarded to all three researchers a few years later. In 1955 Wendell Stanley first crystallized the poliovirus in his Berkeley lab.

The major medical lesson of virology had long been that antibiotics, which kill bacteria, are completely useless against viruses. But immunization had been tested since the time of Edward Jenner in the late eighteenth century and proved to be the only effective technique against viruses. Vaccination works by introducing a weakened or inactivated form of a virus into the body, causing the body's immune system to produce a reserve supply of antibodies against the virus. In theory, if the real virus later infects the body, antibody proteins stand ready to attack the germ.

Now that poliovirus could be grown in cell culture, a vaccine became more feasible. Two groups of researchers had already tried making vaccines from viruses grown in monkeys, but vaccines accidentally caused polio in several children during their 1935 trials. The first to try a vaccine from virus grown in cell culture was Dr. Jonas Salk, who worked for the NFIP. Salk used chemically inactivated viruses in a nationwide field test during 1954, with four hundred thousand children receiving vaccinations. After the results came in, the secretary of Health, Education, and Welfare (HEW) officially licensed the vaccine the following March.

With this stamp of expert approval, all public apprehensions dissolved and the NFIP moved immediately to begin universal distribution. The NFIP even lobbied for federal money to provide free vaccine to the poor, but fortunately did not succeed—for within weeks, reports came pouring in of children who were becoming paralyzed from the vaccine itself, which contained rare virus particles that had survived the inactivating treatment. In other words, fully active polio virus had been injected directly into the bloodstreams of many children. More than two hundred people were hit with vaccine-induced polio the summer of 1955, including eleven who died.

Public celebration turned to horror. The disaster forced vaccine production to stop, and within three months a complete political shake-up hit HEW. The secretary resigned, as did the director of the NIH and various other officials. The vaccine was restarted only after screening procedures were tightened, and later another type of vaccine replaced the Salk version altogether.

On August 1 of 1955, at the time of the Salk vaccine disaster, James Shannon was promoted to director of the NIH. A disciplined and intensely ambitious man with a Ph.D. in physiology, Shannon was known to his associates for his aggressive, even ruthless, leadership style. He had developed grand notions of how science should be restructured through a central authority. The Salk vaccine fiasco handed Shannon the opportunity to refashion a small-time agency into the largest biomedical research establishment in human history. As he retrospectively described his view, "The main deficiency preventing progress was the inadequate funding of research... The difficulty seemed to be in the scaling of the system. There were manpower and resources, but they were too modest in size because of the inadequacy of support funds." It was his "profound conviction that an expansion of the science base for medicine was needed, doable, and should be undertaken with a sense of urgency."[16]

Shannon's aims were well-planned and quite specific: "Success was only possible by breaking out of the confines of the then federal budget for the support of the biomedical sciences... A realistic program would have to provide a continuing expansion of the base for scientists' production and an expansion of physical resources to house the expanding programs. The targets seemed clear."[17] He set about immediately to consolidate his support in both houses of Congress. The chairman of the House appropriations committee, John Fogarty, and his counterpart, Senator Lister Hill, became Shannon's close allies in his bid to spark explosive funding increases for the NIH. With their help and the backing of the Eisenhower administration, Shannon successfully doubled the 1956 NIH budget to $200 million for fiscal year 1957, by far the most radical increase in the agency's long history. He continued expanding the NIH until he retired in 1968, by which time the agency's annual budget exceeded $1 billion. The NIH's growth has continued without letup to this very day, its annual spending of more than $15 billion now making it the powerhouse of biomedical and academic research establishments.

Shannon wanted the NIH to create a huge infrastructure for basic research, but he knew that Congress and the public worried more about the practical questions of human disease. Using the Manhattan Project and the space program as models of heavily funded enterprises built during World War II and the Cold War, he organized basic research for "wars" on disease. Shannon had always disliked the NFIP and the Salk vaccine program for having been funded mostly privately rather than under tight federal control, so he began spending the new NIH's funding and taking over polio research in the United States. His war on polio provided grants that trained a growing field of scientists in studying viruses.

This growing virology program meshed well with the microbe hunters who had long dominated the NIH and helped revive their dwindling fortunes. When Shannon turned to creating a War on Cancer over the next several years, the virologists became his frontline soldiers. And when the NIH got involved in the war on AIDS in the 1980s, virus hunters again took charge. Many of the leading scientists in the war on AIDS, such as David Baltimore and Jonas Salk, launched their careers in the wake of the NIH war on polio.

Since the polio epidemic disappeared in the early 1960s, no other catastrophic infectious plague has struck the industrial nations. Cancer and heart disease have become the prominent examples of noninfectious diseases, mostly affecting those of older age, to which medical science has had to turn for employment. But with Shannon's legacy of a reshaped NIH trapped in a virus program of its own making, microbe hunting was rescued from scientific obsolescence and now has a political stranglehold on research.

SLOW VIRUSES: THE ORIGINAL SIN AGAINST THE LAWS OF VIROLOGY

From the discovery of tobacco mosaic virus through the polio epidemic, scientists have found and legitimately blamed many viruses for a variety of diseases, each having passed the acid test of Koch's postulates. But for every truly dangerous virus, many more perfectly harmless passenger viruses can be found in humans and animals. NIH-sponsored polio research during the late 1950s proved the point. Researchers trying to isolate new strains of poliovirus accidentally found numerous closely related passenger viruses—such as Coxsackie and echoviruses—that, like polio, infected the digestive system. Scientists classified some of these viruses as "orphans"—viruses without corresponding diseases. The virus hunters could not bring themselves to believe microbes could exist without being harmful and expected even these "orphan" viruses would someday find appropriate sicknesses.

When trying to blame a passenger virus for a disease, however, one nagging problem haunts the virus hunter: The laws of virology dictate that the illness will strike the victim soon after infection. When microbes infect a new host, they cause sickness within days or weeks at most. In order to cause disease, viruses need to grow into sufficient numbers to take over the body; otherwise, the host's immune defenses will neutralize the invader and prevent disease altogether. The rate-determining step of such fast, exponential growth is the generation time of the virus. Since the generation time of all human viruses is between eight and forty-eight hours, and since the infected cell produces one hundred to one thousand viruses per day, viruses multiply exponentially, increasing in numbers hundred- to thousandfold per day. Within a week or two, one hundred trillion (10^{14}) cells can be produced—one for each of the one hundred trillion cells in the human body.

Therefore, if scientists wish to convict an innocent virus, they must invent a new property for it that allows the virus to violate the laws of virology. For example, they can hypothesize a "latent period" of months or years between the time the virus invades the body and the appearance of symptoms—hence, a "slow" virus.

However, the slow virus concept has never been reconciled with the short generation time of viruses and the immune system. Once the virus lies totally dormant, an intact immune system will never allow any virus to be reactivated to multiply into numbers that would threaten the host.

For a virus to be reactivated, the immune system first must be destroyed by something else—the real cause of a disease. A reactivated virus would just contribute an opportunistic infection. Thus, there are no slow viruses, only slow virologists.

A conventional virus could, however, be *slow acting* in a defective immune system. Indeed, some exceptional victims suffer pre-existing health problems that prevent their immune systems from reacting decisively against the virus, allowing it to continue growing and damaging the host for a long period of time. This can happen with virtually any type of virus, but it is extremely rare. When such a chronic infection does occur, as with a small percentage of hepatitis cases whose immune system is damaged by alcoholism or intravenous drug addiction, the virus keeps growing abundantly in the body and can easily be found by experimental tests.

Other germs, like herpes viruses, can hide out in some recess of the body, breaking out periodically to strike again when the immune system passes a seasonal low. In both examples, only the weakened immune system of the host allows the infection to smolder or occasionally reappear from hibernation. By contrast, a slow virus is an invention credited with the natural ability to cause disease only years after infection—termed the *latent period*—in previously healthy persons, regardless of the state of their immunity. Such a concept allows scientists to blame a long-neutralized virus for any disease that appears decades after infection. The slow virus is the original sin against the laws of virology.

The slow-virus or latent-period concept, now used to connect HIV with AIDS, can be traced back to the days of the war on polio. The researcher who popularized this modern myth is today an authority for AIDS researchers and one whose career epitomizes the evolution of the virus hunters over the past three decades.

Dr. Carleton Gajdusek is a pediatrician who has worked as a virologist at the NIH for decades. Having spent a great deal of time studying contagious childhood diseases around the world, Gajdusek was sponsored by the NFIP and sent to New Guinea in 1957. There, a doctor with the local health department introduced him to a disease called *kuru*, a mysterious ailment that attacked the brain, rendering the victim increasingly spasmodic or paralyzed until death within months. The syndrome existed only among the thirty-five thousand tribal villagers in one set of valleys, mostly the Fore tribe. Before Gajdusek's arrival, no outsider had ever described kuru, although the Fore tribesmen told him the condition had begun appearing a few decades earlier.

Gajdusek's initial study assumed the disease to be infectious. He reported that the natives routinely cannibalized the brains of relatives for ritual purposes, a practice that they told him had begun around the same time as the arrival of kuru. Gajdusek later explained to one interviewer that cannibalism "expressed love for their dead relatives," and that it also "provided a good source of protein for a meat-starved community."[18] Gajdusek decided that kuru was transmitted by the eating of deceased victims' brains. Yet when he searched for a virus, he ran into a baffling absence of evidence. None of the typical signs of infection could be found in the patients. Their bodies showed no inflammation and no fever, no changes were registered in their supposedly infected spinal fluid, their immune systems failed to react as if any microbe had invaded the body, and those people with suppressed immune defenses had no greater risk of catching the disease. Another scientific group soon arrived from Australia and concluded that kuru might be genetically inherited.

Upon arriving back in the United States, Gajdusek was hired by the NIH to work at its institute for studying neurological disease. While continuing to monitor kuru incidence, he devoted his time to laboratory study of the condition. Word of his discovery of kuru meanwhile made its way to England, where another virus hunter was investigating a sheep disease known as *scrapie*, which involved symptoms of brain degeneration. The English researcher suggested to Gajdusek that kuru might be caused by a slow virus, one with a long latent period.

Gajdusek was immediately hooked by the revolutionary idea, despite his own "misgivings" that genes, toxins, or nutritional deficiencies might be the cause of kuru.[19] Again determined to find an elusive virus, he tried to transmit kuru from victims to chimpanzees. But none of the animals became sick when injected with blood, urine, or other bodily fluids from kuru patients, nor from the cerebrospinal fluid that surrounds the brain, which should have been full of the alleged brain-destroying virus. Indeed, the monkeys contracted no disease even from eating kuru-affected brains—the authentic animal model of cannibalism.

Only one bizarre experiment did work, in which the brains of kuru patients were ground into a fine mush and injected directly into the brains of live monkeys through holes drilled in their skulls. Ultimately, some of the experimental monkeys suffered coordination and movement problems. Surprisingly, though, even this extreme method could not transfer kuru to dozens of other animal species. And no virus could be seen in the brain tissue, even using the best electron microscopes.[20]

At this point, one might expect Gajdusek would have suspected something was seriously wrong with his virus hypothesis. If evidence for the invisible virus could not be found anywhere but in unpurified brain tissue, if it did not elicit any defensive reactions by the body, and if it could not be transmitted in pure form to animals, then probably no virus existed at all. The homogenized brain tissue of dead kuru patients—full of every imaginable protein and other compounds—should in itself be toxic when inoculated into monkeys' brains.

Nevertheless, the sick monkeys convinced Gajdusek and his colleagues he had found a virus. Since he could not isolate it apart from the brain tissue, he decided to study the virus and its structure with a standard experiment: He would define which chemical and physical treatments would destroy the microbe, thereby

78 ■ INVENTING THE AIDS VIRUS

gathering clues about its nature. But to his astonishment, almost nothing seemed to harm the mystery germ. Powerful chemicals, acids and bases, boiling temperatures, ultraviolet and ionizing radiation, ultrasound—no matter how he treated the brain tissue, it still caused "kuru" in his lab monkeys. Further tests also proved that no foreign genetic material, which all viruses require for their existence, could be found anywhere in kuru-affected brains.

Employing the strongest virus-destroying treatments, Gajdusek had failed to render the kuru brain tissue harmless in his experiments. His results lent themselves to one obvious interpretation: No virus existed in the first place, so it could not possibly be destroyed. But Gajdusek clung to his virus hypothesis. Despite his disappointing experiments, he turned the results upside down and argued that the "kuru virus" was actually a new type of supermicrobe or, as he put it, an "unconventional virus." This new virus also needed to act as a slow virus, since long periods of time elapsed between an act of cannibalism and the onset of kuru; he liberally suggested latent periods extending into years or even decades.

At an earlier time, and in another context, Gajdusek probably would have been ignored by orthodox scientists. But he offered this hypothesis to a generation of scientists dominated and impressed by virus hunters. The year was 1965, polio had largely disappeared, and the burgeoning ranks of NIH-funded virologists welcomed any new research direction on which to use their skills. Thus, they embraced Gajdusek's slow virus hypothesis enthusiastically. They listened uncritically when he claimed a similar unconventional virus caused Creutzfeld-Jakob disease, a rare brain disorder that seems to strike mostly Westerners having undergone previous brain surgery (obviously such medical operations might well be suspected as the real cause). Gajdusek proposed slow or even unconventional viruses as the causes of a huge laundry list of nerve and brain disorders, ranging from scrapie in sheep to multiple sclerosis and Alzheimer's disease in humans, and he was taken seriously even though he offered no proof. Entranced, his peers awarded him the 1976 Nobel Prize for medicine, specifically for the kuru and Creutzfeld-Jakob viruses he has yet to find. And the NIH promoted him to head its Laboratory of Central Nervous System Studies.

In the meantime another crucial, if embarrassing, bit of information has emerged as a challenge to Gajdusek's virus-kuru hypothesis. The published transcript of his Nobel acceptance speech, in a 1977 issue of *Science* magazine, included a photo ostensibly showing New Guinea natives eating their cannibalistic meal. The photo is not very clear. When colleagues asked Gajdusek if the photo truly showed cannibalism, he admitted the meal was merely roast pork. According to *Science*, "He never publishes actual pictures of cannibalism, he says, because they are 'too offensive.'"[21] Unconvinced, anthropologist Lyle Steadman of Arizona State University has investigated and directly challenged Gajdusek, claiming "there is no evidence of cannibalism in New Guinea." Steadman, who spent two years doing fieldwork in New Guinea, noted that he often heard tales of cannibalism but when he probed, the evidence evaporated.[22]

Gajdusek, angered by the hint of malfeasance, has insisted that "he has actual photographs of cannibalism, but he would never publish them because they 'so offend the relatives of the people who used to do it.'"[23] This statement contradicts his earlier claims that the tribesman proudly ate their dead relatives out of respect, quitting the practice only in deference to outside pressure from government authorities. For evidence of cannibalism, Gajdusek also cited Australian arrests of tribesmen for the alleged crime—which, as it turned out, were based on hearsay accusations.[24] So perhaps New Guinea natives stand falsely accused of ritual tribalism.

In addition, few people outside of Gajdusek's original research team have ever personally witnessed kuru victims. This means we also depend on his own descriptions and statistics for our knowledge of the disease itself, particularly since he claims cannibalism and kuru both ceased to exist within a few years after his 1957 trip. Phantom viruses, transmitted through phantom cannibalism, cause phantom disease.[25]

Yet Gajdusek has reshaped the thinking of an entire generation of biologists, his seductive message of slow viruses having landed on eager ears. He and the virus hunters inspired by him have built careers chasing viruses and attributing them to latent periods in order to connect them to noninfectious diseases.

SMON, the nerve-destroying disease that struck Japan during the 1960s, became one unfortunate example. Japanese virologists, greatly impressed with Gajdusek's accomplishments, spent years searching for slow viruses they presumed would cause the disease and thereby delayed finding the true cause—a prescribed medication.

Another example of a pointless virus hunt involved diabetes. Beginning in the early 1960s, some scientists tried to blame this noncontagious syndrome on the virus that also causes mumps. The evidence has been pathetically sparse, forcing virologists to point to occasional children who become diabetic after they have also suffered mumps or, if they really stretch their case, to argue that both mumps and diabetes become most common during the same annual season in one county of New York.

Having become soldiers without a war, veteran polio virologists invaded the diabetes field as well, proposing since the early 1970s that Coxsackie viruses may cause the disease. Antibodies against several strains of these harmless viruses, first discovered as by-products of polio research, have been found in a few diabetic children. But between 20 percent and 70 percent of young diabetics have never been infected, and the remainder have already neutralized the virus with their immune systems long before the onset of diabetes. Apparently, an equal percentage of *non*-diabetic children have also been infected with these Coxsackie viruses. Needless to say, none of the above viruses meets Koch's postulates for causing diabetes.

Hilary Koprowski, like Gajdusek, typifies the modern virus hunter.[26] Although Koprowski's virology career began earlier, Gajdusek's work helped rescue Koprowski from the obsolescence that threatened polio researchers after the war on polio. Like so many of his colleagues, he found his newest calling in the war on AIDS.

Koprowski's work on viruses started at the Rockefeller Institute in New York. By the late 1940s he moved across town to the Lederle pharmaceutical company, where he worked feverishly to develop a polio vaccine. By 1954 he had invented one, but Jonas Salk was announcing the field trials for another vaccine, and Koprowski's already-tested product was shunted aside by Salk's public acclaim. Koprowski left Lederle in 1957 to take a position as director of the privately endowed Wistar Institute of Pennsylvania, where he began tests on humans and stepped up the campaign to get approval for his vaccine. By now Albert Sabin had tested his own polio immunization on millions of people in foreign countries, completely overshadowing Koprowski's equally successful but less-promoted vaccine. Nevertheless, Koprowski's day did arrive. His vaccine became the standard used by the World Health Organization in America during the late 1950s and 1960s.

In the meantime he spent several years studying the rabies virus and creating a vaccine against that virus, which attacks the brain and nervous system. But because rabies is relatively rare, Koprowski's vaccine never achieved the stardom of other immunizations. More important, however, his rabies research placed him squarely in the field of neurological diseases just in time to meet up with Gajdusek's kuru work. The news of slow viruses enticed Koprowski with visions of groundbreaking science. He quickly realized that the notion of slow viruses could become a useful tool, allowing him to source slow, noninfectious diseases to viruses, so long believed to be fast-acting agents. He participated as a "program advisor" in Gajdusek's first major conference on slow and unconventional viruses held in 1964 at the NIH headquarters in Bethesda, Maryland. From that point forward, Hilary Koprowski joined the new virus-hunting trend from which he would never turn back.

His first big opportunity to take a crack at slow viruses came at the end of the 1960s. Subacute sclerosing panencephalitis (SSPE), a mouthful of a name for such a rare condition, attacks a small number of schoolchildren and teenagers each year, causing dementia, learning disabilities, and finally death. Doctors first recognized SSPE in the 1930s, and by the 1960s the virus hunters were searching for an SSPE germ. At that time, the most fashionable viruses for research belonged to the myxovirus family, which included the viruses that caused influenza, measles, and mumps. Animal virologists therefore started by probing for signs of myxoviruses. Excitement mounted after trace quantities of measles virus were detected in the brains of SSPE patients, and in 1967 most of the victims were found to have antibodies against measles. The facts that SSPE affected only one of every million measles-infected people and that this rare condition appeared from one to ten years after infection by measles were no longer a problem: Researchers simply hypothesized a one- to ten-year latency period.[27] Little wonder they could also easily rationalize that one virus could cause two totally different diseases.

Koprowski's foray into SSPE research began in the early 1970s. He began isolating the measles virus from dying SSPE victims, a nearly impossible task because their immune systems had long before completely neutralized the virus (some SSPE cases, moreover, had never had measles, merely the measles vaccine). His characteristic patience nonetheless paid off, yielding a tiny handful of virus particles from some patients that could be coaxed to begin growing again, if only in laboratory cell culture. In other patients only defective viruses that were unable to grow had remained so many years after the original measles infection. Rather than concluding the measles virus had nothing to do with SSPE, he employed the new logic of virus hunting to argue that a *defective* measles virus caused SSPE!

Koprowski continued this line of SSPE research for several more years. But in 1985 Gajdusek himself entered the SSPE fray, publishing a paper with leading AIDS researcher Robert Gallo in which they proposed that HIV, the supposed AIDS virus, caused SSPE while remaining latent. With hardly a blink, several leading virologists jettisoned the old measles-SSPE hypothesis in favor of a newly popular, but equally innocent, virus.

Multiple sclerosis (MS), the notorious disease that also attacks the nervous system and ultimately kills, has provided yet another opportunity for the virus hunters. First, they blamed the measles virus starting in the 1960s, since many MS patients had antibodies against the virus. Ten years later others suggested the mumps virus, which is similar to measles. The early 1980s brought the coronavirus hypothesis of MS, the category of virus better known for causing some colds. In 1985, with Gajdusek stealing his thunder for SSPE, Koprowski also published a scientific paper that year in *Nature* with Robert Gallo, in this case arguing that some virus similar to HIV now caused MS. Unfortunately for Koprowski, even this hypothesis was abandoned within just a few years.

PHANTOM VIRUSES AND BIG BUCKS

Most virus hunters prefer chasing real, if arguably harmless, viruses as their deadly enemies. But Gajdusek's "unconventional" viruses—the ones neither he nor anyone else have ever found—have been making a comeback in recent years. Given the abundance of research dollars being poured into biomedical science by the NIH and other agencies, opportunistic virus hunters have been finding creative ways to cash in. One increasingly successful method utilizes modern biotechnology to isolate viruses that may not even exist.

Hepatitis, or liver disease, has yielded profitable virus-hunting opportunities in recent years. Hepatitis can be a truly painful affliction, starting like a flu but progressing to more severe symptoms, including high fevers and yellow skin. At least three varieties seem to exist. Hepatitis A is infectious, spread through unsanitary conditions, and is caused by a conventional virus. Hepatitis B also results from a virus (discovered in the 1960s) and is transmitted mostly between heroin addicts sharing needles, among sexually active and promiscuous people, or in the Third World from mothers to their children around the time of birth.

A third type of hepatitis was found in the 1970s, again restricted to heroin addicts, alcoholics, and patients who have received blood transfusions. Most scientists assumed these cases were either hepatitis A or B, until widespread testing revealed neither virus in the victims. Roughly thirty-five thousand Americans die each year of any type of the disease, a fraction of those from this "non-A, non-B hepatitis," as it was known for years. Today it is called *hepatitis C*. This form of hepatitis does not behave as an infectious disease, for it rigidly confines itself to people in well-defined risk groups rather than spreading to larger populations or even to the doctors treating hepatitis patients. Yet virologists have been eyeing the disease from the beginning, hoping one day to find a virus causing it.[28]

That day arrived in 1987. The laboratory for the job was no less than the research facility of the Chiron Corporation, a biotechnology company located directly across the bay from San Francisco. Equipped with the most advanced techniques, a research team started its search in 1982 by injecting blood from patients into chimpanzees. None of monkeys contracted hepatitis, although subtle signs vaguely resembling infection or reddening did appear. For the next step, the scientists probed liver tissue for a virus. None could be found. Growing desperate, the team fished even for the smallest print of a virus, finally coming across and greatly amplifying a small piece of genetic information, encoded in a molecule known as ribonucleic acid (RNA), that did not seem to belong in the host's genetic code. This fragment of presumably foreign RNA, the researchers assumed, must be the genetic information of some undetected virus. Whatever it was, liver tissue contains it only in barely detectable amounts. Only about half of all hepatitis C patients contain the rare foreign RNA. And in those who contain it, there is only one RNA molecule for every ten liver cells—hardly a plausible cause for disease.[29]

The Chiron team used newly available technology to reconstruct pieces of the mystery virus. Now they could test patients for antibodies against this hypothetical virus and soon discovered that only a slight majority of hepatitis C patients had any evidence of these antibodies in their blood. Koch's first postulate, of course, demands that a truly harmful virus be found in huge quantities in every single patient. His second postulate requires that the virus particles be isolated and grown, although this supposed hepatitis virus has never been found intact. And the third postulate insists that newly infected animals, such as chimpanzees, should get the disease when injected with the virus. This hypothetical microbe fails all three tests. But Koch's standards were the furthest thing from the minds of the Chiron scientists when they announced in 1987 that they had finally found the "hepatitis C" virus.

Now more paradoxes are confronting the viral hypothesis. Huge numbers of people testing positive for the hypothetical hepatitis C virus never develop any symptoms of the disease, even though the "virus" is no less active in their bodies than in hepatitis patients. And according to a recent large-scale study of people watched for eighteen years, those with signs of "infection" live just as long as those without. Despite these facts, scientists defend their still-elusive virus by giving it an undefined latent period extending into decades.

Paradoxes like these no longer faze the virus-hunting research establishment. Indeed, rewards are generally showered upon any new virus hypothesis, no matter how bizarre. Chiron did not spend five years creating its own virus for nothing. Having patented the test for the virus, the company put it into production and began a publicity campaign to win powerful allies. The first step was a paper published in *Science*, the world's most popular science magazine, edited by Dan Koshland, Jr., professor of molecular and cell biology at the University of California at Berkeley. Edward Penhoet, chief executive officer for Chiron, also holds a position as professor of molecular and cell biology at the University of California at Berkeley. The NIH-supported virology establishment soon lent the full weight of its credibility to the hepatitis C virus camp. As Chiron's CEO boasted, "We have a blockbuster product."[30] A regulatory order from the Food and Drug Administration (FDA) to test the blood supply would reap enormous sales for Chiron.

Their big chance presented itself in late 1988 as a special request from Japanese Emperor Hirohito's doctors. The monarch was dying and constantly needed blood transfusions; could Chiron provide a test to make sure he received no blood tainted with hepatitis C? The biotech company jumped at the opportunity, making for itself such a name in Japan that the Tokyo government gave the product its approval within one year. The emperor died in the meantime, but excitement over Chiron's test was fueled when the Japanese government placed hepatitis C high on its medical priority list. Chiron's test kit now earns some $60 million annually in that country alone.[31] By the middle of 1990, the United States followed suit. The FDA not only approved the test, but even recommended the universal testing of donated blood. The American Association of Blood Banks followed suit by mandating the $5 test for all 12 million blood donations made each year in this country—raking in another $60 million annually for Chiron while raising the nation's medical costs that much more. And all this testing is being done for a virus that has never been isolated.

Profits from the test kit have generated another all-too-common part of virus hunting. With Chiron's new income from the hepatitis C test, Penhoet's company bought out Cetus, another biotech company, founded by Donald Glaser, who, like Penhoet, also holds a position as professor of molecular and cell biology at the University of California at Berkeley. And Chiron made an unrestricted donation of about $2 million to the Department of Molecular and Cell Biology at the University of California at Berkeley that generates $100,000 in interest each year.

Unfortunately for Peter Duesberg, who belongs to the same department, his supervisor is yet another professor who consults for Chiron Corporation—and displays little sympathy for Duesberg for challenging modern virus hunting by restricting his academic duties to undergraduate student teaching and by not appointing him to decision-making committees. Such conflicts of interest have become standard fixtures in university biology departments.

The modern biomedical research establishment differs radically from any previous scientific program in history. Driven by vast infusions of federal and commercial money, it has grown into an enormous and powerful bureaucracy that greatly amplifies its successes and mistakes all the while stifling dissent. Such a process can no longer be called *science*, which by definition depends on self-correction by internal challenge and debate.

Despite their popularity among scientists and their companies, "latent," "slow," and "defective" viruses have achieved only little prominence as hypothetical causes of degenerative diseases before the AIDS era. Their hypothetical role in degenerative diseases, which result from the loss of large numbers of cells, remained confined to rare, exclusive illnesses like kuru and hepatitis C.

However, because latent, slow, and defective viruses cannot kill cells, such "viruses" eventually achieved prominence as hypothetical causes of cancer and thus entered the courts of health care and medical research. The next chapter describes the terms under which these viruses were promoted as causes of cancer and how some of these terms were eventually used to promote latent, slow, and defective viruses as causes of degenerative diseases including, above all, AIDS.

CHAPTER FOUR

Virologists in the War on Cancer

DURING THE EARLY PART of the century, while infectious diseases were rapidly declining, a few microbe hunters began to sense the changing of the tide. Cancer was on the rise, if only because people now lived long enough to develop it, and its puzzling nature invited innumerable conflicting explanations. The early microbiologists began applying their tools to the chase of hypothetical cancer-causing germs. Among the first to make the connection was the German Emperor Kaiser Wilhelm II, who addressed Robert Koch in 1905 at a reception in honor of his Nobel Prize: "Mein lieber Jeheimat, nu mal ran an den Krebserreger!" (My dear professor, now you must get the cancer-bug [Berlin dialect].)[1]

Despite the imperial encouragement, the results of cancer microbiology remained wanting and did not impress Hans Dewitt Stetten, the grey eminence of the National Institutes of Health (NIH):

> During the heyday of bacteriology, many attempts were made to find a microbial cause of cancer. Bacteria, fungi, and other micro-organisms, often named after their discoverer, were isolated and proposed as candidates. But none of the

claims withstood the rigorous criteria of bacterial causation enunciated by Koch.[2]

But virus hunting was gradually arriving. As more sophisticated technologies for working with viruses became available, the virologists wished to try their hand at explaining cancer.

However, they faced two bothersome paradoxes in trying to blame viruses for cancer: First, cancer is not contagious but all viral diseases are, so how could a virus cause cancer? And second, the typical virus reproduces by entering a living cell and commandeering the cell's resources in order to make new virus particles, a process that ends with the disintegration of the dead cell. Cancer, on the other hand, is a disease of cells that continue to live. Something goes wrong with perfectly normal cells, and they begin changing their behavior and appearance, refusing to cooperate with the rest of the body. Such abnormal cells eventually begin growing relentlessly, invading nearby tissues and ultimately spreading throughout the host. The patient dies once these increasingly voracious parasites have caused enough disruption. So, if viruses kill cells, how could they possibly cause some cells to grow *too well*?

Amazingly, over the next several decades cancer virology not only rescued itself from this initial quandary and the threat of obsolescence, but even managed to seize control of the entire cancer field. Their answer to the first question was that cancer may very well be infectious if one is only patient enough to wait for the virus to progress from infection to cancer—a period said to be over fifty years for viral leukemia and viral cervical cancer. A very delayed infectivity indeed!

The answer to the second question was to postulate either defective killer viruses, unable to multiply but still able to cause cancer, or a unique class of noncytocidal (non–cell-killing) viruses, the retroviruses, acting as carcinogens. With these sophisticated concepts, "tumor virologists" reached the pinnacle of political success in the 1980s and were well positioned to dominate AIDS research from the start. This is the story of their rise to power—despite having no proof for their germ-cancer hypothesis.

As with any example of science gone awry, cancer virology began with perfectly legitimate observations of rare phenomena. Searching relentlessly, virus hunters did come across a few special types of viruses that cause a tumor in some extraordinarily susceptible animal; however, these are freak accidents of nature.[3] Over the years virologists learned to repeat these accidents in the laboratory's artificial conditions. But only decades later did the virus hunters exaggerate the importance of these early results, seizing upon them as precedents claiming harmless passenger viruses as causes of cancer.

The first known tumor virus surfaced in 1908, when a pair of Danish veterinarians studied leukemia in chickens. Vilhelm Ellermann and Oluf Bang experimented and discovered that only something tiny enough to pass through a bacteria-screening filter—a virus—in causing the same leukemia in newly infected chickens would meet Koch's postulates. The following year, a virologist named Peyton Rous, who worked at New York's Rockefeller Institute, made an even more dramatic finding. When a farmer brought him a chicken with a large, well-developed solid tumor, Rous discovered that some filterable virus from the bird produced amazingly rapid tumors in other chickens within weeks or even days of infection. The Rous sarcoma virus (RSV) is a retrovirus. The hallmark of retroviruses is to not kill the cell it infects. As such, they are potential carcinogens.

But neither of these experiments shook the scientific world, because human cancers are not contagious. They dismissed Rous's virus as some oddity of chickens. Tumor biologists also could not find viruses in the human cancers they studied, and they therefore refused to take seriously the observations of the early cancer-virus hunters.

Several more animal tumor viruses were found during the 1930s. A possible leukemia retrovirus was noticed in certain strains of mice, as was another retrovirus that appeared to cause breast cancer in some mice and that also seemed to pass from mother to child through the milk. Both cancers, however, proved almost impossible to duplicate in the lab and would affect only special strains of mice weakened through generations of incestuous inbreeding, a process long known to cause medical problems in humans and animals, including spontaneous cancers. The same viruses produced no effect when injected into wild mice.[4]

Meanwhile, another researcher at the Rockefeller Institute, Richard Shope, isolated the cause of warts in rabbits, a virus that performed more consistently. A handful of virus hunters became excited when Peyton Rous caused true cancers, rather than mere warts, in rabbits injected with the wart virus and some substance called *tumor promoter*. But this virus, found in wild rabbits, would induce the dramatic tumors only when inoculated with tumor promoter.

In a sense, both sides of the virus-cancer controversy were right. Some viruses could genuinely cause some rare cancers, though only in specially susceptible animals under precise conditions. Yet such exceptions bore no relevance for human and animal tumors in general. Such scattered observations by virus hunters did not sway the cancer investigators. When Franklin Roosevelt signed the 1937 legislation creating the National Cancer Institute (NCI), a report issued by an advisory group of cancer biologists declared without hesitation that "[t]he very exhaustive study of mammalian cancer has disclosed a complete lack of evidence of its infectious origin," and dismissed viruses as "agents that may be disregarded."[5] The report echoed the view common among cancer researchers. With no real evidence on which to stand, the field of cancer virology seemingly faced certain extinction.

As a new federal institute charged with managing the fight against cancer, the NCI turned its main attention to developing radiation and chemotherapy treatments against tumors. Of the twenty-four grants disbursed by NCI during its first five years, only two funded virus research, both relatively small. Ironically, however, over the next two decades the NCI would become the very instrument that kept cancer virology alive. Despite their lack of relevance for human cancer, a few virus hunters managed to secure positions in the new agency. The steady trickle of virus experiments did little to advance a general understanding of cancer, but it did begin attracting a few virus hunters to the field. Their one trump card lay in the Rous virus, which stood out for causing its tumor within days of infection, in contrast to chemicals, radiation, and other factors that required months to produce a few tumors in animals.

One of these new cancer virologists, Ludwik Gross, began his tumor virus work at the Veterans Administration Hospital in the Bronx, New York. Having returned from the Second World War and been turned down for an NCI job, Gross accepted a position at the hospital because they allowed him lab space in the basement for part-time research. He picked up the work first done in the 1930s on a virus suspected of causing mouse leukemias, one that seemed to induce cancer only in the more sickly inbred strains but not in healthier mice. After years of persistent study, he finally isolated a retrovirus in the early 1950s. As a leukemia virus that could cause disease only after months of chronic infection in newborns of certain mouse strains, his finding stirred little interest. But during one of his virus isolation procedures, Gross also accidentally found a virus that caused a much more pronounced tumor of salivary glands in the mice.

These two mouse viruses soon became the foundation upon which a revival of tumor virology was built. Only a couple of years after Gross announced his findings in 1953, James Shannon took over as director of the National Institutes of Health. By this time the NCI had become a branch of the NIH. The sudden cash flow that followed, and the spending priority on polio, uncorked the virus-hunting bottle. Many scientists decided to redirect their careers toward cancer viruses.

At the NCI, Sarah Stewart, a former NIH researcher trained in virus research, had already begun duplicating the work of Ludwik Gross in isolating his two viruses. She discovered that the second virus not only caused tumors in the salivary glands but also induced many other cancers throughout the bodies of her newborn mice and therefore dubbed it *polyoma* (meaning "many faces"). A number of the cancer biologists continued to criticize the virus discovery, but virologists enthusiastically followed her lead. The challenge became obvious: to find a virus that causes cancer in *humans*.

The war on polio provided an unexpected opportunity for finding new viruses. In 1959, the Salk polio vaccine was in wide distribution, and the Sabin vaccine was undergoing large-scale trials in foreign countries. Almost simultaneously, two scientists independently found a new virus in the monkey kidneys in which the poliovirus was being mass-produced for the vaccine—in other words, a contaminant. The virus was native to monkeys and caused cell death in the kidney tissues. Inspired by the polyoma discovery, both researchers injected this virus into newborn hamsters in an attempt to cause cancer, even though neither yet knew of the other's work. To the investigators' excitement, the hamsters did indeed get tumors from the virus. As the fortieth virus isolated from monkey cells used to propagate polio vaccines, it was named *Simian Virus 40*, or SV40.

The new virus was first publicly announced in 1960. Millions of children in the United States and abroad had already been immunized with polio vaccine contaminated with this potentially cancer-causing monkey virus. Another million soldiers had received vaccines for a different disease that had been similarly contaminated. Huge studies tracking vaccinated people soon confirmed no unusual cancer cases among them, but the virus hunters had achieved their victory. In the wake of the near panic over SV40, growing amounts of research dollars were earmarked for cancer-virus study. In 1959, for example, NCI specifically reserved the extraordinary sum of $1 million for the field. The notion that viruses might cause cancer in humans had been firmly embedded in the thinking of the scientific community. NIH investigator Robert J. Huebner spoke for many scientists who had joined the growing polyoma research program, "Wouldn't it be interesting if more tumor viruses turned out to be similar to and spread like the 'common cold'?"[6]

Meanwhile the mouse leukemia virus first isolated by Gross had created a parallel field in the tumor virus hunt. Dozens of scientists rushed to find leukemia viruses in animals and humans. From 1956 to 1970 at least a dozen different viruses were isolated from mouse leukemias by researchers throughout the United States and other parts of the world, even as the NIH were disbursing new grants all over the globe. None of these viruses proved to be any more potent than the first one. Several reports of viruses infecting human leukemic cells also poured in, though none met Koch's postulates. The researchers chasing such human viruses knew how to get public attention: One lab named a virus after its discoverer, Elizabeth S. Priori, giving it the intriguing name "ESP virus."

While achieving only dubious results, the net effect of this research was to draw large numbers of virus hunters into studying cancer. As the war on polio wound down, its soldiers switched to the only medical field left with high expectations of success, bringing with them many harmless human viruses they had isolated as by-products of their polio research. Ludwik Gross and other virologists openly argued that human cancer viruses would soon be found. Albert Sabin and many of his fellow polio virologists attended conferences and listened to the new clarion call. Talk of vaccines against cancer filled the air.

Wendell Stanley, the first scientist to receive a Nobel Prize for viruses, entered the national spotlight as one of the leading lobbyists for a full-scale cancer virus program. At the Third National Cancer Conference, held in Detroit in 1956 and partly sponsored by the NCI, Stanley declared:

> I believe the time has come when we should assume that viruses are responsible for most, if not all, kinds of cancer, including cancer in man, and design and execute our experiments accordingly...
>
> Literally dozens of hitherto unknown human viruses have been discovered during the past year or so [mostly as by-products of polio research]... The discovery of this great array of hitherto unknown viruses coursing through human beings made necessary a special conference devoted to these agents. This conference was held in May, 1956, at the New York

Academy of Sciences under the thought-provoking title of "Viruses in Search of Disease." Thus we have today many more human viruses than we know what to do with; hence there is now certainly no reason to shy away from giving consideration to viruses as causative agents in cancer for lack of the viruses. Actually these recent developments lead one to suspect that there are many more undiscovered viruses present in presumably normal human beings.7

Scientists now had plenty of raw material—many human cancers to explain and a growing list of (evidently harmless) viruses to blame them on. The new NIH money rolled in as Stanley and others beat the drums for a new virus hunt. As had happened so often in the history of microbe hunting, such battle cries ultimately generated medical disasters. But this time the crusade was better financed and organized than ever before.

SLOW VIRUSES TO THE RESCUE

No amount of enthusiasm, by itself, could bridge the giant chasm between viruses and cancer. The handful of cancer-causing viruses found in some animals were considered odd precisely because most viruses kill the cells they infect, rather than making them grow better. And as clinical cancer specialists knew all too well, human tumors rarely contained any detectable virus particles. Nor did they expect to find any, since cancer typically behaves as a noninfectious disease: most tumors develop gradually over years or even decades, rather than striking quickly and affecting large populations, as seen in flu epidemics and other contagious diseases. Against such common sense the virus hunters somehow had to justify their anticipated cancer viruses.

Carleton Gajdusek's sudden popularity in the early 1960s derived largely from the cancer-virus crowd. His hypothetical "slow viruses" presented part of the answer they were looking for—viruses that could supposedly act as slowly as the cancer. The cancer virologists lent their full support to promoting Gajdusek, and he responded in kind. Already at the 1964 scientific conference on "unconventional viruses" hosted at the NIH, he proposed in his introductory presentation some nine human tumors as possibly being caused by slow viruses, including two types of leukemia.

But even this invention would not suffice. Virologists needed some way to rationalize the absence of detectable viruses in tumors and the inability of such hypothetical microbes to kill the infected cells. A full decade before Gajdusek arrived on the scene, a French biologist named André Lwoff had already supplied this missing ingredient: the notion of a dormant virus. As with so many virus-hunting myths, the notion of dormant viruses began with a minor but genuine observation in bacteria that was later twisted into relevance for human cancer.

Lwoff began his microbiology career in the 1920s with the Pasteur Institute in Paris. Over the next twenty-five years he developed better methods for culturing microbes and learning about their nutritional requirements. During the mid-1930s, while his nutrition work continued, he heard about a strange phenomenon being studied at the institute. According to a couple of his peers, certain strains of bacteria could be infected by a virus that would often become dormant. The virus literally went to sleep inside the cell, rather than killing its host and infecting new cells. Then at some later time, seemingly normal bacteria could suddenly burst open, releasing the newly reactivated virus.

Since Lwoff could not come up with a rational explanation, many prominent scientists refused to believe his observations were true. But one year after the end of World War II, Lwoff was challenged to prove his ideas of dormant viruses at a conference in the United States. Returning to Paris with a grant from the U.S. National Cancer Institute, he set up his own research program to study this virus latency. After a series of careful experiments, he proved that the virus could indeed become latent in the infected cell for varying periods of time and would reawaken when exposed to ultraviolet radiation. Soon even the most hardened skeptics were convinced.

The phenomenon was certainly interesting, but it applied to only a few viruses. Most viruses lack the ability to become dormant and must either kill the infected cell immediately or fail altogether. Nevertheless, Lwoff's timing could not have been more perfect for the cancer virologists, and he soon made the connection. From 1953 onward he argued forcefully that cancer resulted from the reactivation of dormant viruses, which would begin to recruit cells to form tumors. His hypothesis struck the right chord with tumor virologists. Ludwik Gross, while in the process of experimentally describing his mouse leukemia virus, echoed the emerging view:

> When inoculated into a susceptible host [mouse], the agent remains dormant, or harmless for its host, until the host reaches middle age. At that time, for obscure reasons, the hitherto latent agent becomes activated, causing rapid multiplication of cells harboring it. This results in the development of leukemia and the death of the host.[8]

Both Gross and Lwoff encouraged the increasingly popular belief that all tumors might be caused by such viruses.

At this point the virus-cancer view ran headfirst into another fundamental problem. In the frenzied drive to isolate tumor viruses from humans, scientists could find no virus that had been active in tumors of a given type. By Koch's first postulate, this would eliminate all such microbes as tumor-causing candidates. But the virus hunt was in full swing, and no virus researcher intended to give up the chance to find the trophy of his career. So rather than abandon cancer viruses in favor of Koch's postulates, the search was on for viruses that could cause cancer without ever multiplying in the tumor.

The leukemia virus discovered by Gross did have a latent period, but it seemed to cause the cancer only after awakening to multiply aggressively in the body. The polyoma and SV40 viruses, on the other hand, caused cancer in hamsters by inserting just some of their genes into infected cells. The products of these genes were sufficient to cause cancer, but were insufficient to assemble cell-killing viruses. Either situation—a reactivated virus that does not kill a cell or active viral genes left behind by a killer virus—could potentially have worked as an explanation for cancer, though only in immune-deficient animals. An intact immune system would cure such cancers just like any other viral disease. But unable to find such tumor viruses in humans, biologists took a huge leap over logic and Lwoff's classic precedent: According to the revised view, viruses could cause tumors long after infection *even while remaining latent.*

Under the spell of this new paradigm, Koch's postulates and most other formal rules of science disintegrated. Now a virus could perform miracles. It could infect a new host one day, remain latent for any arbitrary amount of time, and then cause a deadly cancer without even being present. Moreover, scientists could now pretend that any cancer was infectious simply by blaming it on any virus they found in the patient's body, without fear of being disproved. One would not even need to find the virus to prove its guilt, and if it was found, one had decades of immunity before the virus would cause cancer.

This self-delusion joined hands with the hunt for human leukemia viruses and in the 1960s claimed its first success. The story began with Dennis Burkitt, a British surgeon working at a medical school in Uganda in the late 1950s. He noticed large numbers of children with malignant lymphoma, a cancer of white blood cells. Determined to investigate further, he spent three years conducting surveys of doctors all over Africa, asking them detailed questions about their lymphoma patients. Drawing the points on a map, he found the cancer struck people throughout central Africa, especially along the eastern side. Upon seeing that the risk of getting the disease depended on which climate people lived in, Burkitt proposed that the cancer was contagious, possibly transmitted by insect bites. His idea fit the leukemia virus program splendidly.

News of an obscure disease in Africa, reported by a virtually unknown English medical doctor, tended at first to be ignored. Although this fate greeted his 1958 paper, one doctor back home in London, seeing his opportunity, paid attention. M. Anthony Epstein, working at London's Middlesex Hospital, contacted Burkitt in 1961 and arranged to have sample tissues flown back to England. There Epstein began searching for a virus.

By the end of 1961, word of Burkitt's strange lymphoma and its transmission by insects brought magazine and television reporters to his doorstep. The media had not yet caught up with the new belief in infectious cancer among scientists, and they broadcast this curiosity all over the world. Another source of this news was a young C. Everett Koop, later to become U.S. Surgeon General, who encouraged virologists to study the newly discovered lymphoma after his trip to Africa. As pressure mounted, Epstein struggled to make the tumor cells grow in lab conditions. Succeeding by 1963, he and his new lab associate Yvonne Barr spent more months looking for the virus under the electron microscope. The following year one showed up, a previously unknown herpes-class virus. Once they could find the virus in almost every single culture of cells from Burkitt's lymphoma patients, Epstein and his coworkers officially proposed their virus to be the cause.

This Epstein-Barr virus has since been shown instead to cause mononucleosis, the so-called kissing disease, for which it may meet Koch's postulates. But where the virus causes mononucleosis before the body's immune system has suppressed it, Burkitt's lymphoma strikes an average of ten years after the immune defenses have neutralized the virus. In other words, the virus would cause mononucleosis on its own, but to cause cancer the virus needs help from something else that is available only ten years after infection. During mononucleosis, the virus multiplies actively and infects many cells; during Burkitt's lymphoma, it sleeps soundly in its continuing dormant state. Epstein could only find the virus growing in cells from lymphoma patients that had been cultured outside the body for quite some time. This condition gave the virus a chance to reactivate after arriving in the laboratory, with no immune system to interfere. To resolve this paradox Epstein and others insisted that the virus had a ten-year latent period for causing cancer, but not for mononucleosis.

Because the virus itself can rarely be found in a lymphoma patient, researchers must test whether the blood contains antibodies against the Epstein-Barr virus, indicating the patient was infected sometime in the distant past. Investigators first became excited when they discovered that all Burkitt's lymphoma victims had the antibodies. Upon wider testing, however, they slowly realized that all central Africans, with or without the cancer, also had the antibodies. In the United States, where a small number of people have also developed this lymphoma, roughly half the population has been infected by the Epstein-Barr virus. Apparently, most children catch the virus from their mothers during the first few months of life. Now two more paradoxes raised their ugly heads. Why did the vast majority of infected people never get the cancer, and why is it less common than mononucleosis? And why is an infected African one hundred times more likely to contract the lymphoma than an infected American?

To answer these questions, Epstein and his colleagues resorted to yet another virus-hunting invention: the "cofactor." If Africans face a higher risk of cancer, scientists explain away the problem by hypothesizing that since Africans are also more likely than Americans to be infected with malaria, perhaps malaria helps bring on the cancer. Just like that. Now the virus researchers would like everyone to believe that a cancer requires two separate infections, not just one. To explain away other discrepancies, more cofactors can be thought up.

The American and European lymphoma cases have provided an even bigger blow to the Epstein-Barr virus hypothesis. One-fifth of the patients have no antibodies at all against the virus, meaning they have never been infected. Further, more than two-thirds of the cases have no traces left of the virus in their tumor tissues, not even tiny fragments. What could be causing Burkitt's lymphoma in these people? Something else, according to virologists, that remains unknown. Koch's first postulate—that the suspected cause should be present in all cases of the disease—no longer enters the equation.

Finally, evidence gathered at the level of DNA shows that each patient's cancer originated from a single white blood cell. If virus infection caused cells to become cancerous, one should find every tumor having originated from the millions of infected cells, but each cancer comes from only a single cell. Virus hunters simply cannot explain why all the other infected cells remain normal.

Many scientists have found the above paradoxes too much to swallow. Within just a few years of the announcement of the Epstein-Barr virus, many researchers were already expressing serious doubts about the virus hypothesis of Burkitt's lymphoma. "Today epidemiologists disagree amongst themselves about whether or not Burkitt's lymphoma is an infectious disease," declared a well-respected 1973 textbook.[9] Other prominent scientists have admitted having reservations, switching instead to a chromosomal mutation hypothesis.[10]

Several scientists in the 1960s began proposing that the Epstein-Barr virus also caused a second cancer: nasopharyngeal carcinoma. This cancer, a tumor occurring at the back of the nasal passages, mostly shows up in adults in China, India, and parts of Africa, and among Eskimos in Alaska. The virus also was blamed for this cancer simply because many of these patients have antibodies against the virus. But, as with Burkitt's lymphoma, many of these victims also have never been infected by the virus, while it is dormant in the rest.

So now Epstein-Barr has become a virus that causes at least three diseases, two of them cancers that only appear long after the virus has settled into permanent latency. Despite all doubts, most virologists today thoroughly believe in this virus-cancer hypothesis. It is taught as unquestioned doctrine in college courses and textbooks and employs large numbers of virologists in performing endless experiments on the virus. Epstein has even worked on a vaccine against the virus in order to protect the world from cancer—though the cancer patients hardly need immunization, given that their antibodies have long ago suppressed the virus. After years of work and spending nearly $10 million on research, British scientists announced they would test a new vaccine in late 1993 or early 1994. Once they expand the trials, they will need decades to see if they can prevent cancer.

Despite its failure in terms of public health benefits, the Epstein-Barr virus hypothesis helped accelerate the hunt for cancer viruses. The search specifically for leukemia viruses had grown so dominant that the NCI had set up a special Acute Leukemia Task Force in 1962. Under James Shannon's leadership, the NCI had learned to set up programs that would attract more funding from Congress, making it the largest and most powerful of institutes under the NIH umbrella. The first of these, established during the 1950s, involved a huge effort to develop chemotherapy treatments for cancer; the second, begun in 1962, was a testing program to find potentially cancer-causing chemicals in the environment. The third was built around the leukemia virus group in 1964 and became known as the Virus-Cancer Program, which by 1968 took under its wing all other cancer virus research, including the work of Peter Duesberg, who had then just been appointed assistant professor of molecular biology at the University of California at Berkeley. Illustrating the complete reversal of fortunes on the part of the virus hunters, this third program became the only major NCI effort to determine the fundamental cause of cancer.

The NCI budget, at some $90 million in 1960, jumped to more than twice that figure by 1970. Fueled largely by the Epstein-Barr virus discovery, the Virus-Cancer Program seized the lion's share of this new funding. Its 1971 spending level had reached $31 million, almost equal to the other two cancer programs combined. Thus, cancer virology came to dominate the NIH itself, holding the most powerful position within the entire biomedical research establishment. Some grumbling about this inequality periodically surfaced from nonvirologists, but the growing budgets and accumulating prizes spoke more authoritatively in the politics of science. Even the sheer volume of research papers published by the virus hunters, growing rapidly during the 1960s, tended to drown out all criticism. Yet the rise of virology had only begun.

PRESIDENT NIXON'S WAR ON CANCER

James Shannon's retirement from directing the NIH in 1968 left a decided vacuum at the top of the biomedical research pyramid. In the absence of his firm control, the growth of the NIH temporarily slowed. Although their budget had reached $1 billion the previous year, the spending increases during the subsequent two years ended up being smaller than before.

"After 15 years of soaring affluence, the leaders of American biomedical science were poorly conditioned for austerity," recalled Daniel S. Greenberg, editor and publisher of *Science & Government Report*. The NIH certainly faced no financial troubles whatsoever, for spending was still moving upward. Nevertheless, "The [research] community rang with alarms and doomsday prophecies." The bloated but hungry science establishment and its lobby wanted some way to relive the glory days of James Shannon. "Their decision: maneuver the government into declaring War on Cancer."[11]

After three years of aggressive lobbying by wealthy political strategist Mary Lasker, plus a Senate-created National Panel of Consultants on the Conquest of Cancer, public drum-beating by columnist Ann Landers, self-serving testimony by medical scientists, and even a procession of cancer victims before Congress, the National Cancer Act was passed in 1971 and signed at a large press conference by Richard Nixon two days before Christmas. Some lobbyists had openly boasted this would bring about a cure for cancer by 1976. Others drew the analogy with the moon landing, persuading legislators that the shower of money would work similar miracles for medicine.

In the final analysis, neither benefit materialized. But some $800 million extra poured into the NCI over the next five years, bringing with it equally generous sums for the rest of the NIH. The largesse of the War on Cancer has continued up to the present day. Once again the growth of biomedical research skyrocketed, much of the money being used to train yet greater numbers of new scientists who would themselves become grant dependents. Of all research areas so funded, virus hunting grew the fastest and emerged by the 1980s as unquestionably the dominant force in the science establishment. Its research now fills more than one thousand pages of scientific journal space every month.

The Virus-Cancer Program of the NCI had positioned the cancer virologists to be first in line for the War on Cancer. Prominent spokesmen such as Wendell Stanley, Ludwik Gross, and André Lwoff had kept up the crusade for the growth of this field in the early 1970s. Along the way they were joined by many others, including Robert J. Huebner, a veteran of the war on polio who, until 1968, had run a lab at the National Institute of Allergy and Infectious Diseases, another branch of the NIH. He then transferred to the NCI, where he was given one of a handful of well-funded labs. Having first studied the Coxsackie virus and other spinoffs of polio research, he switched into the cancer field by adding to the growing literature on the polyoma virus. In 1969 he published a key paper amplifying André Lwoff's hypothesis, proposing that all human cancer was caused by latent viruses that awoke to cause tumors when radiation or other insults struck the body.

That same year, Nobel Laureate James Watson joined the cancer virology crusade. As head of the Cold Spring Harbor research facility on Long Island in New York state, he brought SV40 research to the laboratory in 1969. From that point forward, he added his prestigious voice to the chorus of virus hunters. In 1974 he hosted the Cold Spring Harbor Symposium on tumor viruses, the first international cancer meeting held exclusively for virologists. Annual tumor virus meetings have been held there ever since, becoming the most highly respected tumor virus conference worldwide.

Not all virologists held as much enthusiasm. In his 1966 Nobel acceptance speech, Peyton Rous, the discoverer of the Rous sarcoma virus of chickens in 1909, admitted having left the study of tumor viruses altogether for several years after his finding. He had failed to isolate any other tumor viruses and felt the field held little promise. Despite having reentered cancer virology, Rous could only comment by 1966 that "[t]he number of viruses realized to cause disease has become great during the last half century, but relatively few have any connection with the production of neoplasms [cancers]."[12]

Regardless of Rous's skepticism, the very fact that he had won a Nobel Prize for his chicken sarcoma virus helped boost the prestige of the Virus-Cancer Program. The cancer-virus field, boosted by Nobel awards, public advocacy, highly visible scientists, and some landmark discoveries in 1970 (see below), benefited more than any other program under the War on Cancer. Even the man appointed as NCI director to manage this war, Frank Rauscher, was a virologist. This favored position caused some resentment by other scientists. A 1974 report issued by an outside committee outlined the problem:

> First, the committee said, the VCP [Virus-Cancer Program] is too expensive. (It costs about $50 million to $60 million a year and consumes slightly more than 10 percent of the total NCI budget.) Second, the program must be opened up to the scientific community. At present, it is run by a handful of persons who have undue control over large amounts of money, which goes to only a limited number of laboratories. Furthermore, the individuals who award contracts are in a position to award them to each other, which somehow does not seem quite right.[13]

The virus hunters certainly made up a powerful and entrenched clique that increasingly dominated biomedical research. Minor bureaucratic reforms altered the operational details but, as the money continued to flow, their influence only grew. Given this built-in bias, cancer biology was quite likely to search for more viruses.

VIRUSES TO CAUSE CERVICAL CANCER

During the 1960s and 1970s cervical cancer became possibly the single most important virus-cancer project of all time. By blaming the tumor on viruses, tumor virologists have managed to cultivate public interest through a widespread campaign of fear. Readers of the *Los Angeles Times Magazine* opened their March 11, 1990, issue to find disturbing news. A large color photograph of a young, frightened-looking married couple drew one's eye to the ominous title, "Dangerous Liaisons." Several paragraphs down, the story explained further:

> Patty and Victor Vurpillat are infected with a strain of human papilloma virus—HPV—the virus that lurks behind one of the country's fastest-spreading sexually transmitted diseases and is rapidly becoming a prime suspect in the search for the causes of cervical cancer.
> As much as 15% of the population may already be carrying the virus—a fact that many health officials view with alarm...
> As a result, millions of Americans find themselves condemned to a sentence of life beneath the cloud of HPV, carrying in their tissues an incurable and highly infectious virus that may eventually unleash a devastating cancer...
> There are no drugs that can rid the body of the virus, just as there is no vaccine.[14]

Making no attempt to calm public fears, the article and its medical sources instead fanned the flames:

> What's more, some people are spreading the virus unknowingly: It is transmitted by contact with warts, and warts often go unnoticed. Some physicians suspect that HPV may even occasionally be spread indirectly—perhaps on a tanning bed, toilet or washcloth.[15]

According to the *Times*, biomedical authorities wanted far-reaching powers to respond to this supposed crisis:

> HPV infection is rampant among her clients, says Catherine Wylie, who oversees the family-planning program at the H. Claude Hudson Comprehensive Health Center... The spread will continue, she says, until the law requires that partners of people who have HPV be tracked down and treated.

"Our women have sex early because they marry at 16 to 18," Wylie said recently. "As long as this disease is not reportable, and there's no partner follow-up and treatment, I think we're going to have an epidemic of cervical cancer."[16]

For the victims, the diagnosis could be as devastating as the threat of cancer itself. For Patty Vurpillat:

> "It was just awful—not knowing what's going on with your body and if you're going to be OK or not," she said recently. "There's a certain percent chance you're going to be all right. But then, maybe you're not."[17]

In the case of Annie, diagnosed by Dr. Louise Connolly of the Manhattan Beach Women's Health Center:

> "It was horrible, just horrible," Annie remembers, referring to her fear of what Connolly might find. "There you are, spread-eagle, for [nearly half] an hour. None of it really hurts... But every time she'd stop and look at something, I'd think, 'Oh God, oh God, oh God.'"[18]

And for "Nan Singer," whose husband developed genital-type warts:

> Even after she confronted him, her husband was reluctant to see a doctor... Nan felt betrayed and disgusted; their sexual relationship deteriorated. Existing problems in their marriage grew worse...
> [Nan] believes her husband's response to the disease contributed significantly to their subsequent divorce.[19]

The disease in question—cancer of the cervix—is a relatively common tumor that develops slowly and can eventually destroy a woman's reproductive ability or even cause death. As with most cancer, the risk of contracting it increases with age, especially after midlife.

Microbe hunters first began the study of cervical cancer with their microbiological tools in the nineteenth century, when an Italian doctor conducted surveys and found the tumor more often among married women than among nuns. To the eager bacteria hunters, this could only mean that sexual activity was the risk factor for the cancer, which was translated to mean some sort of venereal infection was at fault. A variety of microbes were indeed blamed for causing the disease, including the bacteria that cause syphilis and gonorrhea, as well as mycoplasma and chlamydia bacteria and the trichomonas protozoa.

Virologists entered the cervical cancer field in the mid-1960s, shortly after the Epstein-Barr virus had been isolated and blamed for causing Burkitt's lymphoma. Because Epstein-Barr was a strain of herpes virus, all other herpes viruses immediately became popular among tumor-virus chasers. By 1966 virologists had revived the observation that women with cervical cancer tended to have had more sexual contacts than those without. That same year one lab reported that a higher proportion of the cancer patients had previously been infected by herpes virus than had people without the tumor.

This proved too tantalizing a thread to pass up. Within two years, researchers were able to distinguish two different herpes simplex viruses: type 1 was the most common, causing sores around the mouth, while type 2 caused its sores in the genital areas—including the cervix. The latter became the target for the virus hunters, who proposed it to be the cause of the cancer.

Trying to explain why a tumor would appear only years after the original herpes infection, scientists were forced to construct a new hypothesis. According to this idea, the virus would first infect and kill millions of cells, occasionally making a mistake and mixing with the DNA of the cell and become impotent in the process. In other words, the virus would mutate the genetic code of a few cells, leaving only a piece of the original virus stranded therein. Such cells would survive the infection and eventually grow into a tumor, and years later this leftover piece of the virus could still be detected in the tumor cells.

But as more data accumulated, several embarrassing facts came to light. About 85 percent of all American adults have been infected by this same herpes virus (many without symptoms), including women without any hint of cervical cancer. And scientists consistently found many women with the tumor who had never been infected by the herpes virus. Even among those women with both the cancer and past herpes infection, the leftover pieces of the virus in the tumor cells were always different and inactive, meaning that no particular part of the herpes virus was needed to cause the cancer.

In 1983, desperate but not willing to abandon the herpes virus hypothesis, researchers seriously proposed in the journal *Nature* a "hit-and-run hypothesis—that the herpes virus briefly infects cervix cells in the unsuspecting woman and makes some mysterious, undetectable change. Then it abruptly vanishes, leaving behind no evidence of the infection, so that the tumor can somehow develop many years down the road."[20] This idea threatened to make virus hunters a laughingstock. How could anyone perform experiments to test for a hypothetical event that left behind no evidence? The "hit-and-run" hypothesis nevertheless survived into the early 1990s, by which time scientists quietly retreated out of the herpes virus hypothesis altogether.

Meanwhile, in 1977 a former herpes virologist named Harald zur Hausen, working at the German Krebsforschungszentrum (Cancer Research Center) in Heidelberg, proposed another virus as the agent causing cervical cancer. Human papilloma virus (HPV), the mild virus that causes warts, seemed to him a reasonable possibility based on the observation that cervical warts could occasionally turn into full-fledged cancers.

By the early 1980s technology had become available to detect small DNA fragments of long-dead viruses. Using this technique zur Hausen found broken, leftover pieces of the papilloma virus DNA in the tumor cells of some patients. Soon everyone had joined the new parade, never hesitating to ask if they might be making the same mistake as with the herpes virus.

Indeed, the evidence for the papilloma hypothesis has since

fallen apart. When zur Hausen and his colleagues discovered that at least half the American adult population and, therefore, half the adult women, had been infected by the virus, yet only 1 percent of women develop the cancer in their lifetime, they began to see a discrepancy. Koch's first postulate has also tested the credulity of the cancer virologists, since at least one-third of all women with cervical cancer have never been infected by the virus. The rest of the cervical cancer patients are not all infected with the same strain of papilloma virus; over a dozen different varieties of the virus can be found in these women.

An incredibly long time elapses between infection by the virus (in those who do get infected) and the onset of the tumor. Papilloma virus tends to be contracted by women who are younger and more sexually active—estimated at an average twenty years of age. Cervical cancer, a disease of older age, strikes women in their forties through their seventies. By subtraction, zur Hausen calculates a whopping "latent period" ranging between twenty and fifty years! Nor does the virus reactivate when the cancer appears; in keeping with the revised Lwoff hypothesis of viral latency and cancer, scientists simply assume the virus caused some sort of necessary but not sufficient mutation twenty to fifty years earlier and can therefore remain soundly asleep in the tumor tissue.[21] But this explanation cannot account for several key facts. For one thing, the leftover pieces of the virus cause entirely different, and therefore irrelevant, mutations in the genetic code of each tumor. Also, each cervical cancer grows from one single cell, leading to the obvious question of why all the other millions of infected cervical cells never develop into tumors.

As with virtually all cancers, the dynamics of cervical cancer development simply do not match the behavior of viruses. Papilloma virus causes papillomas, or warts, on young, sexually active adults. These small overgrowths of slightly abnormal cells can appear (or disappear) almost overnight and are not malignant. They typically disappear spontaneously as a result of antiviral immunity. The immune system recognizes the viral proteins and rejects the wart together with the wart virus.

But most cancers, including cervical cancer, are diseases of old age; they develop slowly over many years or decades. Cervical cancer develops from benign *hyperplasias*, meaning excessive growths of nearly normal cervical tissue. Most or all of these hyperplasias regress and disappear, while a few may instead progress further into *dysplasias*, meaning larger growths of abnormal cells. Even such dysplasias are potentially reversible. But the occasional dysplastic growth can give rise to *neoplasia*—meaning "new growth," or cancer. And a percentage of such cancers can even become malignant, invading surrounding tissues and spreading throughout the body. The major feature of cancer progression is that it is irregular, unpredictable, and gradual—quite unlike the rapid and consistent development of warts. Above all, the cancer is never subject to rejection by antiviral immunity, because no viral proteins are ever expressed in cervical cancer. While virus hunters have speculated that wart virus might somehow further the development of cervical cells into cancer cells, the reverse may be true: The active cell growth in dysplasias may simply encourage papilloma viruses to become active. That is exactly what Peyton Rous proposed long before the wart virus was considered to cause cancer.[22]

The final blow to this virus hypothesis lies in the fact that equal numbers of men and women have genital warts, yet rarely do men contract any penile cancers. A cancer virus that can infect both sexes should cause tumors in both sexes equally well, a conundrum that leaves viral epidemiologists perplexed.[23] Perhaps better explanations exist in some of the other risk factors for cervical cancer: Other than aging, two of the most important factors coinciding with the tumor are long-term smoking and oral contraceptive use. Oral contraceptives contain powerful sex steroid hormones that directly regulate the function of cervical tissues and might explain the superficial correlation between cervical cancer risk and the number of sexual contacts a woman has had. In any case, cancer of the cervix is not contagious.

Nevertheless, the virus hunters continue to push for the virus-cervical cancer hypothesis, which today remains one of the most popular and widely accepted among scientists. To help rationalize away some of the paradoxes, they have even revived herpes simplex virus-2 as a cofactor for the papilloma virus—two zeroes that hardly add up. Yet the biotechnology company Digene Diagnostics, based in Maryland, has won government endorsement for its papilloma virus test. Already widely in use, the test is now recommended by medical research authorities for some seven million American women each year, although only thirteen thousand cervical cancers appear each year in this country. The test costs $30 to $150 per person. Given that a woman who tests negative today may become infected tomorrow, there is no upper limit to testing. Many research laboratories are also kept in business with NIH grants to study endlessly every detail of the papilloma virus, and thus scientists would be the last to reevaluate this virus hypothesis. Unfortunately for tens of thousands of women each year, the ongoing media publicity and the tests can have devastating psychological consequences, not to mention the damage from preventive treatments for women who may have little more than harmless warts.

THE HEPATITIS B VIRUS-LIVER CANCER HYPOTHESIS

Another product of the War on Cancer emerged during the 1970s, when the virus hunters took up research on liver cancer. This time their sights focused on the hepatitis B virus.

Most people infected by this virus either experience no symptoms at all or experience a temporary liver inflammation, after which their immune systems clear the virus from the body, leaving behind only antibodies against the virus. In a few cases, however—one out of every one thousand infected people in the industrial world and 5 percent of those infected in Asia—hepatitis B can become a chronic infection that neither escalates to kill the patient nor disappears. Instead, it gradually wears away at the victim, constantly damaging the liver while causing on-again, off-again symptoms. People develop chronic hepatitis for understandable reasons, when their immune responses have deteriorated from alcoholism, heroin addiction, or the malnutrition so common in the Third World.[24]

Scientists first noticed an overlap between hepatitis B virus and liver cancer in the 1970s. Nations with high rates of infection also had many cancer patients. Upon closer inspection, some studies revealed that people with chronic virus infections had an enormously higher risk of eventually developing the tumor. In 1978 a paper was published arguing that chronic hepatitis infection directly damaged the liver enough to cause cancer, and another virus-cancer hypothesis was born. No one bothered to point out, however, the complete absence of any evidence for liver cancer being contagious.

As researchers began jumping onto the new bandwagon, they uncovered data that unraveled the virus hypothesis. For one thing, only a tiny fraction of chronic hepatitis cases ever progressed to the cancer, that fraction being much higher among Asians than among Americans. And unlike in the industrial world, where the cultures of drug abuse and prostitution largely transmit the virus, Asians mostly become infected by their mothers around the time of birth. Since liver cancer in the Third World shows up in people between the ages of thirty and sixty years, virologists simply calculated the latent period between infection and cancer as ranging from thirty to sixty years—longer than the life expectancies of many people. No researcher stopped to ask whether other health risks might also endanger the victim during those many decades, obviating the need to blame a virus.

The case for the virus hypothesis degenerated further when most liver cancers were found in patients who had been infected long ago but were not chronic carriers of hepatitis B virus. Rather than continuing as a chronic infection, the virus had been cleared from the body. Hoping to rescue the virus hypothesis, scientists resorted to an old favorite among cancer explanations: Perhaps the tumor could result from cells in which the virus DNA accidentally combines with a specific gene of the cell to produce a cancerous mutation. But follow-up investigations showed that the pieces of viral DNA did not affect any consistent part of the cell's genetic structure and that most of them were biochemically dead and therefore not producing any viral proteins.[25] This implied that such mutations were random, inconsequential accidents. And as with cervical cancer, each liver tumor arose from a single cell at the start, while millions of other cells had been infected with the virus, producing untold numbers of mutant cells. Why did all these other cells remain normal? No answer has been offered. More important, many liver cancer patients have never been infected by hepatitis B at all; in the United States, at least one-quarter of all these tumor patients have never encountered the virus.

Finally, the virus hypothesis has failed miserably when put to the test of Koch's third postulate. Upon injection into chimpanzees, the human hepatitis B virus does infect and inflame liver tissues, but no liver cancer ever appears. The virus, in fact, cannot cause cancer in any animal.

Hepatitis B infections that do not become chronic cannot possibly cause liver cancer. On the other hand, chronic infections might damage the liver enough to promote the tumor. But the more likely explanation for this noninfectious cancer may lie in the health risks, including drug abuse and malnutrition, that allow chronic infections in the first place. Perhaps these risks in themselves cause cancer. Only a small amount of scientific research has examined diet in connection with this cancer—far too little to be sure.[26]

Despite all evidence to the contrary, most scientists still believe wholeheartedly in the hepatitis B–liver cancer hypothesis. It has even become the primary justification for mass immunization programs against the virus in Asian countries, where people inherit the virus at birth and usually suffer no harm. As two biotechnology experts recently put their argument, "While hepatitis B infection may be asymptomatic, chronic carriers have a high risk of developing hepatic [liver] cancer."[27] After three to six decades, that is. Huge government-sponsored vaccination programs are already underway in several Asian nations. Until recently the cost for immunization was $100 per person, now having declined to $38. Given cooperation by the World Health Organization and various governments, such figures can spell enormous income for biotechnology companies, even as they place strains on the economies of nations like Taiwan and Thailand. More than two million people have been vaccinated, and large field trials are being conducted. Since most of these people have been "vaccinated" by natural infection anyway, soldiers in the War on Cancer cannot explain how adding an artificial vaccine could possibly help. Yet they keep marching on.

KING RETROVIRUS

Even among the modern virus hunters, a hierarchy of sorts has developed over the years. Those studying the most popular viruses—above all, cancer and tumor viruses—receive the bulk of the awards and grant money. The "lesser" virologists understand their place in the hierarchy and display proper reverence for their superiors, while still retaining the confidence of aristocracy relative to the rest of the science establishment.

Since 1970 the most elite circle within virology has belonged to the RNA tumor virus researchers. Since AIDS, RNA tumor viruses were renamed *retroviruses* because most of them are now considered potential immunodeficiency viruses. Even the tumor virus meetings at Cold Spring Harbor, New York, were renamed *retrovirus meetings* in 1992 to accommodate the new view on retroviruses. The expectation that they would cause human cancer has been quietly buried after their reformation to "AIDS viruses." The rise to power of this retrovirus club, numbering roughly a couple of hundred until recently, started as the conquest of a bare handful of scientists whose story begins in the 1950s.

Harry Rubin had spent years as a veterinarian, tending mostly farm animals throughout the United States and Mexico. Having tired of this work, he turned to academic research science and learned the methods of culturing cells and growing viruses at the California Institute of Technology in Pasadena. Wendell Stanley took notice of this aspiring virologist, and in 1958 brought Rubin to his Virus Lab at the University of California, Berkeley. This move took place just as the cancer-virus hunt was ascending among scientists.

Before moving to Berkeley, Rubin had become fascinated by the chicken tumor virus discovered half a century earlier by Peyton Rous—the Rous sarcoma virus (RSV). Most researchers had since moved on to other viruses, largely because they could grow RSV only in live chickens, which was expensive and time-consuming as well as too clumsy for good experiments. Determined to find a better technique, Rubin turned to the culture dish. He soon found a way to grow chicken cells in dishes and then learned how to infect them with RSV. Every cell infected by the virus immediately became a cancer cell, a change that could be seen easily in the dish.

Having achieved this laboratory breakthrough, Rubin began lobbying colleagues to study the Rous virus, which he sensed might contain clues to the role of viruses in cancer. Until 1958 he supervised Howard Temin (died 1994), a young doctoral student equally interested in cancer viruses. Rubin trained Temin in the new methods of culturing RSV, and together they observed some strange behaviors of the virus that convinced them both it was fundamentally unlike most other viruses. Rather than killing cells shortly after infection and then departing, the RSV genome seemed to become part of the DNA of each cell, incorporating itself into the genetic material permanently. This distinctive strategy of replication is why retroviruses do not kill cells—they become part of the cell instead, as genetic parasites.

Now thoroughly possessed with this idea, Temin moved on to establish his own lab at the University of Wisconsin in 1960. There he performed more experiments, confirming that RSV did indeed copy its own tiny RNA genome into DNA before inserting this short piece of DNA into the infected cell's DNA and becoming a permanent resident. But having failed to prove this notion, he faced mild disbelief from some and cautious interest from others when he formally proposed his hypothesis in 1964. He and several colleagues then labored away for the next several years, confident they would prove their point.

Temin finally succeeded in 1970, isolating an enzyme (a protein catalyzing chemical reactions) that did the work of making a DNA copy of the Rous virus' RNA. He announced his finding to an excited crowd of virologists at the International Cancer Conference in Houston, Texas. Because the Rous virus copies its genetic information from RNA to DNA, the reverse of the cell's own process, it was later designated a *retrovirus*.

Where Temin saw vindication, others saw golden opportunity. The quickest of these was David Baltimore, a young associate professor at the Massachusetts Institute of Technology. Baltimore had spent the past several years studying the poliovirus in detail, a remnant of virus research from the 1950s. Like so many of his fellow veterans of the war on polio, he found his research slipping into medical irrelevance as the 1960s wore on and realized he would soon have to enter the cancer-virus field. He had a keen sense of politics, and he made his move just as Temin's announcement opened the door.

The inside joke making the rounds among the top virologists immediately after the meeting went something like this: "Can you guess who took the fastest plane out of Houston? Answer: David Baltimore." This story reflected an important truth. Upon hearing the news of Temin's finding, Baltimore instantly transformed himself into a retrovirus researcher:

> Baltimore confesses that he "jumped the fence" for two days to do the experiment. The virus used was obtained by a phone call to his old friend and NCI project monitor George Todaro.[28]

Baltimore's rush to duplicate Temin's observations paid off. His paper was published alongside Temin's in the prestigious journal *Nature*, and they shared the Nobel Prize in 1975 for the discovery of the retrovirus enzyme, dubbed *reverse transcriptase*.

Several other scientists also rushed to confirm the enzyme's existence. One of the first was a chemist-turned-virologist, Peter Duesberg, another young researcher noticed by Wendell Stanley. In 1964 Stanley hired Duesberg right out of Germany's Max Planck Institute for Virus Research in Tuebingen and into the Virus Lab in Berkeley, where Duesberg promptly went to work studying retroviruses.

Duesberg accepted a position as an assistant professor at the university. Having also formed a friendship with Harry Rubin, he had previously decided to take up the retrovirus field. His research question seemed straightforward: How did the Rous virus cause cancer? The problem, however, had baffled scientists, especially since the virus seemed identical in every respect to many other chicken retroviruses that were entirely harmless. Collaborating with virologist Peter Vogt, Duesberg solved the puzzle in 1970, demonstrating that the Rous virus contained an extra gene that caused cancer. Rous's virus turned out to have been a freak accident of nature, having picked up and mutated part of a gene from the cell that made it a cancer virus: Remove the sarcoma gene—as it is now called—and the virus becomes perfectly harmless.

The Temin and Duesberg discoveries, respectively, launched a new field to the forefront of virus hunting. Soon researchers found that many of the tumor viruses long studied had also been retroviruses, including the breast cancer virus of mice and the leukemia viruses in many animal species. But unlike the Rous virus, few of these others contained special cancer genes. So whereas the Rous virus caused massive tumors within days in almost any chicken, these other retroviruses had to maintain active infections of the body for many months before causing a leukemia, and then only in specially susceptible inbred strains of animals. In short, no retroviruses ever killed cells, and only very rare ones caused tumors in animals. Virtually all retroviruses proved to be benign passenger viruses in animals outside the laboratory.

Even the very few oncogenic retroviruses—those endowed with cancer genes—hardly play a role as carcinogens for two reasons. First, viral cancer genes accidentally acquired are never kept by retroviruses after they are generated because they are entirely useless to the virus—just like a genetic cuckoo's egg. Second, even if a rare oncogenic retrovirus infects an immunocompetent animal, a small tumor will appear within days after the infection, only to disappear again as the animal develops antiviral immunity. Antiviral immunity kills both the virus and all virus-infected cells. As a result, retroviral cancers are extremely rare and very survivable tumors in wild animals. Their statistical relevance as carcinogens is negligible.

Yet there are at least one hundred retrovirologists alive today for every one of the fifty retroviral tumors found since the beginning of the century.[29] Five Nobel Prizes, including that for Peyton Rous, have been given to students of the chicken that died from Rous sarcoma virus in 1910! In addition at least a dozen, including Duesberg, have been elected into the U.S. National Academy of Sciences for their studies of the Rous sarcoma virus and the fifty other rare oncogenic (tumorigenic) retroviruses.

The wave of excitement following the 1970 discoveries of cancer genes and reverse transcriptase helped pass Nixon's National Cancer Act the next year, and retrovirologists quickly rode to power. A 1970 *Nature* editorial accurately predicted that the new retrovirus findings "are likely to generate one of the largest bandwagons molecular biology has seen for many a year... it is especially the case today when cancer is one of the few remaining passwords to the dwindling coffers of the granting agencies in the United States."[30] *Nature* itself jumped on the bandwagon, launching a parallel journal under the title *Nature: New Biology*, its purpose being specifically the publication of retrovirus papers.

As a group the retrovirologists have had more to say about science policies than anyone else, including what directions biomedical research should take and which researchers should get the funding and awards. They have redefined the scientific enterprise and with it our popular culture. Their voices carry enormous weight, and when they choose to blame another retrovirus for cancer, AIDS, or any other disease, the governments of the world and the news media respectfully cooperate.

The next logical step for the retrovirologists was to isolate their first human retrovirus, preferably one that causes cancer. A major effort materialized, but every investigator who tried ended up facing enormous frustration. Hints and echoes of retroviruses would briefly appear, only to vanish upon closer inspection. Many of the experiments suffered from flawed design, while others detected genuine retroviruses that turned out merely to be contaminating animal retroviruses. Scientists should not have been so surprised at the failure, because chronic retrovirus infections are restricted even among animals to sickly inbred strains that have lost natural immunity. Also, retroviruses can be much harder to find in wild animals and humans. But this point was lost on the virus hunters.

Inspired by the breast cancer virus found decades earlier in certain inbred mice, researchers focused much of their energy on the search for a similar human retrovirus. The work began almost immediately after 1970 and continued into the 1980s. In mice, the virus generally passed from mother to offspring through the milk; scientists used this as their starting point. Several studies examined human mothers with breast cancer, failing to see any higher occurrence of the tumor among their breast-fed daughters. Such results hardly discouraged the virus hunters, who promptly turned their high-powered electron microscopes to human milk and samples of breast tumor tissues. A number of reports were published throughout the 1970s by some of the most prestigious investigators claiming to see "virus-like particles."[31] Many such particles were also seen in milk from tumorless mothers, while contradictory reports found no such particles in milk or tumors.

Retrovirologists began applying a battery of increasingly sophisticated technology to hunt down the elusive virus. Some thought they found reverse transcriptase (the unique retrovirus enzyme) in milk and tumor samples, others probed breast cancer tissues for genetic information resembling that of the mouse retrovirus and got some positive signals, and still others checked for virus pieces that might be recognized by mouse antibodies against the mouse virus. Fewer than half of the human breast tumor tissues studied reacted with the antibodies, but this was enough to excite the virus hunters.

Indeed, these findings led to a sensational press conference in October 1971 at the National Academy of Sciences. There, in the middle of an otherwise routine meeting with reporters, several virologists dropped hints they were finding cancer viruses in human breast milk. Sol Spiegelman, one of the first virologists to have jumped on the Temin bandwagon the year before, lived up to his flamboyant reputation by openly suggesting some women should not breast-feed their babies. Peppered with questions, Spiegelman repeated himself: "Look, if a woman has a familial history of breast cancer in her family and if she shows virus particles and if she was [sic] my sister, I would tell her not to nurse the child." Soon one of his colleagues standing beside him piped up, "Why inoculate a child with virus particles? I mean, it's clear." Spiegelman struck a more cautious note warning, "You cannot start a scare like this when we don't really know for sure that this virus particle is the causative agent."[32] Nevertheless, headlines appeared the next day in the major newspapers and on television screaming dire warnings over breast-feeding.

To this day, however, no human retrovirus has ever been isolated from breast cancer, relegating these many expensive research projects to the trash bin of falsely positive results so common in experimental science.

Retroviruses ultimately saw their major impact in reviving the old virus-leukemia program. All leukemia viruses studied in mice and other animals before 1970 offered no insights for understanding human cancer, because they caused leukemia only in a few young, sickly animals under special laboratory conditions. Such viruses did nothing to normal, healthy wild animals. Similarly, such retroviruses could not be expected to affect healthy humans.

But a cat retrovirus isolated in the 1960s, though really no different than other retroviruses, served as the tool for virologists to bridge the gap. Named Feline Leukemia Virus (FeLV) because it had been isolated from a leukemic cat, the virus became the primary object of study by Myron ("Max") Essex, a rising professor at Harvard University's School of Public Health. He picked up this research once others had shown that young lab cats could become leukemic after months of continuously active infection. Outside the lab, however, as many as two-thirds of all cats eventually catch FeLV, quickly and permanently neutralizing the infection with their immune systems. Leukemia among such animals appears only rarely, in four of every ten thousand cats each year. Indeed, because leukemia is a cancer of blood cells and therefore causes immune deficiency, retrovirus infections in leukemic animals may simply be a consequence of acute immune deficiency. But Essex wanted to prove the cat leukemia an infectious disease and had to argue that FeLV could cause the tumor even while remaining latent.

Docile veterinarians and the news media alike have accepted cat leukemia as infectious. The specter of leukemia epidemics among household pets, aggrandized with suggestions of transmission to their human owners (since disproved), has popularized Essex's own nostrum for the perpetual crisis. Having founded his own biotechnology company, Cambridge Bioscience Corporation, Essex has developed a vaccine against the FeLV. One year after approval in 1989, the vaccine had already sold to half the estimated French market of cat owners. Unfortunately for the owners, they have no idea that in most cases their cats already have natural immunity against the virus from natural infection nor that a vaccine can do nothing against a virus that becomes latent anyway. Nor, for that matter, that one-third of all leukemic cats have never been infected by FeLV at all, the same proportion as among healthy cats.

The more important consequence of Essex's research, however, lay in its inspiration of a human leukemia virus search. When Robert C. Gallo arrived in 1965 at the National Cancer Institute fresh out of medical school, his NIH bosses put him to work treating leukemia patients and researching potential new therapies. After several years of unspectacular work, Gallo found his chance to move up in the ranks following Temin's 1970 retrovirus announcement. The glamour of new retrovirus discoveries and of the free-flowing cancer money attracted Gallo to the retrovirus field like many other scientists.

He got his first taste of glory in 1970 when he joined several better-established virologists, including Sol Spiegelman, in chasing retroviruses in human leukemia. They quickly found evidence of the reverse transcriptase enzyme in tissue samples from leukemic patients. During the first week that November, Italian pharmaceutical company Lepetit and the Pasteur Institute sponsored a tumor-virus conference in Paris. Spiegelman seized the opportunity for publicity. The lectures were being given on a stage at the nearby Hilton Hotel, the podium standing in front of a huge curtain that parted in the middle. When the time came for Spiegelman's presentation, he began by solemnly announcing the evidence of retroviruses in leukemia patients.

In the middle of his speech, the curtain suddenly parted and an impeccably dressed bellboy walked up to him, holding a telegram on a silver platter. Spiegelman picked up the envelope, opened it with dramatic flair, and read to the audience the late-breaking news from his laboratory back in the United States. Several more patients had just been tested—with positive results for the virus. Some insiders in the hushed audience of some four hundred—including such colleagues as Temin, Lwoff, and Duesberg—could not help but suspect that the delivery of this news had been planned in advance.

Gallo followed suit with his own leukemic patients. For his own positive results in a few leukemia patients, he was rewarded by being named head of NCI's brand-new Laboratory of Tumor Cell Biology. The year was 1972, and the new department was a product of the lavish War on Cancer funding.

The retrovirus work of Essex had also brought Gallo fully into the virus arena. Gallo's team accelerated the intensive hunt for the first human retrovirus. But his earlier results in competition with Spiegelman turned out to be nothing more than false positives, mistaken observations that were simply lost in the rapidly growing scientific literature. Still, the virus search was stepped up. By 1975 his lab had finally isolated a retrovirus from human leukemia cells. To Gallo's dismay, however, he faced humiliation when he presented the finding at the Virus-Cancer Program's yearly conference. Other scientists had tested his virus and discovered it to be a mixture of contaminating retroviruses from woolly monkeys, gibbon apes, and baboons. Gallo tried to save his reputation, speculating wildly that perhaps one of the monkey viruses caused the human leukemia. This excuse did not fly, and he later described the event as a "disaster" and "painful," admitting that it placed "human retrovirology, and me with it, at a very low point."[33]

Despite Gallo's repeated booms and busts, virus hunting was the fashion, and he doggedly pursued retroviruses for the next few years. In 1980 he finally reported having found the first known human retrovirus. The virus was isolated from human leukemia cells grown for a long time in the lab, with no immune system to interfere or suppress the virus. Gallo's team even had to shock the cells repeatedly with potent chemicals to coax the soundly sleeping virus out of latency. No such virus could be found in a second batch of leukemic cells, but Gallo remained unfazed, giving the new virus a name with strong propaganda value—*Human T-cell Leukemia Virus*, or HTLV.

Gallo's next step was to find a disease for his virus. Having made up his mind it should cause some leukemia, he began scouring the world for a connection to such a cancer. With the help of other scientific teams, Gallo soon found HTLV concentrated among residents of the Japanese island of Kyushu, as well as in certain parts of Africa and among Caribbean people. Among these peoples also happened to exist one type of leukemia, a disease since dubbed *Adult T-cell Leukemia* (ATL). Having found an overlap between his virus and a cancer, Gallo swung the weight of scientific consensus behind his hypothesis, which now ranks among the most popular virus-cancer programs. Even standard biology textbooks now discuss Gallo's hypothesis as unquestioned fact.

But no one should worry about catching this leukemia. By testing the blood supply, the Red Cross counted some sixty-five thousand Americans as having been infected by HTLV, of whom about ninety, or one out of every thousand, have the cancer. Kyushu natives fare little worse, with only 1 percent of infected people developing the leukemia ever in their lives. For that matter, not a single American infected by HTLV through a blood transfusion has ever developed the disease. Conversely, quite a number of people worldwide have this cancer *without* HTLV infection. Indeed, there is not one epidemiological study in which the incidence of leukemia is higher in HTLV-positive groups than in virus-free control groups. Gallo and his colleagues, however, have calculated a means of circumventing this latter problem—by redefining the disease. Doctors may not diagnose patients as having "ATL" unless the victim also has antibodies against the virus; uninfected patients with identical leukemias are categorized under a different clinical name. This little trick handily abolishes one of the gaps between the disease and the virus.34

HTLV researchers can change other rules, too. Having first assumed the virus is spread between adults, scientists calculated a "latent period" of five years between infection and development of leukemia. Soon they adjusted that figure to ten years, then thirty, as they found increasing numbers of healthy carriers of HTLV. Once they decided the virus is transmitted sexually, while the leukemia strikes roughly at age sixty, they subtracted twenty from sixty to generate a forty-year latent period. Then, upon realizing that the virus is actually transmitted from mother to child around birth, the latent period grew to an official forty to *fifty-five years*.35

Even when the leukemia does strike a patient, the virus continues to sleep soundly, forcing doctors to test for antibodies instead of the virus itself. Again, as with cervical and liver cancer, the virologists assume the virus must cause a mutation in each cell upon infection and before entering latency. In this case, however, a virus mutation hypothesis is at least plausible, for the very nature of retroviruses dictates that they combine with the cell's genetic material as soon as they infect it. However, of the millions of cells infected by HTLV, only one ultimately gives rise to the leukemia, the other cells functioning normally as ever. But there is no common leukemia-specific mutation in different viral leukemias—leaving the viral leukemia in search of a nonviral cause.36

Now researchers have granted the virus yet another disease: HTLV-Associated Myelopathy (HAM), a brain disease modeled after kuru and other "slow virus" syndromes. To maintain even a tenuous connection between the virus and HAM, Gallo and his colleagues, including Carleton Gajdusek, have decreed that the disease must be renamed *HAM* when the patient is infected with HTLV. All identical cases without the virus must be diagnosed under their old disease names.37

Given the political power of the retrovirologists within the research establishment, such arbitrary science not only survives but can even be made profitable. Since 1989 the American Association of Blood Banks has required testing the blood supply for HTLV, tacking an extra $5 to $11 onto each of the twelve million blood donations made every year. For scientists holding interests in the biotechnology companies producing HTLV tests, the income is enormous.

Flushed with victory, Gallo did not stop with his first human retrovirus. He isolated a second one in 1982, from a cell line derived from a patient with a different type of leukemia. The old virus became *HTLV-I*, the new one *HTLV-II*. But since that time HTLV-II has been retrieved from only one other patient with a similar leukemia, while plenty of cases have been found without the virus. So although Gallo continues referring to it as a leukemia virus, most other scientists prefer caution as long as the virus has been found in only two patients. Gallo's second virus, much to his chagrin, remains a virus in search of a disease.

THE NO-WIN CANCER WAR

At a press conference on February 4, 1992, an informal group of scientists released a signed statement evaluating the War on Cancer. In this sharply worded presentation, the sixty-eight prominent researchers made several poignant observations:

> We express grave concerns over the failure of the "war against cancer" since its inauguration by President Nixon and Congress on December 23, 1971. This failure is evidenced by the escalating incidence of cancer to epidemic proportions over recent decades. Parallelling and further compounding this failure is the absence of any significant improvements in the treatment and cure of the majority of all cancers...
>
> We express further concerns that the generously funded cancer establishment, the National Cancer Institute (NCI), the American Cancer Society (ACS), and some twenty comprehensive cancer centers, have misled and confused the

public and Congress by repeated claims that we are winning the war against cancer...

Furthermore, the cancer establishment and major pharmaceutical companies have repeatedly made extravagant and unfounded claims for dramatic advances in the treatment and "cure" of cancer.[38]

Two months later, the *Journal of the American Medical Association* reflected this widespread view in an article on the same topic:

> By some estimates, the federal government has spent as much as $22 billion on this effort in the past 20 years...
>
> However, some critics contend that this war is being lost. They argue that too little change is being seen in death rates from many major cancers...
>
> Whatever the case, the fact remains (the American Cancer Society said last week...) that about 83 million persons now alive in this country eventually will contract cancer—"about one in three, according to present rates."[39]

This disaffection with the War on Cancer had begun appearing by the early 1980s, voiced by some of the most prestigious scientists in the business. By that time the public had also lost interest in the program, which had not delivered on its ambitious promises.

Respected science watcher Daniel Greenberg has commented on the early signs of this failure:

> The gusher of new money financed rapid expansion of a previously low-keyed quest for a cancer virus, which in turn might lead to the magic bullet of a cancer vaccine. University scientists were appalled to find that most of the virus money was being dished out to industrial firms, without peer review. An outside inquiry concluded that the virus program, which would soon cost $100 million a year, was intellectually shoddy and unproductive.
>
> It was reorganized to emphasize research by NCI scientists and peer-reviewed university researchers, and became one of

the prime movers of the molecular biology revolution... But the early stumblings of the virus program were duly noted...

In 1975, shortly after stepping down as the senior health official in the Department of Health, Education, and Welfare (HEW), Charles Edwards, a doctor and research administrator, wrote that the cancer program was based "on the politically attractive but scientifically dubious premise that a dread and enigmatic disease can, like the surface of the moon, be conquered if we will simply spend enough money."[40]

Cancer treatment arguably accomplishes little today. This problem is rooted mostly in our lack of understanding the cause of cancer. The War on Cancer budget was narrowly focused by politically powerful virus hunters in their obsessive search for tumor viruses.

If anything, the cancer fight greatly strengthened and consolidated virus hunting, and placed the retrovirologists in charge. At the same time, they needed some new war to revive their popularity after the cancer debacle. Thus, when AIDS appeared in 1981—a textbook example of a noncontagious syndrome—the virus hunters were poised and eagerly waiting to take advantage of another opportunity. The next chapter tells the story of how they seized control of the war on AIDS, mobilizing the entire world behind their latest virus hunt while boosting their own prominence beyond their wildest dreams.

CHAPTER FIVE

■

AIDS: The Virus Hunters Converge

HAD THE AIDS EPIDEMIC struck years earlier, so goes the common belief, medical science would have been unprepared to deal with the crisis. In what would seem to be an amazing coincidence, the crucial technology for confronting this plague emerged just as the first AIDS cases were being documented in the early 1980s. The ability to grow and measure T-cells—a central component of the body's immune system—arrived in the nick of time to see AIDS patients losing their T-cells. The technique for detecting and isolating retroviruses had just evolved to the point that the Human Immunodeficiency Virus (HIV) could be found. And a huge, well-financed establishment of scientific research teams had been set up, ready to gather vast quantities of data on any new disease. Dozens of new biotech companies mushroomed just in time to mass produce HIV tests, HIV vaccines, and antiviral drugs.

But the construction of AIDS as a "contagious disease" caused by a virus had little to do with science and even less to do with luck. "Was the NIH's apparent preparedness for the epidemic an accident?" asked Edward Shorter in his 1987 book, *The Health Century*.[1] No, he concluded, it resulted from the enormous funding of science in preceding years. Shorter hardly knew how right he was. Virus hunters were coasting on their laurels ever since they had won the war on polio. But they had failed to produce any public health benefit in their fifteen-year-old War on Cancer—despite enormous funding and enormous technology. They needed a success because their lease on public support, extended for a few years by the War on Cancer, was wearing. Using their new technology, the virus hunters could now find whatever they wanted in terms of slow, latent, or defective viruses. When AIDS appeared, the tumor virologists, who could not find a tumor virus, were poised to take full advantage of the situation to at least find an AIDS virus.

The NIH, however, did not lead the charge against this new syndrome in the early 1980s; many scientists at first refused to believe it existed. Virus hunters had more experience searching for viruses in older, better recognized diseases such as hepatitis or cancer. The overfunded science bureaucracy was too cumbersome to exploit such a new phenomenon before most people had even heard of it. The NIH proved more effective in mobilizing the scientific community behind a virus hunt and crushing all opposition, only after AIDS had been widely sold as a pandemic threat.

This new syndrome also had little chance of exciting the lay public. Since it was found mostly in male homosexuals and heroin addicts, it was too obscure to concern the average heterosexual person preoccupied with career and family. Even among homosexuals, AIDS was at first denied as an obstacle to their sexual freedom.

Yet, against all the odds, AIDS launched a bandwagon almost overnight. Less than two years after the first five cases were identified, the syndrome officially became the federal government's top health priority. By then the virus hunters had already elbowed past each other to find a viral culprit and HIV was just being discovered. How could this epidemic have mushroomed into a virus-hunting bonanza so rapidly? The answer lies in a parallel branch of biomedicine: public health. To understand this movement's role in blaming AIDS on a virus, one must first understand its history.

Public health seeks to prevent disease rather than treat it and is based on the notion that a healthy lifestyle is not just a matter of personal responsibility, but also a government management imperative. Unlike the academic style of research scientists, public health professionals take a more activist approach to disease—quarantining individuals or populations, seizing control of food and water supplies, conducting mass immunizations, pushing slogans in health campaigns, running aggressive family planning programs, regulating or restricting access to items ranging from cigarettes to dietary supplements, or otherwise targeting anything they believe is a risk factor for disease. Public health experts are inclined to view almost any infectious disease as an emergency.

The federal government officially adopted such a system in 1912 with the reorganization of the Public Health Service (PHS), headed by the Surgeon General. Based largely on the German model, PHS members formed a corps of commissioned officers, complete with uniforms, that dispatched teams to impose quarantines and other crisis-control measures on cities with contagious epidemics.

The National Institutes of Health, one branch of the PHS, became its center for biomedical research. Though originally responding to requests from state governments for epidemic control, the NIH always felt more at home with laboratory research. When James Shannon restructured the NIH in the 1950s to centralize its role in basic science, the agency finally left public health activism altogether.

Its heir had already been born during World War II. In early 1942, anticipating outbreaks of malaria like those in World War I, the Public Health Service established a special unit called *Malaria Control in War Areas* (MCWA). Its primary mission was to prevent spread of the disease among the hundreds of military bases throughout the southern United States, which MCWA accomplished mostly by using pesticides like DDT to kill malaria-carrying mosquitoes.

The creators of the MCWA clearly anticipated their own obsolescence following the end of the Second World War. Within weeks of U.S. entry into the war and before the Surgeon General had officially established MCWA, its soon-to-be leaders were already discussing their long-term prospects. Justin Andrews and Louis Williams, two of the agency's founders, entertained visions of public health management on a national scale, which they preferred over the disbanding of MCWA. So by the time the war was ending in 1945, these MCWA officials convinced Congress to extend their authority to include civilian malaria control. While not terribly glamorous, the new work kept the program alive. The PHS reorganized the MCWA in 1946, creating the permanent Communicable Disease Center (CDC), based in Atlanta, Georgia. The name has been changed several times since 1967, though the initials have remained nearly constant. It is now called the *Centers for Disease Control and Prevention.*

This new agency absorbed all remaining public health activities of the NIH. The CDC enthusiastically responded to all calls for help from states experiencing disease epidemics of various kinds. While malaria remained the focus of CDC energies, its resources were stretched to branch into new diseases, such as rabies and typhus, and even tapeworms. But the CDC wanted to assume full control over the nation's public health system, rather than being relegated to serving state and local health departments on request. At the same time, their disease-control mission was increasingly being regarded as obsolete, prompting serious discussions about abolishing the CDC altogether.

The situation changed in 1949 when the CDC brought on board Alexander Langmuir, an associate professor at the Johns Hopkins University School of Hygiene and Public Health. Langmuir was the CDC's first VIP, bringing with him both his expertise in epidemiology (the statistical study of epidemics) and his high-level connections—including his security clearance as one of the few scientists privy to the Defense Department's biological warfare program. Like the rest of the CDC, he hoped to empower the agency to monitor and exert authority over all epidemics throughout the nation. His dream might have stood little chance of materializing in an age of vanishing infectious disease, but because civil defense ranked high in government priorities at that time, officials of the PHS listened when Langmuir proposed that the CDC develop a comprehensive disease surveillance system to detect the earliest signs of a biological warfare attack. Such an infrastructure could also serve to control hypothetical epidemics—using such techniques as quarantine measures and mass immunizations.

By the start of the Korean War, Langmuir had talked public health officials and Congress into giving the CDC contingent powers to deal with potential emergencies. He shut down the malaria project, freeing millions of dollars to create a special new division of the CDC. In July of 1951 he assembled the first class of the Epidemic Intelligence Service (EIS), composed of twenty-three young medical or public health graduates. After six weeks of intensive epidemiological training, these EIS officers were assigned for two years to hospitals or state and local health departments around the country. Upon completing their field experience, EIS alumni were free to pursue any career they desired, on the assumption that their loyalties would remain with the CDC and that they would permanently act as its eyes and ears. The focus of this elite unit was on activism rather than research and was expressed in its symbol—a shoe sole worn through with a hole. According to British epidemiologist Gordon Stewart, a former CDC consultant, the EIS was nicknamed the "medical CIA."

Every summer since 1951 a new class of carefully chosen EIS recruits has been trained, some classes exceeding one hundred people in size. Although a complete list of EIS officers and alumni was available until the spring of 1993, its members rarely advertise their affiliation; now the membership directory has been withdrawn from public circulation. Over the past four decades two thousand EIS trainees reached key positions throughout this country and the world. Many work in the CDC itself, others in various agencies of the federal government; one of the original 1951 graduates, William Stewart, went on to become the Surgeon General of the United States during the late 1960s. Some have staffed the World Health Organization (WHO), including Jonathan Mann and Michael Merson, the two directors of WHO's Global Program on AIDS, while their fellow agents can be found in the health departments of foreign nations. Several dozen have entered university public health programs as teachers and researchers. Roughly 150 have taken jobs in state or local health departments, closely watching every outbreak of disease. Hundreds have become private practice doctors, dentists, or even veterinarians, while others work in hospitals. Some have joined biotechnology or pharmaceutical companies or have risen in the ranks of major insurance corporations. Some reside within tax-exempt foundations, helping direct the spending of trust funds on medical projects.

A few have obtained prominent positions in the media. Lawrence Altman became a medical journalist for the *New York Times* in 1969 and is now its head medical writer. Bruce Dan joined ABC News as its Chicago medical editor for six years beginning in 1984, the same year he became a senior editor of the influential *Journal of the American Medical Association* (*JAMA*), a position he held for nine years. *JAMA* regularly publishes a section written by the CDC. Marvin Turck has held the title of editor at the University of Washington's *Journal of Infectious Diseases* since 1988. These three men were recruited into the EIS in 1963, 1979, and 1960, respectively—each one years before he entered the media.

Regardless of which career paths EIS alumni take, the vast majority of them retain their contacts with the CDC. Not only do they constitute an informal surveillance network, but they can act as unrecognized advocates for the CDC viewpoint, whether as media journalists or as prominent physicians. And they serve as a reservoir of trained personnel for any CDC-defined emergency. As Langmuir himself described it in 1952, "One of the primary purposes of the Epidemic Intelligence Services of CDC is to recruit and train such a corps of epidemiologists... As a result of their experience, many of these officers may well remain in full-time epidemiology or other public health pursuits at federal, state, or local levels. Some, no doubt, will return to civilian, academic, or clinical practice, but in the event of war they could be returned to active duty with the Public Health Service and assigned to strategic areas to fulfill the functions for which they were trained."[2]

The EIS network has functioned very much as Langmuir first

envisioned, except that it has grown up in the post-contagion industrial world, where infectious diseases have largely become subject matter for historians. The awaited biological attack never arrived. The CDC has nevertheless continued to exploit public trust by transforming seasonal flus and other minor epidemics into monstrous crises and by manufacturing contagious plagues out of noninfectious medical conditions.

SEARCHING FOR EPIDEMICS

During the decades after its founding, the CDC searched for authentic public health emergencies. Tuberculosis was no longer the scourge of industrial nations, measles had largely stopped taking lives, and other potentially fatal diseases ranging from diphtheria to pneumonia ceased striking fear in the hearts of the public. Only polio was left, and by the 1960s it, too, basically vanished. In identifying "epidemics," then, the CDC was forced to attend to continually smaller outbreaks of disease. Before long, experts began defining contagious epidemics on the basis of disease "clusters." Almost any coincidence of two or more closely spaced persons contracting the same disease could qualify as an incipient epidemic, even if they occurred weeks or months apart.

Clustered outbreaks, however, provide no conclusive evidence of an infectious disease. When the bacteria hunters sought to blame scurvy, pellagra, and other vitamin-deficiency diseases on microbes, they mistakenly cited clusters of sick people to argue the diseases were spreading. Likewise, the virus hunters pointed to clusters to support their indictment of viruses for SMON and other noncontagious diseases. Clustering actually reveals very little information. It can reflect several people sharing the same diet, behavior, or environmental hazard of almost any kind, not just common exposure to a germ. Even in cases of truly infectious disease, clusters may only indicate a group of people is susceptible to a sickness for similar reasons, while other people infected by the same microbe will remain healthy—in other words, no epidemic will ensue. If anything, epidemiologists have classically studied clusters of sick people as clues to subtle environmental hazards, not infectious agents. But when public health officials issue ominous warnings about mysterious disease outbreaks, they terrify the public with visions of deadly pandemics.

The most recent examples include the premature panics generated by an imminent Hantavirus epidemic in the United States in 1994. The Hantavirus presumably had jumped species, from mice to American Navaho Indians. But after killing just a few, the virus made peace with the Indians and apparently retired to its mouse reservoir. The epidemic failed to materialize.[3] A front-page article in the *San Francisco Chronicle* reported that CDC "epidemiologists [shown in space suits] across the nation are carefully monitoring the deer mouse population and the level of virus within it." But all that was left to discover of the former "Navajo flu" by the CDC epidemiologists in their space suits were healthy mice in the mountains of California.[4]

In May 1995 the CDC rang the alarm once again, this time threatening with an imminent Ebola Virus pandemic.[5] The deadly killer virus was expected to leave its hidden reservoir in the rain forests of Africa to claim Europe and the United States. The article in *Time* magazine was peppered with "CDC sleuths" in space suits and color electron micrographs of the virus, although the electron microscope cannot take color pictures, and no virus is colored. A CDC virologist suggested the virus could leave the rain forest if "we get a virus that is both deadly to man and transmitted in the air." A European epidemiologist who heads the United Nations' AIDS program echoed the CDC's alarm, warning, "It's theoretically feasible that an infected person from Kikwit could go to Kinshasa, get on a plane to New York, fall ill, and present transmission risk there." But within a month the epidemic had faded away in Africa and not a single Ebola case was reported in the United States or Europe.[6]

A month later the CDC was once again sounding the alarm. In an article entitled "After AIDS, Superbugs Give Medicine the Jitters," the public was warned of an impending crisis in the form of "superbugs."[7] Superbugs are strains of bacteria said to be highly resistant to antibiotics. As usual, the CDC issued its warnings in the form of a chorus of conforming voices recruited from within and without the agency. Robert Shope, professor of epidemiology at Yale, warned, "If we don't gear up to bring matters under control, we could face a new crisis similar to the AIDS epidemic or the influenza epidemic that killed 20 million people worldwide in 1918 and 1919." Ruth Berkelman, deputy director at the CDC, resounded, "If we continue to let this get out of hand, we're setting ourselves up for a major catastrophe... I'm talking about going in for a routine operation and dying from an infection."[8]

Most people have no idea of the more than one thousand outbreaks of disease each year, including colds, seasonal flus, hepatitis, and numerous noninfectious syndromes, all running their course and disappearing, often despite remaining unexplained by scientists. They are natural coincidences between immunodeficiency acquired by some noncontagious risk factors, like drugs, and infections by any one of the ubiquitous microbes, termed *opportunistic infections*. But these many outbreaks provide the CDC with its inexhaustible source of epidemics.

The first genuine success of the CDC emerged from the polio epidemic. Ironically, it was the vaccine against polio, not the disease itself, that provided the opportunity. The Salk vaccine was entering its large-scale testing phase in 1954, and Alexander Langmuir wanted a piece of the action for his fledgling EIS. Insisting on CDC participation in the field trials, Langmuir was able to assign EIS officers around the country to monitor newly immunized children. The EIS aggressively followed up the first cases of vaccine-induced polio appearing in the spring of 1955 by ultimately uncovering the hundreds of victims, who then received national attention over the next several months. The findings of the EIS investigation led to the suspension of the Salk vaccine and to the political shake-up at the NIH that brought James Shannon to power. Although this incident involved neither a natural epidemic nor biological warfare, it built the CDC's reputation as an efficient surveillance agency.

The next major CDC initiative ended less spectacularly, yet the agency emerged untarnished. The spring of 1957 brought news of a flu sweeping nations of the Far East. Influenza is generally a rather benign disease, but CDC officials exploited memories of the deadly 1918 flu epidemic that returned with U.S. soldiers from Europe and killed nearly half a million people. The decision to predict a deadly flu epidemic was arbitrary, considering that thirty-nine flu seasons had since gone by without disaster. Ignoring the fact that circumstances in 1918 differed radically from 1957, the CDC rang the alarm over an imminent Asian flu epidemic. A frightened nation quickly jumped into line. Congress gave Eisenhower a half million dollars, a large sum at the time, into which Langmuir dipped to expand the ranks of the EIS. Seasonal flu did arrive by summer and continued spreading until the following winter. As soon as the epidemic began slowing, public health officers rushed to issue warnings of a second round.

In the end, the CDC and other agencies accomplished little or nothing to slow the epidemic. Large numbers of vaccine doses were crash-produced, mostly after the flu season had finished. The flu itself was probably no worse than in any other year, but the heightened surveillance of the disease, together with the frantic public warnings, helped feed the false impression of a particularly horrible epidemic. Several leading public health experts openly criticized the over-hyped flu scare, and some of them suggested the whole incident merely helped stimulate vaccine sales. But the CDC came out ahead anyway as a heroic group, having gained public acceptance for mass immunization on command. Since the Asian flu, the CDC has regularly produced vaccines of unproven effectiveness for each new flu season and has maintained a permanent flu surveillance program.

With its political standing secured, the CDC began expanding its reach into virtually any disease over which it could gain authority. Collaborations with other biomedical institutions often worked to promote both parties. One such arrangement directly fueled the Virus-Cancer Program. During the early 1960s, EIS personnel were assigned to investigate every cluster of leukemia cases reported anywhere in the country and to search for a virus on the assumption leukemia was infectious. The efforts amounted to little more than a wild goose chase, but in medical circles the repetitive publicity surrounding these random clusters drummed into every scientist's head the notion that viruses must cause cancer. Most researchers, after all, had readily accepted the belief that clustering somehow proved a disease to be contagious. The National Cancer Institute backed this EIS project enthusiastically, and it ultimately benefited through the extra funding it received for chasing cancer viruses. Robert Gallo was one of the young scientists powerfully influenced by such thinking.

Until the advent of AIDS, however, the CDC's most ambitious program—and its most embarrassing disaster—played itself out in 1976. By that time the EIS network of officers and alumni had so widely penetrated hospitals, health departments, and other institutions that, potentially, any minor disease outbreak could easily be detected. In January 1976 five soldiers at Fort Dix in New Jersey contracted a flu. One of them died after overexerting himself against doctor's orders. Such a minor episode met the CDC criteria for a cluster, and the agency sprang into action.

Since 1966 the CDC director had been David Spencer, a medical doctor by training who had experience in various research, public health, and administrative jobs and who had just received an honorary membership to the EIS in 1975. Spencer used a local flu outbreak at Fort Dix as an opportunity to replay the Asian flu public relations victory of 1957, only on a larger scale. Relying on historical precedent, Spencer declared an imminent flu epidemic that would rival the deadly plague of 1918. But what Spencer failed to understand was that Americans in 1976 were much less vulnerable to infectious disease as opposed to the undernourished, immunodeficient people at the end of World War I. The new epidemic was nicknamed "swine flu," based on the belief that pigs were the reservoir for this human virus.

Spencer placed the EIS network on full alert to monitor for cases of swine flu. The large Auditorium A, located in CDC headquarters in Atlanta, became the command center—called the "War Room." Set up especially for this occasion, it contained "banks of telephones, teleprinters, and computers, the hardware for an unprecedented monitoring system which, to work, also required a typing pool, photocopy machines, and doctors sitting at rows of desks in the center of the room."[9] Experts worked around the clock, week after week, chasing down every rumor of flu clusters.

Spencer officially called for the most aggressive emergency immunization crusade in history to be conducted before the flu season arrived. Congress initially favored the idea; not understanding the CDC's bias for infectious epidemics, the naive legislators easily could be manipulated by the CDC's alarmist rhetoric. President Ford appointed a committee that met within two days of Spencer's vaccination proposal and decided to back Spencer's plan, which would run up costs into the hundreds of millions of dollars. The air of panic spread rapidly: "Minutes after the meeting ended, President Ford appeared on national television and called for the vaccination against swine flu of every man, woman, and child in the United States."[10] The plan gained momentum, despite the fact that even the massive EIS surveillance program could not find any more cases of swine flu.[11]

But when early testing showed that the vaccine produced side effects in 20 percent to 40 percent of inoculated people and potentially deadly reactions such as high fevers in 1 percent to 5 percent, insurance companies backed away from supporting the program. With no insurance coverage, Congress became nervous and also began retreating before the plan came up for a vote. Now Spencer faced serious trouble, his whole reputation standing on the line. No longer able to back out quietly, he chose instead to push more aggressively. The word went out to the EIS network to pursue actively any flu-like illness whatsoever.[12] Spencer had to convince Congress that the swine flu epidemic was real.

Meanwhile, another CDC official took note of the swine flu alert:

> By early July 1976 David Fraser, M.D., hoped that a suitable epidemic would soon appear in the United States.

His definition of "suitable" was quite specific; the outbreak would have no known cause; it could present a serious threat to human life and might even have claimed some victims, thus providing the corpses for all-important tissue samples. With every day that passed, his need for that epidemic grew more urgent. He cast his net wide for news that somewhere between Alaska and the Mexican border a mysterious malady had surfaced. He made sure he was never far from a telephone.[13]

Fraser was the head of the Special Pathogens Branch of the CDC, the section charged with investigating infectious diseases with unknown causes. He had been an EIS member since 1971, and he was awaiting two new EIS trainees who would shortly be assigned to his office. He wanted to give them field experience through managing a real epidemic. With the EIS on full alert, a "suitable epidemic" was likely to be found on short order, selectable from the thousand or more disease outbreaks occurring each year in this country.

The first "suitable" choice presented itself in Philadelphia, days after American Legion members had returned home from their July convention. On Monday morning, August 2, after receiving word of a few pneumonia cases, personnel in the CDC's swine flu War Room established contact with Jim Beecham, a brand-new EIS officer barely settling down to his assignment in the Philadelphia health department. The CDC could not directly intervene in the situation without an invitation, and Beecham helped arrange one immediately. Within hours three EIS officers flew to Philadelphia. They were joined by David Fraser the next morning, followed within days by a team of dozens of CDC experts.

State and local health departments had been willing to accept EIS officers on temporary assignment because of their qualifications and training. But as Philadelphia health officials now discovered, this amounted to a Faustian pact. When the CDC personnel arrived, prepositioned EIS members such as Beecham and top health advisor Robert Sharrar stopped obeying local

144 ■ INVENTING THE AIDS VIRUS

authorities and began following orders from the incoming CDC team. Local officials became helpless to stop the tide of events. The CDC seized the initiative, fomenting rumors that this "Legionnaires' disease" was the beginning of the swine flu pandemic. The media proved cooperative; the *New York Times* assigned none other than EIS alumnus Lawrence Altman to cover the story.

With nationwide hysteria rapidly developing, Congress suddenly changed its collective opinion on the swine flu bill, pulling it out of committee and passing the legislation within days. By the time the CDC team officially acknowledged that Legionnaires' disease was not swine flu after all, President Ford had already signed the vaccine bill into law. David Fraser continued managing the CDC investigation for a few more weeks, allowing his new EIS people plenty of training. After testing the patients for infection by a variety of germs, the CDC experts found nothing consistent and packed their bags to leave. The case was declared unresolved and effectively dropped, leaving Philadelphia officials to pick up the pieces.

This cavalier treatment and the one-track focus on infectious microbes so enraged New York Congressman John Murphy that he held hearings on Legionnaires' disease in November. Calling CDC officials to testify on their "fiasco," Murphy humiliated the agency for not having found the epidemic's cause and for ignoring the possibility of noncontagious or toxic causes.[14] "The CDC, for example, did not have a toxicologist present in their initial team of investigators sent to deal with the swine flu epidemic," he fumed at the meetings. "No apparent precautions were taken to deal with the possibility, however remote at the time, that something else might have been the cause."[15] Likely smarting from the attack, David Fraser returned to Atlanta and put laboratory experts to work on the tissue samples collected from Philadelphia. Fraser's own area of expertise lay in bacteria, not viruses, and the researchers under his supervision searched hard for bacteria. Within a few weeks they found one, a harmless microbe that inhabits soil as well as plumbing in most buildings (see chapter 2). Even though the bacterium fails Koch's postulates for causing disease, the CDC cleared its reputation and convinced the unsuspecting public it had discovered the cause of Legionnaires' disease. In the process the CDC created a whole field of study devoted to this bacterium, which now employs a respectable number of scientists.

The swine flu program, on the other hand, collapsed and could not be salvaged. Millions of people received the vaccine starting in October, although many were not told of the possible side effects. Soon, reports of hundreds of cases of paralysis began pouring in, ultimately including at least six hundred cases and seventy-four deaths. The CDC attempted to classify the victims as having died of other diseases. Ultimately, the vaccine's side effects could no longer be hidden, and the expensive scandal cost David Spencer his job as CDC chief. Ironically, the swine flu epidemic itself never materialized; only the CDC's immunization program caused sickness and death.

Allegheny County Coroner Cyril H. Wecht personally investigated some of the vaccine's most unfortunate victims, including several fatalities. In a stinging indictment of this CDC program, he wrote in 1978:

> The government should limit itself to facilitating public programs. Employing high-pressure sales tactics like Madison Avenue mass media promoters to push a program is not commensurate with this objective. Certainly, when people's lives are at stake, cheap politics has no place.[16]

INVENTING AIDS

In the aftermath of the swine flu and Legionnaires' disease fiascos, the CDC diversified into other areas of public health, ones not always tied to infectious disease. In 1980 the agency was restructured into several units, each focusing on different issues, and as a whole was renamed to the plural—the Centers for Disease Control. But as the CDC grew it still preferred contagious diseases as subjects of investigation.

The NIH was likewise beginning to enter uncertain times,

particularly as the War on Cancer was dragging on without any tangible results. The virus hunters had consolidated their position with their so-called cancer viruses, but none had made enough of a public impression to justify their lavish funding. Public patience was beginning to wear thin, and even many scientists were growing critical.

Both the CDC and the NIH, representing the public health and biomedical research establishments, needed a new war to revitalize themselves. Contagious epidemics had proven the most effective at mobilizing public interest, and the medical and health establishments had spent vast sums of money establishing themselves on microbe-hunting foundations. Yet microbe chasers had exhausted their opportunities with virtually every major disease, from hepatitis to cancer and more. Now they had no clear direction in which to march, no significant diseases to conquer. The virus hunters were heavily armed soldiers without a war to fight. Stated Red Cross official Paul Cumming in 1983, "the CDC increasingly needs a major epidemic to justify its existence."[17]

The AIDS epidemic became their salvation. Here was a brand-new plague, too dauntingly unfamiliar to allow criticism of virus-hunting habits and growing quickly enough to compel urgent action. It was an epidemic that allowed no time to think, only to act. The inherent danger of an infectious disease would quickly unite responsible health care workers, scientists, and journalists to stem the possible danger to the health of the general population. Once recognized and taken seriously, it could easily be exploited by the virus hunters of the huge NIH-funded research establishment. But to identify the syndrome and label it contagious, the CDC and its EIS would first have to stake their claim.

That opportunity arrived late in 1980. Michael Gottlieb, a young researcher at the medical center of the University of California, Los Angeles, wanted to study the immune system and began scouring the hospital for patients with immune deficiency diseases. By November Gottlieb was introduced to one such case. The patient, who suffered from a yeast infection that had taken hold in his throat, also had a rare pneumonia that refused to go away. The *Pneumocystis carinii* microbe that caused the pneumonia was known to inhabit the lungs of almost every human on the planet; the disease rarely struck anyone but cancer patients, whose chemotherapy treatments would destroy their immune systems and leave them vulnerable to such normally benign germs. But this young man, in his early thirties, was taking no such therapy. Given his age, he should have been a specimen of perfect health. In any case, this was Gottlieb's chance to try out the brand-new technology for counting T-cells, one subset of white blood cells that participate in the immune system. The patient turned out to have very few T-cells at all, much to Gottlieb's amazement. On the other hand, scientists knew very little about what a "normal" level of T-cells should be, or any other white blood cells for that matter.

The next several months of searching gathered three more such cases of immune deficiencies. All three displayed the same candidiasis, or yeast infection, as well as the *Pneumocystis* pneumonia. And they all had "low" T-cell counts, the only parameter Gottlieb was interested in testing. By April of 1981, he decided he had a new syndrome on his hands. He called up the local public health department to ask for data on any similar patients elsewhere in Los Angeles. The staffer he spoke with, Wayne Shandera, was an active EIS officer trained the previous year. Shandera perked up at the news and found one more such case to add to the list. Now a pattern was emerging: All five men were active homosexuals.

Gottlieb knew precisely what this discovery could mean for his career. As Randy Shilts recorded in his 1987 book, *And the Band Played On*, Gottlieb phoned the *New England Journal of Medicine.* "'I've got something here that's bigger than Legionnaires', he said. 'What's the shortest time between submission and publication?'"[18] The *Journal* refused to bend traditional rules of publication. Frustrated and impatient, Gottlieb turned again to Shandera, who contacted the CDC. He figured this was probably the sort of outbreak that the CDC would be only too happy to publicize without delay.

Shandera was right. James Curran, an official in the Venereal

Disease Division of the CDC, wrote "Hot Stuff. Hot Stuff" on the announcement and hurried it into press with the agency's *Morbidity and Mortality Weekly Report*.[19] Like Gottlieb, CDC leaders could see the political benefits of managing another epidemic on the scale of Legionnaires' disease. On June 5, the report was published, written so as to imply that these five unexplained cases spelled a major new disease. Despite the fact that the five victims had no contacts with each other, the report wasted no time suggesting this might be a "disease acquired through sexual contact."[20]

Buried in Gottlieb's paper was another common risk factor that linked the five patients much more specifically than sex: all five had reported the use of recreational drugs, specifically, nitrite inhalants. Sex, being three billion years old, is not specific to any one group and is hardly a plausible source for a *new* disease.

Once Gottlieb's paper was published, new cases were reported to the CDC, some of whom suffered the rare blood-vessel tumor known as *Kaposi's sarcoma*. The CDC immediately set up a special task force, called *Kaposi's Sarcoma and Opportunistic Infections* (KSOI), to find the cause of this syndrome. All of the known patients had been active male homosexuals who reported using "poppers," the volatile nitrite liquid that had become the rage in the homosexual community for its ability to facilitate anal intercourse, as well as to maintain erections and prolong orgasms. This drug presented itself as the most specific explanation, especially given its known biochemical toxicity. But CDC experts had bigger plans for this illness, which could mobilize public concern only if it were believed to be infectious and therefore a threat to the entire population. The microbe-hunting bias of the KSOI Task Force was set in stone through its composition. Of the dozen or so members, its three leaders came from the venereal diseases section of the CDC, including two EIS officers (Harold Jaffe and Mary Guinan) and James Curran, who became the group chair. Other members specialized in studying viruses or infectious parasites.

Curran and his associates further stacked the deck by allowing only two alternative hypotheses on the agenda: Either this syndrome was a short-lived tragedy caused by a single bad lot of poppers or it was contagious.[21] The task force failed to consider the possibility that the long-term use of poppers might itself cause immune deficiency, a situation analogous to the connection between long-term smoking and lung cancer. The KSOI Task Force strategy was simple. The poppers hypothesis would be thrown out as soon as they could prove the victims had used different batches of the drug; the infection hypothesis would be supported by defining "clusters" of patients. The EIS network would assist the effort with extensive legwork, finding as many patients as possible and tracing their sexual partners. As historian Elizabeth Etheridge has demonstrated, based on later interviews with Harold Jaffe and other task force members, the fix was in: "While many of the patients were routine users of amyl nitrites or 'poppers,' no one in the KSOI task force believed the disease was a toxicological problem."[22]

As expected, no "bad lot" of poppers could be found. The results of the cluster study were equally predictable. The men turning up with such rare and fatal diseases had all spent years in extremely promiscuous homosexual activity, generally involving hundreds or thousands of sexual contacts. They also had "been frequent users of inhaled amyl and butyl nitrite" and "of recreational drugs other than nitrite."[23] Many patients tracked down by the CDC personnel could ultimately be traced through chains of sexual encounters to other immune deficiency patients, especially given their enormous sexual activity over time; "approximately 250 different sexual partners each year."[24] The CDC investigators had their hands full in trying to trace each patient's list of partners, considering the long "latent period" preceding AIDS.[25] Once again the "cluster" method of epidemiology proved its worthlessness, for even noncontagious diseases usually appear in such clusters. Nevertheless, the CDC accepted the clusters as proof of the infection hypothesis and announced their results one year after Gottlieb's first report.[26] Most outsiders began yielding upon seeing the supposedly impressive cluster study.

But one decade later, even the CDC had lost confidence in its hypothesis that clustering could prove AIDS to be infectious:

"Such clusters may be difficult to identify because most persons with AIDS have had contact with many different people. In particular, drug users and homosexual and bisexual men may have had contact with hundreds of partners that they did not know very well."27 In the early 1980s the KSOI Task Force members looked for evidence that the syndrome was spreading to heterosexuals.28 Using hepatitis B as their model, they hunted down every heroin addict and every blood transfusion recipient, including hemophiliacs, who might have conditions vaguely resembling the immune deficiencies in homosexuals. EIS personnel scoured hospitals and monitored local health departments for patients with serious opportunistic infections.29 Within months, one hemophiliac in Colorado and a small handful of heroin users were found with similar problems. The hemophiliac had actually lived much longer than expected, given the severity of his blood clotting disorder; he was dying primarily of internal bleeding but had also happened to contract a *Pneumocystis* pneumonia that caught CDC attention. His pneumonia, and the diseases of the heroin addicts, were immediately rediagnosed to include them in the new immune deficiency epidemic. One young KSOI Task Force member, EIS officer Harry Haverkos, was even sent to Florida and Haiti to study the epidemiology of Haitians suffering from malnutrition—who tended to have different diseases altogether.30 By adding more diseases to the definition of AIDS, all such patients could now be reclassified under the new epidemic.

Having decided the syndrome was a single contagious disease, the CDC now worked to swing the most powerful biomedical and political institutions behind its new war. Support would be hard to gather unless the disease had an easily remembered name; by July of 1982 the CDC decided to call it the *Acquired Immune Deficiency Syndrome* (AIDS). This name also swept under the rug any connection between the syndrome and risk groups, a move favored both by the CDC and the homosexual rights movement.31 "Certainly, the gay groups were putting much pressure on Congress [because of] the emphasis... on AIDS being a gay disease. They wanted the emphasis put someplace else," acknowledged one CDC official.[32] In addition, more federal money had to be appropriated to give this disease more respectability and to attract more experts into this new field. CDC officials soon developed contacts on several congressional staffs, and before long they had won two powerful representatives as allies: Phillip Burton of San Francisco and the powerful Henry Waxman from Los Angeles, who controlled the House committee in charge of health issues. Both congressmen wasted no time in raising a public furor over the immune deficiency syndrome, holding hearings and demanding crash spending programs. Facing little organized opposition, Burton and Waxman succeeded in diverting millions of additional dollars to the CDC and other agencies.

Meanwhile, the CDC courted influence at medical conferences and in journals. They lobbied doctors at every possible opportunity, spreading word of a new epidemic. More important, they put pressure on blood suppliers to screen out homosexuals, or at least people previously infected with hepatitis B, from donating blood. The CDC held meetings with the Red Cross and various blood supply associations, demanding immediate screening procedures. CDC representatives were infuriated when blood bank officials pointed out that the CDC had produced no serious evidence that AIDS was infectious. EIS member and CDC official Bruce Evatt, who worked with the KSOI Task Force, later admitted this to be true:

> CDC was calling shots on almost no evidence—educated guesses rather than proof. We did not have proof it was bloodborne; we had five hemophiliacs and two or three blood transfusion cases. We did not have proof it was a contagious agent; we had epidemiological evidence suggesting it.[33]

That "epidemiological evidence" was little more than the loaded and now discredited cluster studies.

To step up the pressure on medical institutions, CDC leaders used their full array of public relations skills to plant stories on AIDS in the news media.[34] By late 1982 dozens of articles were appearing in national print media, exploding to hundreds per month during the first half of 1983.³⁵ *Time* and *Newsweek* jumped into the act, running cover stories on the mystery disease and hyping up the supposed danger to the general population. *Newsweek*'s cover of April 11, 1983, called AIDS "the Public-Health Threat of the Century."³⁶ Eight months earlier, Dan Rather had broadcast a special segment on AIDS on the *CBS Nightly News*. As the public became fearful, the biomedical establishment began to take notice of the CDC campaign.

The stage was set for the search for an AIDS virus. Bacteria were less favored as potential culprits, given that antibiotics did not control AIDS; besides, virus hunting had become the dominant trend in medical science. The scientists with the appropriate laboratories, resources, and experience mostly worked at the NIH, but their small AIDS research program had so far focused on poppers, finding that homosexuals who had inhaled the most nitrites for the longest times had the highest risk of developing AIDS. Researchers were now beginning to test the chemical on mice, the logical next step.³⁷ Such powerful evidence, however, could no longer budge CDC officials, who had thoroughly convinced themselves AIDS had to be contagious. They began exerting pressure on the NIH to hunt viruses, using every scientific meeting and social occasion to collar researchers.

One of the earliest NIH responses came from its National Institute of Allergy and Infectious Diseases (NIAID), a traditional hot spot for virus hunters. Deputy clinical director Anthony Fauci started up an AIDS research program by early 1983 under his own supervision, readily embracing the CDC's view of AIDS as an infectious disease. Researchers at the National Cancer Institute (NCI) responded more slowly, partly because of their ongoing poppers studies. But by April of 1983 the NCI had established its own AIDS task force, and viruses soon replaced poppers as the focus of research.

Now came the big question: Which virus to blame? Finding one would be the easy part; since AIDS patients were inherently full of infections, virus hunters would almost have too many choices. In his 1981 report on the first five AIDS cases, Michael Gottlieb had offered the first suggestion—the herpes-class cytomegalovirus. This virus had been isolated in the 1950s and had been found to cause a disease similar to mononucleosis, but it was so mild that few people, other than cancer patients whose immune systems had been suppressed by chemotherapy, ever suffered from cytomegalovirus disease. The virus spreads easily and has infected perhaps three-quarters of the adult population, although most people are healthy enough to avoid symptoms. The virus had infected virtually all sexually active homosexuals, including all five of Gottlieb's AIDS patients. Over the next two years, the cytomegalovirus hypothesis picked up steam, attracting researchers in key positions around the country. Part of its popularity derived from the widely accepted belief that two other herpes viruses—Epstein-Barr and herpes simplex 2—could cause cancer. Some scientists even hypothesized that the Epstein-Barr virus itself might cause AIDS. This second hypothesis embraced a strange contradiction, since the Epstein-Barr virus was simultaneously believed to cause Burkitt's lymphoma, a cancer in which it was supposed to make white blood cells grow too well (see chapter 4). To cause AIDS, it would have to kill the very same cells.

While the cytomegalovirus hypothesis slowly gained supporters, retrovirologists also discovered the up-and-coming AIDS research bandwagon. Despite their prestige and Robert Gallo's recent discovery of a human retrovirus thought to cause leukemia, the glow of the War on Cancer was fading fast. The prediction that leukemia was a contagious viral disease simply did not pass muster; no epidemiologist or virologist could convincingly argue that leukemia spreads as an infectious disease. However, despite the retrovirus hunters' need for some disease to blame on a retrovirus, most of them had spent too many years trying to explain cancer to think of anything else. The door to AIDS would have to be opened by a retrovirus hunter outside the NIH academic complex, one not committed to studying cancer.

In stepped Donald Francis. A conscientious objector against the Vietnam War, he received a medical degree in the late 1960s, finished his residency training, and was recruited into EIS in 1971.

Virtually his entire career since that time has revolved around the CDC, in which he has risen to ever more powerful positions. His job history reads like a tour guide, encompassing public health assignments in parts of Africa and the Far East. He gained much of his experience imposing strict, even truly coercive, public health measures—which may not have made much of a medical difference. Since Francis published no controlled studies, improved standards of living, rather than his public health measures, may have reduced infectious diseases. "Years of stamping out epidemics in the Third World had also instructed Francis on how to stop a new disease. You find the source of contagion, surround it, and make sure it doesn't spread," wrote author Randy Shilts.[38] Francis partly demonstrated his methods in 1976 when he was sent to Zaire to control Ebola Fever, one of the innumerable Third World diseases that are constantly appearing and vanishing without explanation:

> When it became obvious that the disease was spreading through autopsies and ritual contact with corpses during the funerary process, Dr. Don Francis, on loan to the World Health Organization from the CDC, had simply banned local rituals and unceremoniously burned the corpses. Infected survivors were removed from the community and quarantined until it was clear that they could no longer spread the fever... The tribes people were furious that their millennia-old rituals had been forbidden by these arrogant young doctors from other continents.[39]

Ebola Fever, as it turned out, had been transmitted primarily through the use of dirty needles in one particular hospital, not through the native burial process. Nor did the CDC and WHO teams accomplish much. According to historian Elizabeth Etheridge, "The epidemic was virtually over before their work began."[40] The guilty hospital had already dosed itself, and the epidemic disappeared spontaneously. Nevertheless, for his stem techniques, Francis was credited by his peers for "singular brilliance."[41]

Francis returned to school in the late 1970s for a graduate degree researching the so-called Feline Leukemia Virus in Max Essex's laboratory at Harvard (see chapter 4). Thus, Francis joined the circle of retrovirus hunters. But he has preferred public health activism over research science and since 1981 has developed the reputation as one of the CDC's most ardent proponents of aggressive health controls over the population. In the early meetings between the CDC and the blood bank associations over possible AIDS transmission through the blood supply, Francis became known for his table-pounding confrontational style.

By the time Gottlieb's report on the first AIDS cases was published in June of 1981, Francis had reached a high position within the CDC's Hepatitis Laboratories Division and had worked for years with the homosexual community in organizing a major hepatitis B study. Upon hearing that these mysterious patients had lost their T-cells, he evidently saw an opening and leapt for it. A mere eleven days after the Gottlieb report—when only five AIDS cases officially existed and only a handful of other possible ones had been reported—Francis placed a telephone call to Max Essex. Randy Shilts described the start of the conversation: "'This is feline leukemia in people,' Francis began." Retroviruses were generally known to prefer infecting white blood cells, including T-cells, he reasoned. Further, in Shilts's words, "Feline leukemia has a long incubation period; this new disease must have long latency too, which is the only way it was killing people in three cities on both coasts before anybody even knew it existed."[42] On that June day, no one could even say for sure that this was even a real epidemic nor had any retrovirus been found in AIDS patients. Yet Francis had already mapped out the entire future of AIDS research: This new syndrome would be contagious, caused by a retrovirus with a long latent period between infection and disease. According to Shilts, "Francis was already convinced."[43] This decision had no basis in any scientific evidence but was destined to shape scientific thinking for years to come.

As soon as Francis had made his decision, he transformed himself into a relentless champion of this retrovirus-AIDS hypothesis.

He doggedly pushed this view whenever someone would lend him an ear and even when no one would. Within a year, KSOI Task Force head James Curran was echoing the Francis hypothesis, as were other key CDC staffers. Max Essex eagerly joined in, helping Francis lobby the NIH to find a new retrovirus. The perfect man for the job was Essex's old friend, Robert Gallo, who headed a huge and well-funded retrovirus lab at the National Cancer Institute. By 1982 both Essex and Gallo were searching part time for an AIDS retrovirus.

But rather than waiting for some new virus to be discovered, Essex decided to use something more readily available. Gallo had already found HTLV-I, the first known human retrovirus, which he believed caused T-cell leukemia after a long latent period. Why couldn't this virus also cause a second disease, AIDS? This would not be the first time virus hunters had blamed a single virus for two or more radically different diseases. In this case, HTLV-I would infect the same T-cells in both diseases. And so Gallo and Essex, in articles published back to back in *Science* in 1983, asserted that HTLV-I can cause AIDS.

Therein, however, lay the problem. If HTLV-I caused infected cells to grow into cancers, it could not also kill those same cells. Indeed, retroviruses had seized the high ground of cancer research during the 1970s precisely because they did not kill infected cells, but rather integrated themselves into the cell's genetic material, and therefore could be thought of as potential cancer-causing agents. Still, Essex's hypothesis, implicating HTLV-I, tickled Gallo's fancy—until he finally noticed the contradiction. Gallo then changed the name of the virus in 1985; for Human T-cell *Leukemia* Virus he substituted Human T-cell *Lymphotropic* Virus, meaning one that favors infecting T-cells. This new name implied neither cancer nor cell killing, thereby maintaining an ambiguity that could allow the virus to cause both diseases at once.

Late in 1982, while Essex and Gallo were reporting many AIDS or immune-suppressed patients who had been infected by HTLV-I, a French retrovirologist named Luc Montagnier was seizing the opportunity to stake his claim on an AIDS virus. Working at the Pasteur Institute in Paris, he cultured cells from a homosexual patient with swollen lymph nodes but no AIDS. Within weeks he isolated a new retrovirus. Not being prone to overstatement, he cautiously named his new find the Lymphadenopathy-Associated Virus (LAV), though certainly hoping it would be accepted as the cause of AIDS. Knowing he faced an uphill battle, he decided to enlist Gallo's help in promoting this discovery. That later proved to be a serious mistake.

As soon as Gallo caught wind of the new retrovirus, he hit the roof. The CDC was putting heavy pressure on him to find an AIDS virus, and he was having trouble gaining widespread support for his HTLV-I–AIDS hypothesis, especially from cancer virologists who hated to lose a leukemia virus. Now a lesser French virologist had beaten him to finding another human retrovirus. Gallo began quietly telling colleagues that Montagnier had made a mistake. Hedging his bets as always, Gallo also generously offered to write the short summary for the beginning of Montagnier's upcoming scientific paper. The unsuspecting French scientist agreed, and Gallo wrote in it that the new virus was closely related to his HTLV-I and -II retroviruses. So while Gallo was denouncing Montagnier's discovery and stepping up his own campaign to make HTLV-I the "AIDS virus," he was also trying to take credit for the new virus.44 Gallo proudly defended his new title, "father of human retroviruses," and lived up to it by adopting all human retroviruses to his HTLV family.

Montagnier's paper was published, and Gallo spent the next several months furiously trying to find the same virus. Finally, by April of 1984 he was ready to announce having found a similar retrovirus, which he unsurprisingly named *HTLV-III*. He had prepared four separate papers reporting his discovery of the virus and its isolation from a number of AIDS patients. Ethical protocol among scientists required that he first publish those papers, allowing his peers to analyze the results before he went to the news media. But Gallo and his employer, the Department of Health and Human Services, pulled a coup d'état on Montagnier by holding a press conference on April 23, more than a week before the papers were to be printed in the journal *Science*. Margaret Heckler, Secretary of Health and Human Services, sponsored the huge event and introduced Gallo to the press corps. Backed by the full prestige of the federal government, she officially declared this new virus was probably the cause of AIDS, a conclusion dutifully reported by the media. By April 24, EIS member Lawrence Altman had dubbed it the "AIDS virus" for the readers of the *New York Times*.[45]

Thus, before any other scientists could review and comment on Gallo's claim, it had been set in stone. The press conference marked a point of no return. Career-minded scientists immediately dropped all other AIDS research, including work on the Epstein-Barr virus, cytomegalovirus, and HTLV-I, as well as all remaining experiments on poppers. From that date forward, every federal dollar spent on AIDS research funded only experiments in line with the new virus hypothesis. Had researchers been politically free to examine Gallo's papers for themselves, they might have objected that some of his AIDS patients had never been infected by the virus. They would have pointed out that no virus had been found in any of Gallo's AIDS patients, but only antibodies against it. Antibodies are traditionally a sign that the immune system has *rejected* the virus. Researchers also could have remembered that retroviruses do not kill cells. For that matter, they might have noticed that Montagnier had found the virus first.

But the CDC had raced to victory. The entire world now knew about AIDS and believed it to be contagious. The news media had begun beating the drums for a war on this syndrome. Hundreds of millions, and then billions, of new dollars began flowing into the CDC and other biomedical research institutions. Most important, the virus hunters had finally reached center stage; not since the polio epidemic had they reveled in the glory of so much public attention. The fear of infectious disease had now been revived on a mass level for the first time in decades, and the lay public had no choice but to trust their appointed experts for answers.

And on the very day of the press meeting, while the rest of the world was struggling to come to terms with the first infectious plague in many years, Gallo quietly filed his U.S. patent application for the virus antibody test. The patent stated under oath that the virus could be mass produced for HIV tests within indefinitely growing, "immortal" T-cells. But according to Gallo's scientific papers the virus caused AIDS by *killing* T-cells.

SCANDAL IN THE ESTABLISHMENT

More than just a politically driven event, the declaration of LAV/HTLV-III as the "AIDS virus" was a sordid affair. The story largely centers around Robert Gallo but also fulfills the worst expectations of the over-funded science bureaucracy. Gallo himself has a history of questionable claims to timely scientific discoveries. Given such a track record, the fact that he nevertheless steadily rose to one of the most powerful positions at the NIH serves as an indictment of federally sponsored research.[46]

Gallo's first attempt to get a piece of the action came in 1970, on the heels of Howard Temin's announcement of finding reverse transcriptase, the retrovirus enzyme that allows it to embed itself in the genetic material of an infected cell. Seeing the chance for a quick and easy way to explain human cancer, Gallo soon declared finding evidence of retrovirus infection in human leukemias. Virus hunters stampeded to confirm his discovery but, to their dismay, could not. Reflecting on this incident, Gallo's colleague Abraham Karpas later observed that "he probably thought that he could tie himself to Temin and Baltimore's wagon which was going to lead to a Nobel Prize within five years. The reason he lost that opportunity to become a Laureate early in the game was because many scientists from around the world, including ourselves, who spent time and efforts trying to reproduce Gallo's 'milestone discovery,' found that it was an uncontrolled artifact."[47] In other words, a false positive.

Gallo further embarrassed himself in 1975 by announcing he had isolated the first known human retrovirus from a leukemia. In his excitement, he did not bother to test his virus carefully. When other laboratories did so, they quickly found it was not a human virus at all, but a mixture of three monkey retroviruses. Caught unprepared, Gallo spent many months trying to argue his way out by insisting that perhaps one of the monkey viruses could cause human leukemia.

In 1980 Gallo was finally credited for discovering a genuine human retrovirus, HTLV-I, which he blamed for a leukemia in blacks from the Caribbean (see chapter 4). But he ran into trouble trying to find the virus in American leukemia patients. At the same time, a Japanese research team reported isolating a human retrovirus from leukemic patients, which they named *ATLV*. After they courteously sent Gallo a sample of the virus to compare with his own, Gallo published the genetic sequence of HTLV-I. The sequence of Gallo's Caribbean virus proved to be nearly identical to the Japanese virus; it contained a mistake identical to one made by the Japanese group.[48] Since all other non-Japanese HTLV-I isolates differed much more widely from the Gallo-Japanese twins, some retrovirologists suggest Gallo may have offered the Japanese sequence as his own.[49] No formal investigation has probed this incident, and Gallo was awarded the prestigious Lasker Prize as the presumed discoverer of the leukemia virus.

Gallo's report of finding a new retrovirus in AIDS patients smacked of similar tactics. Luc Montagnier, of course, was the first to report finding LAV in 1983. Gallo insists he independently found the virus at the same time, but waited nearly a year to test it before releasing his results to the world. The first journalist publicly to question this version of events was Steve Connor, a correspondent for England's *New Scientist* magazine, who wrote an exposé of Gallo in 1987.

Both Montagnier and Gallo published the genetic sequences of their viruses in January 1985, as did a third scientist, Jay Levy, who independently isolated the virus in San Francisco. Several other researchers immediately noticed a suspicious coincidence: The Gallo and Montagnier viruses were so similar to each other that they had probably come from the same patient. Normally, a retrovirus isolated from two different people has mutated, if only in trivial ways, enough to mark the two isolates as distinct. But Gallo's virus was almost identical to Montagnier's. The French researcher had generously sent samples of his virus to Gallo on request, and now Gallo was offering an amazingly similar one as his own.[50] When challenged, Gallo failed to produce any of the other virus isolates he claimed to have. To explain away the similarity, he even proposed that the American and French isolates had come from two patients who just happened to be sexual partners. Finally, in 1991 Gallo publicly admitted in the science magazine *Nature* that the French virus was indistinguishable from his own and has excused his lack of other viruses by weaving tales of laboratory accidents that somehow happened to destroy his dozens of isolates.[51]

Connor's journalistic investigation also revealed a deliberate cover-up. In 1986 Gallo was forced to admit that the photographs of HTLV-III published in his 1984 papers had actually been photos of the French LAV. The switch was discovered after two copies of a letter, written in 1983 by the researcher who photographed the virus with his electron microscope, found their way into the hands of lawyers representing the Pasteur Institute. One copy stated that the virus was indeed LAV, while the other had been doctored to remove that information. Gallo claims to know nothing about the altered letter, and he tried to excuse the switched photo as having been "largely for illustrative purposes"—presumably the usual reason photos are published.[52]

But more recently another hidden fact has come to light. Mikulas Popovic, a Gallo lab associate who co-authored the key 1984 paper announcing Gallo's virus in *Science*, presented an original draft of the paper to the NIH's Office of Research Integrity. In this earlier manuscript, Popovic gave full credit to the French for finding the virus first and showed that the Gallo lab had been able to grow LAV soon after receiving the sample. Those admissions were crossed out in the draft, and in the margins Gallo's handwriting scoldingly declares, "Mika, you are crazy... I just don't believe it. You are absolutely incredible."[53] The published version of the paper contained none of the statements giving credit to the French scientists. With this piece of damning evidence, Gallo has been caught lying about his supposed inability to grow the French virus in his lab.

In 1989 *Chicago Tribune* correspondent John Crewdson joined in the fray with another exposé of Gallo, followed by several more articles. That began an avalanche of scientific fraud investigations originating in the NIH itself, in the National Academy of Sciences, and in Congress. As a result, Popovic was fired from the NIH for fraud, and Gallo himself was convicted of scientific misconduct at the end of 1992. The story has since expanded—Gallo apparently also commandeered the cell line in which he grew the stolen French virus.[54] A sample of the leukemic T-cells, originally named *HUT78*, was sent to his lab for isolating a leukemia virus. Unable to find any retrovirus, Gallo renamed them *H9*, claimed he developed the cells himself, and used them instead to grow HIV. No prosecutions have yet precipitated over this second alleged misappropriation.

Theft seems to be a common problem among Gallo lab personnel. Syed Zaki Salahuddin, another researcher in the lab, pleaded guilty and was fired in 1991 for receiving illegal payments. The money had come from Pan Data Systems Inc., a company founded in 1984 by Salahuddin's wife. Salahuddin had used his authority in the Gallo lab to arrange purchases of supplies from Pan Data, paid for by the NIH budget. For this he received compensation from the company. He even stripped Gallo's lab of viruses and equipment that he handed over to Pan Data for use and resale at below-market rates. Salahuddin was a major author on the 1984 Gallo papers announcing the discovery of the "AIDS virus" and had the habit of referring to himself as "doctor" despite having no such degree. Authorities are also investigating yet another scientist in the lab, Dharam Ablashi, for involvement in the Pan Data scandal.[55]

Another co-author on those Gallo papers, Prem Sarin, soon found himself on trial and was fired by the NIH for embezzlement. When a German company sent a payment of $25,000 for experimental work performed by the Gallo lab, Sarin deposited the check in a special personal account. He later testified he was simply borrowing the money, although he actually used it to pay off personal debts. The check, originally intended for the hiring of a laboratory technician to conduct the desired experiments, had been made out to the initials of the NIH Foundation for the Advancement of Education in the Sciences (FAES). Sarin's own bank account also had the initials *FAES*—which he later claimed represented the "Family Account for the Education of the Sarin Children." A jury convicted him of criminal charges in July 1992.[56]

Gallo sank into still deeper trouble in 1990 through a collaborative project with French scientist Daniel Zagury. In the United States, government scientists are prohibited from participating in dangerous experiments on human beings. Zagury, with the help of Gallo's lab, tested a supposed AIDS vaccine on nineteen human volunteers, some from Africa. Three of the patients died, a fact that Zagury left completely out of his published paper on the experiment. Word of the disaster and cover-up got out after an article appeared in the popular press by *Chicago Tribune* reporter John Crewdson.[57] This in turn led to a major NIH investigation.

Gallo sensed his worsening plight. But as always, whenever he finds himself in a corner, mysterious events take place. A few weeks after the Zagury paper appeared in print, Gallo returned home one August evening from a big dinner to discover the aftermath of a burglary. County police who responded to the call found a baffling scene. "The Gallo family jewelry, silverware, and VCR were in their familiar places, untouched... as police detective John McCloskey told *Science:* 'Not a thing was taken.'"[58] According to Gallo, only one thing had been disturbed—some scientific data sent from Zagury. Gallo eagerly offered John Crewdson, the *Chicago Tribune* reporter, as his first suspect. The police eventually dismissed this idea and dropped their investigation. Several months later, shortly before Gallo was to appear before Congress in one of many fraud investigations, he once again precipitated an unusual but convenient incident:

> The alarm system Gallo bought after last summer's break-in went off in the night. He phoned the Bethesda police, saying he thought Crewdson was again trying to break into his house. The detective bureau concluded it was a false alarm. Despite Gallo's insistence, the police disregarded the complaint.[59]

But Gallo proved not to be the only leading AIDS scientist to offer

Montagnier's virus as his own. The leading English AIDS scientist Robin Weiss in 1985 reported independently isolating an AIDS retrovirus—after Montagnier had also sent him samples of LAV. A British investigation revealed in early 1991 that Weiss's virus also appeared to be identical to the French virus, and Weiss publicly agreed that he might have accidentally contaminated his cultures with LAV.[60]

Both Gallo and Weiss have managed to cash in on their incredible series of "mistakes." Gallo secured the U.S. patent rights for the virus test, while Weiss received the British patent. Facing legal actions by a wrathful Pasteur Institute cheated of its patent royalties, Gallo and Weiss have acted as mutual benefactors. Weiss, for example, managed to be the anonymous peer reviewer on a key Montagnier paper in 1983; by rejecting it, he bought time for Gallo to discover the virus himself.[61]

Other powerfully placed colleagues have rushed to Gallo's defense, either to protect the image of the NIH or to protect the immaculate image of dedicated truth seekers that all scientists enjoy in the open and in the eyes of the public. Several of these researchers have developed such a close alliance with Gallo that they privately call themselves the "Bob Club." Among its informal members has been Gallo's longtime friend Max Essex, the Harvard retrovirologist who studies the so-called Feline Leukemia Virus and who trained Donald Francis. Essex has publicly supported Gallo's claim to isolating HTLV-III. He also shared the 1986 Lasker Prize with Gallo and Montagnier, in his case for relabeling a monkey retrovirus sent him by another lab and calling it his own.[62] Harvard retrovirologist William Haseltine, another "Bob Club" insider, had copied the genetic sequence of HTLV-II, the second known human retrovirus, from a presentation at a science conference. He then published the sequence, unknowingly including a deliberate error planted by the Japanese research team who had actually done the work.[63] Gallo has also found allies among his bosses and other administrators in high NIH positions, many of them helping to stall or water down the investigations.

Naturally, Gallo's 1984 press conference aroused French ire

and precipitated an international legal fight for almost three years. But with support for Gallo in the federal bureaucracy, a deal was worked out by March of 1987. In a public meeting between President Reagan and French Prime Minister Jacques Chirac, the two governments agreed to share credit for the virus discovery. Montagnier's lawyers were silenced for the sake of political compromise, despite the strong evidence supporting their case. That same year a committee of prominent retrovirus hunters met and chose a new, and therefore more neutral, name for the virus: the *Human Immunodeficiency Virus* (HIV). While this name did not discriminate between Gallo and Montagnier, it provided propaganda value by assuming this virus did indeed cause AIDS. The name has stuck, largely because a subsequent letter, published in the journal *Nature* in 1987 and signed by sixteen science stars, including ten Nobel Prize winners, such as David Baltimore, Howard Temin, André Lwoff, Jonas Salk, James Watson, and the director of NIH, backed the decision.

To ensure Gallo's place in the science hall of fame, his old friend Hilary Koprowski, the polio vaccine pioneer, launched a campaign in 1987 to elect Gallo to the elite National Academy of Sciences. Koprowski had worked for years alongside Gallo, chasing slow viruses as the director of the Wistar Institute. Citing Gallo's "brilliant discoveries" and "leadership," he succeeded by 1988, when Gallo joined the ranks of the most prestigious scientific body in the nation.

Koprowski himself probably felt a common bond with Gallo, for he was beginning to face his own troubles. The 1984 Nobel Prize for medicine had honored two European scientists for inventing a biochemical tool known as the *monoclonal antibody*. Upon request, the European researchers had generously sent Koprowski a sample of their cell line, along with a letter warning against any commercial use of the product. In speaking with Cesar Milstein, one of the Nobel Laureate European researchers, Koprowski denied seeing the letter, insisting it had somehow been lost. In any case, Milstein directly reminded Koprowski not to use the technique commercially. Yet, after that warning, Koprowski managed to patent the technique himself. To reassure the angered

Milstein, Koprowski declared that the money was going entirely into scientific research. It was in the form of Koprowski's brand-new biotechnology company, Centocor, which was reaping the profits.[64] Meanwhile, the Wistar Institute's board of directors fired Koprowski as director in 1991. During his last ten years at the helm, he had so mismanaged the institute's finances that its coffers dwindled from tens of millions of dollars to a several-million-dollar deficit. Centocor fared much better; by the end of 1986, Koprowski's own stock holdings in the company already exceeded $15 million in value.

By 1988 Gallo surely believed his position had been secured. But after the Connor and Crewdson articles, the whole stolen-virus scandal reopened in 1990. His career began tumbling, finally leading, on December 30, 1992, to his official conviction on a charge of scientific misconduct. The Office of Research Integrity found that Gallo had falsely claimed he could not grow the French virus in his own lab.

Gallo appealed the decision to a committee of lawyers under the authority of the Department of Health and Human Services. After months of legal wrangling by both sides, the panel shocked observers by raising the burden of proof on the prosecution. Suddenly, the investigators had to prove not only that Gallo had fabricated his results and covered up the evidence, but also that he had consciously planned to do so—as if this scientific review were actually a criminal proceeding. Unable to meet the new standards, the NIH prosecutors were forced to drop the charges, and Gallo was officially "acquitted." Since then, none of the many investigators on the Gallo case have even tried to prosecute the remaining charges, mostly related to the allegedly misappropriated French virus.

But the controversy is not going away. According to columnist Daniel S. Greenberg, "The misconduct case against Robert C. Gallo is showing signs of an afterlife of seething resentment among his detractors and canonization by supporters."[65] Gallo has made many enemies over the years, and many scientists remember the powerful evidence against him. In July 1994 the director of the NIH, Harold Varmus, reluctantly agreed to relocate to France the American royalties for the Gallo-NIH patent of the HIV-antibody test. The issue whether Gallo and Popovic should nevertheless receive their annual $100,000 salary supplements for the patent on the test from the U.S. government remained unresolved. The director's decision to reallocate the royalties to France was based on several years of investigations of Gallo's laboratory by the NIH's Office of Research Integrity and by the Subcommittee on Oversight and Investigations of the U.S. House of Representatives, chaired by Democratic Congressman Dingell. By the end of 1994 the Dingell subcommittee released a 267-page staff report and a 65-page summary report providing overwhelming evidence that Gallo and the NIH had patented Montagnier's virus. The *Chicago Tribune* summarized the report's conclusion in an article entitled "In Gallo Case, Truth Termed a Casualty,"[66] and in a subsequent editorial, "Defending the Indefensible Dr. Gallo."[67] According to well-informed sources, Gallo was asked to leave the NIH in 1995. This happened in the summer of 1995 when Gallo moved to Baltimore.[68]

THE VIRUS SURVIVES

No amount of controversy over the integrity of leading AIDS scientists has weakened the political support for the HIV hypothesis. The CDC, NIH, and dozens of biotechnology and pharmaceutical companies have invested their full resources in this view, making it unchallengeable for all practical purposes.

In the wake of massive spending increases on HIV research, virologists have converged from all fields to stake their claims. Many have taken up HIV research itself, while others have begun reclassifying animal diseases as "AIDS." Animal retroviruses once presumed to cause cancer now suddenly cause immune deficiency, at least in the minds of retrovirologists. Any young animal that will develop a flu or pneumonia when injected with huge quantities of a retrovirus now becomes an experimental model for AIDS. Virus hunters have transformed one strain of Feline Leukemia

Virus into a case of "Feline AIDS" (FAIDS), isolated the "Simian Immunodeficiency Virus" and blamed it for causing "AIDS" in monkeys (SAIDS), and even indicted a mouse retrovirus simultaneously blamed for leukemia as also causing "Mouse AIDS" (MAIDS).

No virus goes to waste. Even Gallo's original HTLV-I–AIDS hypothesis has not died completely. Gallo has proposed that HTLV-I could perhaps serve as a cofactor in causing AIDS, somehow cooperating with HIV when infecting the same victim. Gallo also simultaneously proposed exactly the opposite notion—that HTLV-I might function as the *cure* for AIDS. His logic was simple: If HIV kills T-cells, and if HTLV-I makes them grow more aggressively as leukemia, then the two viruses might cancel each other's effects. Few scientists have bought into either of these hypotheses which, nevertheless, stand mostly unchallenged.

The free-flowing money spent on AIDS has thoroughly reshaped modern science. Virus hunting, nearly discredited by the failed War on Cancer, has now enjoyed a spectacular revival. The CDC has shifted its resources back into managing contagious disease, and it masterminds public campaigns for controlling HIV. The NIH has continued to experience an ever-growing budget. In an era with no serious infectious disease in the industrial world, the otherwise healthy population has regained its fear of contagion. The dangerous public hysteria formerly witnessed with scurvy, pellagra, SMON, and other noninfectious diseases now repeats itself, but on a larger scale.

The next chapter will examine how the HIV-AIDS hypothesis shaped this public hysteria and will present the full evidence against this virus causing AIDS.

CHAPTER SIX

A Fabricated Epidemic

BY THE MID-1980S, a sinister specter had been launched. The media buildup around AIDS, combined with the 1984 announcement of an AIDS virus, had painted a picture of a twentieth-century bubonic plague capable of ravaging our nation and the planet. Now everyone was aware of the deadly disease spreading through the homosexual community.

The scientific and government experts, most prominently including Surgeon General C. Everett Koop, predicted an explosion into the heterosexual population. In early 1987, Koop and the World Health Organization were forecasting that a staggering 100 million people would be infected with the virus by early 1990.[1] Talk of casual transmission became popular once top officials at the CDC and NIH announced HIV could be found in saliva.[2] Evidence that the virus could survive for long periods outside the human body led to nervousness about restaurants and public toilets.[3] Naturally, the fact that HIV was a blood-borne virus spurred discussion of mosquito transmission, including among top AIDS researchers.[4]

AIDS was such a new syndrome that most of its mysteries remained to be solved. Certainly no vaccine, and probably no potent therapy, would be available for several years, by which time hundreds of thousands—or millions—of people would already have died.

In the meantime, it seemed that only public health measures could work. Authorities tried to prevent further spread of the illness by discouraging the major risk activities, those routes most easily transmitting HIV—the most obvious threat was said to be sexual intercourse. Official warnings were always accompanied by reminders that, although the virus was now transmitted by homosexual contact, it would soon follow the usual pattern of infectious diseases by spreading among heterosexuals of all walks of life. Frightening reports of the African epidemic were exploited to paint a picture of our own future; there, whole villages were apparently disappearing as the new syndrome cut a wide swath of destruction among men and women alike. In the industrial world, heterosexual intravenous drug addicts were already passing HIV around by sharing their used syringes. AIDS officials confidently reassured the public of their timely screening and protection of the nation's blood supply, but noted they were too late to save most hemophiliacs.

Ominous statistics hit the news: 50 percent to 100 percent of everyone carrying the virus would die, and the unpredictable latent period between infection and AIDS ranged from five to ten years, during which time the carriers could infect many more people. Once infected, an individual's antibody defense raised against HIV was inexplicably useless, except to alert doctors to the fatal infection. Once the virus was reactivated (for unknown reasons), it proceeded to kill off the body's entire supply of T-cells, the white blood cells regulating the immune response against all other microbes. AIDS victims suffered horribly slow, painful deaths, being eaten alive by pneumonias, yeast infections, cancers, uncontrollable diarrhea, and dementia from brain degeneration. No recovery was possible since the patient was completely defenseless against many diseases normally harmless to a healthy person.

To add a further sense of urgency, AIDS experts supplemented their official estimate of one million HIV-positive Americans with suggestions of two million to three million, plus dire predictions that the number might double every year.

The public response to such news was inevitable. Battle lines rapidly emerged between two political camps—civil rights advocates for the HIV-positives and those championing health rights for the HIV-negatives.

Under the banner call, "Fight AIDS, not people," groups ranging from the militant AIDS Coalition To Unleash Power (ACT UP) to the federal government's National Commission on AIDS insisted that the syndrome be treated basically as a handicap. Although acknowledging that AIDS was contagious, many political activists feared the potential backlash from widespread panic. They preferred to mobilize support for the care of AIDS patients, assiduously avoiding any hint of blame on the victims. As the National Commission on AIDS proclaimed, "HIV disease has a devastating impact on those who are already marginalized members of society... HIV disease could not be understood outside the context of racism, homophobia, poverty, and unemployment."5 Likewise, President Bush admonished that "once disease strikes we don't blame those who are suffering. We don't spurn the accident victim who didn't wear a seat belt; we don't reject the cancer patient who didn't quit smoking. We try to love them and care for them and comfort them."6

The CDC and other agencies deeply involved in managing the war on AIDS continued to warn of an imminent heterosexual epidemic. Activists for HIV were therefore forced to offer some solution to halt the syndrome's spread, but without endangering homosexual liberation; they found an answer in condoms and programs to provide heroin addicts with sterile needles. But many activists, including those in the National Commission, also saw in AIDS much opportunity:

> The HIV epidemic did not leave 37 million or more Americans without ways to finance their medical care—but it did dramatize their plight. The HIV epidemic did not cause the problem of homelessness—but it has expanded it and made it more visible. The HIV epidemic did not cause collapse of the health care system—but it has accelerated the disintegration of

our public hospitals and intensified their financing problems. The HIV epidemic did not directly augment problems of substance use—but it has made the need for drug treatment for all who request it a matter of urgent national priority.7

Another side of the debate operated on the principle of "Better safe than sorry," viewing AIDS in more grand and threatening terms. This alarmism created strange alliances between such individuals as California Congressman William Dannemeyer and former Marxist (head of the U.S. Labor Party) Lyndon LaRouche. Most of these people were convinced the AIDS epidemic was actually far worse than officially acknowledged. They certainly had a rich source of raw material upon which to draw, including frequent quotes and numerical projections by federal officials. A 1985 book written by an investigator at the NIH provides a typical example:

> The AIDS virus shows every sign of being just as deadly as the plague during the Middle Ages. We are on a crash course with reality. This is not a practice run. There is no second chance. AIDS may be to the twentieth century what the Black Plague was to the fourteenth century.
> The alarm must be sounded, loudly and persuasively. If it is not, the conclusion is inescapable: millions may die.8

Believing the population to be on the verge of decimation, a variety of alarmists called for strong public health measures by the government. Their reaction on behalf of the uninfected took on the strenuous tone of Gene Antonio, whose 1986 book *The AIDS Cover-Up: The Real and Alarming Facts About AIDS* became an underground bestseller: "In the pell-mell rush to identify with the plight of AIDS sufferers, compassionate concern for the rest of society has been largely ignored. Permeated with heterophobia, AIDS victim identification hysteria has dangerously impeded compassionate steps being taken to safeguard the health of the rest of society."9 The alarmists generally insisted on mandatory HIV testing, particularly for health care workers and those in AIDS risk groups, as well as infection contact tracing and reportability to government agencies, and they even discussed possible quarantine of infected persons. More than fifty countries, including the United States, adopted immigration or tourism restrictions on infected people, and the Cuban government established a quarantine detention center for its HIV-positive citizens.[10] Alarmists derided the weaker proposals of their opponents, often leaping to the defense of medical workers wanting more safeguards from potentially infected patients.

Yet despite their differences, both sides of the controversy agreed on one thing: More money was needed to fight AIDS—and quickly. Federal AIDS officials were no doubt delighted to hear California Congressman Dannemeyer, in an unusual alliance with Michigan Representative John Dingell for increased medical funding on AIDS, declare enthusiastically:

> The AIDS Prevention Act of 1990 is a pathbreaking piece of legislation in many respects. For the first time, the federal government would make resources available to states, hospitals, high risk clinics, and nonprofit health care facilities to provide "preventive health services" to low income individuals afflicted with a specific disease—AIDS...
>
> This legislation breaks new ground in bringing federal resources to bear on a very specific national health problem—the epidemic of HIV infection. It includes many admirable provisions which, if enacted, would establish sound priorities and provide state and local health officials with appropriate resources to fight this horrible epidemic.[11]

This push for larger AIDS budgets certainly succeeded. Some $7 billion were spent by the federal government during 1994, and well over $35 billion has been spent since the AIDS epidemic began. What are the results of this modern-day Manhattan Project? A staggering one hundred thousand scientific papers so far have been published on HIV and AIDS, a number unprecedented for any other virus. But AIDS investigators have yet to demonstrate that even a single life has been saved by any of their programs. No vaccine exists; condom and clean-needle programs have made no measurable impact on the epidemic; the admittedly toxic drugs AZT, ddI, and ddC, which do not cure AIDS, are the only therapy substitutes available today. Despite projections of wild spread, HIV infection has remained virtually constant throughout the industrialized world ever since it could be tested in 1985, whether in the United States or Europe; the estimated incubation period between infection and disease has been revised from ten months to more than ten years; and the predicted heterosexual explosion has failed to materialize. When a disease can be neither treated nor controlled, nor its course even roughly predicted, some fundamental assumption is probably badly askew.

HIV NOT GUILTY

Twenty years of belief in dormant human viruses causing disease after long incubation periods, plus many decades of hunting animal retroviruses, rendered most biologists utterly incapable of challenging Gallo's 1984 announcement of an AIDS virus. Prestigious awards and new grant moneys awaited scientists who could apply their animal models or "slow virus" concepts to human disease. Researchers also felt insecure about venturing outside their narrow fields of specialization to raise questions in other areas. Epidemiologists assumed clinicians were accurately describing their cases; virologists trusted the statistics of the epidemiologists; the immunologists placed confidence in the virologists' lab experiments; and the computer modeling experts believed them all. Any intrusion into another scientist's domain entailed peer rejection and humiliation.

In this atmosphere of pressure to conform, the lessons of the bacteria-hunting era were easily overlooked. Virtually no one thought to test HIV according to Koch's postulates. These time-tested standards apply even more perfectly to viruses, which are nonliving parasites with no behavioral flexibility, than they do to bacteria, which can sometimes release toxins or adapt to changing environments. The growing mountains of data on HIV were instead interpreted solely to fit the consensus virus-AIDS hypothesis, and researchers forgot the very rudiments of virology itself as they assigned increasingly bizarre properties to this virus. But Koch's postulates do indeed cut to the heart of the issue, exonerating HIV and rendering most AIDS research entirely pointless:

1. *Koch's First Postulate: The microbe must be found in all cases of the disease.* Robert Koch explicitly stated that a causal germ would be found in high concentrations in the patient and distributed in the diseased tissues in such a way as to explain the course of the symptoms. In the case of AIDS, the affected tissues include the white blood cells of the immune system, particularly the T-cells, as well as the skin cells in lesions of Kaposi's sarcoma and brain neurons in dementia. But no trace of the virus can be found in either the Kaposi's sarcomas or the neurons of the central nervous system. Since retroviruses, in fact, cannot infect nondividing cells like neurons, the absence of HIV there is hardly surprising. However, because Kaposi's sarcoma itself has long been synonymous with AIDS, the absence of virus in this cancer seriously undermines the HIV hypothesis.

If HIV were actively infecting T-cells or other members of the body's immune system, cell-free virus particles, known as *virions*, should easily be found with great ease circulating in the blood. This is the case with all classical viral diseases: In a patient suffering from hepatitis B, one milliliter of blood (about five or ten drops) contains approximately ten million free virus particles. Likewise, flu-like symptoms appear only in the presence of one million rhinovirus particles per milliliter of nasal mucous, and one to one hundred billion particles of rotavirus per gram of feces will accompany diarrhea in the patient. But in most individuals suffering from AIDS, no virus particles can be found anywhere in the body. The remaining few patients have at most a few hundred or a few thousand infectious units per milliliter of blood. One paper published in March of 1993 reported two individuals with about one hundred thousand virus particles per milliliter of blood, out of dozens of AIDS patients with little or no detectable virus.[12] Thus HIV behaves as a harmless passenger microbe, only sporadically coming back to life long after the immune system has been destroyed by something else and can no longer suppress the virus.

Even those patients with some detectable virus never have more than one in every ten thousand T-cells actively producing copies of the virus; on average, only one in every five hundred or more T-cells contains even a dormant virus. The abundance of uninfected T-cells in *all* AIDS patients is the fatal, definitive argument against the many false claims for high viral "loads" or "burdens" in AIDS patients.[13] Nothing could ever stop infectious viruses from infecting all susceptible cells in the same body (except of course antiviral immunity). If T-cells remain uninfected, there are no viruses to infect them. The absence of active, infectious virus automatically disqualifies HIV as a player in the syndrome. Microbes can cause serious damage only when infecting the host's cells faster than the body can replace them; T-cells, the presumed target of HIV, are constantly regenerating at much, much higher rates than dormant HIV in the presence of antiviral immunity.[14]

To gain some perspective, one should remember that most people carry inactive forms of several viruses, none of which cause disease while the microbes remain hidden and dormant in the body. Two out of every three Americans carry the herpes virus, and an equal number harbor the herpes-class cytomegalovirus; Epstein-Barr virus, causing mononucleosis ("kissing disease") when active, resides in dormant form in four of every five Americans; and an even higher proportion of people host the papilloma, or wart, virus. If these viruses could cause disease while latent, the absurd situation would arise in which virtually no one would be left to treat the hundreds of millions of sufferers.

HIV is not, of course, behaving differently from other viruses. Upon infecting a new host, a typical virus invades its target cells and begins replicating in large quantities, producing new virus particles that spill into the bloodstream and infect more cells; this is the period during which high levels of virus can be isolated from the patient and the symptoms are strongest. The body's immune system responds to the threat by mobilizing to mass-produce the specific antibody proteins that attack and neutralize the virus particles. As this battle heats up, antibodies are produced more rapidly than the virus, ultimately eliminating active virus from the body. Most viruses are thereby completely destroyed, although some herpes viruses can establish chronic infections by hiding in certain tissues.

Retroviruses, by nature, insert their genetic information into infected host cells, becoming dormant once neutralized by the host's immune system. HIV, like other retroviruses, can achieve high levels of virus when first infecting the body (up to one hundred thousand particles per milliliter of blood), but in most people HIV is then permanently inactivated by the antibodies generated against it. During this brief period of HIV activity, some newly infected people have reported mild flu-like symptoms at most—but no AIDS diseases. But all of these rare cases were male homosexuals from high-risk groups, meaning people who had used recreational drugs that can cause exactly the same symptoms.

Outside this risk group are the seventeen million HIV-positive healthy people identified by the World Health Organization[15] who cannot connect any past disease with HIV infection; they are either surprised or shocked when they find out about being "positive" or are blissfully unaware of it. The reason is that HIV is one of the many harmless passenger viruses that cause no clinical symptoms during the acute infection. By contrast, most people have lasting memories of their mumps, measles, hepatitis, polio, chicken pox, and flus, after which they become "antibody positive" for the respective viruses.

AIDS patients, on the other hand, have generally been infected by HIV for years, not days, before they deteriorate and die. Thus, the virus has long since been neutralized, forcing doctors to test the patient either for the dormant virus or the antibodies against it. This is the operating principle of the "HIV test," which identifies antibodies, and yet ironically stands as proof of the innocence of this virus.

Not all AIDS patients, however, carry even dormant HIV. Antibody-positive patients usually do have some latent virus left over from past infection. But many people dying of AIDS-like conditions, ranging from Kaposi's sarcoma to immune deficiencies and various opportunistic infections, have never been infected by HIV in the first place. The CDC does not include most of these antibody-negative cases in its AIDS figures, rendering these people invisible.

According to the CDC's own statistics, at least 25 percent of all official AIDS cases have never been tested for antibodies against HIV, many of whom might turn out to be negative. Further, the HIV test itself often generates false-positive results, particularly in members of AIDS risk groups who have been infected with large numbers of interfering viruses.[16] Thorough follow-up testing could reveal HIV-negative cases in the official AIDS tally. The scientific literature describes some 4,621 confirmed cases of HIV-free people dying of AIDS diseases, including homosexuals and heroin addicts in the United States and Europe, and central Africans.[17] These dozens of studies generally found that, among any group of clinically diagnosed AIDS patients, many test negative for HIV. But because the CDC ignores virtually all HIV-negative patients, counting only those with the virus as AIDS cases, the total number of such cases may never be known.

Even a "slow virus" hypothesis of HIV cannot explain how uninfected people would develop AIDS conditions. From every angle, HIV fails Koch's first postulate.

2. Koch's Second Postulate: The microbe must be isolated from the host and grown in pure culture. This postulate was designed to prove that a given disease was caused by a particular germ, rather than by some undetermined mixture of noninfectious substances. HIV has been isolated and is now grown continuously in HIV research labs. This rule therefore has technically been fulfilled, but only in some instances.

Since free virus is rarely found in AIDS victims, HIV can be retrieved only from the great majority of them by reactivating the latent form of the virus. Millions of white blood cells must be taken from the patient and grown in culture dishes for weeks, during which time chemical stimulants that shock cells into growing or mutating are added to awaken any dormant HIV from within its host cells. Given enough patience and plenty of repetition of such procedures, a single intact virus can eventually be activated, at which point it starts infecting the remaining cultured cells. Yet even this powerful method does not yield active virus from many AIDS cases that have confirmed antibodies against HIV. Gallo himself faced this intractable problem, a frustrating situation that may have led him to claim Luc Montagnier's virus as his own.

The situation is a mirror image of biological virus isolation that happens every time an uninfected person contracts the virus from an infected host. Natural transmission by unprotected sex has been studied in "discordant" couples, i.e., HIV-free women married to HIV-positive hemophiliacs or HIV-free male homosexuals having HIV-free sexual partners. These studies have revealed a rarely mentioned fact: After neutralizing the virus with the immune response, an HIV-positive person requires an average of one thousand unprotected sexual contacts to pass this virus along just once.[18]

A pregnant mother is a different story; in effect, she provides her child with a nine-month continuous exposure to her blood and therefore has at least a 50 percent chance of passing HIV to the baby. HIV, as with any retrovirus, survives by reaching new hosts perinatally (mother to child), this being five hundred times more efficient than by sexual transmission.[19]

This would explain why the numbers of HIV-positive people, in America as well as Africa, have remained so constant: HIV is transmitted from mother to child just like a human gene. This also reveals the reason for the virus being so widespread and equal between the sexes in Africa—HIV has been passed along from mother to child for many centuries (not through one thousand heterosexual contacts as is commonly assumed).[20]

In the industrial world, HIV can be readily transmitted only among the most sexually active homosexuals, among needle-sharing addicts, and through blood transfusions to hemophiliacs—the routes that so easily transmit numerous other microbes. In short, the very people with tremendous health risks to begin with *also* more easily pass along HIV, making it a surrogate marker for the real cause of AIDS (see chapters 8–10). Therefore, a rough correlation exists between HIV and AIDS diseases, but it is imperfect and misleading.

The extremely low efficiency of sexual transmission explains the failures of Gallo, Weiss, and other leading AIDS researchers in isolating HIV: Even for the most experienced virus hunters, a virus that is not present is difficult to find. Only rare luck or misfortune, depending on one's purposes, and extreme persistence can extract HIV from an antibody-positive person.

The very ability of retroviruses to survive as dormant genes by attaching themselves to human chromosomes has been exploited for the most sensitive HIV assay yet—the Polymerase Chain Reaction (PCR). This incredibly sensitive technique was invented in the mid-1980s by Berkeley biochemist Kary Mullis, who was awarded the Nobel Prize for his discovery in 1993. The PCR is a technology that amplifies even the tiniest amounts of any specific DNA sequence, creating enough copies of the desired sequence for detection and analysis. This amounts to finding the proverbial needle of dormant HIV in a haystack of human DNA. But contrary to statements by some HIV scientists, this is not an isolation of the actual virus and does not fulfill Koch's second postulate. It is only the detection of dormant DNA genomes, or fractions of viral genomes, left behind from infections that occurred years earlier. Nevertheless, scientists and journalists alike sometimes mislabel such exhumations of viral fossils as "new, more sensitive techniques"[21] that somehow prove HIV can be found in an ever-greater portion of AIDS patients. Because a few HIV molecules are technically invisible but millions of HIV molecules are visible, Mullis's PCR technique has become the only practical method to detect viral molecules in all those antibody-positive people in which no virus can be found.

3. Koch's Third Postulate: The microbe must reproduce the original disease when introduced into a susceptible host. The official HIV-AIDS hypothesis declares a 50 percent to 100 percent probability of death from infection. In practice, scientists and medical doctors interpret antibodies against HIV as a sure sign of imminent doom. This notion, of antibodies as a prognosis of death, defies all classical experience with viruses and bacteria. Virtually every microbe causes disease in only a minority of infected individuals, since the majority are usually healthy enough to mount a rapid immune response. Certainly no fatal viral disease is known to cause death in nearly all infected people—except the paradoxical "AIDS virus." Any microbe killing all its hosts would soon destroy itself, even if such could exist in the first place; any germ must be able to reach new hosts before the previous one dies, lest it go down with a sinking ship. Any universally lethal parasite would be, by definition, a suicidal organism. HIV would face even less chance of survival, being extremely difficult to transmit from one person to another, and would thus usually die with its infected host.

Traditional incubation periods, defined as the time between initial viral infection and the onset of disease symptoms, are measured in days or weeks. During this period the virus multiplies into concentrations high enough to cause disease. The process is exponential: Each virus particle infects a single cell, and eight to forty-eight hours later hundreds of new virus particles begin to be produced, each destined to infect a new cell. Flu, common colds, and herpes simplex infections develop with short incubations lasting between a few days and weeks; measles, chicken pox, and rubella have longer incubations of ten to twenty days, while extreme conditions such as hepatitis can take two to six weeks. These delays occur before the body has launched an immune response against the new virus.

Because these delays or latent periods are determined entirely by the generation time of the virus, and the generation time of HIV is about forty-eight hours, we can calculate how soon after infection AIDS should appear. Natural infection only introduces a few viruses into the body. But just one infected cell produces at least one hundred offspring within two days. These in turn will produce one hundred times one hundred within two days. Such exponential or explosive growth will produce *100 trillion* (100,000,000,000,000, or 10^{14}) viruses in just two weeks—enough to infect every single cell in the human body. Therefore, HIV should cause AIDS within a few weeks of infection.

But borrowing from their cancer research, virus hunters officially give HIV ten years between infection and the onset of AIDS—years after antibodies have neutralized the virus. Such latency periods have been invented solely to circumvent Robert Koch's third postulate. But any germ not causing symptoms before being cleared by the immune system should be ruled out as causing disease.

Koch's third postulate insists on reproducing the disease in at least some cases by injecting the allegedly dangerous microbe into a number of uninfected and otherwise healthy hosts. This condition can be tested in one of three ways: infection of laboratory animals, accidental and natural infection of humans (deliberate infection would be unethical), or by vaccination experiments. HIV fails all three tests:

(a) Blood from AIDS patients was injected into several chimpanzees in 1983, before the availability of HIV tests. The animals were infected by HIV, as later evidenced by antibodies against the virus, but in ten years none has yet developed any sickness. Roughly 150 other lab chimpanzees, injected with purified HIV since 1984, have proved that antibodies against the virus are generated within a month of inoculation just as in humans; but again, none has developed symptoms to this very day.[22]

In short, no animal becomes sick from HIV, although monkeys and other test animals do suffer disease from human viruses causing polio, flu, hepatitis, and other conditions.

By the end of 1992 the CDC had reported some thirty-three medical workers as most likely having received HIV accidentally, of whom seven were diagnosed with AIDS symptoms. None of these reports has been confirmed with published medical case histories, although in a 1989 issue of the *New England Journal of Medicine* an informal editorial entitled, "When a House Officer Gets AIDS" was written by a doctor infected by a patient. The article describes only minor weight loss of ten pounds and a "bit" of fatigue as being the doctor's AIDS "complications."[23] This hardly counts as evidence for Koch's third postulate. Nor has the CDC stated whether any of these medical workers have taken the dangerously toxic AZT, the official AIDS treatment, which itself causes immune deficiency (see chapter 9).

(b) During the past decade, more than four hundred thousand AIDS patients have been treated and investigated by a system of five million medical workers and AIDS researchers, none of whom have been vaccinated against HIV. Doctors who have treated AIDS patients were initially admired by their peers and the press for their courage to face a fatal, contagious condition for which there was no cure, no drug, and no vaccine.

But ten years later there is *not even one* case in the scientific literature of a health care worker who ever contracted presumably infectious AIDS from a patient. Imagine what it would have been like if four hundred thousand cholera, hepatitis, syphilis, influenza, or rabies patients had been treated by health care workers for ten years without protection from vaccines and antimicrobial drugs—thousands would have contracted these diseases. This is exactly why we consider these diseases infectious. The complete failure of four hundred thousand AIDS patients to transmit their diseases to even one of their unvaccinated doctors in ten years can mean only one thing: AIDS is not infectious.

However, several thousand health care workers have by now been diagnosed with AIDS, but these individuals belong to the same AIDS risk groups as 90 percent of all AIDS cases—homosexuals and intravenous drug users. And although three-quarters of all health care workers are female, more than 90 percent of these AIDS patients are male, the exact same ratio as with all other AIDS cases.[24] *In other words, medical accidents are not producing the expected AIDS epidemic among unvaccinated personnel in that industry.*

Nor has HIV affected the recipients of blood transfusions, most notably hemophiliacs. Some fifteen thousand hemophiliacs in the United States—about three-quarters of the total—were infected with HIV before screening of the blood supply began in 1984. But also during the past fifteen years, improved medical treatment has doubled their median life expectancy. The virus-AIDS hypothesis would have predicted that now, ten and more years later, more than half of them would have died from AIDS. Instead fewer than 2 percent of these HIV-positive hemophiliacs develop AIDS each year. According to several dozen small studies, this matches the rate of immune deficiencies and death among HIV-negative hemophiliacs, a phenomenon apparently related to hemophilia itself.[25]

(c) The third postulate can be tested in humans through a reverse method. If vaccines or other techniques can be used to provoke the body into neutralizing the microbe with antibodies and the disease is thereby prevented, the germ has been proven guilty experimentally. But since AIDS is found in each patient only after the immune system has already suppressed HIV, the virus plays no role. Most AIDS researchers have conveniently forgotten this important principle and continue to blame the virus when only antibodies against it can be found; others blatantly reverse the logic of the vaccination test, declaring antibodies useless because they do not prevent AIDS.

(d) The acid test of Koch's third postulate would be to infect newborn babies with HIV, because newborns are immunotolerant and thus much more susceptible to a virus than adults. It is known from experiments with animals that a virus is totally harmless if it does not cause a disease in newborns.

It would, of course, be unthinkable to inject HIV experimentally in human babies to test whether it causes AIDS. Yet, exactly this experiment has already been done millions of times by nature to generate most of the seventeen million healthy, but HIV-positive, people living on this planet.[26] Most of these people picked up HIV by natural infection from their mothers.

Indeed, all animal and human retroviruses, including HIV, depend on mother-to-child (perinatal) transmission for survival. Since sexual transmission is extremely inefficient, depending on one thousand sexual contacts in the case of HIV, retroviruses could never survive by sexual transmission. They can only survive by perinatal transmission, which is about 50 percent efficient.[27] Therefore perinatal transmission must be harmless or else the baby, the mother, and the virus would not survive; HIV would be a kamikaze killer—it would kill itself together with its host.

If that were true, one would expect thousands of healthy young American men or women to have HIV but not AIDS. That is exactly what the U.S. Army reports. The U.S. Army tests all applicants and all its young men and women annually and identifies thousands of HIV-positives who are totally healthy. While some of these might have acquired their virus sexually, it is impossible that thousands would have had the 1,000 sexual contacts with HIV-positives or the 250,000 sexual contacts with average Americans (of which only 1 in 250 is HIV-positive) that are necessary to pick up HIV by sexual transmission.[28] Therefore, most of these HIV-positive young men and women must have acquired HIV from their mothers sixteen to twenty years prior to their application to the U.S. Army. The same must be true for most of the remaining seventeen million humans who are healthy and HIV-positive.

The fact that millions have acquired HIV at birth yet are healthy adults is the most devastating argument against the HIV-AIDS hypothesis. It proves that HIV, like all other microbes that are transmitted perinatally or sexually, cannot be fatally pathogenic. Indeed no fatally pathogenic microbe exists in animals or humans that depends either on perinatal or sexual transmission for survival.

No matter how one looks at the HIV hypothesis, it is flawed either in terms of facts or in theory or in both.

(e) Koch's third postulate can also be tested provisionally on human cells in culture. If HIV cannot induce disease in whole organisms, one might at least expect it to kill T-cells grown in laboratory culture dishes, where the concentrations of actively replicating virus are enormously high. Robert Gallo, however, has been able to patent the virus by growing it continuously in immortal T-cell cultures since 1984. The French discoverer of the virus, Luc Montagnier, reported occasional cell death in infected cultures that was stopped by adding antibiotics, which do not affect virus replication but do kill undetected bacterial contaminants. Indeed, the HIV antibody test is made from virus that is mass-produced in T-cells, which grow continuously rather than die. The reports from other labs and biotechnology companies are consistent: HIV grows harmoniously with the cells it infects. The failure to kill T-cells, even under optimal conditions, is the Achilles' heel of the supposed AIDS virus.[29]

HIV typifies a retrovirus in every measurable way. It has the same biochemical structure and infective properties, benignly stimulating some cells to produce more copies of the virus. It has the same amount of genetic information and the same three basic genes as all other retroviruses. It also has six smaller genes, themselves a normal feature of other retroviruses. Although many HIV researchers focus their efforts on studying these "extra" genes as possible AIDS genes, no one gene is unusual and all are needed for virus survival. HIV contains no special "AIDS gene" expressed during the syndrome. However, this does not stop industrious AIDS scientists from endlessly reexamining the genetic sequences for some magical clue to explain AIDS.

HIV clearly fails Koch's postulates. However, virologists should have expected this from the beginning. HIV is, after all, a retrovirus, precisely the kind of virus so benign to its host cells that it had inspired such hope in the War on Cancer, since cancer cells grow and behave uncontrollably rather than die. Retroviruses have never been known to inhibit or kill billions of rapidly dividing cells and could hardly be expected to affect T-cells or otherwise destroy the immune system.

To be the cause of AIDS, the virus would require still more miracles. A number of the AIDS indicator diseases are not opportunistic infections preying on an immune-deficient host, including dementia, wasting syndrome, and the various AIDS cancers—Kaposi's sarcoma, the lymphomas, and, as of 1993, cervical cancer. Altogether these non-immunodeficiency AIDS diseases made up 39 percent of all American AIDS diseases in 1992, and, owing to a new definition of AIDS, 20 percent of all AIDS diseases in 1993 (see Table 1).

HIV would have to kill T-cells while destroying brain neurons it cannot infect and at the same time induce white blood cells and skin cells to grow malignantly. To reconcile these non-immunodeficiency diseases with HIV, AIDS scientists would like to blame even these diseases on immune suppression. But despite years of research, no evidence can be found that the immune system fights cancer cells, which,

TABLE 1

AIDS-defining diseases in the United States in 1992[a] and 1993[a] fall into two classes: immunodeficiency diseases and non-immunodeficiency diseases

Immuno-deficiencies	1992 (in %)	1993 (in %)	Non-immuno-deficiencies	1992 (in %)	1993 (in %)
<200 T-cells	—	79	wasting disease	20	10
pneumonia	42	22	Kaposi's sarcoma	9	5
candidiasis	17	9	dementia	6	3
mycobacterial (including tuberculosis)	12	11	lymphoma	4	2
cytomegalovirus	8	4			
toxoplasmosis	5	2			
herpesvirus	5	3			
Total =	61[b]	80[b]	Total =	39	20

[a] The data are from the Centers for Disease Control (Centers for Disease Control, 1993; Centers for Disease Control and Prevention, 1994).

[b] Over 61 percent and 80 percent are due to overlaps.

In the United States 39 percent of all AIDS cases were non-immunodeficiency diseases in 1992. Owing to the third re-definition of AIDS by the Centers for Disease Control in 1993, that included less than 200 T-cells per microliter of blood as an AIDS disease, about 20 percent of all American AIDS diseases were non-immunodeficiency diseases in 1993. The distribution of AIDS diseases in 1994 was nearly the same as in 1993, since the AIDS definition was not changed that year.

---

after all, are part of the host's own body. In fact, dozens of AIDS patients with Kaposi's sarcoma or dementia have been reported to have normal immune systems.[30] So HIV would indeed have to accomplish many incredible tasks at once. Stranger still, infants with AIDS suffer immune suppression from deficiencies in B-cells, a subgroup of white blood cells altogether different from T-cells.

Since there are no precedents for cell-killing retroviruses and no laws other than Koch's for convicting viruses for a disease, even the HIV orthodoxy admits that their hypothesis stands unproven.[31] However, they insist that Koch's not-guilty verdict of HIV does not prove HIV innocent and that further work will eventually prove HIV guilty.

No matter how convincing the HIV-AIDS paradoxes should be, official AIDS scientists cannot be dissuaded from their virus hypothesis. When forced to answer the above arguments, their imaginations run wild in designing ever-new variations of the same experiments to prove their hypothesis.[32] According to HIV advocate John Maddox, "The remedy is not, of course, to pander to wish-fulfillment, but to redouble effort in the laboratory and the clinic."[33] But these experiments have only proven to this date that the HIV hypothesis is impossible to prove.

INNOCENT VIRUS

According to Koch's postulates, HIV is "not guilty" of AIDS. But this not-guilty verdict is not perceived as innocence by most scientists, particularly by nonscientists, for two reasons:

1. The term *virus* (the Latin word for poison), just by itself, inspires fear. Therefore HIV must be bad. This general prejudice that all viruses are bad is based on the fact that some viruses actually are bad. These pathogenic viruses and microbes are to researchers and to the press what criminals are to detectives—the focus and justification of their existence.

But only a few people know that the great majority of all viruses and microbes cause no disease at all. Such viruses are called *passenger* viruses.[34] They are the most uninteresting of all viruses to virologists, because the standing of virologists in the scientific community depends on the pathogenic potential of the viruses they study. Since passenger viruses do not advertise their presence by causing a disease, most of them go unnoticed, riding with their hosts like a passenger in an airplane. Passengers are the silent majority of animal and human viruses; pathogenic viruses are just the tip of the iceberg.

Passenger viruses infect just enough cells of the host to survive without ever causing a disease. Since passenger viruses keep such a low profile, virologists could not easily detect them until recently, when the technology was developed to detect needles in a haystack. Because a passenger virus neither hurts nor kills, it is the most efficient survivor and hence the most common virus in animals and man.

2. The second reason even scientists consider HIV not innocent in AIDS is the much cited "overwhelming correlation between HIV and AIDS." However, the HIV-correlation argument is not just misleading; it is deceptive on three counts:

First, the overwhelming correlation is not with HIV but with an antibody against it—a difference like day and night. A virus is a potential pathogen, an antibody is a certain antidote.

Second, American and European AIDS risk groups have one common microbial denominator: They have many more microbes and many more antibodies against microbes than the rest of the population.[35] This is because from a microbiologist's point of view, "AIDS risk behavior" is collecting microbes in the process of many sexual contacts with different persons (promiscuity), sharing needles during intravenous drug use, consumption of unsterile drugs, prostitution for drugs, or receiving transfusions for hemophilia. No matter what microbe one chooses—toxoplasma, bacteria-causing syphilis, genital wart virus, human T-cell leukemia virus, cytomegalovirus, one of the many herpes viruses, hepatitis virus, or HIV—it correlates overwhelmingly with risk behavior. In fact, three of these microbes, namely syphilis, HTLV-I, and cytomegalovirus, were considered AIDS causes before HIV, because of "overwhelming" correlations with antibodies against them.[36] However, since HIV was chosen, rather than proved, to be the cause of AIDS in 1984, the correlation with HIV and AIDS became 100 percent—the definition of AIDS. Therefore, the overwhelming correlation is one of the purest examples of circular logic.[37]

Third, the literature includes more than 4,621 clinically diagnosed AIDS cases that are all HIV-free (see appendix C). To cover up this discrepancy with the overwhelming correlation, HIV-free AIDS cases were renamed in 1992 as *idiopathic CD4-lymphocytopenia* (ICL) cases by the CDC and Anthony Fauci, the director of the National Institute of Allergy and Infectious

Diseases.[38] Thus, the "overwhelming correlation" between antibodies against HIV and AIDS is a mere consequence of risk behavior and of the definition of AIDS. It is irrelevant for causation.

The scientific method offers three unambiguous criteria on how to tell a virus that is potentially "guilty by association" from one that is an innocent passenger virus:

1. The time between infection by a passenger virus and the occurrence of any disease, if one occurs, is entirely unpredictable. It could be anywhere from a day to the lifetime of the patient. Since the passenger virus does not cause a disease, the time of infection is irrelevant to the onset of a disease.

2. A passenger virus can be active or passive, rare or abundant, during any disease. Since the passenger does not cause disease, its activity is irrelevant to it.

3. The passenger virus can be present or absent during any disease. Since the virus is not pathogenic, disease can occur in the absence of the passenger virus.

In short, a virus that has been in its host for years before a disease occurs, that is typically inactive and rare during a disease, and that is not present in every case of that disease is not a credible suspect for viral disease. It is an innocent bystander or a passenger virus. HIV meets all of these criteria. Since HIV also fails Koch's postulates, there is no rational basis for the HIV-AIDS hypothesis. In the courts of science HIV must be acquitted of all charges for AIDS—it is an innocent virus.

AIDS NOT INFECTIOUS

In December 1994 *Science* wrote a surprising editorial blaming a newly discovered herpes virus for Kaposi's sarcoma.[39] The surprise was that the AIDS orthodoxy had adopted the view that another virus could cause AIDS. Although this article should have registered as a major heresy among AIDS scientists, it did not. It was received instead only as a "minor sin" because it did not question the central, although tacit, dogma of the AIDS orthodoxy: infectious AIDS. Questioning infectious AIDS is without doubt the ultimate heresy in the AIDS orthodoxy.

The fear of questions about the orthodoxy's most carefully cultivated dogma is understandable, because AIDS does not meet the classic epidemiological criteria of an infectious disease:

1. Infectious diseases do not discriminate between sexes. The first epidemiological law of viral and microbial diseases holds that men and women are affected equally, because no virus or microbe discriminates between the sexes. This law applies to all known infectious diseases affecting large populations. Examples are flu, polio, syphilis, hepatitis, tuberculosis, pneumonia, and herpes—all of which do not discriminate between the sexes nor do they select their victims only from specific risk groups.

By contrast, AIDS selects all its victims from a few, newly established AIDS risk groups: long-term intravenous drug addicts and their babies, male homosexuals using recreational drugs, and hemophiliacs under long-term treatment with commercial clotting factor VIII. Breaking with the sexual equality displayed by conventional infectious diseases, AIDS attacks men ten times more often than women in Europe and the United States. Among men it decidedly prefers homosexuals to heterosexuals. Thus, American and European AIDS is not distributed between the sexes like an infectious disease. (Chapter 8 explains why African AIDS does not discriminate between men and women.)

2. Farr's law: Infectious diseases spread exponentially. Early in the last century the British epidemiologist William Farr first recognized the seasonal rise and fall of microbial epidemics.[40] A new infectious disease rapidly explodes in a population—just as rapidly as microbes are transmitted from person to person. Then it declines within months because it is stopped by the elimination of susceptible victims either by death or more often by natural immunization. In accordance with Farr's law the Hawaiian natives, the California Indians, and the Eskimos were all quickly decimated by European microbes once they had been introduced to them by their European discoverers. But survivors soon became as resistant to these microbes as the Europeans. Likewise, contemporary Americans and Europeans suffer from new, seasonal flu epidemics, following Farr's law to the letter.

Figure 1 shows the exponential rise and fall of a new, seasonal flu epidemic against the backgrounds of several long-established

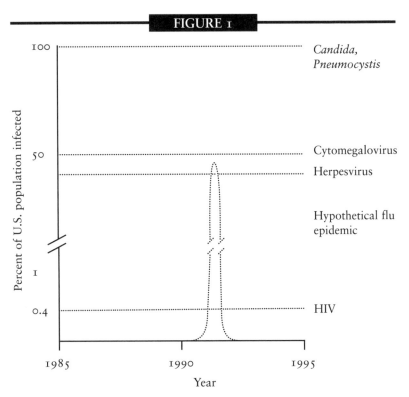

The distribution over time of a new hypothetical flu epidemic, against the background of several parasites long-established in the United States. The long-established parasites differ from a new one in their distribution over time: According to Farr's law the new one rises and falls (or equilibrates) exponentially— the old ones remain steady. Its unchanging incidence in the United States identifies HIV as an old American virus!

microbes. Since the percentage of Americans with herpes virus, cytomegalovirus, and the fungal parasites *Pneumocystis* and *Candida* is constant over time, these are "old" American microbes. Surprisingly, HIV is one of them, because 1 in 250 Americans (0.4 percent) have been "positive" ever since HIV could be detected in 1984. Thus, contrary to its reputation, HIV is an old American virus.

Figure 2A compares the time course of the American AIDS epidemic with that of the American HIV epidemic. The comparison offers another surprise: The HIV epidemic is constant and thus old, but the AIDS epidemic is increasing and thus new. Since the two epidemics follow totally different time courses, the HIV epidemic cannot possibly be the cause of the AIDS epidemic.

In sharp contrast to the bell-shaped curve of a conventional new infectious epidemic, like the flu epidemic shown in Figure 1, the AIDS epidemic increased steadily for fifteen years (Figure 2A). American AIDS gradually spread from a few dozen cases annually in 1981 to more than eighty thousand cases in 1994. It did not explode, as the HIV orthodoxy predicted; neither did it decline, as would be expected from antiviral immunity.[41] Instead of resembling an infectious disease, the time course of the AIDS epidemic resembles the slow progressing epidemics of lung cancer and emphysema in industrialized nations, building up over the years in step with tobacco consumption. These noninfectious epidemics neither rose exponentially nor affected all groups of the population or both sexes equally, nor did they disappear as a result of antiviral immunity or natural resistance.

Thus, AIDS does not meet the classical epidemiological criteria of an infectious disease. The failure of AIDS to meet these criteria destroys not only all hopes of the HIV orthodoxy ever to prove that HIV causes AIDS, but also any other viral or bacterial theories of AIDS.

Despite all these violations of the fundamental principles of virology and epidemiology, the virus-AIDS hypothesis has remained the sole basis for our unproductive war on AIDS. This is as much a scientific as a human tragedy. The reckless rule of the

194 ■ INVENTING THE AIDS VIRUS

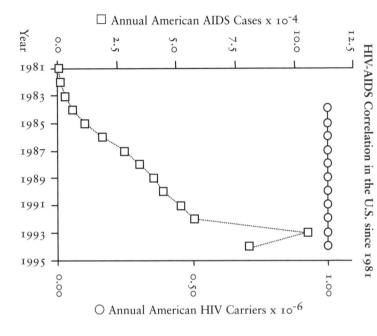

FIGURE 2A HIV-AIDS Correlation in the U.S. since 1981

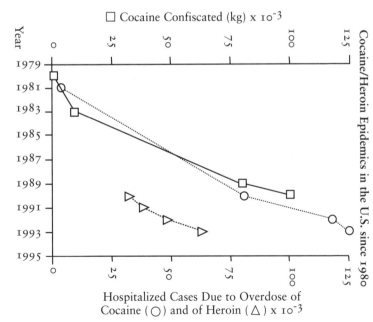

FIGURE 2B Cocaine/Heroin Epidemics in the U.S. since 1980

HIV-AIDS monopoly breaches the most fundamental principle of disease control, "First find the cause, then fight the cause," and closes the door for alternative hypotheses that might be productive.

DEFENDING THE LOW GROUND

After the polio epidemic ended, no new diseases and no fundamentally different viruses were being discovered. To maintain a medical relevance, virologists began connecting known viruses to unexplained diseases—such as cancer or multiple sclerosis. Because these diseases in no way behave as traditional infectious diseases, the virus hunters had to invent new properties for the germs. First, the incubation period of viruses—typically anywhere between one day and three weeks—was allowed to stretch into years. Then antibodies had to be abandoned as a sign of immunity against the microbes. And since the viruses never reappeared during disease, indirect methods of damage had to be postulated.

Nevertheless, all these creative maneuvers merely delayed the inevitable. By the early 1980s, virology was withering from lack of public interest—a fatal weakness when trying to attract new recruits, research money, and federal programs. The public was losing faith in wars on cancer that were never won or wars on diseases that rarely affected the average person.

But AIDS has changed everything, reviving virus hunting as the most glamorous and rewarding branch of biomedical research. To blame HIV for AIDS, virologists had to employ every invention at their disposal, including an ever-expanding latent period, an antibody test, and plenty of paradoxes to keep tens of thousands of investigators busy for many years. The evolution toward these false assumptions had been so gradual, so favored by consensus politics within science, and so shaped by the increasing sensitivity of biotechnology, that most researchers had been lulled into thinking of such rationalizations as normal science. By the time Robert Gallo and other virus hunters had engraved the HIV hypothesis in stone, anyone who dared to raise serious questions appeared truly radical to the rest of the research establishment.

Peter Duesberg first began to ask his colleagues questions about the HIV hypothesis shortly after Gallo's 1984 press conference. The HIV dissidents could see two fundamental problems: HIV was a retrovirus, meaning it should not kill the cells it infected, and the virus could barely be detected even in late-stage AIDS patients. The following year, the NIH awarded Duesberg its Outstanding Investigator Grant, a special seven-year award officially designed to allow free inquiry and latitude for exploring risky new research directions. He took this mandate to heart. As the discussions over HIV continued quietly, he began exploring the issue as a potentially important shift from his usual work on cancer genes and animal retroviruses.

Upon hearing of Duesberg's doubts about whether retroviruses could cause cancer in humans or most animals, the editor of *Cancer Research* invited him to write a special review paper in 1985. Duesberg spent many months compiling the evidence from the scientific literature. While he was working on this piece, the questions about HIV began intruding into his thinking ever more prominently. He finally decided to add a section arguing that HIV could not cause AIDS, citing data that showed HIV was inactive in the body, did not kill T-cells, and could not possibly have a long latent period before inducing AIDS.

He was still writing the paper in 1986 when he took nine months' leave from Berkeley to work in another retrovirus lab at the NIH facility in Bethesda, Maryland. As chance would have it, he worked in the building that housed Gallo's laboratory, though on a different floor. This afforded him many opportunities to test his growing suspicions of the virus-AIDS hypothesis. Not yet realizing Duesberg's intentions, Gallo invited him to be the featured speaker at one of his regular lab seminars. Gallo seemed to enjoy most of Duesberg's talk, which questioned the importance of cancer genes, and did not even become upset when Duesberg threw in a short criticism of the HIV-AIDS hypothesis at the end. Apparently, Gallo thought Duesberg was not really serious, merely dabbling for fun.

But the following weeks brought increasingly tense conversations between them in which Duesberg would constantly raise new questions. One day such a discussion took place in the elevator, on the way to Gallo's lab. Gallo burst into such anger over Duesberg's persistence that he left the elevator on the wrong floor—missing the lab where he had worked for many years! Although Gallo increasingly resisted talking about HIV, several researchers in his lab privately admitted to Duesberg the enormous problem of not finding the virus active in the body. They knew perfectly well something had to give. Rather than abandon HIV, however, they told Duesberg they hoped to explain the problem using "cofactors" or other rationalizations. Naturally, these experiences began confirming Duesberg's suspicion that he had stumbled onto something profound.

Duesberg's twenty-two–page review paper appeared in the March 1987 issue of *Cancer Research*. Colleagues found the section on AIDS especially shocking, privately admitting the importance of the questions about HIV. To this very day, not one scientist has come forward to answer the paper. Traditionally, such deafening silence has been interpreted as a victory for the author, indicating the arguments to be irrefutable. However, despite being unable to find any flaws in the article, no researcher could afford to take on the powerful HIV-AIDS establishment. Unwilling to risk status and career by challenging the growing AIDS research structure, but having no arguments to defend the virus hypothesis, scientists chose the safety of continuing their studies of HIV, claiming that it was at least an "interesting" virus. Some researchers became quite sensitive about the virus hypothesis, reacting angrily to any criticisms.

The *Cancer Research* paper nevertheless generated some interest, and upon invitation Duesberg wrote a guest editorial in *Bio/Technology* that November. Again, no answer. The wide-circulation *Science* soon ran an article on the emerging controversy, placing Duesberg in a rather unsympathetic light. Prompted by Duesberg's letter in response, the editor decided to set up an official debate in this journal, which appeared in July of 1988. Duesberg was on one side, opposing Gallo, Howard Temin, and the epidemiologist William Blattner. Each side offered an opening page and a rebuttal to the opposition's opening page; that was all. *Science* has thereafter refused to publish anything but an occasional letter on the topic, declaring it received as much coverage as it deserved.

Although before this exchange Duesberg still had doubts, he became thoroughly convinced the virus was harmless after seeing this faltering inability to answer his arguments. As he further immersed himself in the AIDS literature, the sheer volume of damning evidence became overwhelming. In a response to the short *Science* debate, he wrote an extended update paper, which after months of fighting he managed to publish in the *Proceedings of the National Academy of Sciences* in 1989. This paper was printed on the express condition that another virologist would respond with an equal rebuttal. Gallo himself promised such but has not delivered as of this date. Once again, no scientist has ever chosen to answer that piece nor to answer Duesberg's subsequent review papers in *Research in Immunology* or the *Proceedings*.

Only a few short, general responses to Duesberg have appeared in other journals: the brief debate forum in *Science*, short exchanges in some 1989 issues of the *Journal of AIDS Research*, terse letters in a May 1990 issue of the *New England Journal of Medicine*, a blatantly *ad hominem* attack in the pages of *Nature* during June of 1990, and a few editorials in 1993. But in December 1994, *Science* published an eight-page article on the "Duesberg phenomenon" by the journal's foremost AIDS journalist. The article acknowledges that "the Duesberg phenomenon has not gone away and may be growing."[42] Although tendentious for the HIV hypothesis, the article made some telling concessions: "(i) According to some AIDS researchers [not all] HIV now [but not earlier, when it was named the AIDS virus] fulfills the classic postulates of... Koch," and (ii) "AZT and illicit drugs, which Duesberg argues can cause AIDS, don't cause the [*sic*] immune deficiency characteristic of that disease," knowing full well that about thirty different diseases are said to be "characteristic of the disease."[43]

From these and excerpts of Gallo's own writings, the standard defense of the HIV-AIDS hypothesis can be reconstructed. None of the most influential AIDS scientists has ever published a definitive defense of HIV, yet when confronted with the paradoxes they all answer with similar arguments. Otherwise, they prefer to ignore the questions.

The arguments for HIV fall into four categories.

1. Arguing for HIV by Ignoring the Facts

The case for HIV as an AIDS virus depends first on bypassing Koch's postulates. The most complete rationale for this is presented by Gallo in his 1991 book *Virus Hunting—AIDS, Cancer, and the Human Retrovirus: A Story of Scientific Discovery*, where he coolly disposes of these time-tested standards:

> Rules were needed then, and can be helpful now, but not if they are too blindly followed. Robert Koch, a great microbiologist, has suffered from a malady that affects many other great men: he has been taken too literally and too seriously for too long. We forget at times that we have made great progress in the last century in developing tools, reagents, and diagnostic techniques far beyond Koch's wildest fantasies...
> Koch's Postulates, while continuing to be an excellent teaching device, are far from absolute in the real world outside the classroom (and probably should not be in the classroom anymore except in a historical and balanced manner). They were not always fulfilled even in his time. Certainly, they did not anticipate the new approaches available to us, especially in molecular biology, immunology, and epidemiology, or the special problems created by viruses. They were, after all, conceived only for bacterial disease, and even here they often fail. Sometimes they are impossible to fulfill; many times one would not even want to try to do so; and sometimes they are quite simply erroneous standards.44

But Koch's postulates consist of elementary logic. Whereas technology is continually being outdated, logic is permanent. Koch's rules, after all, simply restate the germ theory itself in experimental terms. Gallo never tries to explain how logic would change over time; indeed, in this age of ultrasensitive biotechnology, such rules take on more importance than ever in sorting out relevant data from mere trivia. *Nor does Gallo offer any rigorous scientific rules to replace Koch's postulates, leaving HIV science with no standards at all.*

Gallo continues by misstating Koch's postulates, falsely claiming that a germ is required to cause a disease every single time it infects a new host. With most microbes, the majority of infected people or animals experience no symptoms; Koch's test only requires that some animals become sick when injected with a disease-causing germ or that vaccination prevents the illness. Gallo then cites false or misleading examples of germs that supposedly fail the postulates despite causing disease, pretending, for example, that the hepatitis and flu viruses cause no disease in animals. Gallo misses the point that the failure of a given germ to meet Koch's postulate does not call the postulate into question, but rather the germ as the cause of a disease. Or he draws examples from the "slow virus" hypotheses, including measles/SSPE, papilloma/cervical cancer, HTLV-I/leukemia, and Feline Leukemia Virus (see chapters 3 and 4). Or he cites diseases erroneously thought to result from bacteria, such as neurosyphilis (see chapter 2). In reality, all truly viral diseases do fulfill Koch's standards perfectly—yellow fever, measles, polio, chicken pox, herpes, hepatitis A and B, and flu, among others.

Gallo's "these postulates are too old" argument is repeated by English retrovirus hunter Robin Weiss and American CDC official Harold Jaffe: "What seems bizarre is that anyone should demand strict adherence to these unreconstructed postulates 100 years after their proposition."[45] Weiss and Jaffe also forget to explain how logical rules could become outdated and again proceed to misquote Koch and use misleading examples of disease-causing microbes supposedly failing the postulates.

It is generally assumed that stardom in a given field is directly proportional to knowledge: the more famous a person is, the more he knows about his field. However, a star is often born by a coincidence in which the most desirable solution to problems is delivered to the best-prepared audience. To deliver such a popular solution requires a complete knowledge of the politics of science but not of science itself. As we shall see, Gallo and Montagnier fit the formula for scientific stardom in this regard exactly.

Both had studied retroviruses as causes of cancer for more than a decade when AIDS appeared. But neither one had studied other noninfectious causes of diseases, not even other viruses, nor have they treated AIDS patients after AIDS appeared. Retroviruses were their primary investment and their exclusive expertise.

Having persuaded himself to ignore the traditional rules of Robert Koch, Gallo joins with Luc Montagnier in substituting a previously unknown "postulate":

> That HIV is the cause of AIDS is by now firmly established. The evidence for causation includes the fact that HIV is a new pathogen, fulfilling the original postulate of "new disease, new agent."[46]

Superficially, it appears logical to postulate that a new virus would cause a new disease. However, Gallo and Montagnier's argument fails because it ignores a multiplicity of facts:

(i) AIDS is not a disease. Instead, the AIDS syndrome is a steadily growing collection of (currently) about thirty "previously known" (old) diseases (see below). Surprisingly, in view of their notoriety for AIDS, neither Gallo nor Montagnier know the AIDS definition.

It is true, however, that the incidence of AIDS diseases has increased dramatically in the 1980s (Figure 2A) as intravenous drug use has increased and as both the consumption of recreational drugs used as sexual stimulants and the use of AZT as antiviral drug have increased in male homosexuals.

(ii) HIV is not a "new agent." According to Farr's law, a virus is new if the percentage of infected people increases rapidly over time—or "explodes" as the CDC predicted in the early days of

AIDS. A virus is old if the percentage of infected people is stable over time (Figure 1). Since the number of HIV-infected Americans has been an unchanging 1 million since HIV was able to be tested in 1985, HIV is an old virus in the United States (Figure 2A). In order to misjudge the age of HIV so grossly, Gallo and Montagnier must have been unaware of the epidemiology of HIV in the United States and unaware of Farr's law.

Gallo and Montagnier probably assumed HIV is new because it was newly discovered by them. But since the technology used to detect HIV is just as new as the discovery of HIV, there is another interpretation: Gallo and Montagnier discovered a previously unknown but old virus with a new technique. Their claim that HIV is new is just as naive as the claim of an astronomer that a previously unknown star is new because it became detectable with a new telescope.

Since HIV is old in the United States and the epidemic of AIDS diseases is new, HIV is not a plausible cause for a "new" rise of AIDS diseases in the United States.

(iii) AIDS is not an infectious, viral epidemic as Gallo and Montagnier assume. AIDS fails all epidemiological criteria of an infectious disease. Gallo and Montagnier completely ignore the evidence that the new AIDS epidemic could well be the consequence of the new recreational drug use epidemic that started in America after the Vietnam War. Apparently, neither Gallo nor Montagnier were aware of the "lifestyle hypothesis," which originally proposed that AIDS patients were suffering from drug diseases because all early AIDS patients were recreational drug users.[47]

To distinguish between toxic drugs and toxic microbes, Gallo and Montagnier should have investigated whether AIDS is infectious or not. But Gallo and Montagnier completely ignored that AIDS does not meet even one of the classical epidemiological criteria of infectious diseases—possibly because they never considered nonviral causes of disease.

(iv) Considering that hundreds of known retroviruses are harmless passenger viruses, one would have expected that the "leading" retrovirologists Gallo and Montagnier would have explained why they believe that HIV is fatally pathogenic. Yet all that Gallo and Montagnier had to offer in support of HIV pathology was their own credibility.

Indeed Gallo's and Montagnier's reasoning fits their narrow expertise exactly. Two leading retrovirologists agreeing on a retrovirus as the cause of AIDS and ignoring all competing retroviral and nonretroviral explanations. And for the leaders, ignorance is bliss.

2. Arguing for HIV Based on Inappropriate Models

When confronted with the paradoxes of HIV, its defenders simply reach for their bag of virus hypotheses, pulling out on demand a mixture of invented or misinterpreted models. They usually cite viral precedents of three types.

The first comes from the supposed "slow viruses," which are used to justify the long latency period of HIV, but which fall apart in light of the evidence presented above.

The second model suggests HIV reactivation based on authentic prototypes. Herpes simplex virus, for example, can cause lesions even long after the first antibodies against the virus have been produced. However, this can happen only if the virus is reactivated because the original antibodies and anti-viral T-cells have dropped below a safe threshold level. After reactivation the virus multiplies into large numbers just as in the original infection. Using this model, HIV scientists justify both the latent period and antibody test in one breath. But herpes produces the same lesions upon first infecting the body as it does upon reactivation, and antibodies neutralize it both early and late. Herpes can only recur because it hides in certain nerve cells, waiting until some future opportunity when the host's immune function is temporarily reduced. Once the immune system regains strength, the virus is again suppressed and the sores disappear. HIV, on the other hand, is alleged to kill its host only years after being neutralized, and even without reactivating. There is no HIV reactivation and no HIV in most AIDS patients.

The third virus model has been created only since the appearance of AIDS. Some animal retroviruses will cause "AIDS" when injected into hosts of the appropriate species. Simian immunodeficiency virus (SIV), a monkey retrovirus, attracts most of the attention. But these animal diseases can be called "AIDS" only by stretching the definition to extremes. They do not include most of the human AIDS conditions such as Kaposi's sarcoma or dementia. Rather, the animal symptoms usually resemble the flu: The animals become sick within days or not at all, without long latent periods; some animals recover by raising an immune response and never suffer a relapse; and those that die must be injected with large quantities of the virus while very young, before they have developed any immune system at all. In the wild, their cousins retain antibodies against SIV all their lives without ever becoming sick from the virus. These laboratory diseases are, in all respects, very traditional viral flu-like diseases, but HIV scientists rename them "AIDS."[48]

3. Arguing for HIV Based on Evasion

Lacking answers to Koch's postulates and authentic virus precedents, AIDS scientists resort to a variety of excuses. The standard evasions fall into four general categories: the arguments from unknowns, from speculation, from authority, and from irresponsibility.

The argument from unknowns makes the obvious point that scientists never know everything and implies that the HIV-AIDS question is therefore somehow unimportant now, since it eventually will be resolved through more research. According to this argument, the issue is not whether, but how, HIV causes AIDS; paradoxes therefore merely prove that further research is needed and that scientific knowledge will consequently expand, not that the virus is itself in question. William Blattner and Robert Gallo of the National Cancer Institute joined with fellow retrovirologist Howard Temin in using typical arguments from unknowns:

Biology is an experimental science, and new biological phenomena are continually being discovered... Thus, one cannot conclude that HIV-1 does or does not cause AIDS from Duesberg's "cardinal rules" of virology...

Duesberg's descriptions of the properties of viruses [are] in error and [provide] no distinction between knowing the cause of a disease, that is, its etiology ["whether"], and understanding the pathogenesis of this disease ["how"]. There are many unanswered questions about the pathogenesis of AIDS, but they are not relevant to the conclusion that HIV causes AIDS.

The CDC definition of AIDS has been revised several times as new knowledge has become available and will undoubtedly be revised again.[49]

Likewise, Robin Weiss and Harold Jaffe assert:

It is unwise to conclude that because we do not understand the pathogenesis of HIV in molecular detail, it is therefore harmless... So Duesberg is right to draw attention to our ignorance of how HIV causes disease, but he is wrong to claim that it does not.

One need not harp upon molecular quibbles, important though these are for directing research to the prevention or amelioration of HIV infection. To deny the role of HIV in AIDS is deceptive.[50]

It should be clear by now that the questions surrounding the alleged pathogenesis of HIV are too many and too substantial to be dismissed as mere "quibbles." To assert the role of HIV in AIDS is unscientific, particularly since the guardians of the HIV hypothesis have never suggested which standards could prove the virus harmless. Until they propose a scientific experiment that could disprove the HIV hypothesis, they convey the implicit message that they will accept no evidence against it whatsoever.

The argument from speculation is used more often than any other. It uses specialized terms that make it difficult for outsiders to understand, responding to any paradox with one untested assumption after another. For instance, if little or no HIV can be found in the body, scientists propose hidden reservoirs and special routes of infection. If only antibodies against HIV can be found, researchers call them "nonneutralizing" (or ineffective) antibodies and assert that the virus mutates too fast for the antibodies to keep up. If the virus does not make animals sick or kill cells in culture, then researchers claim that the virus somehow makes fine distinctions between humans and chimpanzees, something no other virus can do. All these hypotheses are constantly being disproved or shown to be irrelevant, but the reservoir of new evasions is inexhaustible.

The argument from authority cites the "overwhelming evidence" for HIV, without becoming too specific. In another form, it rebuffs inquisitive epidemiologists for lacking clinical experience while bypassing medical critics for having no epidemiological training. In other words, unless one is an expert in everything, one may not question anything. This response alludes to esoteric scientific data as a reason for critics to remain silent. Blattner, Gallo, and Temin provide perfect examples: "In summary, although many questions remain about HIV and AIDS, a huge and continuously growing body of scientific evidence shows that HIV causes AIDS," and "Thus, we conclude that there is overwhelming evidence that HIV causes AIDS."[51]

The argument from irresponsibility serves as the answer of last resort. In the vein of a "better safe than sorry" warning, such HIV defenders as Weiss and Jaffe assert the weapon of fear:

> If he [Duesberg] and his supporters belittle "safe sex," would have us abandon HIV screening of blood donations, and curtail research into anti-HIV drugs and vaccines, then their message is perilous.[52]

The irony, as will be reviewed later, lies in the danger of the officially approved measures to combat HIV, which are themselves costing lives.

4. Arguing for HIV Based on Antibody Correlations

The three basic arguments outlined above clearly answer no questions. The only positive evidence in favor of the virus-AIDS hypothesis is found in epidemiology, the study of disease epidemics. This field operates entirely by correlation: According to AIDS officials, where HIV goes, AIDS follows. Despite all the sophisticated biotechnology and vast investment in virology, the best evidence for HIV is only by correlation with antibodies present against it. Ironically, the point is made by retrovirologists Blattner, Gallo, and Temin: "The strongest evidence that HIV causes AIDS comes from prospective epidemiological studies that document the absolute requirement for HIV infection for the development of AIDS."[53] Or, as stated by Weiss and Jaffe, "The evidence that HIV causes AIDS is epidemiological and virological, not molecular."[54] Gallo again emphasizes the point in his book, declaring correlation to be "one hell of a good beginning."[55]

What sort of correlations seem so convincing to AIDS officials? The one usually cited first might be called the "geographic overlap." According to Blattner, Gallo, and Temin, "epidemiological data show that AIDS and HIV infection are clustered in the same population groups and in specific geographic locations and in time. Numerous studies have shown that in countries with no persons with HIV antibodies there is no AIDS, and in countries with many persons with HIV antibodies there is much AIDS. Additionally, the time of occurrence of AIDS in each country is correlated with the time of introduction of HIV into that country; first HIV is introduced, then AIDS appears."[56] The three HIV advocates fail to mention, however, that a disease is only recorded as AIDS if antibodies to HIV are also found.

Second, a tighter association is recorded for individual people: Every victim of AIDS has antibodies against HIV, whereas most healthy people do not. This apparently perfect correlation exists in selected surveys that follow people at risk for AIDS. But no national AIDS statistics exist that even document how well HIV compares with AIDS.[57] Clearly, most of the seventeen million healthy HIV-positive humans have yet to develop AIDS.

Altogether fewer than 6 percent (about one million) have developed AIDS in the past ten years.[58] Furthermore, thousands of clinically diagnosed AIDS patients are HIV-free.

A third argument evokes powerful emotional sentiments without much substance and works surprisingly well not only on the lay public, but on scientists as well. When challenged that only people with serious health risks develop AIDS, experts answer with anecdotes, even though the same medical officials will consider anecdotes a worthless type of evidence in any other debate. An anecdotal story is one individual case chosen to prove the absence of other health risks, implying HIV was the only factor that could have led to disease. So, for example, epidemiologists will describe a baby contracting HIV and subsequently developing AIDS. But in a nation of 250 million people, a few anecdotal cases can always be found to support any medical view.

Fourth, AIDS epidemiologists point to their prospective studies, in which the supposedly conclusive proof of the HIV hypothesis can be found. These studies monitor two groups of people over time, one of HIV-positive patients and the other of HIV-negative people in the same age group. According to such reports, the infected people develop AIDS while their uninfected counterparts do not. But all the reports that have also investigated drug use and other noncontagious AIDS risks have found that AIDS correlates with those factors just as well, if not better, than HIV (see chapters 8–10).[59]

Yet, these HIV-AIDS correlations have proven to be the most powerful arguments to scientists and laymen alike. Only a more complete picture can expose the misleading nature of this sloppy epidemiology.

THE OTHER STATISTICS

In one strange sense, officials do refer to some genuine correlations between HIV and AIDS. The syndrome, for example, is rarely found in any nation or individual apart from HIV infection. Indeed, the virus and the syndrome correlate with near-textbook perfection, ironically illustrating the most fundamental problem with the entire virus-AIDS hypothesis—the connection was artificially constructed.

AIDS is a syndrome of about thirty diseases, not a disease. It displays no unique combination of diseases in the patient. Clinically, it is identified by the diagnosis of specific diseases known to medical science for decades or centuries. The CDC has several times increased—but never decreased—the official list of AIDS indicator diseases, most recently on January 1, 1993 (See Table 2). The list now includes brain dementia, chronic diarrhea, cancers such as Kaposi's sarcoma and several lymphomas, and such opportunistic infections as *Pneumocystis carinii* pneumonia, cytomegalovirus infection, herpes, candidiasis (yeast infections), and tuberculosis. Even low T-cell counts in the blood can now be called "AIDS," with or without real clinical symptoms. Cervical cancer has recently been added to the list, the first AIDS disease that can affect only one gender (in this case, women). The purpose behind adding this disease was entirely political, admittedly to increase the number of female AIDS patients, creating an illusion that the syndrome is "spreading" into the heterosexual population.[60] Originally, the AIDS diseases were tied together because they were all increasing within certain risk groups, but today they are assumed to derive from the common basis of immune deficiency. The overlap between AIDS and certain risk groups still holds true but, as pointed out in Table 1, a significant number of these diseases are not products of weakened immune systems.

According to Blattner, Gallo, and Temin, "The CDC definition of AIDS has been revised several times as new knowledge has become available and will undoubtedly be revised again."[61] However, neither the CDC nor other advocates of the HIV hypothesis ever identify the "new knowledge" about HIV that mandates these revisions. It is also remarkable that such "new knowledge" always drives the list of AIDS-defining illnesses upward. Not once has an AIDS-defining disease been subtracted in the light of "new knowledge" about HIV. Irrespective of the undisclosed gains in knowledge about HIV, one thing is clear—the repeated upward adjustments in the definition of AIDS have substantially increased

TABLE 2
Chronology of the CDC's AIDS definitions

Year	Diseases	HIV antibody
1983	*Protozoal and helminthic infections* 1 Cryptosporidiosis, intestinal, causing diarrhea for more than a month 2 *Pneumocystis carinii* pneumonia 3 Strongyloidosis, causing pneumonia, central nervous system infection, or disseminated infection 4 Toxoplasmosis, causing pneumonia or central nervous system infection *Fungal infections* 5 Candidiasis, causing esophagitis 6 Cryptococcosis, causing central nervous system or disseminated infection *Bacterial infection* 7 "Atypical" mycobacteriosis, causing disseminated infection *Viral infections* 8 Cytomegalovirus, causing pulmonary, gastrointestinal tract, or central nervous system infection 9 Herpes simplex virus, causing chronic mucocutaneous infection with ulcers persisting more than one month or pulmonary, gastrointestinal tract, or disseminated infection 10 Progressive multifocal leukoencephalopathy (presumed to be caused by a papovavirus) *Cancer* 11 Kaposi's sarcoma in persons less than 60 years of age 12 Lymphoma, primary, of the brain	not required
1985	13 Histoplasmosis 14 Isosporiasis, chronic intestinal 15 Lymphoma, Burkitt's 16 Lymphoma, immunoblastic 17 Bronchial or pulmonary candidiasis 18 Chronic lymphoid interstitial pnemonitis (under 13 years of age)	required
1987	19 Encephalopathy, dementia, HIV-related	

	20	Mycobacterium tuberculosis any site (extrapulmonary)	
	21	Wasting syndrome, HIV-related	required
	22	Coccidiomycosis, disseminated or extrapulmonary	
	23	Cryptococcosis, extrapulmonary	
	24	Cytomegalovirus, other than liver, spleen, or nodes	
	25	Cytomegalovirus retinitis	
	26	Salmonella septicemia, recurrent	
1993	27	Recurrent bacterial pneumonia	
	28	Mycobacterium tuberculosis any site (pulmonary)	
	29	Pneumonia, recurrent	
	30	Invasive cervical cancer	required
	31	T-cell count is less than 200 cells per microliter or less than 14 percent of the expected level	

References

Selik, R. M., Haverkos, H. W., and Curran, J. W. (1984). "Acquired Immune Defiency Syndrome, (AIDS) in the United States, 1978–1982." *The American Journal of Medicine* 76: 493–500.

Institute of Medicine and National Academy of Sciences (1986). *Confronting AIDS*. National Academy Press, Washington, DC.

Institute of Medicine (1988). *Confronting AIDS-Update 1988*. National Academy Press, Washington, DC.

Centers for Disease Control and Prevention (1992). Revised Classification System for HIV Infection and Expanded Surveillance Case Definition for AIDS among Adolescents and Adults. *Morb Mort Weekly Rep* 41 (No. RR17): 1–19.

Duesberg, P., and Yiamouyiannis, J. (1995). *AIDS*. Health Action Press, Delaware.

---

the American AIDS statistics while HIV infections have remained completely flat since 1985 (see Figure 2A).

The increasing numbers of new AIDS cases until 1993 have largely been products of the artificial AIDS definitions (see Figure 2A). Each alteration in that definition has added, not subtracted, diseases to the diagnostic list. Every time the CDC needs higher rates of new AIDS cases, it expands that definition once again, and more diseases are reclassified into the syndrome. With the stroke of a pen an illusion of the spread of AIDS is created, prominent officials explain the revisions as products of our growing scientific knowledge, and the lay public feels reassured that federal efforts are justified—or perhaps even a little too slow.

One might ask how a doctor would distinguish between an AIDS-related tuberculosis and a traditional one. Clinically, the symptoms are identical, so the CDC has stipulated in its current definition that the tuberculosis must be renamed "AIDS" if antibodies against HIV are also found in the patient. In the absence of previous HIV infection, the disease is classified under its old name, in this case "tuberculosis," and treated accordingly. AIDS, therefore, can never be found apart from HIV infection—entirely by definition!

AIDS officials neglect to mention this crucial fact partly from ignorance, most never having read the definition carefully and in some cases precisely because it shines a disturbing light onto their supposedly perfect epidemiological coincidence between the virus and AIDS. *The observation that AIDS always follows HIV in each nation becomes trivial, since testing for antibodies is followed by a renaming of indigenous diseases.*

The real epidemiological questions, then, must be shifted away from any "correlation" between antibodies against HIV and AIDS to a correlation between HIV and the separate AIDS-diagnostic diseases. Does infection with the virus, independently of any other health risks, lead to an increased risk of contracting pneumonia, cancer, or other diseases? Is HIV new and found in all recent outbreaks of these diseases? Is HIV infection nearly always fatal?

The latter question can be answered most easily. Since the HIV test was made available in 1985, the CDC has officially estimated about one million Americans to be HIV positive, a figure that has not changed with the accumulation of testing data or the passage of ten years (see Figure 2A). Of these, only about four hundred thousand had been diagnosed with AIDS by the end of 1994. But this statistic does not subtract the normal incidence of the thirty AIDS-defining diseases in one million people over ten years. Two-thirds of HIV-positive Americans have not developed any of the AIDS diseases since 1985 (even including the most recent expansion in the AIDS definition).

Nor will most of them do so. The numbers of new AIDS cases have clearly been leveling off for some time now, although different analysts will place the peak at different times. Michael Fumento, the Colorado-based lawyer who gained some media notoriety with his 1989 book *The Myth of Heterosexual AIDS*, draws a curve with its peak in 1987;[62] two epidemiologists, in a 1990 paper in the *Journal of the American Medical Association*, suggest 1988 as the year of leveling.[63] The CDC observed a leveling off in 1994.[64] In any case, a slowly increasing forty thousand to fifty thousand new cases of AIDS—4 percent to 5 percent of the infected subpopulation—have appeared before the 1993 revision of the AIDS definition—hardly the "explosion" that AIDS, as a new infectious disease, was once predicted to show. The enormous gap between HIV-infected people and AIDS patients has induced the CDC to play more tricks with the numbers; at the time of this writing, the CDC is considering lowering its official estimate of one million HIV-positive Americans to a new total of six hundred thousand to eight hundred thousand.[65]

Part of the AIDS scare results from the way the numbers are reported. Rather than giving the numbers of new AIDS cases each year, CDC and other officials use the cumulative total for the current year added to the figures for all years previous, including those victims already dead. So where the annual numbers would remain constant in the first case, the number actually reported to the public grows with each passing year. Such calculation gives the overwhelming but false impression that AIDS is spreading, since the cumulative numbers can only go up. Given enough time, such accounting methods will boost the total AIDS count higher than the number of HIV-positive people. If this method were applied to count the American population, the cumulative number of newborns over several decades would eventually exceed the total number of Americans alive.

The commonly cited 50 percent to 100 percent death rate from HIV has been derived not from national statistics, but from studies on carefully selected cohorts of people. Several ongoing epidemiological studies have for years been observing hundreds, or at most thousands, of homosexual men at high risk for AIDS. Large proportions of the men in these studies have already been infected with HIV. But virtually all the subjects also admit to years of heavy drug abuse, extremely promiscuous sexual activity, and long histories of venereal diseases. Indeed, one major study was specifically organized around homosexual men with repeated bouts of hepatitis B. Researchers calculate the high fatality rate of HIV infection from these health risk groups, casually extrapolating these numbers to average, heterosexual HIV-positives—thus the discrepancy with the higher survival rate among the nation's one million HIV-positives.

The national AIDS figures fall well short of a virus with a nearly 100 percent fatality rate. But rather than abandon the hypothesis, the experts have chosen to revise the parameters of HIV infection. The latency period was originally calculated in 1984 on the basis of tracing sexual contacts, finding homosexual men developing AIDS an average of ten months after their last sexual contacts with other AIDS patients.[66] This "incubation period" has since been stretched to ten to twelve years between HIV infection and disease. For each year that passes without the predicted explosion in AIDS cases, approximately one more year is added to this incubation time. Even this is insufficient; with only 5 percent of infected Americans developing AIDS each year, the average latent period would have to be revised up to some twenty years for 100 percent to become sick.

A deeper look at the disease risk of infected populations reveals stranger paradoxes yet. The probability of developing AIDS varies radically between different HIV-positive populations. Sub-Saharan Africa, with infection rates approaching 30 percent of the population in some areas, has reported only approximately 250,000 AIDS cases to the World Health Organization in the past decade. This stands against six million to eight million Africans infected with HIV since the mid-1980s, whereas more Americans (now over 400,000) have contracted AIDS in a country with only one million HIV-positives. AIDS patients in Zaire, with about three million HIV-infected people, number only in the hundreds; Uganda, internationally considered a model for accurate testing and reporting, had by 1990 only generated some 8,000 AIDS cases out of one million HIV-positives. Roughly 360,000 infected Haitians have produced only a few hundred AIDS patients. In the industrial nations, homosexuals, heroin addicts, and hemophiliacs face greater probabilities of developing AIDS than do HIV-positive individuals without extraordinary health risks. And infants have a much shorter average latent period—two years, as opposed to the ten years in adults. No virus, including HIV, could possibly discriminate so enormously based on such subtle distinctions between its hosts.

HIV would need to perform other miracles to cause AIDS. Virtually all diagnoses of Kaposi's sarcoma are made in homosexuals, not in the other AIDS risk groups. Intravenous drug addicts disproportionately suffer from tuberculosis, Haitians from toxoplasmosis, and hemophiliacs from pneumonias. African AIDS diseases are basically different, manifesting as tuberculosis, fever, diarrhea, and a slim disease, unlike our wasting syndrome. A homosexual with HIV who may develop Kaposi's sarcoma can donate blood for a hemophiliac. But no hemophiliac has ever developed Kaposi's sarcoma from a blood transfusion. Instead he is more likely to develop pneumonia, if he contracts anything at all. Only HIV is common to both victims.

No virus could possibly make such distinctions between its hosts. A more likely hypothesis would blame the health risks specific to each group for their different diseases. If the same diseases can be found on the rise in the same risk groups, but also in people without HIV, then the virus would appear to be a harmless passenger.

The evidence bears this out. Hemophiliacs without HIV develop progressive immune degeneration just like the infected ones.[67] HIV-negative babies of infected mothers develop the same dementia-related symptoms as their HIV-positive siblings. Heroin addicts contract the same pneumonias, herpes infections, weight loss, and tuberculosis with or without the virus, and uninfected homosexuals with Kaposi's sarcoma are now being reported. Outbreaks of pneumonias or tuberculosis in recent years have included as many people without the virus as those with it.

Thousands of central Africans with "slim disease" have now been tested for HIV, and over half are completely negative; given the relatively high cost of HIV antibody tests, most African cases must be diagnosed by symptoms and remain untested for the virus.[68] In the industrial world, upward of one-quarter of all AIDS patients remain untested for the antibodies against HIV, with their doctors merely assuming the virus is present. The existing scientific literature records more than forty-six hundred cases of AIDS-defining conditions in people never infected by HIV.[69] With various AIDS-type diseases increasing in the risk groups even apart from HIV, the virus appears ever less relevant.

All circumstantial evidence aside, the ultimate epidemiological test for HIV would be a case-controlled comparison. In such a study, a large number of infected people would be monitored over time and compared with a large number of uninfected people. They would be matched for age, sex, income, and all other health risks such as drug use. Hemophilia and other medical complications would be excluded. If HIV were truly harmful, the infected group would develop AIDS and the uninfected would not. Scientists would conduct this type of study even before testing Koch's postulates. But no such study testing HIV as an AIDS virus can be found in the more than one hundred thousand studies to date on this virus![70]

When confronted with the whole of the evidence against them, defenders of the HIV hypothesis will sometimes cite studies comparing notorious AIDS risk groups, with and without the virus, to show that only those infected will degenerate and die. But none of the vast number of such prospective studies has actually matched two groups for the health risks that might cause AIDS. They have been designed merely to compare the symptoms of AIDS patients with normal people in the same age group, not to determine the cause of the syndrome. Such studies, their marginal and questionable value notwithstanding, are too often quoted by some researchers as proof of the virus-AIDS hypothesis.[71]

NO AIDS VIRUS AT ALL

Given that HIV fails all standards of scientific evidence as an "AIDS virus," could another, possibly unidentified, virus cause AIDS instead? Such a microbe would have to possess amazing and unprecedented qualities, for AIDS does not behave as a contagious disease at all.

The sexual revolution of the past twenty years has caused increases in all the major venereal diseases, including syphilis, gonorrhea, chlamydia, and genital warts. The same has occurred with hepatitis B. All of these infectious diseases have spread far beyond their original reservoirs into the general population and affect men and women nearly equally.

AIDS, however, has remained absolutely fixed in its original risk groups. Today, a full decade after it first appeared, the syndrome is diagnosed in homosexuals, intravenous drug users, and hemophiliacs some 95 percent of the time, just as ten years ago. Nine out of every ten AIDS patients are male, also just as before. Even the very existence of a "latent period" strongly suggests that years of health abuse are required for such fatal conditions. Among most AIDS patients in the United States and Europe, one extremely common health risk has been identified: the long-term use of hard drugs (the evidence for this new AIDS hypothesis will be presented in chapters 8 and 11). AIDS is not contagious nor is it even a single epidemic.

Tragic deaths, time and money wasted, hysterical public debate over a harmless virus—these have been the fruits borne of a scientific establishment grown too large for genuine science. The creative pursuit of knowledge has been swallowed to satisfy careerism and its voracious appetite for job security, grant money, financial benefits, and prestige. But the monster is twice guilty, for it also destroys or marginalizes those few scientists daring to ask questions. These dissidents against the HIV hypothesis are the subject of the next chapter.

CHAPTER SEVEN

■

Dissension in the Ranks

IN ITS SELF-ORDAINED MISSION to coordinate the war on AIDS, the CDC used its full resources to popularize AIDS as a single, infectious, and terrifying plague. But the agency hardly succeeded in monopolizing interest in the epidemic. Other doctors also took notice of the rising numbers of young homosexual men dying of infections and conditions uncommon for their age group. From the time the CDC advertised its first AIDS cases, the apparently new syndrome invited speculation on its cause.

Those medical professionals who followed the CDC's lead searched for an infectious agent. Michael Gottlieb, the first doctor to report AIDS cases, led a number of virus hunters in suggesting cytomegalovirus. Other well-known viruses, including Epstein-Barr, received growing attention. The retrovirus hunters found themselves torn between Gallo's HTLV-I and the search for a new virus. Still other researchers began thinking of bacteria or even new combinations of several old microbes together all causing AIDS.

The search for the cause of AIDS officially ended with Gallo's 1984 press conference. No American scientist had yet published a single paper on HIV, but most scientists understood the politics and quickly fell into line. Doubts about this virus were relegated to quiet conversations, especially among those researchers whose careers most directly depended on the NIH-CDC medical establishment. Most physicians never even heard any reason to question the official doctrine.

For a few people, however, the press conference settled nothing. Doctors who knew something about the methods of scientific research and who felt a bit more independent of the federal government continued to raise questions. To them, the rush to blame HIV for such a complex and varied syndrome, one that struck people with so many obvious health risk factors, seemed simpleminded. By throwing its weight behind HIV, the AIDS establishment unwittingly spurred some of the alternative thinking it sought to end.

THE EARLY DAYS

Joseph Sonnabend became one of the first to break ranks. Having received a medical degree in his native South Africa, he found his way into basic research upon moving to Great Britain in the late 1950s. There he joined the revived microbe-hunting trend, albeit more from the angle of medical treatment, and began studying the body's immune response against viruses. He focused on interferon, a newly discovered protein that seemed to slow virus infections. Scientists have always placed great hopes in this substance, expecting it to serve as their long-sought magic bullet against viruses and cancer. Both of these dreams have died, but scientists are now trying to revive it for use against multiple sclerosis.

The 1970s brought Sonnabend a temporary chance to conduct his research on interferon and viruses at a medical school in New York. After the money ran out he practiced medicine at a public hospital in Brooklyn. He supplemented his income by working for the city's Department of Health, where in 1978 he briefly became director of the venereal diseases division. In this capacity he encountered many of the "fast-track" homosexuals who constantly needed treatment for their recurring diseases.

Later that year Sonnabend lost both positions. Although he preferred laboratory microbiology, he had little choice but to continue medicine. As a compromise, he decided to continue working on infectious diseases by starting his own private practice in Greenwich Village, New York, treating homosexual men for their venereal diseases. By the early 1980s Sonnabend began seeing AIDS cases, just as similar patients were showing up at the UCLA Medical Center on the opposite coast. He recognized the descriptions in Gottlieb's 1981 report of five such men and immediately conducted research to find the cause. Having seen rising frequencies of venereal disease among homosexual men for years, Sonnabend instinctively reached for the most familiar explanation—that somehow the combination of all these conventional microbial infections caused immune suppression and AIDS.

He went public with his hypothesis by 1982, publishing reports that men with immune deficiencies also had long histories of venereal disease, hepatitis, and even infections by obscure parasites. Meanwhile he started treating his AIDS cases by using antibiotics and other medications directed against the opportunistic infections themselves, including *Pneumocystis carinii* pneumonia. But his views attracted little attention until a publisher suddenly provided Sonnabend the funding to create a scientific journal of his own. *AIDS Research* was thus launched, and the first twelve pages of the first issue, published in 1983, contained a review written by Sonnabend himself. Entitled "The Etiology of AIDS," the article officially proposed what he called the "multifactorial model" of causation. According to this notion, many different infections could have a combined effect that eventually destroys the immune system. He also hypothesized that semen itself—coming in contact with blood when rectal tissues were torn during anal intercourse—might cause immune suppression. Sonnabend opened his review by attacking the CDC viewpoint that AIDS was caused by some new virus, pointing out that no such virus had yet been isolated. Then he turned to his own idea:

> The first issue of this new journal is an appropriate occasion to review an alternative hypothesis regarding the genesis of AIDS. This hypothesis proposes that there is no specific

etiologic agent of AIDS, and suggests that the disease arises as a result of a cumulative process following a period of exposure to multiple environmental factors...

Among homosexual men, it appears that the disease has been occurring in a rather small subset characterized by having had sexual contact with large numbers of different partners... Such conditions were met in New York City, San Francisco and Los Angeles in the 1970s as a result of changes in lifestyle that became apparent in the late 1960s.

The specific factors we propose that interact to produce the disease in homosexual men are: (1) immune responses to semen; (2) repeated infections with cytomegalovirus (CMV); (3) episodes of reactivation of Epstein-Barr virus (EBV); and (4) infection with sexually transmitted pathogens, particularly those associated with immune complex formation such as hepatitis B and syphilis.[1]

In explaining AIDS in Haiti or Africa, Sonnabend argued their diseases might not be new at all and could reflect such factors as "poverty and malnutrition, some tropical infections," while in the case of blood transfusion recipients, "It is well known that blood transfusions are themselves immunosuppressive." In any case, he criticized the CDC assumption of a new AIDS virus in no uncertain terms, specifically taking on their cluster study as not being proof AIDS was a single infectious disease: "That AIDS results from infection with a specific etiologic agent remains a hypothesis... An alternative explanation is that the cases occurred in a relatively small subset of homosexual men who shared a similar lifestyle."

Sonnabend accurately dismantled the assumptions of the virus-hunting establishment, exposing the lack of evidence for AIDS as a single, contagious disease. But his multifactorial hypothesis completely ignored the drug abuse factor in most AIDS patients. Those homosexuals at greatest risk for the syndrome, who had long records of infectious disease, also had used enormous quantities of recreational drugs, especially the alkylnitrites. Sonnabend tended to overlook drugs as a risk factor largely because of his virology background and his experience treating venereal diseases.

He forgot that infectious diseases do not affect everyone equally; probably no germ on earth, from the most common flu virus to the deadly cholera bacterium, causes disease in every infected individual. Only those people whose resistance is lowered for some reason—even a temporary immune deficiency from lack of sleep or other causes—become ill; a healthy person's immune system efficiently suppresses microbes and prevents symptoms, regardless of the number of infections. Multiple contagious diseases, therefore, could not cause immune suppression in a person, but must rather be the result of immune deficiencies for other reasons. Even semen, particularly in the minute quantities that could contact blood in anal intercourse, could not have an irreversible effect on the immune system.

Sonnabend continued making his argument. Soon after launching *AIDS Research*, he published a similar review paper in the *Journal of the American Medical Association*, co-authored with his colleagues and close collaborators, Steven Witkin and David Purtilo. As he stepped up the debate, Sonnabend found himself increasingly crossing paths with Robert Gallo. To counter Gallo's early hypothesis that HTLV-I was the "AIDS virus," Sonnabend tested seventy patients and reported that none of them had antibodies against the virus. He shortly thereafter published a letter to the editor of *Nature* in 1984, following Gallo's press conference announcing "HTLV-III" as the cause, stating that since HTLV-I, -II, and now -III could each be isolated from some AIDS patients, this "suggests that they are more likely to represent opportunistic infections or reactivations from latency."[2]

He made the same point in a 1985 letter published in the *Wall Street Journal*, suggesting that HIV might only be a harmless, opportunistic virus found in some people after their immune systems had already been destroyed. He also acknowledged "the possible role of drugs in the causation of AIDS," an unusual departure from his multifactorial hypothesis.[3] As late as 1988, while he was working on a chapter for a medical textbook, Sonnabend wrote to Peter Duesberg, describing the effects of growing political pressure to swallow the HIV hypothesis: "I just spoke to David Purtilo who does not wish to be on the update—unless a role for HIV can be put in. Steve Witkin also wants a role for HIV, so I'll do it alone."4 By the time the textbook finally appeared, Sonnabend's longtime collaborators had removed their names.

His prestige had been such that James Curran, head of the CDC's KSOI Task Force (which scoured for evidence to prove AIDS an infectious disease), personally consulted with Sonnabend in 1981. Sonnabend takes credit for devising the notion of "safe sex," the use of condoms supposedly to prevent transmission of AIDS or the venereal diseases he believes cause it, which has become popular with public health authorities as a fetish of AIDS prevention. When a press conference was organized in February of 1985 to announce that Gallo's isolate of HIV was suspiciously identical to that of Montagnier, Sonnabend was the man chosen to make the presentation. The FDA used Sonnabend's unorthodox clinical trial (he had dispensed with the time-honored testing rules of double-blind controls and placebos) to approve the aerosolized drug pentamidine for treatment of *Pneumocystis* pneumonia and to set precedent for future licensing.

His most powerful connection has been Mathilde Krim, a colleague who also studied interferon's effects on virus infection. Krim was more than just another scientist; her husband, a Hollywood veteran who founded Orion Pictures, had also been chairman in charge of finances for the national Democratic Party and therefore a consultant to several presidents. Krim herself had been one of the powerful individuals selected for the Senate's Panel of Consultants in 1970, which advised Richard Nixon to launch the War on Cancer. Krim had long befriended Sonnabend, and when he began running out of money to continue his AIDS research in 1982, she stepped in. She organized the American Medical Foundation (AMF) to finance his work, and her clout brought onto the board several important scientists, as well as former president Jimmy Carter's wife Rosalynn. So much money flowed into the foundation that other scientists offered to collaborate with Sonnabend in order to benefit.

Although Sonnabend sometimes enjoyed forays into unfashionable areas of medical research, he clearly stood to lose much by straying too far from the official line. By continuing to question the HIV hypothesis, he unwittingly did precisely that. In 1985, one year after Gallo's press conference, the axe began to fall. The publisher who had financed his journal, *AIDS Research*, suddenly replaced Sonnabend with Dani Bolognesi, a retrovirologist at Duke University. Bolognesi was one of Gallo's closest allies, a member of the informal "Bob Club," and therefore a partisan for the HIV hypothesis. As the new editor, Bolognesi dumped Sonnabend and his supporters, bringing on board his own retrovirus-hunting friends Max Essex and Robert Gallo. The journal's new title became *AIDS Research and Human Retroviruses*, and thereafter it published only papers founded on the HIV hypothesis. Its days of open inquiry were over.

Meanwhile, Mathilde Krim was reorganizing the AMF, negotiating a merger with a more glamorous and better-funded foundation under Michael Gottlieb, the scientist who reported the first five AIDS cases. Gottlieb objected to any doubts about HIV, and Krim ejected Sonnabend from the foundation and its support. Sonnabend found himself isolated, having learned a bitter lesson about challenging a view so cherished by the medical powers-that-be.

At this point Krim stepped in again, playing good cop to Gottlieb's bad cop. She helped Sonnabend establish a new organization for sponsoring research on AIDS treatments, the Community Research Initiative. After more than a year of setup, the group began receiving funds. Sonnabend's criticisms of the HIV hypothesis gradually became muted or were relegated to obscure newsletters. By 1989 he had so sufficiently won his way back into good graces that Krim arranged a public meeting at Columbia University with NIH officials. At the luncheon table, Sonnabend was seated between Sam Broder, Gallo's boss and head of the National Cancer Institute, and Anthony Fauci, director of the National Institute of Allergy and Infectious Diseases. Both were power brokers of the AIDS establishment, to whom Sonnabend had finally become acceptable.

The sanctions have taken their toll. In 1992, when an interviewer asked, "What if HIV doesn't cause the disease?" Sonnabend responded, "Well, I have reluctance in speaking about this, too, because I am a great believer that safer sexual practices are important and that needle sharing is not a good idea."[5] He continued to evade the interviewer's questions about Peter Duesberg, finally declaring bluntly, "There are good reasons why HIV is a respectable candidate. For Duesberg to say that HIV cannot be the cause would mean that he wouldn't want any research to be done on HIV, and that's kind of ridiculous, too. I'll go to great lengths to make sure that I am not confused with Peter Duesberg."[6]

Indeed he does. He joined Duesberg and other dissidents at a May 1992 meeting of HIV critics in Amsterdam, Holland. On the final day of the conference, Sonnabend stunned the participants by issuing a press release attacking Duesberg, on official symposium stationery. The man who once argued AIDS was not infectious now lashed out at Duesberg for saying the same thing and insisted that "his outrageous assertion that safe sex is irrelevant to the spread of AIDS is appalling and may kill people."[7] He even managed to get a few participants to cosign the release. But AIDS dissident John Lauritsen rallied most of the others at the meeting to Duesberg's defense, issuing their own contrary press release. Some of Sonnabend's cosigners switched sides or publicly apologized. Sonnabend himself was seen by witnesses privately apologizing to Duesberg, although he officially denied it later in print.[8] His public attack may have primarily resulted from worries about attending the meeting in the first place.

Even Sonnabend's private medical practice has been changed. Originally, he had been widely known for his vocal opposition to the toxic chemotherapy AZT as AIDS treatment. A 1988 article quoted him as declaring "AZT is incompatible with life," and he refused to prescribe the drug to his own AIDS patients.[9] But he now admits to giving his patients AZT when they request it and no longer lobbies against it.

He has shifted course sufficiently that his old nemesis, Robert Gallo, invited him in 1993 to speak at the NIH. Sonnabend accepted, and his talk on interferon was well received by the believers in HIV. But of his old friend Duesberg, Sonnabend could only comment to an interviewer that "on balance I think [Duesberg has] been bad" for the debate over the virus-AIDS hypothesis.[10] In a letter to *Science* on January 13, 1995, Sonnabend settles in the middle ground. "I may not be supportive of Peter Duesberg's arguments and dogmatism in rejecting HIV as the cause of AIDS, but John Cohen, in citing my criticism, did not make it clear that I continue to believe the issue of AIDS causation still remains open."[11]

His research group has been reorganized as the New York–based Community Research Initiative on AIDS (CRIA), where he conducts research on AIDS treatments and maintains a relatively low profile on the HIV controversy.

At the same time that Sonnabend was first struggling against the growing AIDS virus hunt, another rebel was emerging nearby in New York City—John Lauritsen. Several years later, he would be described as "one of the heroes of the epidemic" by another medical dissident against HIV. "He is not only a top-notch investigative reporter. In his own way he is also a scientist."[12]

Lauritsen has worked in the survey research field since the mid-1960s, where he performed tasks as a market research executive and analyst. Professional survey research, he explains, maintains much higher professional standards than does its academic sister, epidemiology: questionnaires require careful designing, data must be rigorously checked after they are gathered, tables must show all data clearly and completely, and statistics are analyzed critically. He had also co-authored *The Early Homosexual Rights Movement (1864–1935)* and edited an anthology of writings by John Addington Symonds. Lauritsen the scientist and Lauritsen the journalist were both products of an A.B. degree from Harvard's Department of Social Relations.

He first got involved in AIDS after he learned of Sonnabend's work. His attention was focused on the syndrome in 1983, when he decided to spend a week in the library of the New York Academy of Medicine, reviewing for himself the still-small scientific literature on AIDS. The evidence quickly fell into place, strongly suggesting that AIDS was not an infectious disease. Lauritsen now suspected that some lifestyle environmental factor was killing people, not a microbe.

Shortly thereafter, he stumbled across an article describing Hank Wilson, a well-known homosexual rights activist in San Francisco. Wilson was waging a one-man crusade against the use of "poppers," the nitrite compounds inhaled almost entirely by "fast-track" male homosexuals as bathhouse aphrodisiacs and muscle relaxants. The volatile drugs made anal intercourse easier by relaxing the anal sphincter, but also had toxic effects on the blood and other parts of the body. Wilson had taken up this cause after friends who used poppers heavily began suffering swollen lymph nodes, which had led him to research the chemical nature of the nitrites. He founded the Committee to Monitor Poppers in 1981, warning homosexuals of the dangers and lobbying for legal bans on the substance.

Lauritsen began corresponding with Wilson and soon concluded that poppers and other recreational drugs being used in the bathhouses played some role in AIDS and other sickness. As a member of the New York Safe Sex Committee, Lauritsen began circulating warnings about poppers, prompting the group to include the following ending in a 1984 brochure: "Avoid drugs. Shooting up kills. Uppers and downers put a real strain on your system. Pot and alcohol confuse your judgment. Poppers are also dangerous."[13] But the advice fell on deaf ears. No one wanted to give up the popular drug. He then turned to Wilson, and the two of them began organizing a small but nationwide educational campaign that helped push Congress into outlawing poppers a few years later. By February of 1985, Lauritsen was able to publish his first article on AIDS, exposing the CDC's statistical tricks in hiding the association between poppers and the syndrome (as the CDC had been doing since the first reported AIDS cases, part of its campaign to paint AIDS as infectious). The piece appeared in the *Philadelphia Gay News*. As he soon discovered, the widespread hostility to his message meant that he could publish only in the homosexual press, and then only in a small subset of that.

Lauritsen found a journalistic niche freelancing for the *New*

York Native, the largest independent homosexual-interest weekly in the country. Independent it certainly was. Its publisher and editor, Charles Ortleb, had infuriated the CDC and other public health and medical officials when he began questioning the official theory that HIV is the cause of AIDS. Ortleb knew he would not believe the HIV hypothesis when he published Lauritsen's critique on the research that led to the approval of AZT as a treatment for AIDS in record time. Lauritsen also wrote in the *Native* about his own drug-AIDS hypothesis alongside the unfolding story of the HIV debate. By including complete bibliographies, Lauritsen's articles first introduced scientific documentation to science reporting in the nonprofessional literature. In 1987 the *Native* first introduced Duesberg to the gay community with a Lauritsen interview; on October 5, 1992, the paper even put Duesberg on its cover. Over the furious objections of ACT UP officials from New York, the cover called Duesberg "An International Hero" because "Peter Duesberg Bravely Speaks Truth to Power in His Battle Against AZT and HIV Apartheid." In 1988 Ortleb added his own AIDS hypothesis to the list of HIV challenges, postulating that AIDS is caused by Human Herpes Virus 6. Together with staff reporter Neenyah Ostrom he has made this hypothesis the focus of the journal's investigations on the cause of AIDS.

By 1986 Lauritsen had left full-time survey research to allow himself to focus on AIDS. That year he and Hank Wilson produced a small self-published book, *Death Rush: Poppers and AIDS*. In it he made his complete case for the role of poppers and other drugs in causing AIDS, impressively documented with dozens of scientific papers on the subject. He also thoroughly exposed the conflicting interests of homosexual publications and academia in their ties to the poppers industry. He included two pages citing Koch's postulates to argue against HIV as the cause of AIDS.

His articles continued to reflect his own research. In March of 1987, for example, he wrote a devastating attack on a National Academy of Sciences report, pointing to their own admission that HIV is neutralized by antibodies as evidence against the virus hypothesis. But two months after his article was published, he read Duesberg's original *Cancer Research* article. To Lauritsen, it was a stunning confirmation of everything he had suspected. In his own words: "I had never heard of such concepts as 'biochemical activity,' and it clicked. I no longer had any doubt that HIV was not the cause."[14] Lauritsen was referring to Duesberg's argument that a dormant, biochemically inactive virus, like HIV, could not cause any disease, let alone the many fatal AIDS diseases.

The following June an article by Charles Ortleb appeared in the *Native*, excitedly reviewing Duesberg's paper. Ortleb tracked down Duesberg, finding him near the end of his stint at the NIH. Lauritsen immediately caught a train to Bethesda, becoming the first journalist to interview Duesberg.

In preparing for the interview, Lauritsen had phoned the CDC and NIH to pester officials with questions about HIV. Confronting the National Cancer Institute's press officer, he pressed for the definitive proof that the virus caused AIDS. She was unable to answer and deferred until the following day, returning the call to read off a hastily prepared response. Nothing she said directly answered Duesberg's arguments, so Lauritsen raised the obvious issue of Koch's postulates. Her reply serves as the perfect picture of modern virus hunting:

> What are those? I've never heard of them. How do you spell that? Coke? What did you say? Koch? When were those made? [Lauritsen: About a century ago.] Oh, well then, would you say that those apply now?[15]

The approval of AZT as AIDS therapy pushed Lauritsen to take on a new fight. He read the evidence and concluded that such a toxic chemotherapy could do nothing but worsen an AIDS condition, since the drug destroyed the immune system. His investigation led him through a maze of sloppy scientific papers, the federal bureaucracy in trying to release documents under the Freedom of Information Act, and uncooperative researchers. A critical letter to the editor of the *New England Journal of Medicine*, which had published the original AZT trials in humans, yielded Lauritsen nothing but a private response that dodged his facts and airily declared, "I don't know of any noteworthy clinical investigator in the AIDS field who takes your position."[16] Fed up with the closed doors and arrogance of the establishment, Lauritsen wrote several articles on AZT for the *Native* and compiled his information into another book, *Poison by Prescription: The AZT Story*, self-published in 1990. The book remains the most comprehensive critique of AZT available today.

In 1993 Lauritsen self-published another book, *The AIDS War: Propaganda, Profiteering, and Genocide from the Medical-Industrial Complex*. The angry tone reflects his years of struggle. A mix of new material and previously published articles, its 480 pages cover topics ranging from AZT to the death of ballet superstar Rudolf Nureyev from AZT and AIDS. Most of Lauritsen's first interview with Duesberg is printed, along with exposures of the cozy relationships between AIDS organizations and the pharmaceutical industry. Portions even discuss a "program of recovery" from AIDS, focusing on the health risks Lauritsen implicates in causing the syndrome. Mostly, the book is a personal story, documenting the fight against HIV as seen by someone on the front lines.

OTHER ALTERNATIVE VIEWS

In the wake of challenges against the HIV hypothesis by Sonnabend, Lauritsen, and Duesberg, other medical doctors and scientists gradually began joining the chorus of opposition. Some were encouraged to find their open doubts shared by prestigious figures, others had previously felt intimidated in speaking out alone, and a few simply had never given thought to possibilities other than HIV. Not all of these people volunteered their own alternative hypotheses, but all were united in questioning the HIV monopoly in AIDS research and treatment.

For those who did propose alternative causes, the temptation lay in imitating Sonnabend's multifactorial model. AIDS patients not only carried a multitude of opportunistic diseases, but also engaged in extremely promiscuous sexual activity or needle-sharing, behaviors that gave the patients long histories of venereal and parasitic infections. Thus, a researcher could easily blame any of those microbes for AIDS purely on the basis of a heavy overlap between almost any germ and the syndrome. Some, like Sonnabend, chose to blame many or all of the microbes simultaneously, creating a cumbersome and largely untestable notion of AIDS as the consequence of some undefinable combination of diverse germs. Others preferred to implicate one or two specific microbes, sometimes as "cofactors" with HIV; according to this view, AIDS was still a truly contagious disease for which the wrong microbe had been identified.

For a few years, syphilis became the most popular alternative hypothesis.[17] Some superficial associations made this idea seem plausible. The syphilis bacterium, for one thing, had the old reputation as the "Great Masquerader," supposedly being able to imitate symptoms of diverse and unrelated diseases. Neurosyphilis—brain rot—had achieved legendary proportions in this regard and seemed to parallel the symptoms of AIDS dementia. The standard test for syphilis infection, moreover, turned out to be less reliable than previously thought, generating false-negative results in people who had been infected. Improved testing revealed high percentages of AIDS patients with prior syphilis. And AIDS education had taught everybody that AIDS, just like syphilis, was a sexually transmitted disease.

On the other hand, a hard look at AIDS quickly dispels any connection between the two. Neurosyphilis, as we discussed previously, most probably never really had anything to do with syphilis bacillus (see chapter 2). It never appeared during the original syphilis infection, instead manifesting only after the common treatments of the day—mercury, antimony, and arsenic. (Mozart is said to have been one person so treated until his early death.) Poisoning has often been blamed for late-stage "syphilis" symptoms, including the many conditions that earned syphilis its image as a masquerader. Indeed, carefully monitored syphilitics have proven to have normal life spans in the absence of toxic treatment.

Exotic symptoms aside, syphilis behaves no differently from any other microbe in that the disease itself results from, not causes, immune deficiency; as we have noted, healthy immune systems easily suppress any microbe and prevent sickness. Even the syphilis bacterium can do little damage in an otherwise healthy person (see chapter 2). For thousands of years before the age of antibiotics, most people survived syphilis without lasting consequences and without any treatment. This microbe, further, has no latent period between infection and disease, which contrasts the years required for AIDS to develop. Finally, AIDS is not a contagious disease, as evidenced by its tight restriction to risk groups.

Another bacterial hypothesis of AIDS was evolving at the same time the syphilis proposal was gathering supporters. In 1986, a virologist named Shyh-Ching Lo first reported finding a new virus in several AIDS patients. He performed some of these experiments at the National Cancer Institute, where colleagues scoffed. In a noble attempt to meet Koch's postulates for causing AIDS, Lo went on to grow the virus in cultured cells and then infected four monkeys—all of which died of a wasting disease within months. But at that point he ran into trouble. "Lo had a tough time getting his findings published. 'I forget how many journals turned us down,' he says. One colleague put the figure at more than half a dozen."[18]

Lo had some protection from other virus hunters because he worked at the Armed Forces Institute of Pathology, a military research facility entirely independent of the NIH-funded establishment. Nevertheless, he could not publish his results until 1989, and then only in a relatively obscure journal. By that time he had further identified the nature of his "virus," discovering he had actually been working with a mycoplasma, a tiny bacterium that prefers to hide inside cells. He named his find *Mycoplasma incognitus*, reflecting the fact he had originally confused it with being a virus. Lo finally began receiving applause for his discovery the following year, once it was endorsed by Luc Montagnier, the French discoverer of HIV.

Although Lo tested his mycoplasma using Koch's postulates, his microbe hunting enthusiasm overran his scientific sense. The mycoplasma had in reality failed the test. He could not find the bacterium in many AIDS patients, thereby falling short of the first postulate. And the infected monkeys, while wasting away and dying, developed nothing like the wide spectrum of AIDS diseases nor did their conditions have a latent period. Thus, the third postulate also eliminated the mycoplasma as a candidate. Mycoplasmas have been textbook subjects for decades; they cause roughly one-third of all human pneumonias and frequently contaminate the cell cultures of unsuspecting researchers. Unlike viral pneumonias, the mycoplasma pneumonias can be treated with tetracycline and other antibiotics. Mostly, these microbes function as opportunists, preying on people with weakened health. And since AIDS, as this book shows, is not infectious at all, it could not be caused by this mycoplasma or any other microbe. The failure of tetracycline to cure AIDS drove the last nail into the coffin of the mycoplasma hypothesis.

With regard to Gallo and the HIV dogma of AIDS, Lo did have poignant comments. In a letter to *Policy Review* in 1990, he and his supervisor wrote that "to commit oneself exclusively to a particular agent and completely rule out any other possible role of a different microbe, may... result in a greater loss of AIDS victims."[19]

SPREADING DOUBTS

Peter Duesberg's entry into the HIV debate in 1987 suddenly changed its scope, particularly with his insistence that the virus clearly had nothing whatsoever to do with AIDS. Faced with such a compelling and uncompromising argument, scientists could no longer easily ignore dissension. Several prominent researchers chimed in with their own doubts about HIV, although they cautiously avoided naming alternative causes for AIDS, preferring simply to question official dogma. Despite their own impeccable credentials, some of them quickly ran into the same political pressures that had plagued other dissenters.

Albert Sabin became the first to follow Duesberg into the fray.

Following his days working on the polio vaccine, he had retired to the NIH as a consultant with his own office. The position was granted to him because of his honored status, having been a member of the National Academy of Sciences since the early 1950s and one of the most respected virologists in the world. His sometimes gruff and forceful personality had even helped enhance the respect his peers afforded him.

In 1987, while still on leave at the NIH, Duesberg was asked to give a lecture in honor of the Fogarty fellowship supporting his NIH research. He chose to speak about his recent paper in *Cancer Research* that criticized the HIV hypothesis. Sabin was one of many NIH people filling the lecture room. Duesberg had barely finished his speech when Sabin leaped to his feet. He headed straight for the microphone, seizing the podium as if to throw it.

"I think the views of a person like Dr. Duesberg are terribly, terribly important," he bellowed, "and we must pay attention to them."[20] Turning to the whole question of whether AIDS would actually spread to the general population, his voice took on an angry tone. He denounced the panic-ridden projections of a heterosexual epidemic. "This is not the population where you find AIDS. We have known this for almost 10 years and the pattern has not changed. I am astonished by the hysteria. This is absolute madness." He thundered along, no one in the room daring to interrupt. "These are irresponsible statements without any scientific foundation... I don't want to be a psychiatrist and try to figure out why these things are said in the absence of evidence, but unfortunately they are receiving a great deal of publicity."[21]

Sabin's years of virus hunting now came into play. He had worked with truly Nobel Prize–quality disease-causing viruses, including polio, which induced symptoms only when flooding the body in high numbers. "Presence of virus doesn't mean anything in and of itself," he reminded the audience, "because virologists know that quantities count." This meant, he concluded, that HIV itself, being extremely rare in AIDS patients, should be difficult to pass along between people. "The basis of present action and education is that everybody who tests positive for the virus must be regarded as a transmitter and there is no evidence for that." Finally, he threw barbs at the virus hunters who spent all their time investigating the genetic details of HIV, never asking whether it had been proven to cause anything. "Up to the present time, all that beautiful knowledge about the molecular biology of the virus isn't helping us at all to deal with it."[22]

Sabin spent twenty minutes at the microphone, nearly as long as Duesberg himself. The added comments touched off excited rounds of questions and discussion, Sabin's own personality magnifying the charged atmosphere. He was now fired up enough to fight back against the one-sided media coverage of AIDS and arranged a press conference the following month at the Third International AIDS Conference in Washington, D.C. Duesberg was asked by Sabin to participate, but Duesberg had not been invited to the AIDS conference and thus could not attend. Sabin therefore held the meeting himself.

But after that occasion, he was never again heard defending Duesberg or questioning HIV. Confronted by his peers, Sabin may have reconsidered his strong spontaneous stand at Duesberg's Fogarty lecture for two reasons. First, the grand, emeritus poliovirus pioneer now earned many reflected glories from HIV research, as a virus spokesman and consultant. Second, having no tenure or other protection, Sabin's emeritus position at the NIH was subject to the whims of intolerant superiors, ones who did not enjoy being embarrassed by a scientist with his prestige. Duesberg worked at a university, a more difficult target for NIH retaliation; Sabin was more directly vulnerable. Until he passed away in 1993, Sabin declined to speak out against HIV again. When called by Duesberg and several reporters, he cited failing health and lack of familiarity with the AIDS literature as reasons. Shortly before his death, Sabin had made peace with the virus-AIDS establishment; true to his reputation, he wrote a last paper dealing with the problems of making an AIDS vaccine.

Duesberg's next outspoken supporter did retain a safer university position. Walter Gilbert, a professor of molecular biology at Harvard, had won the Nobel Prize for chemistry in 1980.

Considered one of the more important Nobel awards in recent years, Gilbert had won it for inventing the modern technique for sequencing, or reading, the genetic material DNA.

Upon reading Duesberg's *Cancer Research* paper in 1987, Gilbert was immediately fascinated. He told a reporter, "It is good to have it [the HIV hypothesis] questioned and argued. I absolutely do consider it a valid debate."[23] Specifically, he argued from the time-tested principles of virology that Duesberg "is absolutely correct in saying that no one has proven that AIDS is caused by the AIDS virus. And he is absolutely correct that the virus cultured in the laboratory may not be the cause of AIDS. There is no animal model for AIDS, and where there is no animal model, you cannot establish Koch's postulates."[24] The arguments against HIV are so strong, according to Gilbert, that "I would not be surprised if there were another cause of AIDS and even that HIV is not involved."[25]

Gilbert has made the *Cancer Research* paper required reading for his graduate students, using it as an illustration of how skeptical thinking ought to work in science. This he considers his most important message. As he sees it, "The community as a whole doesn't listen patiently to critics who adopt alternative viewpoints, although the great lesson of history is that knowledge develops through the conflict of viewpoints, that if you have simply a consensus view, it generally stultifies, it fails to see the problems of that consensus; and it depends on the existence of critics to break up that iceberg and to permit knowledge to develop."[26]

With his honors and awards, Gilbert remains fairly immune from political repercussions of his public statements. Thus, he can continue to criticize HIV though he does not take an activist role in the debate.

Another Nobel Laureate sympathized with HIV dissidents in 1991. Having received the prize in the early 1980s, Barbara McClintock was finally vindicated after decades of scientific isolation. She had discovered transposons, small genes that periodically jump from one spot to another in the DNA of various organisms. Her long struggle to gain acceptance for the concept has since become legend, her findings hailed as one of the momentous discoveries of biology since World War II. Even in the popular literature, she now stands as a symbol of tireless dissent against an intolerant scientific establishment.

McClintock's years of pioneering research were performed at the Cold Spring Harbor research labs in New York, headed by Nobel Laureate James Watson, where she remained all her life. This placed her in the right spot to meet Duesberg in 1991. That May, shortly before Duesberg left for the Cold Spring Harbor facility to attend Watson's annual conference on retroviruses, he received a telephone call from the elderly McClintock. She said that a colleague at Harvard, asked by Duesberg to review the draft of an update paper on AIDS, had sent her a copy. She loved it and even thought he should make it stronger and more forceful. Would he meet with her at the conference?

After arriving, Duesberg had an opportunity one morning to break away. He found McClintock in her office, and the two of them hit it off immediately. She told him stories about her own conflict with majority scientific opinion. In those days, she laughed, her observations on "jumping genes" were dismissed by her male colleagues. "Isn't it just like a woman," they would say, to propose such a silly idea?

She reminisced that science itself had become huge and thoughtless. Most researchers, she emphasized, prefer "knitting" together raw data rather than interpreting it. Thus, a "deluge of information" tends to swamp out genuine science. Such people are perfectly happy merely gathering data, and they uncritically accept "tacit assumptions" that force real thinkers to fight an uphill battle.

Turning to Duesberg's paper, she offered some minor points of advice but agreed wholeheartedly that the epidemiology of AIDS did not fit the pattern for a contagious disease. By the end of their two-hour conversation, she had wished him the best of success.

But her own energies were already failing. The following year, Duesberg saw her again at the same retrovirus conference. This time McClintock suffered from a weakened condition, a consequence of her advanced age. Walking with a crutch, she had little time except to say hello and mention that she was seeing a doctor. She would never have the opportunity to speak out publicly against HIV, for she passed away that fall.

James Watson himself took up an interest in the HIV debate at the 1992 retrovirus meeting. He had won the Nobel Prize in 1962 for discovering the structure of DNA, the genetic material. Always one to recognize promising trends in science, he had started the tumor-virus meetings at Cold Spring Harbor in the late 1960s, just as the War on Cancer was about to emerge.

Watson had the habit of making transient appearances at his conferences, greeting colleagues according to their unofficial social status. This time he spoke with Duesberg, and the two struck up a conversation about the HIV debate. On this subject Watson was short-tempered; he had previously told a reporter that Duesberg had no "convincing evidence" against the HIV hypothesis."[27] Now he confronted Duesberg with his skepticism: If AIDS is not infectious, why do hemophiliacs get it? Duesberg pointed out that hemophiliacs actually began living longer since roughly the time HIV infected three-quarters of them. Watson was startled. "If that's true, I'll call a special meeting here at Cold Spring Harbor," he declared. His curiosity aroused, he invited Duesberg for a private meeting at his office.

Again Watson demanded answers, still suspicious. "Where is your evidence? You say all these things without data." Duesberg objected, mentioning some of the evidence he had uncovered in the scientific literature. Watson then wanted to know why he had not published in the *Proceedings of the National Academy of Sciences*, the journal in which all members of the academy have an automatic right to publish. At that point Watson learned about a paper Duesberg had failed to get printed in the *Proceedings*, one that reviewed the evidence that drug use causes AIDS.

Genuinely shocked on learning this, Watson now wanted copies of all the correspondence between Duesberg and the journal's editor. "Send me everything," he insisted at the close of their half-hour meeting. He promised to look into the matter without delay, planning to exert whatever influence he could to stop this act of censorship. Although Duesberg sent the material, he has not heard back from Watson, and his paper remains unpublished.

Duesberg had a brief written exchange with another dissident of sorts, Manfred Eigen, one of Germany's most revered Nobel Laureates. Based at the prestigious Max Planck Institute, Eigen decided in 1989 to offer himself as an arbiter of the argument over HIV. In a paper entitled "The AIDS Debate," he reviewed existing evidence and formulated mathematical models in his analysis.[28] In the end, he chose a compromise solution. He offered that HIV did help to cause AIDS but needed some sort of cofactor to finish the job. Even this modest concession proved too much for the AIDS establishment. Allegedly, Eigen originally submitted his review to the prestigious journal *Nature*, from which it was rejected; certainly, Eigen's stature was too great normally to publish in the lesser-known German journal *Naturwissenschaften*, where his paper finally appeared. Duesberg's response was published a few months later in the same journal, taking issue with Eigen's attempt to save a role for HIV. Eigen countered with a series of rationalizations to explain away the puzzles of the HIV hypothesis. Duesberg ended his reply on a philosophical note:

> Eigen feels that in the absence of scientific proof for the hypothesis, "It is dangerous to state 'This ends the fear of infection'... because it may trigger wishful thinking."
>
> By contrast I will not accord the virus-AIDS hypothesis any more respect or concern than I would any other unproven hypothesis, as for example, the hypothesis that we are going to be invaded by the Martians and hence must build an interplanetary defense system. The burden of proof... is on those who propose a hypothesis, not on those who question it.[29]

Eigen has not publicly spoken further on the HIV debate.

Perhaps the most unexpected defection from the orthodox HIV establishment has been the discoverer of HIV himself, Luc Montagnier. He dropped his announcement in the midst of the Sixth International Conference on AIDS, the huge gathering of scientists and reporters in June of 1990. That year the meeting was held in San Francisco. To everyone's surprise, he used his allotted presentation to declare that HIV could not itself be enough to cause AIDS. The virus needed a cofactor, and he had already chosen a candidate—Shyh-Ching Lo's mycoplasma!

That evening, television broadcasts carried the news internationally. Headlines screamed the new hypothesis the following morning. "Almost all researchers working on AIDS said Montagnier was out on a limb," recalled *Science* a few months later.[30] Robert Gallo's reaction was particularly furious: "Since 1984 we've established enough evidence that there is a single cause for this disease. There is no evidence that anything else is needed."[31] Gallo's book, published the next year, bore down hard on Montagnier for breaking ranks. "This surprising view, which has been chiefly presented in press conferences [this from Gallo, who first announced his discovery of HIV at a press conference], has given, and may do so for a while, added longevity to confused and confusing (to others) arguments that HIV is not the primary cause of AIDS... In short, he has lent *some* support to Duesberg" [emphasis in original].[32]

A 1991 *Science* article mentioned one of the direct consequences of such unapproved behavior:

> But Montagnier has had difficulty getting his new work published. One paper, for example, was rejected last year by *Nature*.
> "I have high resistance from the virologists, and high enthusiasm from the mycoplasmologists," Montagnier says.[33]

The reasons behind his sudden shift, however, never made the news. The story actually began several months before the announcement, in the fall of 1989. A Canadian scientist had brokered an arrangement between Duesberg, Montagnier, and

Research in Immunology, a journal published by the Pasteur Institute in Paris. The journal would print a complete debate about the HIV hypothesis between the two scientists. The two sides would volley arguments back and forth by facsimile machine, stopping at a maximum of twenty-five hundred words each. Duesberg was chosen to submit the opening round.

In November, after several revisions, Duesberg launched his first installment. He had used fourteen hundred words, more than half his total, defining as many of the issues as possible. Summoning arguments from virology and epidemiology alike, he raised such points as the absence of active virus in AIDS patients, the long latent period, and the extreme bias of AIDS for males. Having laid out a rather overwhelming case, he ended the round with two very tough questions: "What proves that AIDS is infectious? If so, what proves that it is caused by HIV?" Then he, and the journal, waited for a response.

And waited. And waited. Attempts to contact Montagnier only received brush-offs, the French scientist constantly claiming to be preoccupied with other temporary matters. Finally, *Research in Immunology* decided to wait no longer. They published a slightly modified version of Duesberg's original installment in their January issue, with a written promise to publish Montagnier's answer at whatever future date he would submit it. But no such response ever arrived.

Instead, everyone found out what Montagnier had been up to by March, when he published a startling and obviously rushed paper in the Pasteur Institute's other journal, *Research in Virology*. This paper actually marked the first time he announced his cofactor hypothesis of AIDS, preceding the San Francisco AIDS Conference by three full months. He had miraculously discovered that cultured cells infected with HIV, which normally died in his laboratory, grew perfectly well when given the antibiotic tetracycline. HIV itself was unaffected by the treatment, so he inferred that some undetected bacterium had been killing the cells. In fact, he concluded the hidden microbe must have been a mycoplasma. He may well have been right, for mycoplasmas commonly contaminate cell cultures, cannot easily be seen, and are killed by tetracycline. Indeed, this sort of contamination is so common that no laboratory ever publishes such a trivial observation as a scientific paper.

The paper became Montagnier's opportunity to announce his cofactor hypothesis, a point he drove home in the last sentence of the paper: "Further experiments are presently being undertaken to isolate and identify the microorganism and to investigate its role in HIV-induced pathogenicity."34 For those who knew about his abortive debate with Duesberg, Montagnier indirectly gave away the reason for his sudden change of direction—Duesberg's arguments had apparently changed his mind about HIV. Articles and interviews covering Montagnier's June surprise at the AIDS conference quoted him repeating several of Duesberg's arguments, including the low levels of HIV in the bodies of AIDS patients, the latent period, the large number of infected people who never develop AIDS, even the inability of retroviruses to kill cells. But Montagnier never mentioned Duesberg's name.

More recently, Montagnier privately admitted to a colleague that he has tested hemophiliacs for several years, finding the same immune suppression in HIV-negative individuals as in their HIV-positive counterparts. But Montagnier has neither published nor officially announced this study.

The decision to back a cofactor hypothesis of course allows Montagnier to move easily toward or away from the HIV hypothesis at any time. Depending on the pressures exerted, it seems that he has indeed vacillated. In any case, Shyh-Ching Lo has enjoyed a revival of his fortunes now that Montagnier has chosen to work with him on his *Mycoplasma incognitus*. And AIDS officials have been forced to handle one more annoying dissident.

THE DISSIDENTS ORGANIZE

The ranks of HIV dissidents continued growing steadily. Inevitably, they united to present their common message, a move that took place in the spring of 1991. The man who organized this opposition, Charles Thomas, Jr., had all the right credentials. As a

244 ■ INVENTING THE AIDS VIRUS

professor of biochemistry at Harvard University, he had pioneered studies of how the body synthesizes proteins. But he found the years of NIH-financed science too intellectually restrictive. Thomas was motivated by his libertarian political values to leave the government-funded academic setting, opting to use his personal finances to conduct research. He moved to San Diego, California, and started the nonprofit Helicon Foundation, as well as his own small biotechnology company, Pantox.

Thomas read about Duesberg's *Cancer Research* article and decided to provide a focus to the growing ranks of dissidents by launching the newsletter *Rethinking AIDS* (renamed, since fall 1994, *Reappraising AIDS*). Thomas especially deplored the lack of controlled studies comparing HIV-infected people with those uninfected. Amidst writing a steady stream of letters to editors and prominent individuals, he drafted a statement that remained carefully neutral with respect to alternative hypotheses of AIDS, yet conveyed the skepticism of many scientists about HIV:

> It is widely believed by the general public that a retrovirus called HIV causes the group of diseases called AIDS. Many biomedical scientists now question this hypothesis. We propose that a thorough reappraisal of the existing evidence for and against this hypothesis be conducted by a suitable independent group. We further propose that critical epidemiological studies be devised and undertaken.

Thomas recruited scientists from all over the world to affix their names to the statement. Within weeks, he already had more than two dozen signatures of biomedical researchers with solid credentials garnered from the United States, Europe, and Australia, as well as a smattering of professionals in other fields. Most held academic positions. The membership, however, did reflect the political pressures inside science: Most had some form of protection from the virus-hunting establishment, whether because they worked in entirely unrelated fields, were near or past retirement, or, like Thomas, were self-employed.

By early June, Thomas had sent the statement as a letter to *Science*; the editor responded within days, reassuring him that "If we decide to publish it, we will be in touch with you before publication."[35] The statement fared no better at such prestigious journals as the *New England Journal of Medicine* and *Lancet*. The editor of *Nature* did call back, promising to print it, but nothing ever happened. In 1991 only *Christopher Street*, an independent homosexual-interest monthly run by the *New York Native*'s Charles Ortleb, was willing to print the letter. Realizing this would be a long-term fight, Thomas established a group around this statement, the Group for the Scientific Reappraisal of the HIV/AIDS Hypothesis. The Group had grown to some forty members by the end of 1991 and swelled to more than one hundred signatories after the 1992 International AIDS Conference, at which cases of AIDS without HIV infection were announced. By the beginning of 1995, more than four hundred people had joined, including scientists, physicians, nurses, lawyers, journalists, teachers, students, and nonprofessional observers.

In February 1995 Thomas's letter was finally published by *Science*.[36] The letter in *Science* was soon followed by another in the German-based international journal *AIDS-Forschung*.[37]

Of the dissidents so far discussed in this chapter, only John Lauritsen and Kary Mullis have joined the Group. But others who did sign on brought some rather impressive credentials. One of the best known for speaking out on the HIV debate, Robert Root-Bernstein, independently developed his suspicions about the virus shortly after Gallo's 1984 press conference, years before Duesberg published his *Cancer Research* paper. Barely out of graduate school with a degree in the history of science, Root-Bernstein was awarded the MacArthur Prize fellowship—a five-year "genius grant"—in 1981. This afforded him the opportunity to work alongside polio vaccine pioneer Jonas Salk, followed by a professorship at Michigan State University in physiology.

Inspired by Duesberg's outspoken challenge against HIV, Root-Bernstein eagerly added his own energies to the debate. He had always shown a rebellious streak in his science, the very reason for his MacArthur Prize. His 1989 book, *Discovering*, revolved around the theme that large, well-funded science tends to stifle genuine innovation. By early 1989 he had begun corresponding with Duesberg and other critics of the HIV hypothesis. Scouring the scientific literature, Root-Bernstein found hundreds of cases of AIDS-like diseases dating back throughout the twentieth century. These data he extracted into a letter published in the *Lancet* in April 1990, showing that Kaposi's sarcoma had not been as rare as supposed before the 1980s. The next month he fired off in rapid succession several more papers on the history of other AIDS diseases, all of which the same journal now rejected. Ultimately, he was forced to compile the remaining data into a paper published in a smaller French journal.

He also began documenting the explosive increases in immune-suppressive risk factors since the 1960s, including venereal and parasitic diseases, and drug abuse. This material, and a bevy of arguments against the HIV hypothesis, formed the basis of several more papers submitted to an amazing array of biomedical journals. His major 1990 paper "Do We Know the Cause(s) of AIDS?" clearly laid out the stakes: "It is worth taking a skeptical look at the HIV theory. We cannot afford—literally, in terms of human lives, research dollars, and manpower investment—to be wrong... the premature closure of inquiry lays us open to the risk of making a colossal blunder."[38] By 1993 he had written a book incorporating all of his extensive research, entitled *Rethinking AIDS*.[39] He was also a founding member of Charles Thomas's Group.

Nevertheless, peer pressure left its mark on Root-Bernstein. In a 1990 interview taped for a British television documentary, the following exchange took place:

Q: Do you think HIV causes AIDS?
A: I don't—absolutely not... I believe that HIV by itself cannot cause AIDS.[40]

But by the 1992 meeting of HIV dissidents in Amsterdam, he had signed Joseph Sonnabend's press release condemning Duesberg, a

Dissension in the Ranks ■ 247

move for which Root-Bernstein later seemed apologetic. His book, *Rethinking AIDS*, also contained a different tone than in the past:

> I believe that Duesberg is wrong in ignoring the role of HIV in AIDS... I posit that at the very least HIV... can have just as serious and potentially as deadly effects as cytomegalovirus, toxoplasmosis, or *Pneumocystis carinii* pneumonia.[41]

The book also makes a special acknowledgment of Sonnabend, whose multifactorial model of AIDS as a product of repeated venereal infections has begun shaping Root-Bernstein's own view. The book, however, delves further into his own developing hypothesis of AIDS, the autoimmune model. According to this idea, specific combinations of microbes, if they infect the body all at once, might trigger a chain reaction in which the immune system is fooled into attacking itself. Root-Bernstein includes HIV as one of the infections that might start the process.

The autoimmunity hypothesis, however, suffers several fatal flaws.[42] For one thing, autoimmune reactions have been poorly documented in any disease, not to mention AIDS. In fact, they may never occur in an otherwise healthy person. Moreover, the immune system works so well precisely because it has built-in (but poorly understood) safeguards that prevent it from attacking its own host body; the immune system's inherent function is to attack only foreign particles. For an invading microbe to induce a self-destructive immune response would be a contradiction in terms. Even if an autoimmune reaction could somehow take place, AIDS would have a latent period of days, not years. Further, the AIDS diseases against which the immune system provides no defense anyway—including the cancers, dementia, and wasting disease— cannot be explained by this model, or any other, that only accounts for destruction of the immune system. And if AIDS did result from autoimmunity, it would have spread out of its original risk groups into the general population years ago, rather than striking men nine times out of ten. Root-Bernstein himself admits these problems.

Soon after the appearance of his book, Root-Bernstein distanced himself from Thomas's Group, seeking the middle ground between the pro- and anti-HIV camps. In January 1995 he started a letter to *Science* by expressing his gratitude for being identified as a "Duesberg critic" and then proceeded to endorse Duesberg's drug-AIDS hypothesis, writing: "The fact that HIV is remaining within high-risk groups characterized by immunosuppressive risks (for example, disease, drugs, malnutrition, and blood products) argues in favor of performing such tests."43

Harry Rubin, the retrovirology pioneer, Lasker Prize recipient, and member of the National Academy of Sciences who trained Howard Temin and who has been a mentor and close friend of Duesberg since the 1960s, has spoken out against the HIV hypothesis since 1987. Rubin's instincts about retroviruses were shaped by his changing views of biology over the years; since the early 1970s he had drifted away from the field, precisely because simple agents such as viruses hardly seemed to contain the answers to complex problems such as cancer. He told the interviewer in a 1990 British television documentary:

> I don't think the cause of AIDS has been found. I think [in] a disease as complex as AIDS that there are likely to be multiple causes. In fact, to call it a single disease when there are so many multiple manifestations seems to me to be an oversimplification.44

Always cautious, Rubin nonetheless clearly stated, "I don't necessarily agree with everything that Peter [Duesberg] is saying. But I *do* support his questioning the simplistic idea that this very complex syndrome is caused by this one virus."45 Writing in Duesberg's defense, he sent letters to both *Science* and *Nature* in 1988, both of which were printed. Since that time, Rubin has not been able to have similar letters published. He also rallied to Duesberg's side at a 1988 "conference" sponsored by Mathilde Krim's

American Foundation for AIDS Research (AmFAR). The two Berkeley colleagues faced an ambush of hostile virus hunters and media reporters at the Washington, D.C., meeting, yet they boldly made their points. Rubin himself leans toward a multifactorial hypothesis, one that includes drug abuse as one of many potential health risk factors that could cause AIDS over time.

British epidemiologist Gordon Stewart, another founding member of the Group, has run into roadblocks against questioning the HIV hypothesis. Stewart also favors a multifactorial model of AIDS, but his argument with HIV focuses on the failure of AIDS to spread out of its original risk groups, an indication that no one microbe causes the syndrome.

After a struggle, he was able to place a letter in *Lancet* in 1989. But virtually all attempts to speak out thereafter failed, despite Stewart's predictions of the size of the AIDS epidemic continually proving far more accurate than the wildly exaggerated estimates of AIDS officials. *Lancet* itself rejected two more letters by Stewart. A paper sent to *Nature* in early 1990 took months of review before the editors rejected it. As Stewart's predictions began coming true, *Nature* went on to refuse publication three more times, an embargo that continues today. A paper submitted to the *British Medical Journal* met with instant rejection, though with the suggestion that they might print a shorter letter. Stewart complied, but his second attempt met with equal indifference. A comprehensive review of Stewart's AIDS models finally appeared in *Genetica* in 1995, a small but open-minded journal published since 1919 in Holland.

Harvey Bialy, the research editor of *Nature* subsidiary *Bio/Technology*, is a graduate of the University of California at Berkeley, an associate professor at the University of Miami, and another early member of the Group. Bialy's interest focused on Duesberg's arguments after the 1987 *Cancer Research* paper, and he invited Duesberg to publish an editorial in *Bio/Technology* late that year. When *Science* attacked Duesberg a few months later, Bialy wrote a forceful letter to the editor demanding fairer coverage. This led to a news article that revived interest in the controversy just when many virus hunters were hoping Duesberg would fade away. Duesberg then wrote a letter to the editor, but *Science* instead published a brief written debate between Duesberg and Blattner, Gallo, and Temin (see chapter 6). Bialy has sometimes opened the pages of his own journal to other AIDS dissidents and has given lectures critical of the HIV dogma. He explained his own view of the epidemic to the *Sunday Times of London*:

> The [HIV] hypothesis has become all things to all people. It violates everything we previously knew about virus disease, and allows any kind of therapy, any kind of research, to generate research bucks. What kind of science continues to place all its marbles, all its faith, all its research bucks, in such a theory? The answer I keep coming back to is that it has nothing to do with science; the reasons are all unscientific. We have taken sex and equated it with death, and into that mixture we have thrown money. What an ugly stew.[46]

Bialy has faced uphill battles, even at his own job, to keep dissent alive. In 1993 he invited Duesberg to write a standard-length paper for publication in *Bio/Technology*. Bialy was partly overruled, and the paper was cut down to a small fraction of its former length. When the paper finally appeared in August, it had been printed as a "last word" but with an unnecessary disclaimer that "The views expressed here are the author's own, and not necessarily those of *Bio/Technology*."[47] Even the column by editor Douglas McCormick expressed mixed feelings for publishing Duesberg's carefully documented paper, admitting that "we enter the fray reluctantly" because "we think that Duesberg is wrong in his conclusions" and because of Duesberg's debating style.[48] But McCormick deplored that Duesberg was denied his right of reply after a personalized challenge of "his drug hypothesis" by *Bio/Technology*'s sister journal, *Nature*.

Other top names have joined the Group, many criticizing the HIV hypothesis before Charles Thomas began organizing.

Beverly Griffin, director of the Virology Department at London's Royal Postgraduate Medical School, wrote a review in a 1989 issue of *Nature* arguing that "the burden of proof for HIV as a deadly pathogen" rests squarely on "those who maintain that HIV causes AIDS." She also unflinchingly brought up "the pressures of silence imposed by the establishment (including journalists and journals)."[49]

The editor of *American Laboratory*, Frederick Scott, seconded Duesberg's questions in an April 1989 editorial. There he proposed that nutritional deficiency might contribute to causing AIDS, particularly zinc deficiency. Citing the microbe-hunting mania that once controlled research and treatment of scurvy, beriberi, and pellagra, he argued that AIDS might prove to be a tragic parallel, another noncontagious syndrome falsely blamed on a microbe.

Kary Mullis, another former graduate student from Berkeley, achieved international fame for inventing the Polymerase Chain Reaction (PCR) a few years ago. This, ironically, is the sensitive detection technique used by AIDS officials to claim they can find HIV in almost every antibody-positive AIDS patient. Mullis refuses to buy this argument: "I can't find a single virologist who will give me references which show that HIV is the probable cause of AIDS... If you ask a virologist for that information, you don't get an answer, you get fury."[50] Asks Mullis, how could a dormant virus cause fatal AIDS? Biochemistry demands that every biochemical reaction is a consequence of an equivalent biochemical action. How could a virus that can be seen only after a billion-fold amplification be responsible for the fatal biochemical "reactions" that kill AIDS patients?[51]

But even Mullis's logic cannot penetrate orthodox AIDS-think. For example, take the response of a prominent AIDS researcher to Mullis's case against HIV. The incident was a television debate in New York on May 23, 1994, in which Duesberg used Mullis's arguments against HIV. The AIDS researcher's response was a rather unprofessional question, "Isn't he [Mullis] the surfer?" Obviously, in the mind of this mainstream scientist, surfing is not compatible with serious science. Indeed, Mullis is a Trojan horse to the AIDS establishment, adored for his invention of the only technique to detect at least a gene of the elusive AIDS virus, but feared for his outspoken criticism of the virus-AIDS hypothesis.

For his PCR invention, Mullis has won the 1993 Nobel Prize for Chemistry, making him the third Nobel Laureate to question the "AIDS virus" and the first to belong to the Group for the Scientific Reappraisal of the HIV/AIDS Hypothesis. Many scientific colleagues had not previously realized that Mullis questioned HIV's significance, and they now are becoming seriously unnerved by his comments. Although many journalists refuse even to mention his dissenting view, Mullis continues to hammer the AIDS establishment with his outspoken criticisms:

> Where is the research that says HIV is the cause of AIDS? We know everything in the world about HIV now. There are 10,000 people in the world now who specialize in HIV. None have any interest in the possibility HIV doesn't cause AIDS because if it doesn't, their expertise is useless.[52]

Australian medical professor Eleni Papadopulos-Eleopulos has independently questioned the HIV hypothesis since 1988. In June 1993 she and her colleagues from the University of Western Australia in Perth published an article in *Bio/Technology* that even shocked the HIV dissidents.[53] Their paper proved the HIV test thoroughly unreliable, producing up to 90 percent "false-positives" and relying on standards that differ between countries and even between official AIDS laboratories of the same country.[54] It outraged even those faithful to the HIV hypothesis that the fate of thousands of lives, every day, are determined by a test that cannot be trusted. The Papadopulos group has since become the most outspoken medical team to challenge the HIV hypothesis.[55]

Hundreds of other professionals have now lent their names to Thomas's statement, all agreeing on the need to re-open the HIV hypothesis for testing. Many of the scientists propose their own ideas of what causes AIDS. But by far the most compelling case can be made for the notion that long-term drug use is the culprit in most AIDS cases. The growing evidence for this hypothesis is the subject of the next chapter.

CHAPTER EIGHT

■

So What Is AIDS?

LOS ANGELES, CALIFORNIA, 1980: The man of thirty-three years being examined by Dr. Michael Gottlieb is deteriorating quickly. His fever refuses to go away, as do an active cytomegalovirus (CMV) infection in the blood and liver problems. Soon his immune system collapses to the point that native microbes, ones that have lived at peace with him for more than three decades, begin eating away at his body. *Pneumocystis carinii* and *Candida*, germs that normally reside in all humans and most mammals, now take over the patient; the former germ grows into a severe pneumonia, the latter establishes a thick yeast infection that begins choking his throat. By May 3 of the following year, the young artist has died, the autopsy revealing a CMV infection in his lung that was hidden by the *Pneumocystis* pneumonia.

This patient gained the dubious distinction of being the first officially recorded AIDS case in history, one of the five reported by the CDC in June 1981. Gottlieb had dutifully noted the man was an active homosexual who admitted using "poppers," the aphrodisiac nitrite inhalant so popular in the homosexual bathhouses and discos of major cities.[1]

Kenya, Africa, several years later: The hospital that the foreign woman enters is considered better than the few clinics in surrounding areas. She needs the best care medicine can provide.

Only thirty-nine years old, she has just arrived from Zaire, desperate to find treatment for her lung condition. It had begun with a relatively innocent cough and an unexpected drop in weight. Soon her coughs began bringing up blood. Tuberculosis is the diagnosis of the Kenyan doctor, but the patient has a strong allergic reaction to the drugs he prescribes. Her condition progresses from bad to worse, adding diarrhea, uncontrollable fever, swollen lymph nodes, and anemic blood disorders to her list of symptoms. But while the tuberculosis takes over, the *Pneumocystis* and *Candida* microbes also residing in her body remain perfectly hidden, causing no complications. She represents in every way the typical African AIDS patient.

The woman's husband, staying in the same hospital, suffers something entirely different and more unusual. Doctors assume he must have transmitted AIDS to his wife, though his diseases bear no similarity to hers. He has some sort of pneumonia, as well as a *Candida* yeast infection in his mouth and lesions of Kaposi's sarcoma, a blood vessel tumor, on his now-irregularly-pigmented skin. For African patients this tumor appears so rarely, it is almost totally unknown. He loses weight to a relentless diarrhea and is constantly fighting off episodes of gonorrhea. He knows he is on his deathbed.

Oddly enough, their children have no such medical troubles.[2]

According to the public health officials directing our war on AIDS, the male homosexual in Los Angeles and the Zairian couple all suffered the same disease. But did they? Each person was affected with radically different diseases—a *Pneumocystis* pneumonia, a tuberculosis, a Kaposi's sarcoma—conditions that in the past would never have been connected by medical doctors. The only common factor between these patients was the presence in each of antibodies against HIV. At least, that is the presumption; Gottlieb's first AIDS case was never actually tested, since the virus had not yet been discovered. And African AIDS patients are routinely diagnosed for AIDS without ever conducting an HIV test.[3] A glance at the statistics proves that AIDS is not one, but several, totally different epidemics and is thirty, in part totally different, diseases under one name (see Table 1). The global AIDS empire is held together only by its name and the hypothesis that it is caused by HIV.

So What Is AIDS? ■ 257

TABLE 1

AIDS Statistics*

Epidemics	American	European	African
AIDS total 1985–1991	206,000	66,000	129,000
AIDS annual since 1990	30–40,000	12–16,000	~20,000
HIV carriers since 1985	1 million	500,000	6 million
Annual AIDS per HIV carrier	3–4%	3%	about 0.3%
AIDS by sex	90% male	86% male	50% male
AIDS by age, over 20 years	98%	96%	?
AIDS by risk group:			
male homosexual	62%	48%	
intravenous drugs	32%	33%	
transfusions	2%	3%	
hemophiliacs	1%	3%	
general population	3%	13%	100%
AIDS by disease:			
Microbial	50% *Pneumocystis* pneumonia 17% candidiasis 8% mycobacterial disease 3% tuberculosis 5% toxoplasmosis 8% cytomegalovirus 4% herpesvirus	75% opportunistic infections	fever diarrhea tuberculosis slim disease
Microbial total	62% (sum > 62% due to overlap)	75%	about 90%
Nonmicrobial	19% wasting 10% Kaposi's 6% dementia 3% lymphoma	5% wasting 12% Kaposi's 5% dementia 3% lymphoma	
Nonmicrobial total	38%	25%	

*Status as of 1992; see P. H. Duesberg, "AIDS Acquired by Drug Consumption and Other Noncontagious Risk Factors," *Pharmacology and Therapeutics*, 55 (1992): 201–277.

Epidemics in different parts of the world are the same if the same diseases are observed in the same groups of people. For example, if lung cancer occurs in smokers in America and in Africa and if the sex ratio of these lung cancers reflects the sex ratio of smokers, the cancers of both continents are part of the same epidemic.

But the statistics reveal the unbridgeable gaps between the "AIDS" epidemics in America and in Africa. Since 1985, official estimates have placed the number of HIV-positive Americans at around one million, of which some 206,000 had developed AIDS by the end of 1991 and about 400,000 by the end of 1994.[4] Nine of every ten cases occur in men.[5] Most AIDS victims are older than twenty years and a few are infants, but virtually no teenagers have been affected. Male homosexuals make up 62 percent of American AIDS patients, intravenous drug users and their children another 32 percent, and hemophiliacs and other blood transfusion patients remain at 3 percent. The balance of 3 percent represents the CDC's "other categories," which is the normal low background of AIDS-defining diseases in the general population of America.[6]

While nearly two-thirds (62 percent) of American AIDS diseases do fit the popular image of opportunistic infections caused by microbes taking advantage of decimated immune systems, the remaining one-third (38 percent) do not (see Table 1). Kaposi's sarcoma, dementia, weight loss, wasting disease, and lymphoma can even strike people with healthy immune systems. These are nonmicrobial and noncontagious diseases whose causes are not dependent on the immune system (see also Table 1, chapter 6).

The African picture stands in sharp contrast. Also tested for HIV since 1985, six to eight million[7]—eight times as many Africans as Americans—are infected, yet the entire continent has produced fewer AIDS cases: 129,000 by 1992 and exactly 345,639 by December 1994.[8] Women overall are diagnosed as often as men (see Table 1).[9] No particular age group seems to be singled out by the syndrome, nor can risk groups be easily defined by sexual activity or identifiable health risks. Despite the universal presence of *Pneumocystis* and *Candida* microbes in Africans, as in all world populations, these germs do not dominate the

So What Is AIDS? ■ 259

African AIDS statistics as they do in the industrial world. Instead tuberculosis and the fevers and diarrheas associated with parasitic infections show up most commonly. Even their "slim disease" appears to be a different sort of wasting condition than is found in the United States or Europe (see Table 1). The African, but not the American, wasting diseases are associated with parasitic infections. And Kaposi's sarcoma, which now strikes 10 percent of American AIDS victims, appears in only 1 percent of African cases.[10] The pulmonary Kaposi's sarcoma, a lung cancer that manifests itself in a third of all American Kaposi patients, has never been diagnosed in Africans nor ever in the United States or Europe before AIDS.[11]

To find the cause of AIDS, therefore, one must define the health risks common to each separate group. Since this syndrome is not spreading outside of any AIDS risk group, the causes must be noninfectious; a contagious disease, by definition, spreads into the general population, as do all microbes. As witnessed in the past, noncontagious causes of disease can include medically prescribed drugs (as was the case with SMON), vitamin or other nutritional deficiencies (as with scurvy, pellagra, and beriberi), or long-term recreational drug use. For example, long-term use of tobacco causes lung cancer and emphysema, and long-term use of alcohol causes liver cirrhosis.

The infectious AIDS paradigm cannot explain why (1) in Africa AIDS is not new and is not infectious; (2) in the United States and Europe, most AIDS cases do reflect an independent increase in opportunistic infections, Kaposi's sarcoma, weight loss, and dementia, but one that has coincided tightly with the explosion in heavy drug use (this is why AIDS is restricted to risk groups, male homosexuals using sexual stimulants for years and intravenous drug users); and (3) hemophiliacs and blood transfusion recipients are not dying from HIV. Instead hemophiliacs suffer from immunosuppression caused by the long-term transfusion of blood products and from immunosuppressive treatments with anti-AIDS drugs like AZT. Transfusion recipients die from diseases that necessitated the transfusions regardless of the presence of HIV.

DRUG USE AND AIDS—THE SAME EPIDEMIC?

Virtually everyone's life has been directly impacted by the drug-use epidemic—the only new health risk of the Western world since World War II. Most people in the industrial world either have tried an illicit drug or know others who have. Just one or two generations ago, high schools spent their time trying to control cigarette smoking in the rest rooms; in those same rest rooms today, students can find a laundry list of recreational drugs for smoking, swallowing, snorting, or even injecting.

The 1960s gained the reputation as the decade of freely available drugs, especially marijuana and psychedelics. But in reality, the widespread escalation in drug use began largely during the Vietnam War, about a decade before the appearance of AIDS. Much of the explosion has taken place only in recent years. Overall drug arrests in the United States totaled approximately 450,000 in 1980, according to the Bureau of Justice Statistics, and the total was up to 1.4 million by 1989.[12]

Heroin-related arrests roughly tripled during the 1970s, corresponding with a jump in the number of heroin overdose victims. In the decade 1976 to 1985, the number of addicts admitted to hospitals doubled and then doubled again. During 1985, some 580 injection drug addicts died in hospitals, increasing to 2,483 such deaths by 1990. Between 1992 and 1993 heroin-related hospital emergencies increased 44 percent, based on data from the Department of Health and Human Services (see Figure 2B, chapter 6).[13]

The situation with cocaine looks even more grim. More than five million Americans had tried the drug by 1974, but eleven years later this figure had jumped to twenty-two million. Currently, about eight million Americans are regular users of cocaine.[14]

By the mid-1980s "crack," an addictive, smokable form of cocaine, evolved into an epidemic among poor young adults of minority groups.[15] A recent national household survey of drug use found that one million Americans between eighteen and twenty-five years of age had used crack during the previous year. Unlike injected cocaine, which is predominantly consumed by men, the use of crack is widespread among both men and women.[16]

The Drug Enforcement Administration confiscated about 500 kilograms of cocaine in 1980, 9,000 in 1983, 80,000 in 1989, and 100,000 in 1990—a total increase of 20,000 percent in one decade. During that same time the number of cocaine overdose victims admitted to hospitals also exploded, from slightly more than 3,000 in 1981 to more than 120,000 in 1993—a 4,000 percent jump. And direct cocaine-related deaths increased more than tenfold from 1980 to 1990 (see Figure 2B, chapter 6).

Law enforcement agencies seized some two million doses of amphetamines in 1981 but caught some ninety-seven million doses just eight years later in 1989.

Alkylnitrites, used primarily as aphrodisiacs, became popular during the 1970s, escalating into a "popper craze" in the 1980s.[17] The National Institute on Drug Abuse estimated that by 1980 some five million Americans were inhaling the drug at least once a week. By 1978 the once-tiny poppers industry was already grossing $50 million in annual profits, a figure that continued to climb.[18] Poppers manufacturers even became the largest source of advertising revenue to such homosexual magazines as the *Advocate*, which in turn ignored the efforts of some public health authorities and activists to warn homosexual men of the dangerous effects of poppers.[19]

Naturally, one might expect major health problems in the wake of this drug explosion. If the timing of the AIDS epidemic—following on the heels of the drug epidemic—was no coincidence, then one should also find the spread of AIDS following the spread of drug use.

Not only did the drug-use epidemic take off shortly before AIDS appeared, but it hit hardest among precisely the same risk groups. The parallels are astounding. Both AIDS and drug use, for example, are concentrated in younger men. Between 1983 and 1987 the death rate among American men ages twenty-five to forty-four increased by about ten thousand deaths per year, the same as the average number of AIDS deaths per year in that time period. But also during the 1980s deaths from drug overdoses doubled in men of exactly the same ages, while deaths from blood poisoning—an indirect consequence of injecting drugs—quadrupled. During that same period, AIDS deaths sharply increased among New York injection drug addicts, as did deaths from blood poisoning or other pneumonias—both at exactly the same rate.[20]

Ninety percent of all AIDS cases occur in men. But nine of every ten people arrested for possession of hard drugs are also male. Even the age distributions coincide perfectly. Men between the ages of twenty and forty-four make up 72 percent of AIDS cases, just as they make up 75 percent of people arrested or treated for use of hard drugs.[21]

What can be said of drug use in the AIDS risk groups?

The fact that injection drug users make up one-third of American AIDS cases, more than 130,000 by the end of 1993, should give pause for thought. Consider how that number breaks down. This figure includes three-quarters of all heterosexual AIDS cases and more than two-thirds of all female AIDS cases.[22] More than two-thirds of all babies with AIDS are born to mothers who inject drugs. Even 10 percent of the hemophiliac AIDS cases inject drugs. These statistics incorporate only self-reported drug injection, for they cannot confirm such illegal habits in people who will not admit to them. And more important, most drugs are inhaled or taken orally, not intravenously.[23] The CDC, however, does not ask AIDS patients about nonintravenous drug use. It is more concerned about possible HIV contamination on the injection equipment—hence the "clean needle" programs. But heroin or cocaine itself is most likely more dangerous than the dirty needles through which it is passed.

The remaining AIDS cases occur mostly among male homosexuals, the group that originally defined the epidemic. But the homosexuals who get AIDS form a special subset—sexually hyperactive and often promiscuous men, the so-called fast-track homosexuals. Their lifestyle emerged during the 1970s together with the new drug-use epidemic in the bathhouses, discotheques, and sex clubs. These men accumulated hundreds or even thousands of sexual contacts within just a few years. Venereal diseases and exotic parasites spread like wildfire. Infectious diseases ranging from the flu to hepatitis B became commonplace, and heavy doses of antibiotics were taken by many each night before sex, just to prevent unsightly sores or acne.[24]

Such extreme sexual activity cannot be done on a cup of coffee alone or even on natural testosterone. The fast-track lifestyle required liberal drug use—stimulants to get going, poppers to allow anal intercourse, downers to unwind afterward. Several drugs, combined with alcohol and marijuana, became par for the course of an evening, a routine that would go on for years. One homosexual man, a math professor in New York who has witnessed the fast-track scene, described the situation in a 1993 letter to Duesberg. The letter is a testimony to the high-risk lifestyle behind AIDS:

> From my experience in the New York City and Fire Island gay communities I can testify that more than a thousand (an ever increasing number) of my acquaintances have been diagnosed with HIV/AIDS over the past decade. Unfortunately some 250 (an estimate, it could be greater) of these are now prematurely dead...
>
> I have a list of my friends and acquaintances who died under the HIV/AIDS diagnosis. There are 150 names on the list... The remarkable thing about the people on this list and the hundreds of people living with an HIV diagnosis who presently come in and out of my life, sometimes daily, sometimes weekly, is that they almost all have a drug (recreational and medical) use and an alcohol use history of duration of often more than ten years...
>
> Most of the people on my list abused some, if not all, of the following drugs used recreationally: alcohol, amylnitrite, barbiturates, butylnitrite, cocaine, crack, ecstasy (XTC), heroin, librium, LSD, Mandrex, MDA, MDM, mescaline, methamphetamine, mushrooms, PCP, purple haze, Quaalude, Seconal, special K, THC, tuinol, and Valium.
>
> Most of the people on the list hosted many diseases and some of these diseases more than once. The following microbial diseases or microbes were common: *Candida* albicans,

chlamydia, cytomegalovirus, cryptosporidiosis, Epstein-Barr virus, gonorrhea, giardia, hepatitis A or B or C or D, herpes simplex (both 1 & 2), herpes zoster, gay bowel syndrome, scabies, venereal warts, and other parasites. In almost all of these cases the diseases were contracted before an HIV+ diagnosis.

I know that my acquaintances ingested large amounts of various antibiotics, antifungals, and antiparasitics. Some used antibiotics before going out for sex as prophylaxis against sexually transmitted diseases. These antibiotics were routinely given to them by gay doctors familiar with the fast-lane scene. Of course, after HIV diagnosis the overwhelming majority of these people used antibiotics, antifungals, antivirals (AZT, ddI, ddc, d4T, acyclovir, ganciclovir, etc.), as a matter of course, in various combinations over varying intervals of time...

At gay discos, both in New York City and on Fire Island, the use of recreational drugs is prevalent. The most common drugs are cocaine, ecstasy, poppers, and special K. On weekends on Fire Island drug dealers hawk their goods on the beach and on the walks as well as announce their hotel room numbers. Drug consumption among the fast-track gays is "de rigueur."

I emphasize that my remarks on drug usage are my observations or they were related directly to me by the individuals involved. They are not judgments...

As a result of these observations I am inclined towards the Duesberg drug-AIDS hypothesis.[25]

One Texas doctor, while studying AIDS risk factors among his patients in the early 1980s, discovered some of the dangerous practices in the bathhouses. "As an example, one of the drugs used was the readily available ethylene chloride," he wrote. "I was curious as to how this could be utilized for a 'high' until it was explained to me that a group formed a circle, saturated a towel, and then passed it from person to person for deep inhalation, which certainly seemed an excellent way to transmit disease to me."[26]

On the West Coast, AIDS activist William Bryan Coyle now battles the AIDS and HIV dogma. He has painted a similar picture of the fast-track life:

So What Is AIDS? ■ 265

These were the gayest of years! But the question is: Were they a bit too gay? In deciding to "party" and celebrate our newly obtained freedom, where would the limits be set? How much partying? How many cocktails? Would that be just on the weekends or seven days a week? These decisions were, of course, up to each individual; however, the tendency by most to want to be social and sexual would lead most gay men to either a gay bar, a gay dance club, or "the baths"... For many the choice became regular visits to the infamous bathhouses, whether it was the main plan for an evening or the finale to an assorted night of gay bars and discos. Whatever the case, it was usually routine to use one or more "mood-elevating substances" to enhance this social/sexual experience. Substances frequently chosen included cocaine, Quaaludes, amphetamines, LSD, MDA, amylnitrite, and, of course, marijuana and alcohol. The combined "recipe" for an evening might possibly involve four or five of these, and in this depressed state the sexual exposure to one or more person's germs would occur and an increased tendency toward indiscriminate additional promiscuity due to distorted judgment capabilities...

Some men would use poppers 30 to 40 times while dancing and then additionally at home or at the baths during their post-disco sexual liaison.

As the market grew and "bootleg" amylnitrite was now available in half-ounce screw-top glass bottles, it was not uncommon to be in a disco where someone had either accidentally or deliberately spilled a quantity on the dance floor, intoxicating everyone in reach...

So many of the poor souls deteriorating so rapidly with AIDS had gone from illegal/recreational drug abuse, directly into multiple daily prescribed drug abuse [such as AZT]... I, for one, will not be another statistic.[27]

Coyle credits ending his drug use, his objection to AZT, and paying careful attention to his diet to help control yeast infections for his gradually improving health. He has even found the energy to write his own book, currently in progress.

266 ■ INVENTING THE AIDS VIRUS

Larry Kramer, the volatile homosexual rights and AIDS activist who founded the AIDS Coalition To Unleash Power (ACT UP), has himself criticized the excesses of "fast-track living." A playwright and author by profession, he used his 1978 novel *Faggots* to lament the emptiness of anonymous homosexual activity. His book described the intense sexual promiscuity in the bathhouses, a lifestyle that could never be separated from the endless drug use on which it depended. Indeed, Kramer specifically listed many of the most popular drugs:

> MDA, MDM, THC, PCP, STP, DMT, LDK, WDW, Coke, Window Pane, Blotter, Orange Sunshine, Sweet Pea, Sky Blue, Christmas Tree, Mescaline, Dust, Benzedrine, Dexedrine, Dexamyl, Desoxyn, Strychnine, Ionamin, Ritalin, Desbutal, Opitol, Glue, Ethyl Chloride, Nitrous Oxide, Crystal Methedrine, Clogidal, Nesperan, Tytch, Nestex, Black Beauty, Certyn, Preludin with B-12, Zayl, Quaalude, Tuinal, Nembutal, Seconal, Amytal, Phenobarb, Elavil, Valium, Librium, Darvon, Mandrax, Opium, Stidyl, Halidax, Caldfyn, Optimil, Drayl.[28]

Years passed before AIDS forced the homosexual community as a whole to acknowledge Kramer's point.

Medical physicians and researchers have also described the drug problem rampant among many homosexuals. A surprising guest editorial appeared in a 1985 issue of the *Wall Street Journal*, cowritten by a journalist and a Washington, D.C., doctor, Cesar Caceres. The two authors cited official CDC AIDS statistics, as well as Caceres's own patients, to argue that drug use was so universal among AIDS patients that HIV could not be considered the syndrome's primary cause. AIDS patients, they protested, have "pre-existing immune damage" from years of drug use, without which AIDS cannot occur. In a direct challenge to the AIDS research establishment, they rhetorically asked, "Since drug abuse can severely damage the immune system, why has AIDS been identified primarily with sex, especially sex among homosexuals?"[29]

Joan McKenna, an AIDS therapist from Berkeley, California,

described similar drug use patterns among one hundred homosexual men in her medical practice: "We found... nearly universal use of marijuana; a multiple and complex use of LSD, MDA, PCP, heroin, cocaine, amyl and butyl nitrites, amphetamines, barbiturates, ethyl chloride, opium, mushrooms, and what are referred to as designer drugs."[30]

John Lauritsen and Hank Wilson noted that "Leaders of People With AIDS, who have known hundreds of PWA's, state that most of them were heavily into drugs, and all of them used poppers," and that the owner of a prominent homosexual sex club in New York candidly admitted, "I really don't know anybody who's had AIDS who hasn't used drugs."[31]

Large-scale studies of fast-track homosexual volunteers confirm these descriptions. An early CDC study, interviewing more than 400 homosexual men recruited from venereal disease clinics, counted 86 percent of them as using poppers frequently. Another study of 170 such men found that 96 percent admitted inhaling poppers regularly, while most had also used cocaine, amphetamines, lysergic acid, and methaqualone; many had also taken phenylcyclidine, ethylchloride, barbiturates, and heroin. A study of more than 350 homosexual men from San Francisco discovered that more than 80 percent used cocaine and poppers, with a majority simultaneously consuming other hard drugs. And a similar Boston study of more than 200 HIV-infected homosexual men revealed that 92 percent inhaled poppers and 75 percent used cocaine, in addition to the usual laundry list of drugs. Among male homosexual AIDS patients, more than 95 percent typically admitted to popper inhalation; by comparison, fewer than 1 percent of all heterosexuals or lesbians used poppers. In these and other studies (see chapter 5), HIV-positive men had always used more drugs than had uninfected men, and sexual activity was tightly linked to heavy drug use.[32]

In 1993, everyone in a group of 215 male homosexual AIDS patients from San Francisco reported the use of nitrite inhalants, in addition to cocaine and amphetamines. Moreover, 84 percent of these men were on AZT.[33] A parallel study from Vancouver showed in 1993 that virtually every male homosexual AIDS patient had used nitrites, cocaine, amphetamines, and AZT.34 Recreational drugs—including cocaine, amphetamines, and again AZT—were also the common denominator of all male homosexual AIDS patients from a group in Vancouver, Canada.35

Drugs have also brought babies into the AIDS epidemic. A small percentage of the total AIDS cases, infants tend to suffer from their own peculiar spectrum of AIDS symptoms such as bacterial infections and mental retardation. These symptons read like the profile of "crack babies" and is no coincidence. In his book *And the Band Played On*, Randy Shilts revealed which babies were getting AIDS. "Whatever the homosexuals had that was giving them Kaposi's sarcoma and *Pneumocystis*," he noted ominously, "it was also spreading among drug addicts and, most tragically, their children."36 Except that these young victims did not get Kaposi's sarcomas, lymphomas, or various other diseases common to homosexual AIDS cases. Two-thirds of these children have had mothers who inject drugs; some large percentage of the rest have mothers snorting cocaine or otherwise using noninjected drugs. But only a few studies have reported identical syndromes among babies of drug-using mothers, regardless of HIV infection.37 Even the scientific jargon of medical studies cannot hide the tragedies of unborn junkies:

1. At the University of California in San Francisco, the mental development and the coordination of eight HIV-infected and six uninfected infants were observed from six to twenty-one months of age. The mothers of each group were HIV-positive and had used intravenous drugs and alcohol during pregnancy.38 The degree of retardation of the infants correlated directly with maternal drug consumption: the more cocaine, morphine, and heroin their mothers had used during pregnancy, the more retarded and ill were their children.

2. Another medical school observed that the psychomotor indices—a measure of coordination—of infants "exposed to substance abuse in utero" were "significantly" lower than those of

non–drug-using mothers regardless of whether the mothers were HIV-positive or not. The researchers concluded that maternal drug use during pregnancy, not HIV, impairs children.39

3. Ten HIV-free infants born to intravenous drug-addicted mothers had the following AIDS-defining diseases: "failure to thrive, persistent generalized lymphadenopathy, persistent oral candidiasis, and developmental delay."40

4. One HIV-positive and eighteen HIV-free infants born to intravenous drug-addicted mothers had only half as many lymphocytes (white blood cells) at birth as normal controls. At twelve months after birth, the capacity of their white blood cells to proliferate was 50 percent to 70 percent lower than that of white blood cells from normal control infants.41

Yet the AIDS establishment and the news media have exploited these AIDS babies as proof the symptom is contagious, ignoring the drug connection in these unusual infants.

Injection drug addicts, male homosexuals, and the children of drug-injecting mothers constitute 94 percent of all AIDS patients. Thus, the correlation between heavy drug use and AIDS is far better than between HIV and AIDS. Drugs are biochemically active, and hence psychoactive, every time they are taken—the reason for their popularity. But HIV is inert and dormant in persons with and without AIDS.42 And although thousands of HIV-free AIDS cases have been described in the medical literature,43 possibly indicating hundreds of thousands more, no study has ever presented a group of AIDS patients genuinely free of drug use or other AIDS risks such as hemophilia.44

Taken together, these facts imply a central role for drug use in AIDS. But there are also experimental reasons to indict these drugs as causes of AIDS. Indeed, each of the major AIDS-risk drugs shows evidence of toxicity that could destroy the immune system or cause other AIDS diseases.

AIDS THROUGH CHEMISTRY

Medicine first pioneered the use of alkylnitrite compounds in the 1860s. Because the substances relaxed muscles and dilated blood vessels, they helped patients with heart diseases such as angina. These liquids were carried in tiny glass vials that would be broken open to inhale the powerful fumes and thus gained the nickname "poppers." Using only these tiny amounts, terminal heart patients never lived long enough to report dangerous health effects.

During the 1960s, male homosexuals discovered the aphrodisiac effects of nitrites. Receptive anal intercourse became less painful because the anal sphincter (muscle) would relax; therefore, receptive men used far more of the drug than did their insertive partners.[45] Nitrites also helped maintain erections and intensified orgasm, and some users even claimed a euphoric "high." The cost at first seemed little more than a brief rush and often a headache. The interest in poppers for sexual purposes soon turned into a stampede, the drug becoming a staple of bathhouse and discotheque life. Bottles of the drug could be purchased in sex shops under such brand names as "Rush," "Ram," "Thunderbolt," "Locker Room," "Climax," "Discorama," and "Crypt Tonight."[46] As described in one research paper, "Common settings in which these agents are used include the bedroom, parties, backrooms of pornographic bookstores, pornographic theaters, bars, and dance floors. Some users have told us that a few discotheques use special lighting effects to indicate that they are about to spray nitrite fumes over the dance floor."[47] According to Lauritsen and Wilson, "With regular use, they become a sexual crutch, and many gay men are incapable of having sex, even solitary masturbation, without the aid of poppers."[48] Nitrite manufacturers, however, managed to sidestep most federal controls by labeling the substance as a "room odorizer," and the "popper craze" took off during the 1970s.[49]

Few chemicals are more toxic than nitrites. Sodium nitrite, a much weaker, related compound used in tiny amounts as a preservative in meats, has been regulated for years as a potential cancer-causing agent. The alkylated nitrites (poppers), on the other hand, react more violently with almost anything. Upon mixing with water, as in the human body, these nitrites form the unstable nitrous acid, which in turn destroys any biological molecules within reach. The nitrites and their breakdown products have long been known to scientists for their ability to mutate DNA, a point recently verified by direct experiment.[50] In addition, nitrites are some of the most powerful cancer-causing chemicals in existence.

In contact with living cells, nitrite inhalants are cytotoxic (cell killing), which means they either poison or kill cells including, of course, the blood-forming cells and the epithelial lining of the lungs. Since these are among the fastest growing cells in the body, they will also be among the first cells to be in short supply if the sources are intoxicated. This is the reason that nitrites cause anemia, immunodeficiency, and pneumonia in experimental animals and humans.[51] In view of the toxicity of nitrite inhalants, a prescription requirement was instated by the FDA in 1969.[52] Moreover, the FDA limits nitrites as food preservatives to fewer than 200 ppm because of direct toxicity and because "they have been implicated in an increased incidence of cancer."[53]

By 1986 a statistical "AIDS link"[54] to nitrite inhalants had become so convincing to public health officials that the sale of nitrites was banned by the United States Congress in 1988 (Public Law 100–690)[55] and by the "Crime Control Act of 1990."[56] However, there is no report that nitrite bans are ever enforced or that nitrite warnings are taken seriously.[57]

On the contrary, the medical establishment turns a blind eye to drug toxicity in its single-minded pursuit of HIV with safe sex and clean needles.[58] For example, *Science* described nitrite inhalant-AIDS links as just another "hatched" theory in December 1994.[59] As a result, it comes as no surprise that nitrite use continues to remain popular and has even sharply increased recently, particularly among male homosexuals.[60]

The reactivity of nitrites easily compares with such toxins as carbon monoxide, the gas that suffocates its victims when a car engine is allowed to run in a closed garage. Carbon monoxide

272 ■ INVENTING THE AIDS VIRUS

destroys the hemoglobin in blood, preventing oxygen from reaching the body despite normal breathing. Nitrites do the same, a process that can be fatal if too much is inhaled at one time. At the height of the "popper craze," for example, a number of overdose victims arrived in hospital emergency rooms with as much as two-thirds of their hemoglobin chemically destroyed. Or to look at nitrites from another angle, a single dose can saturate the person using it with up to ten million nitrite molecules per cell in the body, leaving plenty of opportunity for damage.[61]

But the important question is whether inhaling the drug at sublethal doses for several years can eventually destroy the immune system or cause cancer. Recognizing the universal popularity of nitrites among homosexual men in 1981, the CDC was forced to consider this drug as one possible explanation of the emerging AIDS epidemic. However, the infection-minded CDC officials missed the point of the hypothesis by only searching for a single "bad batch" of poppers that might have temporarily caused a few sicknesses. It did not even occur to them that nitrites could be toxic by themselves. Therefore they searched for a contaminated or bad batch of nitrites. When that couldn't be found, they dismissed the hypothesis altogether. The CDC also assumed the effects would show immediately after using poppers, not after years of abuse, the way lung cancer and emphysema follow only after years of smoking tobacco. Naturally, no such contaminated batch could ever be found, and the CDC dismissed the hypothesis altogether and thereafter focused its search entirely on infectious agents.

Not all scientists dropped the idea so easily. Some continued testing the proposal that long-term exposure to all nitrites might cause AIDS, and they found some suspicious associations. Kaposi's sarcoma, the blood vessel tumor, grabbed some attention for its direct link to the poppers. This AIDS disease almost entirely affected homosexuals (as homosexuals were by far the major consumers of nitrite inhalants) and left heroin addicts, their babies, hemophiliacs, and other AIDS victims untouched. Often the Kaposi's tumor appeared on the face and upper torso and in the lungs of its victims, precisely where the nitrite fumes concentrated the heaviest during use.[62] (See chapter 11.) Before the age of poppers, nobody, not even Moritz Kaposi, had ever diagnosed a pulmonary Kaposi's sarcoma, which is a lung cancer.

Researchers also discovered that the risk of the tumor was directly proportional to an individual's total lifetime exposure to poppers, regardless of how many venereal or other contagious diseases the person had caught. Interestingly, they estimated that seven to ten years of exposure would, on average, produce AIDS—roughly the same as the supposed "latent period" of HIV.[63]

Time has borne out the nitrite hypothesis of Kaposi's sarcoma. Early public health warnings about the drug's potential effects convinced many homosexual men to stop inhaling it. By 1984 only 58 percent of homosexual men in San Francisco said they used the drug on a regular basis, dropping to less than half that number by 1991. In parallel, the incidence of Kaposi's sarcoma also steadily dropped as a proportion of AIDS cases, from half of all AIDS reports in 1981 to only 10 percent by 1991.[64] This has been the only AIDS disease to decrease this way, a change so shocking that the CDC itself briefly considered the possibility, in early 1991, that Kaposi's sarcoma might be a disease completely independent of AIDS and not caused by HIV. In the end, they retained this tumor on the list of AIDS diseases, correctly assuming few people would pay attention. Reports have now also emerged of young homosexual men with this tumor who have never been infected by HIV, but who do admit to having used poppers.[65]

Because the existence of HIV-free AIDS is the most direct threat to the HIV-AIDS hypothesis,[66] only a few such cases have been published in professional journals after the HIV hypothesis had become national dogma in 1984.[67] However, for offering an alternative AIDS virus some AIDS researchers have been allowed to report Kaposi's sarcoma cases free of HIV. For example, a CDC researcher was quoted in the *San Francisco Examiner* for the discovery of "20–30 men who have Kaposi's sarcoma but no HIV." Asked for comment, Professor Marcus Conant, one of the University of San Francisco's many AIDS specialists, admitted to the *San Francisco Examiner*: "At that point you have to say: Well maybe it could be something else—what could that something else be?"[68]

Indeed such concessions proved to set dangerous precedents for the HIV-AIDS hypothesis, because numerous AIDS researchers now felt free to follow the CDC example in reporting their HIV-free AIDS cases at the International AIDS Conference in Amsterdam in 1992. Realizing the imminent danger, the HIV establishment quickly found a new term for such cases, *idiopathic CD4-lymphocytopenia*, a term that even docile AIDS scientists and reporters had difficulty spelling.[69] And the HIV-free Kaposi cases from the CDC's Dr. Peterman never did appear in a professional journal.

In 1993 two health care workers, one working at the Public Health office in San Francisco, the other at Stanford University, told Duesberg, under the condition of anonymity, that they were directed not to report HIV-free AIDS cases as AIDS. Even though the respective patients were from AIDS risk groups and were clinically just like HIV-positive AIDS patients, their diseases were recorded by their old names, i.e., pneumonia, Kaposi's sarcoma, tuberculosis—rather than as AIDS.

In December 1994 the dam surrounding the HIV-AIDS hypothesis sprang another leak. Again a "new virus" was claimed to cause Kaposi's sarcoma in HIV-free male homosexuals—as the *New York Times* and *Science* reported simultaneously.[70] And all of a sudden it was okay that eleven out of twenty-one Kaposi's sarcoma patients who had the new virus were free of HIV, considered to be the sole cause of Kaposi's sarcoma in the previous ten years![71] Apparently the virus-AIDS establishment can accept another AIDS virus in a squeeze, but not a nonviral cause of AIDS. Virus-free AIDS would be a monumental embarrassment for the current AIDS establishment, with far-reaching consequences for prevention, treatment, and education. But another AIDS virus could be absorbed by AIDS educators and therapists with only minor adjustments.

The toxicity of nitrites to the cells of the lung and the immune system also explains the proclivity of male homosexual nitrite users for pneumonia, which is the most common AIDS disease in the United States and Europe[72] (Table 1). The added toxicity of cigarette smoke explains why, among otherwise matched HIV-positive male homosexuals, cigarette smokers develop pneumonia twice as often as nonsmokers.[73]

Intrigued by the poppers connection, researchers in several laboratories began independently testing long-term exposure in rats or mice to see if the drug could also cause immune deficiency. One CDC research team deliberately used a low dose and carried out the experiment for only a few weeks, finding some side effects but claiming no damage to the rodents' immune systems (see chapter 5). But several other labs used higher doses that resembled the heavy recreational use by homosexuals, and their experiments all showed clear destructive effects on the immune system, especially after a few months. In 1983, however, the CDC publicized only its own mouse study, claiming this as proof nitrites were really harmless. The other studies remained hidden in a monograph entitled *Health Hazards of Nitrite Inhalants*, which was published in 1988 by the National Institute on Drug Abuse.[74]

Even today, most AIDS scientists have not even heard of the nitrite research. In lectures or informal conversations they show curiosity or amazement that human beings would inhale such chemicals. In 1993 the editor of *Science* magazine, the most popular journal among researchers, privately expressed to Duesberg his astonishment at the pervasive use of nitrites among homosexual men with AIDS, a group he previously thought had only HIV as a risk factor. Even Gallo, the "nation's leading AIDS researcher," as he used to be called by the press, at a May 1994 meeting organized by the National Institute on Drug Abuse in Gaithersburg, Maryland, privately asked his old friend Duesberg, "Tell me, what are these guys using poppers for?"

Heroin is another AIDS-risk drug with a long history of serious health effects, though not as well studied as the nitrites. Some of this information even dates back to the time opium was smoked, rather than injected as heroin. Descriptions of health troubles in drug users date back as far as 1909, often following waves of addiction.75 Persistent drug users have shown loss of white blood cells, the cornerstone of the immune system, as well as lymph node swelling, fever, rapid weight loss, brain dysfunctions and dementia, and a marked vulnerability to infections, the classical consequence of immune deficiency.76 Addicts who inject heroin have classically died from pneumonias, tuberculosis, and other opportunistic infections, as well as from wasting syndromes, all precisely the same "AIDS diseases" they suffer today.77

Yet the medical orthodoxy disregards the abundant literature on the health hazards of heroin addiction, while defending the unproductive HIV-AIDS hypothesis. In an effort to blame HIV for the ills of heroin addicts, *Science* recently quoted a dedicated AIDS researcher as saying, "Heroin is a blessedly nontoxic drug."78 The AIDS establishment regards heroin and all other illicit psychoactive drugs simply as psychological catalysts of risk behavior rather than chemical health hazards on their own. Under the influence of psychoactive drugs, safe sex and safe recreational drug use with "clean needles" are abandoned in favor of "risk behavior," which is thought to risk infection by HIV, the cause of all evil.79

Cocaine consumption has escalated both in the numbers of consumers and in the dosage consumed. Once mostly inhaled or smoked, it is now often injected intravenously to achieve a higher concentration in the body. Long-term cocaine addicts often develop lung problems, weight loss, and fever, and have proven unusually susceptible to tuberculosis, an AIDS disease.80

In the United States the epidemics of cocaine-use–related diseases and AIDS diseases track so closely together that not even the experts can tell them apart. Narcotic toxicologist W. D. Lerner, of the University of Alabama, outlined many similarities between the bronchitis and pneumonias of cocaine addicts and AIDS patients in an article, "Cocaine Abuse and Acquired Immunodeficiency Syndrome: A Tale of Two Epidemics."81 Says Lerner, "Outwardly very dissimilar, they share a number of similarities." But Lerner regrets that "with few exceptions, the treatment of drug problems in the United States occurs, both figuratively and literally, a long distance from the major medical centers... the issue is not a lack of clinical cases but, more likely, a lack of scientific reporting of these cases." Lerner pointed out in 1989 that the United States federal government supported just sixty postgraduate research fellowships to investigate the diseases of the eight million American cocaine addicts, but nearly one fellowship for every one of the more than ten thousand AIDS patients recorded that year.[82] The National Commission on AIDS, appointed by the federal government, likewise documented the many overlaps between the drug and AIDS epidemics in *The Twin Epidemics of Substance Abuse and HIV*. The commission reported in 1991 that 32 percent of American AIDS patients are from groups that use intravenous drugs such as heroin and cocaine.[83] The commission concluded that AIDS was not spreading into the general population, a point recently echoed in a book entitled *Sex in America: A Definitive Survey*.[84] And the *Journal of Psychoactive Drugs* ran an article in 1993 conceding that it was not able to tell apart two "Entangled Epidemics: Cocaine Use and HIV Disease."[85] In 1994 even the *New England Journal of Medicine* acknowledged "Intersecting Epidemics—Crack Cocaine Use and HIV Infection."[86]

A new epidemic of tuberculosis has emerged among cocaine and crack addicts within the past few years.[87] In its press statements the CDC first assumed the outbreak resulted from the spread of HIV.[88] But upon testing these new tuberculosis cases, it found only a minority of them infected with the virus.[89] Backed into a corner, the CDC smoothly turned the tables by announcing that a new tuberculosis epidemic, parallel to AIDS, was now surfacing—and would soon threaten the general public! For decades, however, a significant percentage of the population has been infected by the tuberculosis bacterium, more than 90 percent of whom never become ill.[90] Populations in the industrial world no longer develop symptoms from tuberculosis because their immune systems are optimally tuned by their high standards of nutrition and health care. Cocaine users, however, particularly those who have become "homeless" after long-term addiction, seem to have a special inability to fight off disease, forming in fact a subculture of persons with immune systems defective from malnutrition and chronic intoxication.[91]

Although cocaine and heroin may be "blessedly nontoxic drugs" in the pages of *Science*, they are unforgivingly toxic for unborn children.[92] During pregnancy every minute counts for the developing fetus; there is no time to waste on drugs. A completely developed adult can afford to spend a few days in the gutter on his favorite recreational drug without lasting damage, but an unborn child cannot take off even a day from its extremely busy schedule of growing new cells and new organs every day.

It is for this reason that babies born to drug-addicted mothers may develop irreversible AIDS diseases in less than nine months—before they are even born.[93] In adults the grace period of cocaine and heroin addiction that leads to irreversible diseases is much longer, about five to ten years.[94] This is euphemistically called the "latent period of HIV" by the AIDS establishment.[95] The "innocent victims,"[96] born to cocaine- and heroin-addicted mothers, have severe mental retardation and other birth defects as well as bacterial diseases—regardless of whether or not they are infected by their mothers who have HIV.[97]

In fact, modern studies that look at heroin and cocaine addicts with tuberculosis, pneumonia, weight loss, oral thrush, chronic diarrhea, and other AIDS diseases typically find that half of them (or more) have never been infected by HIV, yet all are dying of the same conditions.[98] Sampled from places as diverse as New York and Baltimore, as well as France, Germany, Sweden, and Holland, injection drug abusers manifest pneumonia, tuberculosis, T-cell depletion, and premature death even without HIV.[99] Even the death rates of HIV-positive and -negative addicts are the same—they die at an average age of thirty years.[100] The only common denominator of the high morbidity and mortality of intravenous drug users has been the drug use itself, irrespective of HIV infection. A glance at some medical reports confirms this view exactly:

1. Among intravenous drug users in New York that represent a "spectrum of HIV-related diseases," HIV was observed in only

So What Is AIDS? ■ 279

twenty-two out of fifty pneumonia deaths, seven out of twenty-two endocarditis deaths, and eleven out of sixteen tuberculosis deaths.[101]

2. Pneumonia was diagnosed in six HIV-free and in fourteen HIV-positive intravenous drug users in New York.[102]

3. Among fifty-four prisoners with tuberculosis in New York state, forty-seven were street-drug users, but only twenty-four were infected with HIV.[103]

4. In a group of twenty-one long-term heroin addicts, T-cells declined during thirteen years from normal levels to the low levels typical of AIDS, but only two of the twenty-one were infected by HIV.[104]

5. Thrombocytopenia (a deficiency of blood clotting units) and immunodeficiency were diagnosed in fifteen intravenous drug users on average ten years after they became addicted, but two were not infected with HIV.[105]

6. The annual mortality of 108 HIV-free Swedish heroin addicts was just as high as that of 39 HIV-positive addicts, i.e., 3 percent to 5 percent, over several years.[106]

7. The reactivity and the concentration of lymphocytes were depressed as a direct function of the number of drug injections not only in 111 HIV-positive, but also in 210 HIV-free drug users from Holland.[107]

8. The same lymphadenopathy, weight loss, fever, night sweats, diarrhea, and mouth infections were observed in forty-nine HIV-free, and in eighty-nine HIV-positive, long-term intravenous drug users from New York.[108]

9. Among intravenous drug users in France, lymphadenopathy was observed in forty-one and AIDS-defining wasting disease in

fifteen HIV-positives. Exactly the same conditions were diagnosed in twelve and eight HIV-negatives, respectively.[109] The French addicts had used drugs for an average of five years.

10. Among intravenous drug users in New York with active tuberculosis (an AIDS-defining disease), nine were HIV-negative; among "crack" (cocaine) smokers with tuberculosis, three were HIV-negative[110]—excluding HIV as a cause of AIDS.

11. A note in the British medical journal *Lancet* documents five American intravenous drug users with AIDS-defining immunodeficiency but no HIV.[111]

12. The percentage of HIV-positives among thousands of intravenous drug addicts living in Germany was exactly the same as among drug deaths (10 percent to 30 percent, depending on the region that was sampled).[112] Thus, HIV does not contribute to the mortality of drug addicts.

13. Contrary to expectations, a European AIDS study found in 1995 that the median age at death of HIV-positive intravenous drug addicts was even a bit higher, i.e., thirty years, than that of HIV-free addicts, i.e., twenty-nine years.[113]

The steadfast belief in the HIV hypothesis by public health authorities has created problems in trying to control AIDS. The Swiss city of Zurich recently learned the hard way when city officials reserved a park in the center of the city, Am Platzspitz, as a free zone for heroin and cocaine addicts. Each drug abuser was provided sterile needles for injection on a daily basis in order to prevent the spread of HIV. Much to the surprise of government officials and the news media, the addicts have continued to develop their standard pneumonias and other diseases at the usual rate. If anything, the provision of sterile needles actually encouraged further drug abuse, thereby promoting AIDS. But public health officials, convinced they had done their job in fighting the epidemic, referred to the continued pneumonias as "AIDS-like diseases" so as to imply the drug addicts were now dying of something entirely different.[114]

Amphetamines are also becoming a popular recreational drug among AIDS risk groups, particularly homosexuals. Amphetamines are synthetic adrenaline derivatives that were originally given to German pilots and tank commanders to fight off fatigue and anxiety during World War II;[115] Hitler was addicted to them in his bunker in Berlin and had to use barbiturates to find sleep after each amphetamine "high." The drug has now been discovered by both heterosexuals and homosexuals as an agent on which to "cruise" through a night at discos without fatigue. Swallowed as pills in the past, the drug is making new rounds in a crystal form that can be smoked—"ice." Complete with all the addictive problems of crack cocaine, these amphetamines are causing a range of symptoms, from the loss of motor coordination found in Parkinson's disease to psychoses and sudden, radical weight loss. The latter qualifies as the wasting disease of AIDS.[116]

But even ice cannot begin to compare with the devastating effects of "crystal," the street name for methamphetamine. One of the cheapest and most powerful stimulants available, "it raises sexual cravings to new, superhuman levels" and is now becoming an uncontrollable epidemic in the homosexual community.[117] "Crystal is a gay person's drug and a gay community concern," states one official at a Los Angeles drug treatment facility. Many snort the drug in powder form, while others inject it intravenously or use it as an enema. Crystal drives its users to unparalleled heights of intense sexual excitement and frenzied behavior, coupled with periodic crashes of equal horror and the gradual development of psychoses. Overdose victims are beginning to show up. "We're just starting to see heavy usage types in our emergency rooms in New York City,"' says one medical worker, who also notes that "life expectancy for those intravenously injecting crystal is two years." What about those who last longer on the drug? According to the head of a French AIDS foundation, "There is ample evidence to suggest that crystal accelerates premature progression to full-blown AIDS in people dealing with HIV infection. Studies have shown that crystal eats T-cells for breakfast, lunch and dinner."[118] The crystal epidemic is so new that its impact on the AIDS epidemic is probably just beginning to be felt.

Science, however, has little working knowledge of the long-term effects of heroin, cocaine, amphetamines, nitrites, and other common recreational drugs. Ultimately, we can determine whether these drugs truly cause AIDS only by exposing animals, such as mice, to these drugs for several months at a time. Except for preliminary studies with nitrites, no such experiments have ever been done. AIDS research dollars have been plowed entirely into studying HIV, leaving the comparatively tiny field of drug toxicity with virtually no support at all. Most illegal drugs have been given to mice or rats in only a single dose by researchers, who are looking for the short-term effects. Until researchers can perform long-term experiments, the mechanism of how drugs cause AIDS will never be understood. Certainly the evidence above strongly implies drugs consumed at mind-altering concentrations could more easily cause AIDS than can a latent, biochemically inactive virus that is present in one out of one thousand T-cells.

Two more potential risk factors also need to be investigated. As increasing numbers of homosexuals entered the fast-track during the 1970s and 1980s, infections by viruses, bacteria, and other parasites skyrocketed. Antibiotics became the panacea; pop a few pills, and one could return to the bathhouses to risk another infection. "A typical medical history would include dozens of cases of VD [venereal disease] in the decade before the 'AIDS' diagnosis," writes John Lauritsen. "Each case of VD would be treated with stronger and stronger doses of antibiotics. Some doctors gave their gay patients open prescriptions for antibiotics, advising them to swallow a few before going to the baths. One popular bathhouse in New York (now closed) sold black market tetracycline on the second floor, along with all kinds of street drugs."[119]

Tetracycline certainly topped the list of favorite medical drugs, both for treatment and even to prevent new infections—often being taken before visits to the discotheques or sex clubs. Perhaps the only survey of this phenomenon, which "interviewed the patrons of a gay bar in Memphis, Tennessee," found that "over 40 percent of the men surveyed responded that they 'routinely' treated themselves with prescription antibiotics."[120] In some cases this would reach an extreme, as seen in certain of Joan McKenna's patients in Berkeley, California: "I have histories of gay men who have been on tetracycline for 18 years for the possibility of a pimple! I guarantee you their body chemistry isn't normal."[121] A less specific antibiotic than penicillin, tetracycline interferes with the body's normal metabolism. Doctors include with prescriptions a warning to stay out of the sunlight, for this antibiotic stops the skin from repairing sunburns. Used over the long term, it can also cause immune suppression.[122] The same holds true for steroids and erythromycin, also widely prescribed to treat or prevent venereal disease in homosexual men.[123] Possibly the worst side effects of antibiotics result from killing helpful bacteria, such as *E. coli*, that reside in the body. Many people using antibiotics for long periods find yeast or other fungal infections moving in to replace the dead bacteria.

The more toxic drugs came into play when used to treat diarrhea caused by intestinal parasites, such as amoebae. Homosexual men would receive such compounds as Flagyl and diiodohydroxyquin, the latter related to dioquinol, the drug that caused SMON.[124] And to prevent *Pneumocystis carinii* pneumonia, sulfa drugs such as bactrim and septra are now prescribed, which have serious side effects including nausea, vomiting, and diarrhea, as well as folate deficiency, which results in anemia. In AIDS patients thrombocytopenia (a shortage of platelets needed for blood clotting), rash, and hematologic toxicity are often observed.[125]

Malnutrition, another potential AIDS risk factor, also plagues the drug addict, who spends money on drugs rather than on a complete diet. Protein and zinc deficiencies have been described among many drug users, but the nature and importance of these dietary problems has never been researched. In general terms, malnourished people do face a high risk of immune deficiencies and pneumonias. Protein- and vitamin-deficient diets are found in much of the Third World and existed throughout Europe immediately following the havoc of World War II. Under such conditions, opportunistic infections do run rampant.

If recreational drug use and its associated risks have produced 94 percent of the American AIDS epidemic, how can we explain the remaining 6 percent? Half of these extra AIDS victims caught HIV through blood transfusions, a point that fuels the popular belief in AIDS as a contagious disease. But a closer look at these patients reveals some surprising facts, ones that confirm AIDS is neither infectious nor a single epidemic.

AIDS AND THE BLOOD SUPPLY

1. Emergency blood transfusions: Tax rebellion dominated the late 1970s, and Paul Gann epitomized that theme. Working with his long-time friend Howard Jarvis, a crusty California state legislator, he organized the citizens' crusades against rising tax burdens. The years of political agitation finally paid off in 1979, when California voters overwhelmingly passed Proposition 13 to limit property taxes. Excitement spread around the country, and dozens of states began following suit.

Gann had achieved folk hero status in the eyes of millions of taxpayers. Yet even his fighting spirit could not withstand the ravages of age. A worsening heart condition forced him into the hospital by 1982, when he was seventy years old. His heart disease was so bad that doctors made the decision to operate, creating five separate bypasses of the heart during the long operation. Large volumes of blood had to be transfused to make up for the losses. Gann slowly recovered enough to leave the hospital, but by the following year he returned with blocked intestinal arteries. Again, bypass surgery was required.

Gallo's patented test for finding antibodies against HIV became universally available by 1985, after which HIV-positive blood was screened out of the nation's supply. Several years later, ongoing complications and increasing political pressure to find AIDS in heterosexuals[126] prompted doctors to test Gann. As chance would have it, he was positive, probably having received the virus in one of his previous transfusions. The announcement devastated Gann psychologically, who believed he must inevitably die of AIDS. Dismay and anger fired up his old combativeness, and despite old age and ailing health he launched yet one more campaign, the last of his life. Hordes of loyal supporters gathered signatures, placing Proposition 102 on the 1988 California ballot. The measure called for stronger public health controls to prevent the spread of HIV. In a close vote that November, the proposal went down to defeat.

Gann himself fared little better. The next year he again wound up in the hospital, this time with a broken hip, and he was a poor candidate for recovery at seventy-seven years of age. He was immobilized for several weeks, and his condition steadily deteriorated. A severe pneumonia took over his lungs, refusing to disappear, until he finally died.

Media headlines blared the news that Gann had succumbed to the "deadly AIDS virus," reminding the public that the disease could strike anyone, homosexual or heterosexual alike, including the decidedly conservative Gann. Few news reports bothered describing his unhealthy condition, including cardiovascular disease. Nor did they remind readers that his seventy-seven years precisely equaled the average life expectancy of American men.

Gann's death typified the health situation for blood transfusion recipients. Amazingly, even health care workers rarely seem to know the survival statistics of such patients: Half of all blood recipients die within the first year after transfusion.[127] Naturally, this risk does not apply equally to all patients. The very old, the very young, and the most severely injured bear the brunt of death. Transfusions, after all, are not given to *normal*, healthy people. These patients have undergone traumatic medical problems and require the blood transfusions to stay alive after risky surgery for cancer, bypasses, or hip replacements. In the case of an organ transplant, the patient is given special drugs designed specifically to suppress the immune system and thereby reduce the possibility of organ rejection. And the blood itself is foreign material, overloading an already-stressed immune system in proportion to the amount transfused. Transfusion recipients die of many complications, not the least being opportunistic infections that prey on weakened immune systems, as for example Gann's pneumonia.

Among AIDS patients, those who caught HIV through blood transfusions do not suffer Kaposi's sarcoma, dementia, or several other major diseases found in the homosexual or injection drug–using cases. Instead they develop the pneumonias and other conditions typical of such patients as Paul Gann, with or without HIV. No evidence has shown that death rates from blood transfusions ever increased from HIV transmission, nor has anyone demonstrated that death rates declined again once the virus was screened out of the blood supply. One 1989 CDC study reported that among hundreds of transfusion patients, those with HIV died no more often than the uninfected during the first year—the official "latent period" between HIV infection and AIDS for such patients! In short, no new epidemic of disease has affected transfusion recipients in recent years, nor do their diseases belong under the same heading as AIDS in homosexual men or heroin addicts. In 1981 the CDC's KSOI Task Force searched frantically for transfusion recipients to reclassify as having AIDS (then known as "KSOI") only for the propaganda value, as a truly contagious disease would have spread through the blood supply. AIDS has not. But by redefining the standard diseases of transfusion patients as "AIDS," the CDC has left the specter of infection as an indelible impression on the public mind.

2. Long-term transfusion as prophylaxis of hemophilia: Lacking key components that allow blood to clot, hemophiliacs have long faced poor prognoses. Depending on the severity of the disorder, any damage could potentially cause unstoppable bleeding, externally or internally. Hemophiliacs in the past constantly needed blood transfusions, which only added to the problem, although the difference could hardly be noticed against the background of early death. As recently as 1972, hemophiliacs had a median life expectancy of only eleven years.

So What Is AIDS? ■ 287

Then an innovative product changed their lives permanently: Scientists invented a method of extracting from normal blood the proteins that hemophiliacs are missing. Known as *Factor VIII*, this blood component can be injected prophylactically on long-term schedules by hemophiliacs and restores most of the clotting ability they lack. Fewer hemorrhages are now occurring, and the median life expectancy has more than doubled, reaching twenty-seven years by 1987.[128]

The clotting factor brings a price tag, and not just in financial terms. Where hemophiliacs once died from internal bleeding, they now gradually develop immune deficiencies as they get older. Commercial Factor VIII itself seems to be part of the problem: With or without HIV infection, hemophiliacs lose immune competence according to the cumulative amount of Factor VIII consumed.

However, when the clotting factor is highly purified, the immune system remains healthy. Cost, unfortunately, bars many hemophiliacs from using the purified Factor VIII. Hemophiliacs treated with commercial Factor VIII consequently develop some opportunistic infectious diseases in the long run, particularly pneumonia and yeast infections. Those with HIV, who are counted as AIDS cases, get these same pneumonias, while they are unaffected by the Kaposi's sarcoma, lymphoma, wasting disease, and dementia that afflict homosexuals or heroin addicts who have AIDS. And as would be expected if these hemophiliac diseases were not caused by HIV, those with hemophilia-AIDS are on average at least ten years older than the rest—ten extra years of clotting factor and blood transfusions.[129]

Ryan White provides a case in point. The young Indiana teenager became a national symbol of heroic battling against AIDS after his school expelled him as a threat to the other students. His family's lawsuit eventually prevailed, and a court order forced the school to accept him back into the classroom. The ruling was based on the fact that HIV is difficult to transmit. The news media kept a periodic spotlight on White's life, and when he became sick and was hospitalized by 1990, the story splashed across the front pages as implicit proof the deadly virus could kill even the healthiest of people. White's death in April drew so much attention that entertainers Elizabeth Taylor and Michael Jackson attended his funeral. Although the news media portrayed the death as the tragic end to White's long fight with AIDS, the doctor never publicly confirmed that the death certificate actually attributed the cause of death to AIDS.

A phone call to the Indiana Hemophilia Foundation to check the details generated a very different story. A foundation representative directly familiar with White's case was asked of what specific AIDS diseases White died. Only internal bleeding and hemorrhaging, liver failure, and collapse of other physiological systems were listed. These conditions interestingly happen to match the classical description of hemophilia, none being listed as peculiar to the AIDS condition, but the representative did not seem to know that. It was then acknowledged that White's hemophilia condition was more severe than the average, requiring him to take clotting factor every day near the end. On top of all that, White had taken AZT, the former toxic cancer chemotherapy now prescribed as AIDS treatment. Hemophiliacs, needless to say, are particularly vulnerable to the internal ulcerations induced by such chemotherapy. Thus, only media hype transformed White's death from a severe case of hemophilia, exacerbated by AZT, into AIDS.

Those hemophiliacs whose diseases are reclassified as AIDS tend to have the severest clotting disorders in the first place, needing more Factor VIII and transfusions to stay alive. On the other hand, hemophiliacs have less to worry about than ever before. Of the twenty thousand hemophiliacs in the United States, some three-quarters were infected by HIV through the blood supply a little more than a decade ago. Yet during that same time period, clotting factor doubled their life expectancies, and very few are diagnosed with AIDS each year. HIV has made no measurable impact on the well-being of hemophiliacs, except for those who are treated with the highly toxic "anti-HIV" drug AZT.[130] (See chapter 9.)

AIDS IN THE THIRD WORLD

Public health officials never cease predicting the spread of AIDS out of narrow risk groups and into the general population. This line has become less believable with each year-end CDC report showing no such spread. So the public health experts resort to an old standby. For a picture of our future, they say, look to the Third World, where AIDS has already spread into the heterosexual population.

For instance, Thailand. The past few years have brought headlines and news stories on the impending doom in that poor country, where three hundred thousand people are infected with HIV. A disaster for Thailand, but bad for us as well. Several Thai cities host a flourishing sex industry, where men from Europe, the United States, and Japan meet to indulge themselves in abundant prostitution. This sex tourism supposedly could bring the AIDS epidemic back to our countries in force, finally triggering the long-awaited explosion out of the risk groups.

News photographers cannot publish enough pictures of Thai prostitutes, and no estimate of the potential danger is considered too large. In all the hustle, however, reporters forget to mention the grand total of AIDS cases in that country: as of 1991, only 123 individuals had been so diagnosed, rising to 1,569 cases by the middle of 1993. This amounts to only one-half of 1 percent of the 300,000 HIV-positives. Even more shocking, these Thai AIDS victims fall into very strict risk groups. Half of them are either male homosexuals or injection drug users. The other half hold down jobs as "sex workers," more commonly known as prostitutes, among whom drug use is hardly uncommon. Tuberculosis and pneumonia rank as the most common AIDS diseases in this handful of people. So much for an explosive Thai epidemic.

Africa, on the other hand, has been touted as a disaster already in progress, the ultimate example of what can happen in the industrial world if CDC guidelines are not heeded. In a continent with six million to eight million HIV-positives, whole villages are said to have disappeared while burdened economies are strained to the breaking point by massive death. Hospitals allegedly can no longer handle the AIDS load.

Careful inspection yields a different picture. For one thing, African population growth is higher than for any other continent—3 percent per year—a figure that belies the supposed devastation by AIDS. Since the AIDS epidemic began, the entire continent of Africa has reported only 345,000 cases by December 1994, fewer than in the United States. This reduces to an annual AIDS rate of about 0.5 percent of the HIV-positives developing the syndrome, compared to ten times that rate in the United States. Nor is this a product of extreme underreporting. The Ugandan AIDS surveillance system, considered internationally a model for the rest of Africa, provides similar numbers. Medical clinics seeing many HIV-positives commonly find very few AIDS cases. Another confirmation comes from Felix Konotey-Ahulu, a medical physician and scientist visiting London's Cromwell Hospital from Ghana. In early 1987 he toured dozens of cities throughout sub-Saharan Africa, trying to size up the AIDS epidemic. Upon returning, he wrote a scathing editorial for *Lancet*, criticizing news media coverage of the situation:

> If one judges the extent of the AIDS in Africa on an arbitrary scale from grade I (not much of a problem) to grade V (a catastrophe), in my assessment AIDS is a problem (grade II) in only five (possibly six, since I was unable to obtain a visa for Zaire) of the countries where AIDS has occurred...
>
> The phrase "possibly a considerable underestimate" has appeared many times in articles and broadcasts all over the world whenever a colossal figure is attached to the extent of AIDS in Africa...
>
> If tens of thousands are dying from AIDS (and Africans do not cremate their dead), where are the graves?[131]

Konotey-Ahulu made a point of visiting hospitals featured in the Western press as hotbeds of the AIDS epidemic, but consistently found very few AIDS cases. Nevertheless, many African doctors

So What Is AIDS? ■ 291

themselves participate in building the myth of the AIDS pandemic. *Spin* reporter Celia Farber discovered the reason during a recent trip to Africa:

> Many believe that the statistics have been inflated because AIDS generates far more money in the Third World from Western organizations than any other infectious disease. This was clear to us when we were there: Where there was "AIDS" there was money—a brand-new clinic, a new Mercedes parked outside, modern testing facilities, high-paying jobs, international conferences. A leading African physician... warned us not to get our hopes up about this trip. "You have no idea what you have taken on," he said on the eve of our departure. "You will never get these doctors to tell you the truth. When they get sent to these AIDS conferences around the world, the per diem they receive is equal to what they earn in a whole year at home."
>
> In Uganda, for example, WHO [World Health Organization] allotted $6 million for a single year, 1992–93, whereas all other infectious diseases combined—barring TB and AIDS—received a mere $57,000.[132]

To a large extent, the myth of an African AIDS epidemic grew out of a report in the late 1980s entitled *Voyage des Krynen en Tanzanie*. Written by French charity workers Philippe and Evelyne Krynen, it dramatically summarized their findings of devastated villages, abandoned homes, growing numbers of orphans, and a sexually transmitted AIDS epidemic that threatened to depopulate the Kagera province of northern Tanzania. As the heads of *Partage*, the largest AIDS charity for Tanzanian children, the Krynens told a story that the news media could not resist, one that is still repeated today. The vivid images helped shape the Western impression of an AIDS problem out of control.

But after spending a few years working with the people of the Kagera, the Krynens changed their minds. To their own disbelief, they discovered no AIDS epidemic in the region at all. The "sexually transmitted" disease somehow completely missed the prostitutes while it killed their clients; the exact same prostitutes work the towns today. Whatever caused AIDS in these clients did not affect these hardy prostitutes. Then the Krynens discovered that more than half their "AIDS" patients tested negative for HIV. The empty houses turned out to be additional homes owned by Tanzanians who had moved to the city. And the final blow came from the "orphans" themselves, who turned out to be the consequences of the Tanzanian social structure; the parents typically moved to the cities to earn money, leaving the grandparents to care for the children. "There is no AIDS," Philippe Krynen now states flatly. "It is something that has been invented. There are no epidemiological grounds for it; it doesn't exist for us."[133] He also describes how the epidemic is created for media consumption:

> Families just bring [children] as orphans, and if you ask how the parents died they will say AIDS. It is fashionable nowadays to say that, because it brings money and support.
> If you say your father has died in a car accident it is bad luck, but if he has died from AIDS there is an agency to help you. The local people have seen so many agencies coming, called AIDS support programs, that they want to join this group of victims. Everybody claims to be a victim of AIDS nowadays. And local people working for AIDS agencies have become rich. They have built homes in Dar es Salaam, they have their motorbikes; they have benefitted a lot...
> We have everybody coming here now, the World Bank, the churches, the Red Cross, the UN Development Programme, the African Medical Research Foundation, about 17 organizations reportedly doing something for AIDS in Kagera. It brings jobs, cars; the day there is no more AIDS, a lot of development is going to go away...
> You don't need AIDS patients to have an AIDS epidemic nowadays, because what is wrong doesn't need to be proved. Nobody checks; AIDS exists by itself.[134]

Exaggeration involves more than the numbers; the epidemic itself is manufactured. None of the African AIDS diseases is new.

So What Is AIDS? ■ 293

Many common Third World diseases are confused with AIDS even if they are not part of its official definition. The WHO definition for African AIDS includes "slim disease," a composite of weight loss, diarrhea, and fever, plus such conditions as persistent coughing, skin problems, swollen lymph nodes, and some opportunistic infections like tuberculosis. This list reads like a summary of indigenous African health problems. Malaria, the leading killer in the Third World, produces fever and other symptoms frequently misdiagnosed as AIDS. Tuberculosis, also a common killer, presents another problem, as described by a Nigerian medical professor: "The serologic demonstration of HIV infection in patients with tuberculosis in Africa is very important because it aids the separation of seropositive from the seronegative patients since such a separation may be impossible in all cases on clinical grounds."[135] According to a Ugandan doctor treating AIDS cases, "A patient who has TB and is HIV-positive would appear exactly the same as a patient who has TB and is HIV-negative. Clinically, both patients would present with prolonged fever; both patients would present with loss of weight, massive loss of weight, actually; both patients would present with a prolonged cough, and in both cases the cough would equally be productive. Now, therefore, clinically I cannot differentiate the two."[136]

Konotey-Ahulu has illustrated what a complete mess has been made by the AIDS definition:

> Immunosuppressive diseases, of course, there always have been in Africa and elsewhere before antiquity was born... I have clinical photographs from 1965 of a Ghanaian man who looked exactly like some of the AIDS patients I saw in Africa recently. The man who was like a skeleton (from gross weight loss) has severe nonbloody diarrhoea (more than twenty bowel actions a day); he had what looked like fungus in the mouth (candidiasis), skin changes (dermatopathy), periodic fever and cough—all the classical features of African AIDS... The patient (according to relatives) had literally

consumed on average one and a half bottles of whisky every single day for the previous 18 months before admission. We found it difficult to believe the story but there are photographs today showing a complete reversal in 1966 of the physical signs and symptoms, including the diabetes, when hospitalization cut short his alcohol supply and active treatment was administered, with gradual protein calorie buildup and pancreatin supplements.[137]

Konotey-Ahulu had also seen the effects of reclassifying traditional diseases under the AIDS umbrella. From his medical practice in Africa, he recalled that "[b]efore the days of AIDS in Ghana there was a death a day (more in the rainy and harmattan seasons) on my ward alone of thirty-four beds." Listing dozens of fatal diseases ranging from tuberculosis to various cancers, he remarked sarcastically, "Today, because of AIDS, it seems that Africans are not allowed to die from these conditions any longer."[138]

In a 1989 letter to *Lancet*, four Tanzanian doctors reported examples of another source of confusion—the misdiagnosis of diabetes as AIDS:

> Some of the reasons why diabetes may be confused with AIDS are illustrated in these case histories. Weight loss is often marked in newly presenting diabetic patients in Africa, fatigue may be a prominent feature, frequent visits to the toilet may be misinterpreted as indicating diarrhoea... Skin lesions, especially fungal infections, boils, and abscesses, are often present in newly presenting diabetic patients, and these could also mislead observers.
>
> In tropical Africa febrile illnesses are frequently attributed to malaria. Now in certain places AIDS is the fashionable diagnosis, made by the public and doctors. Many patients with treatable and curable illnesses may now be condemned without proper assessment. Public and medical education on AIDS should stress that symptoms such as those described are not unique to AIDS, and that even if a person presents with

clinical AIDS the possibility of coexisting problems such as diabetes should not be overlooked.[139]

So how can doctors tell the difference between AIDS and other conditions? Only by testing for antibodies against HIV! Thus, HIV has no connection with disease, and no new epidemic exists. Several large studies recently published findings that among thousands of randomly selected Africans with standard AIDS diseases, fewer than half were HIV-positive.[140]

As one nurse working in Tanzania put it, "If people die of malaria, it is called AIDS. If they die of herpes, it is called AIDS. I've even seen people die in accidents and it's been attributed to AIDS. The AIDS figures out of Africa are pure lies, pure estimate."[141]

Like everywhere else, AIDS in Africa seems to encompass at least two independent epidemics. Konotey-Ahulu and some other doctors insist that one major risk group is composed of urban prostitutes. As in Thailand, these women supply the goods for a "sex tourism" market. European and American men bring money to such countries as the Ivory Coast to purchase time with these "international prostitutes," who themselves travel from surrounding countries to compete for customers. The same modern jet travel that has made such trade possible has also brought another plague to African cities: recreational drugs. Authorities are becoming frustrated with the rising levels of cocaine and other substances being imported into the cities, creating all the attendant problems since the mid-1980s. Injection drug use remains uncommon, but cocaine and heroin are commonly smoked. The little evidence that emerges from Africa indicates that only those urban prostitutes sinking into the drug epidemic are developing AIDS.[142] Certainly this is true of Thailand.[143]

A completely separate epidemic seems to affect rural Africans, this one with no identified risk group at all. Some reports suggest a correlation between AIDS and malnutrition, which has long been known to cause such conditions. Doctors observe that AIDS patients who eat least often, or whose diets are skewed by food availability, suffer the most rapid decline in health. Other doctors attribute some of the sickness to "voodoo death" syndrome, the term for illnesses induced psychologically. According to one nurse, "We had people who were symptomatically AIDS patients. They were dying of AIDS, but when they were tested and found out they were negative they suddenly rebounded and are now perfectly healthy."[144] Needless to say, sanitation rarely exists in rural Africa, and clean water supplies are rare or nonexistent.

Whatever does cause early death among Third World populations, nothing appears to be new in Africa.

Both sorts of AIDS epidemics may have affected Haitians as well. The country hosts an active sex trade in the cities, while virtually all of the Haitians who arrived in the United States suffered some degree of malnutrition. Tuberculosis has topped the list of their AIDS diseases; Kaposi's sarcoma can hardly be found. Although Haitians still form a risk group for AIDS, the CDC has for years reclassified them under other AIDS risk categories, the reason they are no longer mentioned as a separate group in AIDS statistics.

The widespread belief in the HIV hypothesis has yielded tragic ironies. AIDS control programs in African nations, funded by outside governments, provide little but fear. Konotey-Ahulu's 1989 book reproduces a photograph of a Ugandan child, his filthy clothes ripped in tatters and his bony frame revealing the rampant hunger in his war-torn land, holding up the condoms given him by public health experts. With solutions like this, Africa's burdens are likely to continue crushing the little hope that remains.

Only one group of AIDS victims has not been explained thus far. Three percent of American AIDS patients fall under the CDC's "other" exposure category, having no identifiable risks for catching HIV. This hardly rules out hidden drug use, but some of these cases must result from the expansive definition of AIDS. Each year, people with fairly random backgrounds develop an occasional pneumonia, yeast infection, or hepatitis, any of which will be rediagnosed as AIDS if the person coincidentally also has antibodies against HIV.

To gain some perspective of what the "other" category really is, it helps to make a brief calculation: 3 percent of the 50,000 to

75,000 annual American AIDS patients translates into 1,500 to 2,250 people. They come from the reservoir of one million HIV-positive Americans. Since the average American dies at about 80 years, at least 1 out of 80 (1.25 percent), or 12,500, HIV-positive Americans die per year. The CDC claims about 1,500 to 2,250 of these annually for HIV; they are the "other" category even if they do not practice risk behavior. Viewed this way, the CDC claim seems modest considering that probably even more must die from the thirty common diseases that are now called *AIDS* in HIV-positive people.

Many HIV-positive people, whether they have symptoms or not, would normally not die of AIDS, but do so anyway. The reason lies in their treatment, AZT, one of the most toxic substances ever chosen for medical therapy. This drug is now creating a scandal that may soon explode as the most embarrassing in the history of medicine. The evidence that AZT actually causes AIDS, and the story behind its unethical approval, are told in the next chapter.

CHAPTER NINE

■

With Therapies Like This, Who Needs Disease?

CHERYL NAGEL'S DREAM was on the verge of becoming tangible reality in late October of 1990. She and her husband Steve had wanted a child of their own for a long time. Now her flight was arriving deep in the heart of Eastern Europe, Romania. Steve could not take the time off, so her mother accompanied Cheryl to the remote city of Timisoara. Cheryl felt out of place, having traveled so far from her suburban home just outside Minneapolis; back home Steve cooked for a restaurant, while she worked as a realtor's assistant. But when they heard the news of turbulence in that country and then of the orphanages full of desperate children, the Nagels knew where they had to go.

Arriving turned out to be the easy part. Touring the surrounding area, Cheryl and her mother soon met Lindsey, a baby girl only several days old. She had been born in the small coal-mining town of Petrosani, nestled deep in the Transylvanian Alps. She had been given up by her impoverished mother, who was already burdened by caring for three older daughters. But Cheryl found the adoption process absurdly difficult, considering the desperate economic and social condition of the country. Constant shortages of the littlest things, even light bulbs, could cause delays, and a phone call overseas took four hours to connect before she could hear the reassuring sound of her husband's voice. After two weeks of paperwork, bureaucratic stalling, and struggling with a strange language, Cheryl returned to the United States to recoup her energy. Returning to Romania within a few weeks, however, she overcame the final hurdles and retrieved the two-month-old baby.

Lindsey was a happy, healthy child, her slightly small size reflecting the norms for her original family. The Nagels took her for a complete checkup with a clinic near Minneapolis. The doctor's battery of tests included one for HIV. To everyone's astonishment, Lindsey was confirmed positive. Upon investigation, the Nagels discovered that Lindsey's birth mother did not have the virus. This left only one possible source—the blood transfusion Lindsey had received (despite having had nothing more than a brief ear infection) in Romania's backward medical system, where the method was carelessly used as a treatment for almost any illness. Lindsey still seemed a picture of health. However, the Nagels were now told she had the deadly "AIDS virus."

Then the nightmare began. Steve and Cheryl agreed to treat Lindsey prophylactically, that is, to delay the onset of symptoms as long as possible. They were referred to a specialist at the Children's Hospital in Minneapolis, where the doctor examined Lindsey and found no symptoms at all. No infections, no abnormalities, nothing. "She is [a] very bright, smiling and happy girl," noted the doctor, who nevertheless decided to head off a potential AIDS pneumonia immediately. Lindsey was prescribed Septra, to be taken three times each week.

The drug is known by dozens of brand names, including Bactrim. Septra is a sulfa drug, a remnant of the era before penicillin and the other antibiotics. Sulfa drugs do not target invading microbes as narrowly as the antibiotics and so have become notorious for their side effects. According to the *Physician's Desk Reference*, Septra can cause "nausea, vomiting, anorexia," and "bone marrow depression," and also includes "rash, fever, [and] leukopenia" among its side effects.[1] Even the drug's manufacturer, Burroughs Wellcome, strongly recommends against using Septra for more than two weeks, in children or adults. Young Lindsey, however, would take the drug for some nine months.

When the Nagels brought their daughter back a week later, the specialist announced that Lindsey's T-cell count was perfectly normal. Nor had any infections shown up. But given the HIV infection, the doctor wanted to slow the presumably inevitable appearance of AIDS. This time she prescribed the only drug approved for AIDS therapy—AZT (a chemotherapeutic drug designed to kill growing cells; see below). Lindsey began swallowing a total of 120 milligrams of the drug every single day, in addition to her Septra.

AZT stands for azidothymidine, a drug often marketed under the names Zidovudine or Retrovir. As with Septra, AZT is produced by the pharmaceutical giant Burroughs Wellcome. Both drugs have toxic effects. But compared to the sulfa drug, AZT amounts to poison. The doctor herself admitted some of the effects in her medical report of the visit, stating that Lindsey's "mother was explained the side effects of Zidovudine which are primarily bone marrow suppression with anemia, sometimes nausea and vomiting and rarely the cause of other symptoms like skin rash."[2] If anything, this understated the effects. AZT kills dividing cells anywhere in the body—causing ulcerations and hemorrhaging; damage to hair follicles and skin; killing mitochondria, the energy cells of the brain; wasting away of muscles; and the destruction of the immune system and other blood cells. Children are affected more severely, because many more of their cells are growing than in adults.[3] Amazingly, AZT was first approved for treatment of AIDS in 1987 and then for prevention of AIDS in 1990.

Totally unaware of the toxicity of this controversial drug, the Nagels faithfully fed their daughter the AZT syrup four times a day. At their next visit the following month, the doctor strangely began praising Lindsey's "improvement."[4] Upon reflection, the Nagels grew puzzled. What "improvement" could the doctor have meant, since Lindsey had suffered no medical problems at all before the treatment began? In fact, their daughter was already changing for the worse. Despite gaining slightly in weight, she was beginning to fall behind the proper growth rate for her five months of age. She was also losing her appetite, feeling too sick to drink her milk.

This process continued for months. Lindsey developed no infectious diseases, but her appetite continued to decline. The doctor acknowledged that the child was falling further behind the normal growth curve. By the time Lindsey reached her first birthday on October 15, 1991, her adoptive parents began to lose patience. The doctor seemed to believe the lack of growth had more to do with HIV infection than with any drug effects of Septra or AZT. As the Nagels began reading up on the officially recognized "side effects" of these drugs, their uneasiness turned to outright anxiety when they found perfect descriptions of their daughter's condition. Becoming suspicious of their doctor for not admitting or discussing these side effects, Steve and Cheryl took Lindsey to Dr. Margaret Hostetter at the University of Minnesota clinic. They felt overwhelmed and wanted clearer advice.

Dr. Hostetter possessed all the poise and confidence of her twin appointments as a university professor and director of the clinic's infectious disease program. Exuding an urbane but authoritative charm, she at first appeared much more professional and much friendlier than the other physicians. "Thank you for referring this lovely family to me," she wrote to Lindsey's first pediatrician after her November visit.[5] Running a complete battery of tests, she decided to take Lindsey off the Septra immediately. But the Nagels did notice that the doctor seemed to blame Lindsey's weight loss on HIV, rather than on drug side effects.

As soon as the Septra prescription ended, Lindsey began rebounding. Within one month her weight had again increased, though hardly back to normal, and her appetite recovered slightly. Amazingly, Dr. Hostetter completely missed the point. She had increased Lindsey's AZT dosage at the same time as ending the Septra, so at the Nagels's next visit she credited the baby girl's improvement to the AZT. In fact, she discussed plans to increase the AZT yet again. Even the experimental drug ddI, another powerful drug similar to AZT and just approved by the FDA, started cropping up among the doctor's suggestions.

After the rapid side effects of the sulfa drug had disappeared, the slower toxicity of the extra AZT began taking over. Lindsey stopped improving, and her weight, though still rising slowly, could no longer keep up with the normal growth rate for her age. She remained at the bottom end of the healthy weight range. By March she virtually stopped growing altogether. Her parents, fending off an increasing nervousness with each passing month, nevertheless kept up the daily syrup-feeding routine. The doctor praised Lindsey's nonexistent progress at each visit.

A few weeks later, the doctor had stretched the Nagels's patience by pressuring them to put Lindsey on ddI (a chemotherapy like AZT; see below). The young girl's T-cell levels were dropping, she said, and new drugs might help combat the deadly HIV. Investigating for themselves, the Nagels discovered that all children normally start with more than three thousand T-cells per microliter at birth, declining to about one thousand before adulthood.[6] Lindsey's counts were coming down near the standard, healthy rate. Naturally, the Nagels refused ddI therapy. But now they were reconsidering AZT as well.

The tension finally erupted a few days after Lindsey's second birthday on October 15, 1992. Steve and Cheryl woke up one night to the tormented screams of their daughter. Racing into her room, they found her sitting up and tearfully clutching her legs. The muscle pains were unbearable. Leg massages, Tylenol—they used anything that would allow Lindsey to sleep again. The same thing happened the next evening. And the next one. Night after night, the pain returned with ruthless consistency to deprive the entire family of sleep for a whole month. The Nagels recognized precisely what was happening to their daughter: Based on their own study, they had already learned that AZT produces muscle wasting as one of its "side effects."[7]

By chance they stumbled across an article discussing Peter Duesberg's dissent against AZT treatment for AIDS. Upon tracking down his phone number and calling, they received an earful about the drug's toxic effects. From there the Nagels talked with several other scientists dissenting against the HIV hypothesis. By early November the picture had become clear. The day they received a letter from Duesberg with scientific documents on AZT and on the shaky evidence for an AIDS virus, the Nagels stopped feeding their daughter the drug. Lindsey's changes took even her parents by surprise:

> After Lindsey was off AZT, she became a "new" child almost overnight.
> She started sleeping much better, including longer hours... Her muscle cramps went away.
> She started eating at least 2–3 times as much every day as she had ever eaten before.
> She would now drink milk, and especially around other youngsters, would drink as much as 6 ounces at a time. She would never drink milk before unless we added chocolate syrup, not a very nutritious drink...
> She displayed a much calmer demeanor. Lindsey was described almost as "hyperactive" by several people, including maternal grandparents who babysat a lot. This was a night and day difference! Lindsey, before, could not sit still for 5 minutes, and was seemingly agitated all the time...
> After seeing our nutritionist for only 2 months, and ridding Lindsey's body of toxic effects of being on AZT and Septra, Lindsey, now at 27 months, had an upswing on the chart. Her weight has been going up ever since. Now for the first time in 21 months, Lindsey is at 24 pounds, and is back on the chart at the 10th percentile.[8]

Dr. Hostetter knew nothing about Lindsey's being off AZT. The Nagels contacted the physician to demand an open discussion about the drug's merits at their next visit, to take place in early December of 1992. They were caught completely off guard by the doctor's reaction:

> Dr. Hostetter looked very tense... We were verbally attacked, as if we were 5 years old, and how dare we question

her opinion, let alone the use of AZT! She told us how lucky we were that Lindsey had tolerated AZT so well, and had not needed to go on ddI up until now. Then, Dr. Hostetter drew a large diagram on the black board, and told us (as she reminded us that she had told us all of this before) which cells AZT affects and which ones it definitely does not affect. If one of AZT's main side effects is bone marrow toxicity, how does a doctor know which cells the AZT will affect? (How does the AZT know?)... After our "lecture," Dr. Hostetter gave us her 20 minute sales pitch for AZT.[9]

The parents felt too intimidated by the meeting to let the doctor know they had ended the AZT treatment. In a letter written one week afterward to the Nagels's private physician, Hostetter noted that Lindsey had grown remarkably well during the previous two months, and warned that "we, unfortunately, might well see a return of Lindsey's previous failure to thrive were we to discontinue this drug."[10]

When the Nagels finally informed the doctor in writing and switched Lindsey to a chiropractor and nutritionist, Hostetter's mood turned downright ugly. Her response letter thundered a stream of dire warnings:

> As we have discussed repeatedly, AIDS is a fatal disease... To take Lindsey off Retrovir now will, I am afraid, hasten her decline and death.
> As parents, you are responsible for your child's health and life... Running away from qualified medical care will not help you, and it will certainly jeopardize Lindsey's life. You must take Lindsey to a qualified M.D. immediately.[11]

Hostetter followed up the letter with an angry call to the Nagels's chiropractor—on New Year's Eve. "She wanted to warn our chiropractor that she had no right to be seeing Lindsey," recalled Cheryl. "She also said that there are foster homes to provide care for children who were in Lindsey's predicament! (Living with parents who wouldn't give their daughter AZT.)"[12]

Hoping to stall Dr. Hostetter and to get a second opinion, the Nagels took their adopted daughter to another physician referred by Hostetter. But all he gave them was the same opinion. He recommended they restore Lindsey's treatment, and his nurse-practitioner called AZT a "wonder drug," a term even its manufacturer, Burroughs Wellcome, has never dared use.

Lindsey remains off AZT and all other toxic drugs. Her healthy growth pattern continues, she suffers no unusual diseases, and she is developing normally. Two years after suffering from AZT-induced leg cramps in 1994, she became a budding star in a local ballet school. And on October 15, 1995, Lindsey celebrated her fifth birthday—with HIV and without AZT—in excellent health. According to public health officials, she should already have died of AIDS because babies with HIV are supposed to survive only about two years.

Not everyone is so fortunate. In 1987 three years before Lindsey was born in faraway Romania, Doug and Nancy Simon brought their daughter Candice into the world, in a town south of Minneapolis. Their daughter certainly seemed healthy enough, but by the time she reached one and a half years of age, the doctor discovered she had antibodies against HIV. Investigation traced the infection to her mother, who had contracted the virus from her husband. He, in turn, had contracted it from a blood transfusion several years earlier. None of them suffered from AIDS.

The Simons took Candice to the Minneapolis Children's Hospital, the same one where Lindsey Nagel would be given Septra and AZT a couple of years later. Candice, too, became a victim of AIDS medicine. Doctors there prescribed interferon, a powerful anti-metabolic drug that shuts down cell function.[13] They later added AZT to her regimen, a treatment that, unlike in Lindsey's case, would last three and a half years. The constant testing added to the parents' sense of being overwhelmed: X-rays, blood samples, brain scans. For a while, Candice appeared to handle the therapy without too many problems.

Then her condition took a nosedive. Her appetite declined to dangerously low levels. The hospital became almost a second home, and by late 1992 she could no longer leave her bed. A new symptom, hauntingly reminiscent of Lindsey's AZT poisoning, took effect: "When the pain hit she would double over in her bed like a safety pin and, wild-eyed, grab her ankles until it eased."[14] Soon the doctors found malignant cancer spreading tumors throughout her stomach area. For the pain they prescribed morphine, then surgically cut the nerves to her intestines. Candice could no longer eat on her own, and the doctors began feeding her directly to her blood through intravenous tubes. Though five years old, she had lost control of her intestines and had to wear diapers.

In June of 1993, only three days before she turned six years old, Candice died painfully. Nearby, Lindsey Nagel had already stopped AZT seven months earlier and was recovering her health spectacularly. But Candice continued the drug right up to the end. Now both her parents take AZT as well.[15]

The Nagels know of the Simons's situation and consider themselves lucky for not having followed through on their daughter's AZT treatment.

THE DEATH AND RESURRECTION OF AZT

The virus hunters have always aspired to the glories of their predecessors, the bacteria hunters. Medicine still takes credit for eliminating bacterial diseases with antibiotics such as penicillin. These drugs attacked their bacterial targets with tremendous specificity, meaning they did little direct damage to the host's body. Antibiotics became known as the "magic bullet" for bacterial infections. Fire them into the body, and kill only invading bacteria.

But for viruses the problem was different. These are nonliving microbes, made of proteins, DNA or RNA, and sometimes even a tiny membrane—molecules all made entirely by human cells inside a human body. How could any drug possibly discriminate between the production of proteins and DNA made for viruses and those made for their human hosts? Despite never-ending searches for "magic bullets" against viruses, the efforts have produced little but failure. In principle, an antiviral drug may never be possible. The only solution offered has been vaccination, which prevents viruses from entering cells.

The 1975 Nobel Prize for Medicine, awarded for Howard Temin's discovery of the protein "reverse transcriptase" (see chapter 4), popularized this unique retrovirus enzyme. Many virus hunters switched into chasing retroviruses, and the reverse transcriptase protein took on mythic proportions. It did, after all, copy the virus' genetic information from RNA molecules "backward" into DNA, this new copy integrating somewhere into the genetic DNA structure of the infected cell. Normally, the cell keeps its genetic material in DNA, copying selected genes into RNA as needed. This "reverse" feature of the retrovirus protein inspired virus hunters to make it their key target for "magic bullet" drugs. At least in diseases caused by retroviruses, they speculated, some effective drug could be found. Once AIDS was blamed on HIV, a retrovirus, the race was on to find a drug that could inhibit the viral reverse transcriptase.

Drug development since World War II had also been heavily shaped by cancer research. Cancer, too, fueled ambitions among doctors to find "magic bullets" that could destroy the cancer tissue without killing the host. First came surgery, the attempt simply to cut out the tumor; this method has serious limitations. Radiation also became popular, based on the hope that tumors could be burned away by X-rays or other high-energy beams before destroying the body, but radiation therapy has mostly proved disappointing. Chemotherapy, using powerful cell-killing drugs, came into vogue during the 1950s. Starting in World War I, researchers observed the destruction of blood cells by mustard gas, the chemical warfare agent used to hideous effect in the trenches of Europe's battlefields. A few attempts to use this drug against cancer turned up with minimal results, largely because mustard gas was so toxic to the patient.

Shortly after James Shannon took over the NIH in 1955, he instituted several major research programs to attract vast new budgets from Congress. The largest of these became the Virus-Cancer Program, which ultimately converted itself into the war on

With Therapies Like This, Who Needs Disease? ■ 309

AIDS. The second largest project aimed to develop chemotherapy agents to treat cancer. The 1950s and 1960s therefore saw a proliferation of drugs designed to kill growing cells. At first, the goal seemed straightforward: Since cancer is made of persistently dividing cells, find a drug that prefers to kill cells that grow. The biggest problem with this concept lay in the body's own tissues, many of which replenish themselves constantly with rapidly growing cells. Therefore cancer patients undergoing chemotherapy experience devastating side effects, including hair loss; muscle wasting; severe weight loss due to intoxication of the intestines and benign intestinal microbes; anemia and the need for blood transfusions; and destruction of the immune system, composed mostly of white blood cells. Decades before the appearance of AIDS, chemotherapy patients *often* died of the same *Pneumocystis carinii* pneumonia that later killed young homosexual men.[16]

AZT was invented under this program in 1964. Jerome Horwitz, heading a lab at the Detroit Cancer Foundation and financed with an NIH grant, created a chemically modified form of a DNA building block. Every time a cell divides, it must copy its complete genetic code, allowing one copy for each new cell. Genetic information is stored as a sequence of four "letters" in long chains of DNA, known as *chromosomes*. Each building block of DNA is linked to the one before it, almost like train cars. But Horwitz's altered DNA building block, azidothymidine (AZT), surreptitiously enters the growing DNA chain while a cell is preparing to divide, and acts as a premature "caboose," blocking further DNA building blocks from being added (see Figure 1). In short, the cell cannot copy its DNA sequence and dies trying. AZT was the perfect killer of dividing cells. However, when he tested the compound on cancer-ridden mice, it failed to cure the cancer.[17] Horwitz was so disappointed he never bothered publishing the experiment and eventually abandoned that line of research. The drug must have killed the tumors, which contain dividing cells, but it so effectively destroyed healthy growing tissues that the mice died of the extreme toxicity.[18] The drug was shelved, and no patent was ever filed.

FIGURE 1
Human DNA Sequence

Human DNA is a string of 10^9 A, T, C, and Gs linked in a specific sequence.

a) normal DNA synthesis

chain continues

b) DNA synthesis with the T-analog, AZT

chain terminates

Each of the four building blocks (nucleotides) of human DNA has two links. But AZT, an analog of the nucleotide T, has only one link. Therefore it stops DNA synthesis and kills the cell.

Twenty years later, Gallo's 1984 press conference announcing HIV as the "AIDS virus" set in motion a new hunt, this time for a "magic bullet" drug to act against the virus. The federal government had promised treatment, and it had to deliver. Some virus hunters, including Jonas Salk, scurried to invent an HIV vaccine. Others searched for an antiviral drug and turned to the cancer chemotherapy program for already-developed chemicals. The fastest way to put a drug to market would be to select one that had finished some testing in the past.

Burroughs Wellcome became the pharmaceutical company positioned at the right place and the right time. One of the giants in the industry, the British-based company maintains a relatively unusual corporate structure as a mostly nonprofit, charitable institution. Most of its profits are paid, not to stockholders, but as grants and donations to biomedical research institutions. By throwing so much money around, Burroughs Wellcome has bought enormous influence throughout government and universities, especially through its American branch. A large number of scientists and physicians had developed informal ties to the company, having been paid to test its pharmaceuticals many times over the years.

The company's head researcher in the United States, David Barry, recognized the opportunity after Gallo's press conference. Barry knew his way around the federal bureaucracy in getting a drug approved. He had originally worked at the FDA during the 1970s as a virologist. His research had focused on the flu viruses, and he occasionally dabbled in retroviruses after they became popular. Upon switching to Burroughs Wellcome, he paid more attention to herpes viruses, also a hot research item. He brought to his new job a vast network of connections with fellow virus hunters and top FDA people.[19] Upon hearing the official call for anti-HIV drugs, Barry turned to the company shelves for previously rejected substances. If one of these could be approved, the company would save vast sums of research and development money. The political pressure for a quick solution played in his favor.

The key lay in winning FDA approval, which counted for more than mere permission to sell. The agency bans most potential drugs, automatically suppressing the competition and granting treatment monopolies for approved drugs. This monopoly alone can be worth hundreds of millions of dollars to the pharmaceutical company holding the patent. Back in the days when snake oil could freely be sold as a nostrum, drugs would sell only according to the reputation of the producer and their effectiveness against disease. Now the public depends on, and trusts, FDA screening procedures.

Barry selected a handful of drugs and quietly forwarded them to a couple of Burroughs Wellcome's former collaborators. One of them was Dani Bolognesi, a veteran retrovirus hunter and professor at North Carolina's Duke University, who not only knew Barry but also was so close to Gallo he belonged to the "Bob Club." Bolognesi tested the substances in his laboratory, checking whether they would prevent HIV from multiplying while infecting cells in the test tube. One of the drugs clearly proved most potent against the virus— *compound S*, as it was code-named. Its real name was *AZT*.

Bolognesi then referred Barry to Sam Broder, the man in charge of Gallo's laboratory at the National Cancer Institute. Broder had joined the NIH in the early 1970s, just as Gallo's star was beginning to rise. Broder made his career testing and developing cancer chemotherapy, but he also allied himself to Gallo and thereby practiced a bit of virus hunting himself, soon becoming a full member of the "Bob Club." Politically savvy, he could see by the early 1980s that the time had come to switch his emphasis from cancer to AIDS and immediately after Gallo's press conference he mobilized NIH researchers to find a drug. According to Bruce Nussbaum, "The hallmark of Broder's operation was... simple: Find a drug that had been tested for a previous disease. Make sure it had a big corporate sugar daddy behind it. Push the bureaucracy like hell to move it along. And talk it up. Talk it up."[20]

Broder's tenacity made him a perfect advocate for AZT; Barry realized that Broder, if properly recruited, would aggressively push through the bureaucracy to get AZT approved. So Barry sent compound S to Broder late in 1984, who discovered its powerful effect on HIV and waxed enthusiastic. Broder was so completely hooked, he soon became known as "Mr. AZT."

Barry, Broder, and Bolognesi together published their laboratory experiments on AZT. They reported that only a tiny concentration was needed to block the virus from multiplying. Of course, this would mean nothing if the same dose of AZT would also kill the T-cells in which the virus grew, in which case it would destroy the immune system before the virus supposedly could. Further tests gave an answer that sounded too good to be true: At least *one thousand times* as much AZT was needed to kill the T-cells as to stop the virus.[21] This theoretically meant doctors could use small doses of the drug to stop HIV without seriously damaging their patients' immune systems. No one bothered to check this fantastic result. The Burroughs Wellcome and NIH researchers somehow had to explain their success, and they billed AZT as a compound that specifically attacked reverse transcriptase, the retrovirus enzyme. In other words, they quickly declared, they had finally found a "magic bullet."[22]

AZT, however, did not really attack reverse transcriptase

directly. It only did what it had been designed to do originally—stop the synthesis of DNA. Since reverse transcriptase copies retroviral genes into DNA, the drug certainly interfered with its normal function. But the infected T-cell, meanwhile, produces its own DNA. Every time the cell divides, it must copy one hundred thousand times more DNA than the small virus, giving AZT one hundred thousand chances to kill the cell for every opportunity to block the virus. Since retroviruses can make viral DNA only in cells making their own DNA, the drug could not possibly attack the virus without also killing the cell, casting suspicion on the Bolognesi-Broder experiments. Recent studies conducted by smaller laboratories have tested AZT on other samples of T-cells, finding that the same low concentration that stops HIV also kills the cells. According to these studies, the real cell-killing dose is one thousand times lower than that reported by Broder, Barry, and Bolognesi. AZT is definitely toxic, indiscriminately killing virus-infected and uninfected T-cells alike. Broder and his collaborators have never corrected their original reports, nor have they explained the huge discrepancies between their data and other papers. To this date the *Physician's Desk Reference* quotes the low toxicity of AZT reported by Broder, Barry, Bolognesi, and colleagues in 1986, although the real toxicity of the drug is one thousand times higher according to more than six independent studies published since.[23]

They also overlooked two even more fundamental problems with their lab experiment: (1) The virus against which Broder and colleagues tested AZT was actively growing in the test tube. But in the body of an infected person, antibodies neutralize HIV years before AIDS appears, if it comes at all. In persons with antibody against HIV, the virus is inactive, not making any viral DNA at all. Thus AZT in a human being cannot attack the virus anyway, for it has already become dormant. It can attack only growing human cells. (2) AZT, like all other chemotherapeutic drugs, is unable to distinguish an HIV-infected cell from one that is uninfected. This has disastrous consequences on AZT-treated people: since only 1 in about 500 T-cells of HIV antibody-positive persons is ever infected, AZT must kill 499 good T-cells to kill just one that is infected by the hypothetical AIDS virus. This is called a very bad therapeutic index in pharmacology! It is a tragedy for people who already suffer from a T-cell deficiency.

A toxic chemotherapy was about to be unleashed on AIDS victims, but no one had the time to think twice about its potential to destroy the immune systems of people who might otherwise survive. The pressure was on to find a drug. Barry used this as leverage when he began quiet negotiations with key FDA officials, arguing that AZT should be rushed through the approval process with reduced testing requirements. Broder was doing his bit, championing the drug through every channel of NIH power at his disposal. FDA officials relented and agreed to help the drug through in order to save time. Given the toxicity of AZT, Burroughs Wellcome would need every break it could get to win approval.

Barry and Broder were the right team to get that break. Says Nussbaum in *Good Intentions*: "David Barry was the puppet master, and his favorite marionette was Sam Broder. While Broder was charging around promoting AZT at the National Institutes of Health, Barry was working quietly behind the scenes orchestrating a whole panoply of actors who would ensure the drug's ultimate commercial success."[24]

Broder rushed AZT through its Phase I trials, the tests to determine its toxicity in humans. FDA cooperation allowed him to cut corners, making the drug appear to have minimal side effects. Now they were ready for the Phase II study, to see whether the drug would actually fight AIDS symptoms.

THE AZT COVER-UP

Double-blind, placebo-controlled studies form one of the cornerstones of medical science. This rigorous gold standard puts any promising new treatment to the ultimate test: When applied to humans, does it really work? If properly structured, such a study throws out the prejudices of the researchers and yields the bottom line. A group of people with the appropriate disease is carefully selected, then secretly divided into two subgroups matched for every important characteristic. To test a therapy for tuberculosis, for example, both groups would contain the same number of tuberculosis patients. One group is given the treatment, the other a placebo—a "sugar pill," meaning a sham treatment that appears identical to the therapy itself. This removes any interfering effect of patient psychology or actions. And the study is conducted in a double-blind fashion, so that neither the patients nor the doctors know who is receiving treatment and who gets placebo, until the experiment is finished.

Under normal circumstances, AZT's Phase II trial would have been such a double-blind, placebo-controlled study. But the intense political pressure to approve an AIDS drug, enhanced by fast-spreading rumors in the homosexual community of AZT's powerful benefits, forced FDA officials to take shortcuts. Although the study was finally published as if it had been a double-blind, placebo-controlled test, it most definitely was not.[25] The drug's toxicity inevitably unblinded the study within weeks, its effects on patients being painfully obvious.[26]

David Barry structured the entire study from beginning to end. He tapped Burroughs Wellcome's informal network of scientific collaborators, selecting twelve medical centers around the country for participation. By providing $10,000 per study patient to each clinic involved, he induced a whopping fifty-one researchers to jump on board, a group heavily weighted with old virus-hunting peers of Barry's. Just having that many well-connected medical scientists helped swing the political balance in his favor, and it locked in their own loyalties to AZT. Even Michael Gottlieb, who reported the first five AIDS cases, joined in. There was hardly a medical institution left in the country that was not involved and that could have offered an independent second opinion. Barry chose Margaret Fischl, a virologist at the University of Miami, to head the experiment.

Thus, Burroughs Wellcome not only coauthored (Drucker, Nusinoff-Lehrman, Segreti, Rogers, Barry), but also paid for the licensing study of its own product. But nobody seemed to mind this blatant conflict of interest—not the many non-Burroughs Wellcome researchers on the study; not the NIH, which cosponsored the study; not the FDA; not the editor of the *New England Journal of Medicine*, which published the study.

A total of 282 AIDS patients was recruited, roughly half being put on AZT and the other half receiving the placebo. The trial, conducted in 1986, was scheduled to treat each patient for six months. After four months the announced results seemed stupendous—so amazing, in fact, that the study had to be aborted early. Fischl and her associates decided they could not ethically continue to withhold such a wonderful drug from the placebo group. Nineteen placebo recipients had died during the study, compared to only one member of the AZT group. Forty-five in the placebo group developed opportunistic AIDS diseases, versus only twenty-four in the AZT group. And while the T-cell counts of the placebo patients continued to decline, the AZT group saw a temporary surge in their T-cells. Results like these could propel almost any drug to FDA approval.

But even an inspection of the officially published data reveals some grim problems. The study does not indicate that Fischl and colleagues sorted their patients according to use of such recreational drugs as heroin or poppers. Since most were homosexual men, this could complicate matters if, for example, the placebo group contained more heavy drug users. Fischl herself also admitted that an undocumented number of the patients were allowed to take other medical drugs during the study, a factor that introduced another wild card.

When pioneer AIDS researcher Joseph Sonnabend, from New York, "first read the AZT study report, he had a lot of questions but the first one had nothing to do with AZT: Why had so many placebo patients died? 'I was suspicious of the study from the beginning because the mortality rate was simply unacceptable,' he said. 'My patients were simply not dying in those sort of numbers that rapidly.' Sonnabend had an added difficulty. The causes of death provided to the FDA did not match those in the research Fischl had written for the *New England Journal of Medicine*. 'Sloppy research,' Sonnabend said."[27]

Still, taking the results at face value, a shocking picture of AZT toxicity emerges. Sixty-six AZT recipients suffered "severe" nausea—a category that would have been mentioned only if this was clinically serious—as compared to twenty-five in the placebo group. All AZT users saw their muscles waste away, while only three placebo recipients suffered this symptom. And a full thirty in the AZT group survived only with multiple blood transfusions to replace their poisoned blood cells, compared to five similar cases among the placebo users. The less-publicized "side effects" of AZT more than abolished its presumed benefits.[28]

A follow-up study on those same patients found that Fischl's neat picture mysteriously vanished once everyone was put on AZT. Within months, the death rate of the original AZT test group rapidly caught up to the former placebo group. After a year, one-third of both groups had died. Fischl, "the Queen of AZT,"[29] and her coworkers shrugged off these new results, suggesting that AZT's miraculous effects somehow wore off after a few months.

Or perhaps the benefits never existed in the first place. A flood of previously concealed information has surfaced since the trial, all showing that it became unblinded from the start. The controls completely broke down.

The doctors certainly found out quickly who took AZT and who did not, because AZT induces serious destruction of blood cells and the bone marrow that produces them. Bruce Nussbaum, in his 1990 book *Good Intentions*, described the mood in the trial's first month:

> A move to stop the trial began immediately. The toxicity of AZT was proving to be extremely high, much higher than indicated by Sam Broder's previous safety trials. PIs [Principal Investigators] began to worry that AZT was killing bone marrow cells so fast that patients would quickly come down with aplastic anemia, a murderous disease. This was terrifying to many PIs. "There was enormous pressure to stop," recalls Broder. "People said, 'My God, what's going on, we're getting these anemias, what's going on?' We never saw this level of anemia before."[30]

For those doctors who may have missed AZT patients vomiting

up blood, the routine blood tests gave away the secrets. Michael Lange was one of the researchers in the trial, interviewed for a 1992 British television documentary:

> I don't think [the trials] were really blinded, because when you take AZT, your red blood cells increase in size... You can notice that on an ordinary blood count, and since blood counts were monitored and the information fed back to patients, this information was available to the investigators.[31]

The patients, needless to say, often found out what they were taking by such clues. But they had other methods. For one thing, the AZT and placebo pills tasted different at the study's beginning. When doctors finally caught some patients tasting each other's pills, they fixed the problem. This came too late, of course, for full damage control. The patients who missed this opportunity discovered other ways around the controls. According to Christopher Babick, an AIDS activist with the People With AIDS Coalition:

> During the Phase II trials, we received many phone calls in our office from individuals who wanted to determine whether or not they were using the placebo or actually receiving AZT. There were three laboratories in New York which would analyze the medication. We would refer individuals there. If, in fact, they were on placebo, they would make arrangements to acquire the drug AZT. Oftentimes they would share it with individuals who were in the trials, thus really rendering the Phase II trials unblinded.[32]

The patients had bought the early rumors of AZT's incredible healing powers, and they really did not want to take a placebo. Some of the placebo group secretly did use AZT, explaining the presence of its toxic "side effects" among those patients. The AIDS activist group Project Inform, originally an opponent of AZT, tried to dislodge the trial's internal working papers to confirm that the "placebo group" members with toxicity symptoms had used AZT; despite invoking the Freedom of Information Act, they never could get the documents released. The trial's organizers pulled one final stunt to help AZT succeed. The original plan had called for each patient to participate for six months.

But long before the six-month "double-blind, placebo-controlled trial" was over, the "blinded" researchers saw that the AZT group was doing better than the placebo group. How did they see this, if the study was blinded? The researchers could monitor the tally of AZT versus placebo either by AZT's toxicity or by something else.

As soon as the tally appeared to favor AZT over the placebo, the FDA oversight committee aborted the trial. Insisting they were acting on ethical considerations, the organizers immediately provided AZT to all patients. Patients spent an average of about four months in the original study, some less than one month. The final analysis included all patients, with projected guesses to fill the gaps in the data. As the follow-up study later observed, the death rate among the original AZT group quickly caught up to the former placebo group.33 Had the trial not been unblinded, or had the FDA chosen to wait the full six months, the relative death rates would have looked radically different. In any case, Fischl's pretense of double-blind controls smacks of dishonesty.34

Once the controls broke down, the study began to unravel. While some "placebo" recipients were actually taking AZT, some of the "AZT" recipients were being taken off the drug. Many of the patients simply could not tolerate AZT, and the physicians had to do something to save their lives. "Drug therapy was temporarily discontinued or the frequency of doses decreased... if severe adverse reactions were noted," admitted Fischl in the fine print of her paper. "The study medication was withdrawn if unacceptable toxic effects or a [cancer] requiring therapy developed."35 This astonishing slip reveals that the doctors did indeed know who was using AZT. But never did Fischl tell how many "AZT" patients were taken off the drug, nor for how long.

Other patients dropped out of the trial altogether. Some 15 percent of the AZT group disappeared, possibly including patients with the most severe toxic effects. Fischl and her collaborators never bothered accounting for the loss, fueling the suspicion that they could have even dropped the sickest patients themselves.

This is a likelier possibility than it first sounds. Author John Lauritsen succeeded in obtaining documents released under the Freedom of Information Act and found many examples of incomplete or altered data. Causes of death were never verified, as by autopsy, and report forms often listed "suspected" reasons.[36] Naturally, Fischl and colleagues tended to assume that diseases in the placebo group were AIDS-related, while assuming diseases in the AZT group were not. The symptom report forms looked even worse. Mysterious changes appeared, often weeks after the initial report for a given patient, including scratching out the original symptoms. The unexplained tamperings generally had no initials indicating approval by the head researcher. Other symptom reports were copied onto new forms but often lacked the original form for comparison. And on some forms reporting toxic effects of AZT, the symptoms were crossed out months later.

During the trial, an FDA visit to one of the test hospitals in Boston uncovered suspicious problems. "The FDA inspector found multiple deviations from standard protocol procedure," an FDA official later commented, "and she recommended that data from this center be excluded from the analysis of the multicenter trial."[37] Months after the trial had finished, the FDA finally decided to inspect the other eleven centers. By then much of the evidence had been lost in the confusion. Far too many patients had been affected by test rule violations, and the FDA ultimately chose to use all of the data, good or bad, including data from the Boston center. One FDA official let the cat out of the bag on the hopeless mess: "*Whatever the 'real' data may be*, clearly patients in this study, both on AZT and placebo, reported many disease symptoms/possible adverse drug experiences."[38]

Other than allegedly reducing death, the Phase II trial made two other claims on behalf of AZT: (1) It raised the T-cell levels of immune-deficient AIDS patients and reduced the number of opportunistic infections they suffered. All testing violations aside,

AZT can temporarily raise T-cell counts. So can various other poisons and even severe bleeding after a long period. When some tissue is attacked by a toxin, or blood is lost due to an accident, the body tends to overcompensate for the loss by producing too many replacements—as long as it can.[39] At some point even the ability to replace white blood cells becomes overtaxed and the T-cell counts collapse downward, exactly as observed in the Fischl study. A temporary increase in T-cells does not necessarily indicate the patient is improving.[40] (2) AZT blocks DNA production, not only in human T-cells or retroviruses, but also in any bacteria that might exist in the body. Thus, it can act as an indiscriminate antibiotic, killing opportunistic infections while destroying the immune system. Even Burroughs Wellcome had previously billed the drug as an antibacterial. This effect could explain the lower number of such infections in the AZT group. But the effect lasts only a short time; once the body's immune system is devastated by AZT, the microbes take over permanently.

Ignoring all the chaos, the FDA approved the drug on the basis of this experiment. Apparently, the strategy of involving many medical researchers from many institutions had paid off. There was only one critical voice questioning a therapy that infiltrates a DNA chain terminator into human bodies indefinitely, the voice of the retired bacteriologist Seymour Cohen:

> The severe toxicity of AZT to bone marrow, as well as unexpected interactions of other drugs with AZT, indicate the importance of knowing more about the effects of the compound.
>
> We ask therefore, Which normal cells are severely damaged? Is the damage reversible or irreversible? Are the cells killed and the chromosomes fragmented, as one might expect from a termination of DNA chains? Are AZT and DDC mutagenic, or possibly carcinogenic? These questions have not yet been answered, to my knowledge."[41]

Several leading scientists, even virologists, also felt uneasy about the whole affair. But they preferred to remain silent or restricted their concerns to informal comments to the press. For example, Jay Levy at the University of California at San Francisco had been one of the first scientists to isolate HIV. A *Newsday* article described his comments on the drug: "I think AZT can only hasten the demise of the individual. It's an immune disease," he said, "and AZT only further harms an already decimated immune system."[42] Even Jerome Groopman, one of the Phase II participating scientists, harbored serious reservations. The head of a research group at a prominent Boston hospital, he quickly discovered the effects of AZT on his patients. "When Groopman gave it to 14 patients on a compassionate basis, only 2 were still able to take it after 3 months. 'We found it nearly impossible to keep patients on the drug,' Groopman says."[43]

Sam Broder, on the other hand, never seemed to entertain a second thought. "When the Wright Brothers took off in their first airplane it probably would have been inappropriate to begin a discussion of airline safety," he nonchalantly told the Presidential HIV Commission in 1988.[44]

But Martin Delaney, founder and head of the AIDS activist group Project Inform, San Francisco, was furious:

> The multi-center clinical trials of AZT are perhaps the sloppiest and most poorly controlled trials ever to serve as the basis for an FDA drug licensing approval... Because mortality was not an intended endpoint, causes of death were never verified. Despite this, and a frightening record of toxicity, the FDA approved AZT in record time, granting a treatment IND [investigational new drug] in less than five days and full pharmaceutical licensing in less than 6 months.[45]

David Barry had already negotiated behind closed doors with the FDA for rapid approval. He held a strong bargaining position, given the political climate.[46] But even that took too long for him, and he demanded special permission for Burroughs Wellcome to sell AZT while waiting for the official approval. FDA officials scrambled for an answer, dredging up a permit known as the *Treatment IND*. This method had almost never been used. Within days the technicalities were ironed out, and Barry had his permit to sell.

Next, he had to get the official permission. He wanted it fast, and based on less scientific data than normally required. Again the FDA complied, cutting the process down to several months. Even AZT studies on mice were dropped from the requirements. The final hurdle lay in a meeting of an advisory committee of scientists and doctors, whose recommendation would likely determine AZT's fate. The panel met for a single day in January of 1987. The dice were loaded in his favor, for two of the eleven panel members were paid consultants for Burroughs Wellcome.47 The FDA granted special permission for those two researchers to remain on the committee with full voting powers.48

Dozens of scientists from the Phase II trials showed up to argue their case, packing the room with virtual cheerleaders. They spent hour after hour flashing huge quantities of data past the committee, some of it so new that no one had had the time to review it beforehand. The follow-up results on the patients, showing higher death rates after everyone went on AZT, were cleverly buried in an avalanche of confusing statistics. Dazed, the members of the committee began to feel anxious that something had gone wrong in the testing process. Then Barry played his ace: a high-ranking FDA official, Paul Parkman, showed up and spoke, despite not having been scheduled to do so. After only a minute of suggesting most of the panel's concerns could be addressed, Parkman closed with a dramatic statement: "I think we can probably arrive at a plan that will satisfy people here."49 Suddenly, the arguments stopped, and the mood shifted from opposition to support for AZT. FDA officials had never before interfered in these meetings, and the entire committee was shocked. "Did you hear that?" the panel chairman said to an associate. "He's telling us to approve it."50

Few in the room knew that Parkman was a personal friend of Barry; they had once worked together on virology projects. The

324 ■ INVENTING THE AIDS VIRUS

panel ended up recommending AZT, with only the chairman voting against it. Burroughs Wellcome quickly patented the drug, something no one else had ever bothered to do.

The FDA endorsement could seem a cruel joke perpetrated by heartless AIDS scientists. Patients on AZT receive little more than white capsules surrounded by a blue band. But every time lab researchers order another batch for experimentation, they receive a bottle with a special label (Figure 2). A skull-and-crossbones symbol appears on a background of bright orange, signifying an unusual chemical hazard. The label appears on bottles containing as little as 25 milligrams of AZT, a small fraction (1/20 to 1/50) of a patient's daily prescribed dose. The adjoining warning on the label reveals secrets not conveyed to the unwitting patient:

FIGURE 2

AZT product label, Sigma Chemical Co.

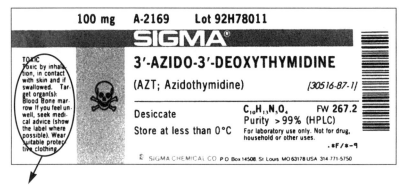

TOXIC
Toxic by inhalation, in contact with skin and if swallowed. Target organ(s): Blood Bone marrow. If you feel unwell, seek medical advice (show the label where possible). Wear suitable protective clothing.[51]

DDI AND OTHER DNA CHAIN TERMINATORS CLAIM A PIECE OF AZT'S ACTION

A different pharmaceutical giant, Bristol-Myers Squibb, produced another DNA chain terminator, ddI. The company sorely wanted to pull this drug off the shelf and into production, hoping to get a piece of the action from Burroughs Wellcome. Sam Broder, after working in the mid-1980s to promote AZT, was only too happy to help along any such chemotherapy. He performed lab experiments on cultured cells, again wrongly trying to argue that the drug blocked HIV production more effectively than it killed T-cells.

Bristol-Myers began sponsoring research on AIDS patients. But they performed no controlled study, never comparing ddI's effects to placebo in matched groups.[52] The studies did, however, reveal a couple of additional toxic reactions not produced by AZT. It can cause fatal damage to the pancreas, and it can destroy nerves throughout the body.[53] On an experimental basis, a number of doctors began giving ddI to thousands of AIDS patients who could not tolerate AZT. Hundreds of unexplained deaths occurred among these patients, but the FDA managed to quell growing concerns.[54]

The FDA advisory committee, meeting to vote on ddI's approval, convened in July 1991. On that day, the panel reviewed the sloppily gathered data on the drug, which was compared to unmatched and untreated AIDS patients from years earlier. On this questionable basis, the committee was told ddI worked as well as AZT. Given the astonishing lack of a controlled study, the panel leaned against approval. That is, until FDA director David Kessler personally intervened on behalf of ddI and pressured the committee to "be creative."[55] The members changed their minds, voting to license the drug, albeit with restrictions. Doctors could prescribe it only for patients they felt did not benefit from AZT, leaving ddI as a secondary treatment and the second drug ever to win "fast-track" FDA approval.

But even four years after its approval for human consumption, Anthony Fauci, director for AIDS research at the National Institute of Allergy and Infectious Diseases (NIAID), stated to the *New*

326 ■ INVENTING THE AIDS VIRUS

York Times: "ddI had never been compared with a placebo in a large study."56

Since that time, the AIDS establishment has backed yet other DNA chain-terminating chemotherapies, including dideoxycytidine (ddC), a drug also developed by Jerome Horwitz in the 1960s and now marketed by Hoffmann-La Roche. The FDA has approved ddC, but only for use in combination with AZT or ddI.

THE CONSENSUS DISSOLVES

In the years following AZT's approval, a flood of studies on AIDS patients have poured forth, illustrating frightening toxicity. None have included placebo groups, an omission rationalized by ethical concerns that patients should not be denied such a miracle drug. But the numbers speak for themselves.

Two years after the end of the Fischl Phase II trial, a group of French physicians working at the Claude Bernard Hospital in Paris published another study on hundreds of AIDS patients. All used high-dose AZT for an average of seven months. One-third of the patients experienced a worsening of their AIDS conditions, and a slightly higher percentage developed new AIDS diseases. By nine months, one of every five patients had died, a rate far higher than in the Fischl study, which also used the high dose. "The bone marrow toxicity of AZT and the frequent need for other drugs with hematological [blood] toxicity meant that the scheduled AZT regimen could be maintained in only a few patients," wrote the authors. This has matched other findings; in most studies, half of any group of people suffer an immediate reaction so severe that they must stop. The French doctors cast a cloud of pessimism, noting that "in AIDS and ARC patients, the rationale for adhering to high-dose regimens of AZT, which in many instances leads to toxicity and interruption of treatment, seems questionable."57

In England, one group of researchers described the medical consequences of AZT in thirteen patients; all thirteen developed severe anemia (Mir and Costello). A subsequent Australian study reported the consequences of treating more than three hundred patients for one to one-and-a-half years. More than half developed a new AIDS disease during the first year, and exactly half needed blood transfusions to survive. Nearly one-third died. A Dutch study found still deadlier results: After little more than a year, most of their ninety-one patients needed blood transfusions and almost three-quarters died. The Dutch researchers despaired at prescribing AZT, warning that most of their patients simply could not stay on the drug for loss of blood cells.[58]

A new complication surfaced in 1990. The National Cancer Institute analyzed the status of AIDS patients who had participated in Broder's Phase I trial and uncovered the fact that of all the people using AZT for three years, half were developing lymphoma.[59] This is a deadly cancer of white blood cells, akin to leukemia but forming solid tumors in the body, and also happens to be included on the official list of AIDS diseases blamed on HIV. Since AZT was killing and damaging those same white blood cells, the drug stood out as the likely culprit. Virus hunters rushed to the drug's defense. Some massaged the statistics to lower the lymphoma rate, while others turned the news completely upside-down by claiming AZT actually helped patients live longer—long enough to get lymphoma![60] Paul Volberding, one of the leading organizers in the Phase II trial, told one interviewer, "So we see the lymphomas as an unfortunate reflection of our success at this point rather than a reason for real caution."[61]

Certainly AIDS officials can hardly be accused of caution when it comes to AZT. Nor does their explanation wash, since only 3 percent of all AIDS patients tend to get lymphoma as their AIDS-defining disease[62]—not 50 percent as in Broder's AZT trial. AZT, furthermore, has given evidence of cancer-causing abilities when tested.[63]

A few small studies have tested the reverse to see what happens to AIDS patients who stop AZT use. In one group of four patients who developed massive blood cell loss weeks after starting AZT, three recovered after the doctor took all four off the drug.[64] Another group of five patients suffered muscle wasting, a symptom that disappeared in four cases only a couple of weeks after stopping AZT; two of these patients lapsed back into the condition after restarting the drug.[65] In the most dramatic such experiment, a doctor took eleven of his worsening AIDS patients off AZT. The immune systems of ten immediately rebounded, and several continued improving.[66]

Yet no amount of warning data could dissuade AIDS officials from abandoning their "antiviral" compound. Having won approval for treating AIDS patients, Burroughs Wellcome and the NIH moved to have AZT recommended as a preventive drug, for HIV-positive people without symptoms. This time Anthony Fauci directed the experiment as a project of NIAID, the NIH division he headed. Burroughs Wellcome again financed much of the study, paying hospitals for participating, and several of its consultants again joined in. Margaret Fischl and many other Phase II researchers signed up, and Paul Volberding secured the top position as organizer. But now a stupendous number of scientists were recruited: The final paper mentioned 130 authors, which Volberding called "a partial list." The investigators read like a who's who among leading virus hunters and medical doctors involved in AIDS research. With that many prominent researchers involved, few colleagues remained to act as independent reviewers. The study's political success was virtually guaranteed, regardless of its outcome.

Volberding and his colleagues enrolled more than thirteen hundred HIV-positive healthy persons from AIDS risk groups, none of whom had AIDS diseases. The subjects were divided into three groups—placebo, high-dose AZT, and, because of growing worries about the drug's toxicity, a low-dose AZT group. Protocol 019, as Fauci designated it, quickly degenerated into a repeat performance of the Phase II trial. More cancers, including Kaposi's sarcoma, occurred in the placebo group, hinting that more users of poppers or other recreational drugs may have ended up there (see chapter 8), biasing the results in favor of AZT. The double-blind controls broke down again, a fact that was covered up publicly. But in the text of the paper, Volberding acknowledged that the drop-out patients tended to come from the AZT treatment groups, removing some of the sickest victims. Having also anticipated sharing of AZT between patients, he had their blood tested for the actual presence of the drug. Nine percent of the "placebo" group were caught with traces of AZT, while almost 20 percent of the AZT groups showed no evidence of ever having used the drug.

The study was terminated early, after patients had been treated for an average of one year. The final analysis showed the AZT groups with fewer AIDS diagnoses than the placebo, but the toxic "side effects" of AZT swamped this small difference. The low-dose group had as many sick people as the placebo group, although their blood disorders and immune deficiencies were not called "AIDS." The high-dose group suffered by far the most, having dozens of deathly ill patients.[67] By calling diseases in the placebo group "AIDS," while avoiding that diagnosis for the AZT groups, Volberding successfully won an FDA recommendation for using the drug on healthy HIV-positive people.

In 1994 Volberding published a stunning aftermath to Protocol 019. The T-cells of 29 percent of the men in the placebo group had increased gradually over two years, while those of the AZT recipients had decreased.[68] It is probable that under clinical surveillance the 29 percent whose T-cells increased, despite the presence of the alleged T-cell killer HIV(!), had given up or reduced their recreational drug use.

After studying patients for another five years, Volberding, the father of AZT prophylaxis, came to a new conclusion about AZT: "Zidovudine [AZT]... does not significantly prolong either AIDS-free or overall survival."[69] Stated otherwise, hundreds of thousands of healthy people had taken AZT for five years for no "significant" reason, assuming that DNA chain termination is indeed "not significant."

In an article entitled "Early Intervention: An Idea Whose Time Has Gone?" the *New York Native* writes on Volberding's latest insight: "The same group of people that has continued to insist that 'early intervention' with AZT is necessary and beneficial—despite data showing that people who take AZT earlier also die earlier and that their quality of life is so diminished as to negate completely any alleged benefits from AZT—have now published research showing that, after all, AZT does not prevent progression to 'AIDS' or delay death. The magnitude of this about-face cannot be overstated."[70]

Not totally convinced by Volberding's original trial, other researchers put together two long-term studies on AZT's preventive effects. An American research group sponsored by the Department of Veterans Affairs ran a two-year trial comparing patients who used AZT before symptoms (the "early" group) to those using it afterward (the "late" group). These scientists found that the early group actually died slightly more often and a bit faster than the late group, but the differences were small. They concluded AZT showed no survival benefits whatsoever when used for prevention.[71] The news hit the stock market with force, knocking down the value of Burroughs Wellcome shares some 10 percent in one day.

British and French scientists organized a similar study, known as the *Concorde trial*, while Volberding's study was still in progress. The Concorde study treated two groups with AZT, one before AIDS symptoms (the early group) and the other after (the late group). Only people without AIDS symptoms were recruited into the study, the late group receiving a placebo until after they contracted AIDS. Apparently, the researchers were seeing enormous toxicity as the study progressed, for midway through, a minor crisis erupted. The scientists became divided over whether to continue or abort early. At a meeting behind closed doors, an audience member secretly recorded Chairman Ian Weller as he voiced the increasing concerns: "If there is benefit [to AZT therapy], is it maintained, or will it wear off? In which case we may do more harm than good."[72] The study organizers voted to continue, albeit nervously.

After each patient had participated for three years, the researchers came out in 1994 to announce publicly that they could find no difference in survival between the early and late treatment groups. In reality, the early AZT group had done worse than estimated. The death rate in the AZT group was 25 percent higher than in the control group—hardly a recommendation for AZT prophylaxia.73 The double-blind controls again seem to have dissipated, for symptom-free patients could easily know they were on AZT by its potent toxicity. Many of these AZT patients could no longer tolerate the nausea, vomiting, and anemia, but they did not have the courage to confront their doctors. So, according to at least one report, "They have thrown their tablets down the toilet."74 This would artificially lower some of the apparent toxicity in the early group.

But the news of no positive benefits did stun the AIDS establishments in all countries, sending various officials scrambling for excuses to explain away the Concorde results. This study has provided the heaviest blow yet against AZT, and the first signs of retreat are beginning to emerge. Based on a preliminary report on the Concorde study, on June 25, 1993, an NIH panel formally announced new guidelines for AZT use, recommending that doctors and patients use more caution. "AZT has benefits, but we are admitting that it is not as good a drug as we thought it was," said the committee chairman.75

Further bad news came from America. One investigation of AZT, as prophylaxis against AIDS dementia, showed in 1994 that—contrary to expectation—there was twice as much dementia in AZT-treated homosexuals than in untreated counterparts.76 Also in 1994, a large American study reached an even more damning verdict on AZT prophylaxis. It found that HIV-positive hemophiliacs had 2.4 times higher mortality and a four-and-a-half–times higher AIDS risk rate than untreated HIV-positive hemophiliacs.77 It may not have been just a coincidence that Sam Broder resigned, apparently, at the height of his career, as director of the National Cancer Institute in December 1994.78

In April 1995 an American study found that AZT treatment doubled or quadrupled the risk for HIV-positive male homosexuals to develop *Pneumocystis* pneumonia.79 In July 1995 the *British Medical Journal* published that AZT prophylaxis reduces the time to death of HIV-positive AIDS patients from three years without AZT to two years with AZT.80

The emerging concerns about AZT treatment are summarized

dramatically in a letter from a German doctor to the editor of *Nature*:

> To the Editor: As a hospital doctor I come face to face every day with the disaster that Gallo and his colleagues have brought about. In the case of each patient with tuberculosis, each patient with herpes zoster, each patient with toxoplasmosis or cytomegalo viral infections, I am confronted with the thought that if these patients were HIV positive, they would, as things currently stand, have to undergo anti-viral therapy. The substances available are pure chemotherapeutic agents, which means that in treating them I bring about the very illness I seek to bring under control. In effect, this means leading the patients to their deaths. As a result of the AIDS-virus hypothesis, things have now reached the stage where treatment of the disease itself gives rise to the bleak prognosis for the disease.
>
> CLAUS KOEHNLEIN, M.D.
> Kiel, February 28, 1995

In his response to Koehnlein, John Maddox, the editor of *Nature*, wrote on September 20, 1995:

> But it seems to me that there are two separate issues—is AZT dangerous in itself and does HIV cause/not cause AIDS? Only physicians such as yourself can establish the first point, but it seems to me that by now there must be a great deal of clinical data on which you and your colleagues could draw to reach a substantial conclusion that could be published.
>
> You say that the Ho and Wei papers we published in February [*sic*—it was January] are unconvincing because their work is based on the "AIDS-virus hypothesis," but how can you dismiss their finding of very large quantities of virus in the blood of people with HIV infection? And the temporary effectiveness of the protease inhibitor, whose design is specifically determined by the sequence of HIV, used in their study?

With Therapies Like This, Who Needs Disease? ■ 333

I should add that a haemophiliac relative of my wife died of AIDS this year. He was infected before 1984 and diagnosed with antibodies against HIV in 1985. His first symptoms of AIDS appeared about 1989.

American gay activists from ACT UP San Francisco who used to conform with the HIV orthodoxy have recently also begun to protest AZT therapy with violence. Crashing the ten-year anniversary party of Martin Delaney's AIDS organization, Project Inform, and turning over tables at the plush Hyatt Regency Hotel in San Francisco, protesters shouted into the faces of Delaney and his guests Larry Kramer and Anthony Fauci: "Tony Fauci, you killed our friends! This is where the murder ends!"

According to *Spin* reporter Celia Farber, the protesters picked the occasion because "Project Inform, they insist, has become so entrenched with authoritarian, establishment, old-boy-network views on AIDS that it has betrayed the community."[81]

In a press release the ACT UP protesters listed their complaints of May 6, 1995:

> The past decade of human tragedy has shown us that trying to kill the AIDS virus with high priced drugs like AZT and ddI harms the people who take them. These compounds make people sick, are not created to be taken for prolonged periods of time and are immunosuppressive. Enough is enough! This on-going circus of death must be questioned...
>
> Government sponsored clinical trials of drugs are obviously not meant to save our lives. These trials are meant to document the drug's effect on laboratory markers that have little, if any, correlation to the health or survival of people with HIV disease. People with HIV need to concentrate on activating their cellular immunity in order to control the opportunistic infections that threaten our lives. Fauci knows this, has admitted it and still does nothing... With 270,000 dead from AIDS and millions more infected with HIV, you [Fauci] should not be honored at a dinner. You should be put before a firing squad.[82]

On July 22, 1995, even the establishment media sent out a signal of distress. The *New York Times* published the following letter condemning AZT studies:

> To the Editor: The recent study casting doubt on azidothymidine's [AZT] alleged therapeutic benefits for carriers of the human immunodeficiency virus (news article, July 16) [from the *British Medical Journal* cited above] contrasts with the majority of AZT studies, in which the drug is claimed to be beneficial.
>
> The best way to resolve the AZT disagreement might be to gather as many of the original articles as possible to see if the experiments were done well. In AIDS research, funding sources can also be illuminating. Here is what I found after reviewing more than 25 studies on AZT:
>
> Evidence of AZT's inefficacy and toxicity has been around a long time, well before the 1994 Concorde studies or the 1992 Veterans Affairs cooperative study. Negative data on AZT were published in the *Lancet*, the British Medical Journal, in December 1988. Those data were not highly publicized.
>
> While the absolute number of studies casting doubt on AZT is small, they tend to have two things in common: good experimental design and "independent" funding.
>
> The more numerous studies supporting AZT's benefits tended to use inappropriate experimental designs and very short follow-up times.
>
> Moreover, these studies were financed, at least in part, by the drug's maker, the Burroughs Wellcome Company.
>
> TIMOTHY H. HAND
> Atlanta, July 17, 1995[83]

Considering the faithful commitment of the *New York Times* to the HIV orthodoxy since 1984, the publication of this letter assumes outstanding significance. In this era of centralized, government-sponsored science, an article against politically correct science can be fatal for a journalist. In such a climate the publication of a letter is a journal's last resort of expressing a dissenting opinion.

AZT, known for decades as a failed and toxic cancer chemotherapy, was resurrected for political reasons and rushed through the FDA's first fast-track approval. By its very nature, such a drug could only worsen AIDS, if not cause some AIDS diseases by itself.[84] One experiment after another, despite flaws, has confirmed the drug's toxicity in humans, yet only now is the AIDS establishment slowly backing down. The virus hunters bring tremendous political and financial momentum behind each of their projects, and AIDS treatment is no exception.

PREVENTING HIV INFECTION— THE LAST STAND OF THE AZT LOBBY

The recent growth of opposition to AZT may save lives in the future, but it is coming too late for some victims. Convinced of the drug's alleged success, the AIDS establishment has aggressively promoted AZT wherever it could despite the drug's poor performance. Often the treatment is paid for by one federal program or another, creating an indirect subsidy of Burroughs Wellcome by the taxpayers. In 1992 at least 180,000 people worldwide took the drug every day.[85]

Frustrated with its failures to cure AIDS, and then even to prevent it, the AZT lobby has concentrated on a last front to save its drug: the prevention of infection. As usual in the rush to "save lives" there was no time for theory. To prevent HIV infection, a drug would have to shut down all cell growth in the body for several weeks. This is because retroviruses like HIV depend on cell division for reproduction and therefore infection. If only a few cells continue to divide, the defense against HIV would be useless. But to achieve a complete shutdown of cell division, so much AZT must be administered that survival is impossible. Even the highest doses ever prescribed would not suffice. Given the choice between lethal doses of AZT that could prevent infection but would likely kill the patient and not to use AZT at all, the AZT establishment chose to compromise. By treating with the known doses, most patients would survive long enough to obscure all drug-mediated diseases by HIV-mediated diseases that are "expected" to occur late after infection.

The early months of 1989 brought an unusual notice posted in the buildings of the NIH. Entitled "HIV Safety Notice," it announced a new policy established by the director himself. Any employee of the NIH who underwent accidental exposure to HIV, as for example by a needlestick injury, would be offered preventive AZT. According to the notice, "The advisory group in providing their recommendations emphasized that administration of AZT should be initiated as soon as possible, preferably within hours following the exposure."[86] Only first aid for the injury itself would precede AZT. Numerous medical institutions have since adopted this policy, and a 1993 report in *Lancet* revealed the practical application. The paper described a doctor who was accidentally stuck by a needle and thus exposed to HIV-infected blood. The doctor began taking AZT within the hour, and continued for six weeks—too quickly even to do an HIV test.[87] Thus, medical workers may use AZT even if they never become infected with HIV.[88] But this doctor became HIV-positive despite the toxic prophylaxis.

A second and more disturbing announcement reached the public in the summer of 1989. NIAID, the NIH division under Anthony Fauci, declared it would be conducting trials of AZT on pregnant mothers infected with HIV. A drug that interferes with growth can lead only to physical deformities in babies developing in the womb. The study, financed through the NIH budget, ironically recruited mothers who had been injection drug addicts. Apparently Fauci believes heroin addiction poses less of a threat to children than does HIV; some of their babies, moreover, may never even contract the virus from their mothers but will receive AZT anyway. Following his lead, the French joined Fauci's trial, and the British government, in 1993, began on its own a study of AZT effects on HIV-positive babies.[89]

However, to prescribe a known mutagenic drug to a pregnant woman was a risky departure from the foremost medical principle, "First, do no harm." According to AIDS reporter Celia Farber in April 1995:

> Though AZT is widely claimed to have been deemed "safe" in pregnant women, in fact, the FDA did not think so and never allowed pregnant women to take it prior to this study, primarily because it was classified as mutagenic. Another mutagenic drug, Thalidomide, was prescribed as a sedative throughout Great Britain and Germany in the 1950s but was never approved in the U.S. Thalidomide was responsible for over 10,000 birth defects in children born to mothers who had taken the drug during gestation. Many of the infants were born with missing or partially missing limbs. From that point on, no potentially mutagenic chemical was to be taken by pregnant women, for any reason. AZT in pregnant women represents a radical break in this tradition.[90]

In February of 1994, Fauci's American-French trial on pregnant women was abruptly terminated. Fauci and his collaborators claimed victory because AZT had reduced "maternal HIV transmission rate by two-thirds"—from 25 percent without treatment to 8 percent with AZT treatment.[91] This euphemism was chosen for the net result that out of 180 babies born to AZT-treated mothers, 13 had been found HIV-positive, compared to 40 out of 184 born to placebo-treated mothers.[92] In other words, to save 27 babies (17%) from HIV infection, 180 mothers and 153 of their unborn infants (who either did not pick up HIV from their mothers or picked it up despite AZT) were first treated for six to twenty weeks every five hours with 100 mg AZT and then again intravenously during delivery. In addition, the newborn babies were given 2 mg AZT every six hours for the first six weeks of their lives.[93]

In view of the possible genetic damage from AZT, Fauci acknowledged "long-term follow up of all of the children... is essential to learn more about the risks and benefits of the treatment beyond these encouraging early results." Recommendations on treatment were said to be "pending developments of consensus on the balance between known benefits and unknown risks."[94]

There is of course a double irony in this apparent caution. First, the benefit of being HIV-free is currently not known, because there is no proof that HIV causes AIDS.[95] Second, the risk of AZT is certainly not "unknown"—thirty years after it was first developed to kill human cells for cancer chemotherapy.

After declaring victory against HIV transmission, the double-blind controls were officially broken prematurely and AZT was offered to all mothers.[96] Clearly, the 124 primary and secondary authors of the maternal transmission study achieved "consensus" on playing down the "adverse experiences" of babies on AZT, acknowledging only that the level of "hemoglobin at birth in the infants in the Zidovudine group was significantly lower than in the infants in the placebo group."[97] However, the "neutropenia, high bilirubin levels, and anemia" reported prior to final publication[98] were not documented in the consensus paper.[99] The AZT-induced neutropenia, the medical term for a critical shortage of the majority of immune cells in the blood, could very well be the explanation for the AZT-mediated reduction of HIV transmission. Since HIV replicates in blood cells, and blood cell synthesis is inhibited by AZT, it is no surprise that HIV is less likely to be transmitted if the cells in which it replicates are killed by AZT, both in the mother and the unborn child.

An editorial in *Lancet* did not share Fauci's optimism: "The most worrisome aspect is the possibility of long-term adverse effects on children exposed to Zidovudine (AZT) during fetal life, especially since the vast majority would not have been infected anyway."[100] Indeed, the as-yet-unpublished side effects of the American-French study confirm this sinister projection. According to Farber's article: "There were two birth defects: One had extra digits and a heart defect, the second had an extra digit on both hands. The report concluded that neither case was related to AZT therapy."[101]

A formal request by the *New York Native* for an official account of the unpublished birth defects was denied by the Assistant Secretary for Health, Dr. Philip R. Lee. Lee advised the reporter on January 6, 1994, to "sue the federal government."[102]

But a study from outside the United States provided a clearer picture: Eight spontaneous abortions, eight "therapeutic" abortions, and eight serious birth defects including extra digits were recorded among babies of 104 HIV-positive pregnant women treated with AZT.[103]

THE TRUST IN MEDICAL AUTHORITY BREAKS DOWN

Long-term survivors of AIDS know better than to use AZT. Michael Callen was diagnosed with full-blown AIDS in 1982 before HIV had even been isolated. Given little time to live, he discovered Joe Sonnabend and switched doctors. Callen had participated in the fast-track homosexual scene for a decade, including sex with more than three thousand partners and the attendant drug abuse. His lifestyle changed radically on Sonnabend's advice, although he began taking enormous amounts of antibiotics and sulfa drugs. Because of his cleaned-up lifestyle and his ongoing refusal to take AZT, Callen lived twelve years with an AIDS diagnosis until he died with pulmonary Kaposi's sarcoma in 1994. He told his story in his 1990 book *Surviving AIDS*, along with the stories of several other long term survivors who tend to avoid AZT. For that matter, the CDC estimated that one million Americans had HIV by 1985, but two-thirds of those have not developed AIDS at all in the past ten years. Most HIV-positives have never received AZT.

In New York, Michael Ellner runs a self-help group to help AIDS patients live. Named HEAL (Health-Education-AIDS Liaison), it strongly advises members against AZT. And a 1990 article in *Parade* magazine profiled thirteen AIDS cases who had survived their diagnosis for five years. They rejected AZT as counterproductive. "It's incredible, isn't it, that the drug designed to save you can also kill you," says Mike Leonard, a survivor. "It can make you anemic, and you end up having to get blood transfusions."[104]

In London HIV-positive male homosexuals at risk for AIDS formed a survivor group called "Continuum." In August 1993 there was no mortality during 1.25 years in all 918 members of that group who had "avoided the experimental medications on offer" and chose to "abstain from or significantly reduce their use of recreational drugs, including alcohol."[105] Assuming an average ten-year latent period from HIV to AIDS, the virus-AIDS hypothesis would have predicted at least 58 (half of 918/10 x 1.25) AIDS cases among 918 HIV-positives over 1.25 years. Indeed, the absence of mortality in this group over 1.25 years corresponds to a minimal latent period from HIV to AIDS of more than 1,148 (918 x 1.25) years. As of July 1, 1994, there was still not one single AIDS case in this group of 918 HIV-positive homosexuals.[106]

Other individuals began to take their health in their own hands rather than rely on medical authority for the "treatment" of HIV. In terms of notoriety, the list is led by one of the nation's top basketball stars, Earvin "Magic" Johnson. In November 1991, Magic proved to be HIV-positive when he applied for a marriage license. Magic was totally healthy until AIDS specialists Anthony Fauci, from the NIH, David Ho, now director of the Aaron Diamond AIDS Research Center in New York, and Magic's personal doctor advised AIDS prophylaxis with AZT. Magic's health changed radically within a few days. The press wrote in December 1991: "Magic Reeling as Worst Nightmare Comes True—He's Getting Sicker." Only after he began taking AZT did Magic's health begin to decline. He "had lost his appetite and suffered from bouts of nausea and fatigue" and complained, "I feel like vomiting almost every day."[107]

But then suddenly Magic's AIDS symptoms disappeared—and so did all further news about his AIDS symptoms and treatment. Had Magic's virus suddenly become harmless, or was Magic taken off AZT? No paper would mention whether Magic was taken off AZT. Nobody knew, except those who joked, "There is no magic in AZT, and there is no AZT in Magic." Indeed, it is very unlikely that he could have won the Olympics in 1992 on AZT, considering his strong reactions to the toxic drug in 1991. The silence of the AIDS establishment seems to confirm this assumption. Nothing would have been a better advertisement for the troubled AIDS drug than having returned AIDS patient Magic to an Olympic victory. But no such announcement was made. At last Magic broke the silence himself. After a "motivational" AIDS talk in Tallahassee, Florida, in the spring of 1995, Magic responded to a teacher that "He had been taking AZT for a while, but has stopped."[108] The media preferred not to mention the news.

About six years earlier another young man fought the battle of his life, having been discharged by the U.S. Navy for being HIV-antibody–positive. Raphael Lombardo has won his one-man campaign against the Navy and the AIDS establishment all by himself. His letter proves that true science does not depend on institutional authority:

> To: Dr. Peter Duesberg
> From: Raphael Sabato Lombardo
> Date: May 30, 1995
> Subject: Life without AZT !!!!!!!!!!!!!!!!!!!!!!!!
>
> Dear Dr. Duesberg,
>
> My name is Raphael Sabato Lombardo, 33 years old and from Cape Coral, FL. I am writing in regards to the enclosed magazine article from this month's issue of *Men's Style*. I was thrilled to read that there was someone in the medical profession who shared the same views I've had for so many years.
>
> I am an HIV positive individual. I learned of my HIV status while in boot camp in the U.S. Navy back in 1985 (I could have very possibly been HIV positive 7 years before that). The Navy wanted to discharge [me] and others and dishonorably at that. Feeling my constitutional rights were being violated, [I] and several others dragged the U.S. Navy into the Federal District Courts of Washington, D.C. for one of the very first AIDS litigation cases ever. I was acting

spokesperson for the group. The enclosed newspaper articles will give you some insight into exactly what transpired during this time.[109]

Going back to my small hometown after all of this publicity was not easy. Remember, this was 1985, a time when HIV was called the HTLV III virus and anything and everyone associated with it meant complete and utter doom (physically, spiritually, societally, politically, etc.—or so they thought)! After discharge, my parents and family insisted I come home and finish school. Education was always stressed in our household and looked upon as the only means of moving ahead in life. My parents, who were at great risk of losing their Mom and Pop Italian deli business (which still exists today) also wanted me home so I could do the most important thing, maintain my health. Although met with discrimination and much verbal and physical abuse as well, I did go home and received my bachelor's in business from the University of South Florida-Ft. Myers. The past 6 years I have been working as a field auditor for the largest and oldest newspaper/magazine circulation auditing firm in the world (The Audit Bureau of Circulations—ABC). I love the work and the job is 100% travel which has afforded me the opportunity to see and experience all that this great country has to offer. It was the love, encouragement and support of my family that pulled me through and the faith in our Lord that sustains us all.

Myself and the other recruits (those who are left) still remain a closeknit group. The bond will forever exist. Several have died of AIDS and several have AIDS. As for myself I've remained completely asymptomatic thank God! To be honest, in regards to HIV, I haven't seen a doctor since the day I was discharged. While in the Navy, we were subjected to incompetent Navy doctors who often gave us inaccurate medical results. As a result, I came to trust no one in the medical profession. I decided to take things into my own hands. I spent countless hours in the medical library at the Bethesda Naval Hospital which is where we were being held and did research on one's immune system and all AIDS information available up to that point in time. Since no drugs had yet been approved by the FDA, there were no forms of treatment available. I came up with my own form of natural healthcare which I follow to this very day. I guess you could say that the General Nutrition Center—GNC, Reebok step, weightlifting and good pasta is what keeps me going.

Shortly after discharge, AZT was approved by the FDA. My family and friends wanted me to jump on the bandwagon immediately! I can't explain why, but I outright refused. There was this inner voice that kept telling me, and continues to tell me, to just stay away from medication. Even back then I had a feeling that taking this medication and going on drug experimental trials would do nothing more than provoke the onset of the disease. Again, this feeling was based not on medical data or research, just an inner gut feeling. I guess you could say my spirit guides or guardian angels have been working overtime. By not going on medication, my family and friends felt I was exhibiting the same "ignorance" and "foolishness" that got me into this mess in the first place. We had countless heated arguments over this, but I told them my mind was made up and that was that—period. We Italian men can often times be quite stubborn! Actually, my dad is the only one who agrees with me. That is reflective in our conversations which last no longer than a couple of seconds. He only has 2 questions for me. First, are you still eating a lot? Second, are you still hitting the gym? If I answer yes to both of these questions, then he knows I'll be OK. Sounds like such a simple philosophy for such a complex virus, but Jesus was such a simple man and people make him out to be so complex.

I learned about "love" and "relationships" in the underground gay subculture in New York's West Greenwich Village while in my teens. Unfortunately, very unfortunately, that was all that was available to gays at that time. I would have much rather have asked out someone my own age for a date and taken a nice drive down to Fort Myers Beach or

Sanibel Island like all kids my age did, but society wouldn't hear of it. Society still wouldn't hear of it. I hope to change all that.

During those years of experimenting, exploring and even rejoicing in my God given sexuality, I did the bathhouse scene, the "Saint" parties, the S&M sex clubs, the backroom bar scenes, the group sex, etc. I guess you could say that sexually, I did it all. I was curious, knew exactly what I wanted to do and experience, and did just that. Something I'm proud of? No! It's just the way it happened. Again, this was all society felt, and still feels, gays are worthy of. While I was part of the "gay scene" in this respect, I always felt I wasn't at all in other respects.

At about the same time as my Navy situation, I began hearing more and more of guys I had dated in N.Y.C. who had died or were dying of AIDS. I speak of approximately 2 dozen friends (that I am aware of, there's probably more) who have died of AIDS from 1985 to 1995. They are all gay men (except for 1 woman). These men were also very much into recreational drugs (steroids, poppers, marijuana, cocaine, ecstasy, etc.). They ranged in age from mid twenties to mid forties. I don't know at what point they started using the drugs such as AZT, ddI etc. I found out my HIV status while I was in the Navy and didn't even know I was being tested and had not experienced any signs or symptoms of the disease. I don't know if these other friends of mine had already progressed to ARC and full-blown AIDS before finally deciding to get tested and go on medication or they took it upon themselves early on to have the test done before experiencing any symptoms and then progressed from simply testing HIV positive and then progressing to ARC, full-blown AIDS and eventually death. My personal suspicion is that these individuals were not aware of their HIV status until they started experiencing physical complications. My friends who were sick and died since the late eighties were taking mega doses of AZT (approximately 12 pills a day). I hear that dosage has been greatly reduced. My friends today take several pills of AZT daily. I'm not sure what the dosage is for any other drugs that they're on.

In regards to the woman I mentioned, she was a heterosexual, and in her late twenties. I am not certain how she contracted the disease. She was married with a set of twins that were merely a few years old at the time of her death last year. I believe she suffered approximately 3 years and was on AZT and several other drugs for most of that time. An unfortunate tragedy! Her husband and children test negative.

I started to ask myself why I wasn't developing any of the classic symptoms? I literally sat down and made a chart of the similarities between myself and all of these guys who had become sick. I've duplicated it here for you:

	Raphael	Friends
oral sex (giving)	great amount	great amount
oral sex (receiving)	great amount	great amount
anal sex (giving— no rubbers)	moderate amount	uncertain
anal sex (receiving— no rubbers)	limited amount	uncertain
fisting (giving)	several times	limited
fisting (receiving)	never	uncertain
deep mouth to mouth kissing	very heavy	moderate to heavy
rimming (giving)	very heavy	moderate to heavy
rimming (receiving)	very heavy	moderate to heavy
poppers	never	heavy use
marijuana	never	moderate to heavy
cocaine	never	moderate to heavy
special K	never	moderate to heavy
ecstasy	never	moderate to heavy
alcohol consumption	don't drink	moderate to heavy
smoking	zero	moderate to heavy
steroid use	never	heavy
weightlifting	great amount	great amount

nutrition	excellent	pretty good
vitamins	heavy use	uncertain
sleep habits/rest	excellent	fair
AZT	zero	heavy, heavy use
ddI	zero	heavy, heavy use
other experimental drugs	zero	heavy, heavy use

Looking at this, the only commonality with myself and the others is the sex. In regards to the drug issue, I'm probably the only gay male who could answer the way that I have (those answers still hold true for me today a decade later). As I said, my spirit guides or guardian angels have certainly been working overtime for me.

My reason for not succumbing to the ill temptations of drugs is simple. I never, ever had the desire or curiosity to try them. Where as with sex, the desire and curiosity was there and I went out and did exactly what I was looking to do. But with drugs, that desire or curiosity was never there. I have my 2 older sisters to thank for that. You see, I lived and spent the first 10 years of my life in the slums of New York City—East Harlem. This was at one time a very close-knit Italian neighborhood that my grandparents settled in when they immigrated from Italy. In the early 1970's, the neighborhood started to change drastically (crime, drugs etc.). My sisters and I (I have 3 great sisters) attended the Catholic grammar school up the street from the tenement building in which we lived. When we would leave our building in the morning we would more often than not be greeted by junkies rolling around in the gutter, shooting up their drugs, clothes all torn, battered and bruised bodies, etc. It was horrifying! My 2 older sisters would shield my younger sister and I when passing by. Each day my older sisters would say, "You see that's just what happens to you when you try drugs." That is all it took. Those words stayed with me for life. As a result, the desire or curiosity to experiment with drugs was never there for me. I guess something could be said for scare tactics.

With regards to HIV, I've always sensed that drugs, or lack of them, has played a big part in keeping me going while so many others have been less fortunate. Another thing I'd like to add is that as a workout enthusiast, I've never experimented with steroids, which unfortunately runs so very, very rampant amongst gays and in my opinion is ravaging the gay community. Amongst other things, it severely compromises one's immune system. To me, there's nothing wrong with good old fashioned, honest hard work.

Several months ago, *USA Today* ran a story about a talent agency out in California which last year opened a modeling division which strictly promotes HIV models. The name of the agency is the "Morgan Agency" and it's located in Costa Mesa, CA. The owner of the agency is Mr. Keith Lewis. Mr. Lewis wanted to dispel the myth that HIV individuals are all emaciated looking people on their deathbeds. He named this division "Proof Positive" and within one year it has been the fastest growing division in his talent agency. Some big name advertisers (such as Nike) have used his models. He feels this segment is going to boom. Well, after reading this article, I wrote Mr. Lewis and sent him some recent snapshots (which I've enclosed for you). Well, Mr. Lewis called me a few weeks ago. He said he couldn't think of anyone who embodies the spirit or philosophy any better than me and would love to have me as part of his "proof positive" family. He's started to promote me to advertisers immediately. I certainly hope something comes of that.

According to the article I've read, it sounds as though you've had a pretty rough time of things in trying to gain support in the medical community and gay community as well. I just wanted to let you know that I share the same views and sentiments as you. If you have any questions at all or would like to contact me for whatever reason or if I could be of any help to you, feel free to contact me. At the time of discharge I said that if the Good Lord sustains me for 10 years, then I would once again come forward and open myself up to the glare and scrutiny of the public eye and media and serve as an inspiration to millions!

This year, 1995, marks the 10 year anniversary of my Navy situation, a milestone in many, many ways.

Respectfully,
Raphael Sabato Lombardo

THE STORIES OF THOSE WHO BELIEVED IN AZT

Not all AIDS victims are fortunate enough to question medical authority. The resulting tragedies can sometimes turn into a media circus promoting the HIV hypothesis. Of all the cases hyped up for their AIDS scare value, the Florida woman who supposedly caught AIDS from her dentist has become the most notorious.

Kimberly Bergalis: The story began in late 1986, in the small town of Stuart on Florida's Atlantic coast. David Acer, a dentist who had begun his private practice five years earlier, felt a bit under the weather and saw a physician. Acer was also an active homosexual, a fact that led him to seek an HIV test. The result came back positive. Although disturbed by the news, he still felt reasonably healthy and saw no reason to stop his dental practice, nor apparently his fast-track lifestyle.

One year later he experienced worsening symptoms and a visit to his doctor confirmed the diagnosis: full-blown AIDS. A Kaposi's sarcoma covered the inside of his throat and his T-cell count had fallen dangerously low. Both symptoms suggested the extensive use of poppers and other drugs so common in the homosexual bathhouse scene. Acer could see his life slowly wasting away. He continued practicing dentistry while remaining discreet about his sexual life and failing health, making sure to follow the standard guidelines for protecting his patients from infection.

That December, in 1987, he pulled two molars from a nineteen-year-old college student, Kimberly Bergalis. At the time he had no idea the business major would one day be touted as his hapless victim.

The story picks up again in May 1989, when Bergalis developed a transient oral yeast infection. Later that year, during the emotional stress of preparing for an actuarial exam for the state of Florida, she felt some ongoing nausea, and she became dizzy during the test itself. Afterward, the symptoms disappeared. But a brief pneumonia that December sent her to the hospital, where the doctor decided out of the blue to test her for HIV. As chance would have it, she had antibodies against the virus.

Up to this point, none of her occasional diseases differed from the common health problems many HIV-negative people encounter. But the positive HIV test changed her whole attitude, as well as her medical treatment. Within three months the CDC had heard of her case, possibly aided by the presence of several EIS members in the Florida health department, and sent investigators to probe further. The CDC team included such EIS members as Harold Jaffe, Ruth Berkelman, and Carol Ciesielski. Bergalis denied any intravenous drug use or blood transfusions and insisted she was a virgin. During the prolonged examination, the CDC officers stumbled across David Acer's positive HIV status and made the connection to Bergalis. Before the HIV hypothesis of AIDS, no medical expert in his right mind would ever have entertained the slightest thought that a dentist with a Kaposi's tumor and a patient with a yeast infection had anything in common. But in the era of AIDS, doctors tended to discard common sense. That the dentist and patient both carried a dormant virus was enough.

Excited by its discovery, the CDC boldly advertised its results in its weekly newsletter, the same one that nine years earlier had broadcast the first five AIDS cases. The July 27, 1990, issue prominently featured their amazing leap of logic—that the dentist must somehow have infected Bergalis. Naturally, the CDC's speculation leapt straight to the front pages and prime-time television news broadcasts.

Acer died in early September 1990. Bergalis meanwhile sought medical care at the University of Miami, where she was treated with an unidentified "experimental" method. Certainly this was the appropriate place for such therapies. Margaret Fischl, the head of the Phase II AZT trial, worked at that medical center, which had served as one of the twelve facilities sponsored by Burroughs Wellcome for the study. So Bergalis was prescribed AZT.[110]

Suddenly she started a precipitous decline in health. In an angry letter, she partly acknowledged her symptoms resulted from the toxic drug:

> I have lived through the torturous ache that infested my face and neck, brought on by AZT. I have endured trips twice a week to Miami for three months only to receive painful IV injections. I've had blood transfusions. I've had a bone marrow biopsy. I cried my heart out from the pain.[111]

This represented only the beginning. Her yeast infection worsened and became uncontrollable, she lost more than thirty pounds, her hair gradually fell out, her blood cells died and had to be replaced with transfusions, and her muscles wasted away. Her fevers hit highs of 103 degrees, and by late 1990 her T-cell count had dropped from the average of 1,000 to a mere 43. She looked just like a chemotherapy patient—which she now was.

The CDC saw its golden opportunity in the Bergalis case. It publicized a second report on the Bergalis case, announcing its belief that four of Dr. Acer's other patients had also been infected by him, and even surveyed the patients of other HIV-positive doctors and dentists—suggesting that all HIV-positive patients had also been infected by their doctors. Such CDC-funded organizations as Americans for a Sound AIDS Policy (see chapter 10) aggressively promoted public fear with these speculations. A media feeding frenzy resulted, with every major television talk show, and every national magazine, running scare stories.[112] The CDC's relentless publicity had its expected effect: By mid-1991, more than 90 percent of the public believed HIV-positive doctors should be forced to inform their patients of their status, and a clear majority favored banning such doctors from medical practice.[113] Many doctors, angered by the publicity campaign, "accused the federal Centers for Disease Control of unduly alarming the public."[114]

With Therapies Like This, Who Needs Disease? ■ 351

The CDC certainly had an agenda behind its campaign. In July of 1991, the agency issued a set of proposed rules that would require doctors to follow extraordinarily burdensome measures, supposedly to protect their patients from HIV infection. By hyping up the Bergalis case, the CDC had created enough public panic and backlash to favor its proposed regulations. To dramatize the point, Bergalis was brought in to testify before a stunned Senate in October of 1991. Her muscles largely destroyed by AZT, she had to be brought in a wheelchair. Her furious testimony, whispered into the microphone, made a powerful emotional impact on the attentive congressmen and the television audience.

Congress soon passed a new law requiring the states to adopt the CDC guidelines—or else begin losing federal funds. When the medical profession resisted the new rules, the Occupational Safety and Health Administration (OSHA), which works closely with the CDC, stepped in with parallel rules of its own. On threat of criminal prosecution, laboratory and medical workers must now follow incredibly restrictive regulations on their practices and equipment, and must deal with extra bureaucratic red tape.

Blaming her deteriorating condition on the latent virus supposedly passed on by her dentist, Bergalis sued the Acer estate. She received a $1 million award, plus unannounced compensation from the dentist's insurance company. She parceled out the money to a variety of friends, family members, and AIDS organizations, and told her father to purchase "a new, red Porsche and deliver it to my aunt with a large bow on top."[115] Had she known better, she could have instead sued Burroughs Wellcome.

Bergalis died in December 1991 at twenty-three years of age, having taken AZT for up to two years. Her death became the ultimate symbol of the deadly powers of HIV. No one pointed out that, according to the HIV hypothesis, the virus should take ten years to kill its victims, particularly someone like Bergalis with no other risk factors. She had died within four years of her initial visit to Dr. Acer. As her symptoms would indicate, AZT must have killed her instead.

In December 1992, another former patient of Dr. Acer tested

positive for HIV, but had no symptoms. Two months later, eighteen-year-old Sherry Johnson began taking AZT. She has since begun wasting away, admitting she periodically feels sick.

The CDC continued to exploit the Bergalis story as proof of the risk of doctor-to-patient HIV transmission. Some eleven hundred of Acer's two thousand former clients volunteered for HIV tests. Seven of these were positive, including Bergalis, two of them having standard risk factors for AIDS. That left five people who supposedly caught the virus from Acer. Expanding its search, the CDC tested almost sixteen thousand total patients of some thirty-two HIV-positive doctors around the country, finding eighty-four infected patients. Though admittedly baffled by how HIV could pass from doctors and dentists to the patients, the CDC nonetheless advertised the alleged threat. Curiously, when confronted with an unexpected outcome for an unproved test, the CDC did not proceed with caution. It published its findings in July 1990 without further verification.

Apart from HIV being a harmless virus, the evidence that this virus has ever been medically transmitted remains dubious. Based on their own research, insurance companies concluded that the HIV strains in the five patients were different from that in Acer, meaning each caught it from a different source.[116] A study out of Florida State University has backed this conclusion.[117] Even the CDC acknowledged this evidence, though it still preferred to believe the dentist had infected Bergalis. But the CDC's own numbers give away the reality. An estimated 1 million Americans have HIV, in a total population of 250 million. Thus, 1 in 250 Americans have the virus. Five HIV-infected patients of Dr. Acer, out of 1,100 tested, comes to 1 in 220, virtually identical to the national average. So does the proportion of HIV-positives from the patients of the 32 doctors, which works out to 1 in 188. These HIV-positive patients merely represent random samples from the general population.

And where did these people get the virus? As suggested in chapter 6, HIV is probably transmitted much as other retroviruses, from mother to child during pregnancy. There is no evidence that

Kimberly Bergalis's mother has never been tested for HIV antibodies, nor that the mothers of Dr. Acer's other patients were tested. Perhaps Kimberly carried the harmless virus for twenty-three years.

The CDC's theory that AIDS was transmitted from Dr. Acer to his patient began to crumble in the mainstream press in 1994 when an investigative reporter researched the alleged victims of Dr. Acer. "He found weak evidence, shoddy science, and the work of a very accomplished malpractice attorney."[118]

The report first casts doubt on the time course of AIDS transmission from Dr. Acer to his patients. "She developed AIDS just two years after the surgery, and only 1 percent of HIV positive patients develop the full-blown disease that quickly."[119] The investigation disclosed that one of the six other patients that Acer presumably infected had visited the dentist's office only once for a cleaning by a hygienist, not by Acer himself.[120] The report further calls into question the exclusive reliance of the CDC and the malpractice attorney of the "Acer six" on the DNA fingerprinting technique to match Acer's virus with those of his patients. This same technique had also been used to determine that the NIH researcher Gallo had claimed HIV obtained from his French rival Montagnier as his own. Several experts have directly challenged the DNA fingerprinting that linked Acer to his patients, claiming that instead Bergalis's virus matched other HIV strains much more closely.[121] In view of this, a writer in the *New York Times* commented, "The CDC owes it to the public to reopen [Acer's] case."[122]

The re-investigation of the "Acer six" provides unknowingly yet another reason why the "CDC owes it to the public to reopen [this] case": It supports the hypothesis that AIDS is caused by recreational drugs and AZT. Only three of the "Acer six" have developed AIDS, and every one of them was on drugs: Bergalis was on AZT; a thirty-year-old male was involved with "drug dealers, and a homosexual relationship"; and another male was a "notorious crack head."[123]

While on AZT, Bergalis once told a reporter she hoped to also get dideoxyinosine (ddI), another experimental AIDS drug. This drug and ddC, two products of cancer chemotherapy research, work in precisely the same way as AZT. Chemically altered building blocks of DNA, they enter the growing chain of DNA while a cell is preparing to divide and abort the process by preventing new DNA building blocks from adding on (see Figure 1). So, like AZT, ddI and ddC kill dividing cells and have similar toxic effects. They destroy white blood cells and therefore can cause AIDS. The only difference between ddI, ddC, and AZT lies in how easily each is absorbed into the body; people who absorb one evidently may not be equally affected by the other.

Alison Gertz: Both ddI and ddC have begun to claim their victims. In 1988, twenty-two-year-old New York socialite and aspiring graphic artist Alison Gertz entered the hospital for a fever and diarrhea. At some point the doctor decided to test for HIV and found antibodies against HIV. Gertz's transient illness was rediagnosed as AIDS. She had not injected drugs, although her wilder days at Studio 54 bespoke the cocaine and other free drugs available to patrons. A process of elimination traced her infection to a one-night stand with a bisexual male—six years earlier. The announcement left her feeling depressed, but she began a lecture circuit at high schools and colleges, admonishing students that AIDS could come from a single sexual encounter. Television talk shows followed, as did the cover of *People* magazine and Woman of the Year for *Esquire*. Even the World Health Organization circulated a documentary featuring her story.

Gertz started AZT treatment in 1989. The 1990 *People* magazine profile recounted the consequent disaster:

> Last October she was hospitalized with a severe allergic reaction to AZT. When doctors called for a lung biopsy, Ali balked. "I told them if they put me to sleep, I'd never wake up," she says. "My strength was gone." Released after 17 days, she recuperated at home, where her mother and girlfriends took turns nursing her around the clock. "They'd help

me to the bathroom, feed me, see that I didn't fall in the shower," says Ali. "My knees were so bony, I had to sleep with a pillow between them."[124]

The doctors switched her to the still-experimental ddI, which Gertz apparently did not absorb as well and thus allowed her partly to recover. She mixed the powder in her drink twice every day. Her immune system and general health declined, though more slowly. "Gertz remains susceptible to infections like thrush, a fungus that frequently affects the mouth," stated the *People* article. "She has lost 30 lbs. since last summer, naps each afternoon and continues to visit her doctor every 10 days."[125] Ultimately, the ravages of the chemotherapy took her life in August 1992, the news media advertising her death as AIDS-related. She was only twenty-six.

A backlash is now rising against the toxic and irrational treatment approaches to AIDS. In 1993, during the Ninth International AIDS Conference in Berlin, Germany, medical reporter Laurie Garrett was interviewed on the *MacNeil-Lehrer News Hour*. She described the growing discontent among scientists and patients alike:

> Most drug trials were terminated early. The AZT trial was terminated early, ddI, ddC, and so on, and people were allowed as soon as there was any sign that something showed promise to jump out of the placebo arm and get into the treatment arm...
>
> Dr. Anthony Pinching, who was really the leader of most of the clinical research related to AIDS in the United Kingdom, gave a very important speech this morning. I think if he had given this precise same speech a year ago, he would have been booed off the stage, and this morning, he was applauded heavily. And what he basically was saying was we have no idea what drugs work. We have no idea what we're doing in treatment, and it's time to return to the use of placebo trials. He went a step further and said that at least in Europe a lot of AIDS activists and patients now agree, because they're

shocked to find out that the drugs they've been taking, thinking they would be helpful, might even be hurtful.[126]

Arthur Ashe: This lesson almost saved the life of the late Arthur Ashe, the tennis star and one-time Wimbledon champion who died in 1993, supposedly of AIDS. Ashe's medical problems surfaced in 1979 with a heart attack, despite his young age of 36. In December he underwent quadruple-bypass surgery. His chronic heart condition continued plaguing him, and by 1983 he had double-bypass surgery. A blood transfusion during either one of the operations may have carried HIV.

His heart condition and its complications nagged him for several years. Then in 1988 he entered the hospital for toxoplasmosis, a protozoal disease relatively uncommon in humans. The germ resides in cattle and household pets, and in 17 percent to 50 percent of the U.S. population, but most people never succumb to the disease because of healthy immune systems. This also happens to be one of the many diseases on the AIDS list, so the doctor tested and found Ashe to be HIV-positive. Although his toxoplasmosis soon disappeared, Ashe was pronounced an AIDS victim. His disease was retroactively blamed on HIV, not on his heart condition.

Yet his condition hardly seemed contagious. Neither his wife nor his daughter, born three years after his second transfusion, ever developed any AIDS conditions. Indeed, his immune system must have neutralized HIV quite effectively, as Ashe never transmitted the virus to his family.

His daily medicine intake expanded to a virtual pharmacy. He continued to take several drugs for his heart problems, one to lower cholesterol by interfering with liver function, another to slow down the heartbeat, and three others, including nitroglycerin, to lower blood pressure. To these his doctors added a spectrum of antibiotics, all with mild to serious side effects, to prevent the possibility of opportunistic infections. Ashe took Cleocin to fight further toxoplasmosis, nystatin to slow down yeast infections, and toxic pentamidine to stave off *Pneumocystis* pneumonia. Two other drugs were prescribed against possible brain

With Therapies Like This, Who Needs Disease? ■ 357

seizures. Eventually his daily regimen included some thirty pills, only a few of them vitamins.

But just as soon as Ashe received his AIDS diagnosis in 1988, his doctor pushed him into taking AZT. He started on an unbelievably high dose, nearly double the seriously toxic levels used in the Phase II trial. His doctor only gradually lowered the dose over the next four years. "I refuse to dwell on how much damage I may have done to myself taking the higher dosage," Ashe later admitted.[127]

In early 1992 he established an acquaintance that came close to rescuing him. A close friend arranged a series of meetings with Gary Null, a New York–based radio talk show host and nutritionist. Null introduced Ashe to the evidence of AZT's toxicity and against the HIV-AIDS hypothesis, desperately trying to convince him to halt the therapy. For the next ten months, Ashe "wrestled with the possibility of breaking away from the medical establishment to seek alternative treatment for AIDS," according to one columnist. Ashe never met Peter Duesberg, but became familiar with his arguments. "He read everything; he studied what we gave him and asked lots of questions," recalled Null.[128] In October, Ashe announced the lessons he was learning in a column he wrote for the *Washington Post*: "The confusion for AIDS patients like me is that there is a growing school of thought that HIV may not be the sole cause of AIDS, and that standard treatments such as AZT actually make matters worse. That there may very well be unknown cofactors but that the medical establishment is too rigid to change the direction of basic research and/or clinical trials."[129] But psychological pressure stopped Ashe short from rejecting AZT. As Null stated, "He wanted to do it, but he would say, 'What will I tell my doctors?'"[130]

In his 1993 book, *Days of Grace: A Memoir*, Ashe openly acknowledged his interest in alternative AIDS hypotheses:

> But AZT was controversial in other ways. A gift from heaven to many desperate people, it was poison to others. Developed for use in cancer chemotherapy to destroy cells then in the process of actively dividing, AZT was only later

applied to AIDS. Some scientists believe that AZT, which relentlessly kills cells but cannot distinguish between infected and uninfected cells, is as harmful as AIDS itself. After all, HIV is present in only 1 of every 10,000 T-cells, which are vital to the immune system; but AZT kills them all. Dr. Peter Duesberg, the once eminent and now controversial professor of molecular and cell biology at the University of California, who bitterly disputes the notion that HIV causes AIDS, has called AZT "AIDS by prescription."

Dr. Duesberg argues that the use of recreational drugs, not sex, led to AIDS. It is well known that many gay men used— and many of them continue to use—drug stimulus in sexual activity or to facilitate intercourse. "Natural and synthetic psychoactive drugs," he has argued (drugs such as cocaine, amphetamine, heroin, Quaaludes, and amylnitrites and butylnitrites, or "poppers"), "are the only new pathogens around since the 1970s and the only new disease syndrome around is AIDS, and both are found in exactly the same populations."[131]

Ashe faithfully summarized the main points against the HIV hypothesis and for the drug-AIDS hypothesis and explained the deadly effects of AZT and the flaws of its Phase II trial. "Some tolerate [AZT] for a while, then must give it up. Still others cannot tolerate it at all," wrote Ashe. "To my relief, I tolerate AZT fairly easily."[132] With that rationalization, he sealed his fate.

During 1992, his doctors placed him on ddI. Each morning he sprinkled the powder on his cereal, in addition to the AZT pills he swallowed throughout the day. By this time he was wasting away rapidly, his underweight frame hidden by loose clothes. He began rotating in and out of the hospital. January of the following year brought more bad news: Now he had a serious case of *Pneumocystis* pneumonia that his poisoned immune system could no longer fight off. He never recovered. On February 6, 1993, he breathed his last.

The list of celebrity AIDS patients who died on AZT for their belief in medical authority includes ballet star Rudolf Nureyev, who died in 1993, Randy Shilts, the author of the bestseller *And the Band Played On*, who died in 1994, and many more.

As a thoroughly politicized epidemic, AIDS began with a falsehood and ended in tragedy. Virus hunters in the CDC-directed public health movement first made the new syndrome appear contagious. Virus hunters in the NIH-funded research establishment then blamed AIDS on a retrovirus. And virus hunters in the NIH, CDC, FDA, and pharmaceutical industry exploited the situation by resurrecting failed cancer chemotherapeutic drugs for AIDS treatment. In the crisis atmosphere created by the CDC, which allowed no time to think before acting, such toxic drugs as AZT, ddI, and ddC could bypass the normal review procedures and achieve a sanctified monopoly status. The final results have been an unnecessary death toll and an artificially expanding AIDS epidemic.

To make all this possible, the virus hunters from all fields first had to join forces. They have used their combined influence, often behind the scenes, to mobilize the government, media, and other institutions behind a global war on AIDS. Few outsiders have realized just how coordinated the whole strategy has been. The story behind this war, and how its leaders are actively suppressing dissent, is told in the next chapter.

CHAPTER TEN

■
Marching Off to War

THE AIDS EPIDEMIC HAD, by the mid-1980s, already become the salvation of the virus hunters. Out of the ruins of the Virus-Cancer Program had emerged the virus-AIDS program of the NIH. Cancer chemotherapy research had given birth to the AIDS treatment program, including AZT. The CDC had recovered its shaky reputation after the swine flu fiasco and had won a renewed mandate to pursue contagious diseases. The lay public had bought the entire "slow virus" notion, paradoxes and all. Thus, the virologists had maintained their powerful grip on the biomedical research establishment, and the retrovirus hunters had further secured their position at the very top.

But AIDS differed from Legionnaires' disease or the old, intractable problem of cancer. This syndrome had surfaced recently enough to retain its novelty, and it was definitely growing steadily rather than disappearing. Thus, the AIDS epidemic presented the first real opportunity in many years to revive virus hunting on a grand scale. The angst of infectious diseases had originally spawned microbe-chasing wars, the enemies being invisible but deadly germs whose epidemics fed the public's terrified imaginations. Frightful images of polio patients in iron lungs and killer viruses lurking in Africa's rain forests now both haunted and titillated the public. Richard Preston's bestseller *Hot Zone* and *The Coming Plague* by AIDS journalist Laurie Garrett stoked the public's fear of deadly viruses while capturing its imagination. And the Dustin Hoffman movie *Outbreak* in 1994 appeared like a dress rehearsal of the CDC's Ebola Virus epidemic of 1995 (see chapter 5). Fortunately, in both cases, catastrophes were narrowly averted only through the heroic efforts of CDC officials. In *Outbreak*, CDC officials would even fight a two-front war to stop deadly viruses. True to its reputation, the military was ready to blow up an infected community to stop deadly viruses, but thoughtful CDC officials would contain both military brutality and microbial threats with strategic quarantines, saving lives, and banning microbes all at once.

While most people had never gone as far as Howard Hughes in his constant hand-washing paranoia, the fear of germs had sustained the glory days of virus hunting. However, not since polio had the public been so anxious over a "deadly virus," wondering whether it would strike via kisses, one-night stands, mosquito bites, or public toilet seats. For the virus hunters, the time had come to declare war.

By 1986, one man stood out as the obvious general to lead such a war. The veteran polio virologist who had shared the 1975 Nobel Prize for finding the "reverse transcriptase" protein of retroviruses, David Baltimore, had ascended to enormous power among scientists. He had brandished his award to win a large network of allies throughout the research establishment and even in political and financial arenas. The War on Cancer during the 1970s, studies on the immune system by the early 1980s, AIDS in the mid-1980s—each allowed Baltimore to enhance his standing as professor at the Massachusetts Institute of Technology (MIT).

Nor did it hurt to establish connections with financial clout. When biotechnology mogul Edwin C. "Jack" Whitehead offered $135 million to universities, Baltimore seized the chance to establish the Whitehead Institute in 1982. Over the furious objections of many faculty members, he rushed the proposal through and managed to affiliate the new research institute with MIT, from which it lured key professors. Baltimore was named director.

Insiders, awed by his enormous influence, referred to him as "the Pope." By the mid-1980s, any ambitious scientist knew the importance of befriending him, and he always reciprocated; those who fell on his bad side faced trouble. Peter Duesberg, for example, publicly raised questions about the scientific validity of cellular cancer genes, and Baltimore retaliated for several years by blocking Duesberg's election to the National Academy of Sciences.[1] Upon hearing that Duesberg had also become a serious candidate for Germany's highest science prize, the Paul Ehrlich Award, Baltimore interceded with an opposing recommendation for his own friends. Whereas a scientist's typical nomination letter includes a few pages of detailed justification, Baltimore's terse letter resounded with his imperial tone:

> Appropriate nominees for the 1988 and 1989 Paul Ehrlich and Ludwig Darmstaedter Award are: Drs. Robert Weinberg of the Whitehead Institute and Michael Bishop of the University of California, San Francisco. Dr. Bishop also worked very closely with Dr. Harold Varmus and they really should be honored together.[2]

Hilary Koprowski, a close ally of both Robert Gallo and Baltimore, sat on the prize committee and apparently made sure the panel quickly changed its opinion.[3] Duesberg was displaced, while his longtime research collaborator, Peter Vogt, alone received the 1988 prize for their joint research on the Rous sarcoma virus. Although Baltimore's nominees did not themselves win the Ehrlich Prize, Weinberg did receive the General Motors Award in 1987, and Bishop and Varmus jointly won the Nobel Prize in 1989. Baltimore's endorsement was widely recognized as a major factor behind those awards. Such maneuvers also serve to marginalize critics of establishment science. Awards serve to distinguish the darlings from the dissidents of the scientific establishment.

THE BLUEPRINT FOR WAR

In 1986, Baltimore stood at the apex of biomedical research as the most influential of all retrovirus hunters. Thus, when key members of the National Academy of Sciences decided to launch a war on AIDS, they sought him out to lead the charge.

The strategy was simple: a committee of prominent scientific figures would issue a report, outlining a program of increased funding to win over researchers in almost every field, all under central supervision. Responding to public fear of a new, sexually transmitted disease, it would mobilize the whole population regardless of political views. In other words, it would create a national consensus, uniting scientist and nonscientist alike behind the new agenda. Even more ambitious than the wars on polio or cancer, this program would authorize extraordinary measures that might normally meet serious resistance. The research establishment, already by far the largest in history, would expand even more rapidly; federal spending programs, including Medicare, could also grow in size; public health officials could implement emergency controls; and even United Nations agencies and foreign governments could get their share of the largesse.

The Institute of Medicine and the National Academy of Engineering cosponsored the project. Funding came from such notable sources as the Carnegie Corporation of New York, the John D. and Catherine T. MacArthur Foundation, the Andrew W. Mellon Foundation, and the Rockefeller Foundation. Twenty-three prestigious scientists were divided between two panels, with David Baltimore chairing the Research Panel and cochairing the Steering Committee that supervised the whole process. The committee featured Nobel Laureate Howard Temin as well as Paul Volberding and Jerome Groopman, two central figures in the AZT Phase II trials (see chapter 9). David Fraser, the CDC's Epidemic Intelligence Service (EIS) member who had led the Legionnaires' disease episode in 1976, sat on the Health Care and Public Health Panel (see chapter 5). Another EIS graduate, J. Thomas Grayston, sat as chairman of the Epidemiology Working Group. More than one hundred other advisors were listed as participating, reading like a veritable who's who of virus hunting. Robert Gallo and various "Bob Club" members were involved, including Max Essex and William Haseltine—but none of them were elected committee members. Two major CDC officials participated, as did five graduates of the EIS, including Donald Francis. AZT-manufacturer Burroughs Wellcome and ddC-producer Hoffmann-La Roche each sent a representative.

The committee had been given a sweeping mandate to mobilize the entire nation behind the war. As instructed by the National Academy of Sciences, "The committee shall evaluate methods whereby the ultimate goals of controlling and combating the disease may be achieved... The committee shall prepare a report outlining a strategy (or strategies) whereby these concerns can be addressed. The report shall contain recommendations for its implementation directed to the Executive Branch, the Congress, the research community, those who treat patients, state and local governments, corporate leadership, and the public."[4] Naturally, anyone not cooperating with the committee's goals would be labeled counterproductive, if not irresponsible, or dangerous.

All the key players in the AIDS virus hunt had their hands in the project, whether from the NIH, the CDC, or the politically well-connected pharmaceutical companies. The outcome was predictable. After two public hearings and a series of private meetings, the committee released its report in August 1986. Entitled *Confronting AIDS: Directions for Public Health, Health Care, and Research*, the book became the bible of the entire AIDS establishment, its guidelines universally adopted as a blueprint for war. The report made recommendations in four areas:

1. *A broad-based research agenda:* The committee boasted about the discovery of HIV and "its unambiguous identification as the cause of AIDS" as a supposed triumph of heavily funded research.[5]

Obviously the reader was expecting from such a high-ranking committee to receive the definitive scientific evidence for the "unambiguous identification" of HIV as the cause of AIDS. But the best the "blue ribbon committee," as it was labeled by the press, could offer were Gallo and Montagnier and an *ex cathedra* statement that appeared in the 1988 edition of *Confronting AIDS*: "**The committee believes that the evidence that HIV causes AIDS is scientifically conclusive**" [boldface in the original].[6] The committee had no more to offer than its belief in Gallo, Montagnier, and other HIV discoverers.

Starting with this assumption, the report went on to outline an unrestrained agenda of "future research needs."[7] The plan offered something for virtually everyone: Molecular biologists could study the genetic structure of HIV, while biochemists would analyze viral protein functions and crystallographers would examine protein structures. Virologists would inspect every detail of the infection process and develop more tests for HIV, and animal researchers would experiment with mice and chimpanzees alike. Epidemiologists would not only monitor HIV infection in the population, but would also receive vast sums of money to follow cohorts (risk groups) of infected people as they lived or died. Pharmacologists would have their hands full developing a spectrum of drugs to attack the virus, the committee specifically suggesting AZT, ddI, and ddC, among others. Of course, vaccines against the virus would have to be invented. Even the social scientists could join in, studying the risk behaviors for transmitting HIV and trying to understand what psychological barriers might prevent the public from adopting the official AIDS doctrine.

The common denominator was that all AIDS research should be predicated entirely on HIV. Even with new money pouring in, no room would be allowed for alternative hypotheses. The committee casually listed "possible cofactors" that might contribute to AIDS, all implying infectious agents: cytomegalovirus or other microbes, genital ulcers as possible transmission routes for germs, and infection by HIV a second time. Psychological stress and diet were added as possible minor contributors. Drug abuse received no mention at all, effectively blocking any research in that direction.

Every recommendation of *Confronting AIDS* has been slavishly followed. Whereas a few scientists privately questioned the HIV hypothesis after Robert Gallo's 1984 press conference, this 1986 report and its 1988 sequel bared enough teeth to squelch most remaining doubts. The report also suggested that drugs being tested for AIDS treatment need not rely on all the proper controls, especially placebo groups, a recommendation later used in approving ddI and ddC. Further, the committee's call for an increase to $1 billion in federal AIDS research money by 1990 was nearly met by the NIH alone. And, as recommended, much of that money has been used to attract scientists into AIDS research, as well as to train larger numbers of new scientists who add to the growing demand for grant money.

2. *Public financing:* The committee estimated the staggering costs of medical care for each AIDS patient and decided that "it appears likely that future financing of AIDS care will necessarily involve substantial public programs and funds."[8] To make this as open-ended as possible, the report declared that "[t]he committee believes that society has an ethical obligation to ensure that all individuals receive adequate medical care."[9] The meaning of "adequate" was never defined, but since HIV infection was considered inherently fatal and untreatable, this meant providing toxic drugs such as AZT or simply comforting the patient as he died. The committee wanted "to ensure that all persons at risk of infection, seropositive, or already ill could make provision for or otherwise be assured that their potential health care costs will be covered."[10] The high price for one year on AZT, in terms of both health and money ($2,000 per year wholesale, about $10,000 retail), ultimately weighed heavily in those calculations.

The key to getting AIDS to qualify for such coverage lay in classifying patients as being disabled. This designation has allowed Medicaid to cover 40 percent of AIDS patients, and Medicare to pay for a much smaller fraction.

3. *Public health measures:* Any actions to slow the virus were considered justifiable, even if they caused hysteria, encouraged drug

use, or impinged on civil liberties. The public health effort directly conflicted with the medical care program, since AIDS patients had to be classified as disabled to qualify for medical coverage, whereas public health controls depended on classifying patients as infectious and therefore deadly to others. The AIDS establishment was constantly issuing such mixed signals, stretching its public credibility. But in the meantime people have tacitly learned that the AIDS virus is only attacking "homosexuals" and "junkies," and that one is not at risk if one does not practice anal intercourse or does not inject drugs.

In the early days of the Public Health Service, decades before the founding of the CDC, federal officers were frequently dispatched to various cities in the midst of disease epidemics. Tuberculosis, bubonic plague, and yellow fever still swept through periodically but were becoming less common and taking fewer lives as the population became healthier. Hoping to stall epidemics, federal public health agents tried to seize emergency powers by quarantining patients, restricting travel, and taking control of water supplies. But the local citizens and governments usually disagreed with such tactics and resisted attempts at control. In time, serious contagious epidemics disappeared altogether, but not merely because of public health measures. The decline of these epidemics occurred during an era of improved nutrition, the construction of plumbing systems, and rising standards of living.[11]

The memories of popular opposition have remained etched in the minds of public health officials, reminding the CDC of its limits. On the other hand, AIDS presented the first opportunity in many years to revive the old public health campaigns. *Confronting AIDS* recognized this potential, recommending a two-step program of education and widespread HIV testing.

The committee gave examples of what it meant by "education." This included advocating use of condoms and sterile needles for the injection of illicit recreational drugs, targeted not only at the AIDS risk groups but at the population as a whole, on the assumption that everyone was endangered by this supposedly infectious disease. Because pregnant mothers can also transmit HIV to their children, the committee noted that "the Centers for Disease Control advises women at risk of HIV infection to consider delaying pregnancy."[12] Education also meant damage control—in case anyone began questioning the HIV hypothesis of AIDS. In reviewing a survey of men asked about the HIV test, the report expressed dismay at one of the results: "The most disturbing finding from the survey was the number of subjects (a majority) who believed that a positive antibody test somehow conferred immunity, that they had successfully 'fought off' the virus."[13] Education would have to change such commonsense views.

The report recommended easily available HIV testing on a voluntary basis. It also raised the possibilities of reporting these results to central agencies, tracing the sexual contacts of infected people, and quarantining HIV-positives, though it dared not explicitly endorse such measures. But stronger controls were not ruled out. "There may be need, however," the committee stated, "to use compulsory measures, with full due process protection, in the occasional case of a recalcitrant individual who refuses repeatedly to desist from dangerous conduct in the spread of the infection."[14] In other words, infected people who do not willingly follow public health guidelines could be forced to do so.

The committee effectively opened the door to more radical proposals. Donald Francis, the EIS graduate who had played a central role in blaming AIDS on a retrovirus, has formulated such explicit plans. In 1984 he had already summarized the goals of many CDC leaders with a proposal entitled "Operation AIDS Control." He later spelled out these ideas in a 1992 speech to his fellow CDC officials, using the audacious title "Toward a Comprehensive HIV Prevention Program for the CDC and the Nation."[15]

Referring directly to the mandate provided by *Confronting AIDS*, Francis called for five major steps to expand the CDC's authority. First, he wanted the CDC to receive a special status, making it immune from accountability to the voters. "The United States needs to establish a separate line of public health authority that allows for accountability, yet is protected from extremist interference. Perhaps the Federal Reserve is an example to emulate...

Specific legislation should be promulgated to protect CDC from political interference with necessary public health practice."

Second, he proposed "guaranteed health care" for HIV-positives, mostly to lure infected people out of hiding. "If we are going to be successful in identifying HIV infected persons through testing programs," he said, "the necessary incentive must be guaranteed health care financing."

His third point called for condoning drug use, using the logic that providing drugs to addicts would prevent sharing of dirty needles. "Following a more enlightened model for drug treatment, including prescribing heroin, would have dramatic effects on HIV and could eliminate many of the dangerous illegal activities surrounding drugs." Francis even called this "safe injection." But what if heroin itself causes AIDS? If that were the case, taxpayers would be financing the death of addicts.

Fourth, he advocated heavier federal intervention in producing vaccines.

Finally, Francis issued a call to consolidate public health authority in central hands. "Establish clear chains of responsibility," he insisted. "The CDC needs to reestablish its leadership role in HIV prevention. Prevention requires close coordination, training, and financial support of state and local health departments." This would subject all public health functions in the country to the CDC's control.

Francis then revealed how these powers would be used to manage HIV. The CDC would develop a central registry with the identities of all infected people, gathered from every imaginable source. "Whether through hospitals, doctors' offices, sexually transmitted disease clinics, jail health clinics, or whatever, routine testing should be strongly recommended to all patients... the concept of routine voluntary testing for *everyone* should be aggressively promulgated as the standard of medical practice" [emphasis in original]. The sexual contacts of all HIV-positives would be traced and registered as well—not a simple task, considering that fast-track homosexuals typically report hundreds or thousands of sexual contacts.

To push the CDC's "educational" programs in schools, Francis proposed overriding local authority to bypass any resistance from parents. "If, in the opinion of those far more expert than I, schools cannot be expected to provide such programs, then health departments should take over, using as a justification their mandate to protect the public's health."

Little wonder that Francis boasted of "the opportunity that the HIV epidemic provides for public health." Letting down his guard a bit, he revealed the virus hunter within. "The cloistered caution of the past needs to be discarded. The climate and culture must be open ones where old ideas are challenged. Those who desire the status quo should seek employment elsewhere... This is the epidemic of the century, and every qualified person should want to have a piece of the action."[16]

4. Parallel efforts abroad: Finally, the *Confronting AIDS* committee recommended extending parallel efforts to other nations, because "infectious diseases know no national boundaries."[17] This meant scientific research collaborations with foreign scientists as well as public health programs in their countries, including condom distribution. The Agency for International Development (AID) has largely picked up the tab, giving millions of dollars to central Africa.

Confronting AIDS specifically called for increased aid to the World Health Organization (WHO) for public health measures. That same year, the WHO released a book outlining its action plan for contagious disease control, *Public Health Action in Emergencies Caused by Epidemics*. This book was considerably more explicit, describing "quarantine," "mass immunization," "restrictions on mass gatherings," and "restrictions on travel" (including the formation of a cordon sanitaire) as options when the WHO intervenes in a nation's epidemic—measures that would be more politically ticklish to carry out in the United States.[18] Third World citizens, after all, tend to have less power to object to such actions.

The committee proposed that a special commission be established to oversee the implementation of the war on AIDS. This was carried out in the form of the Presidential Commission on the HIV Epidemic, established in 1987 and lasting for one year. The Presidential Commission gathered testimony and issued a report, which merely affirmed the guidelines in *Confronting AIDS*. In 1988 the National Academy of Sciences set up a second committee, which included David Baltimore and EIS graduate Donald R. Hopkins as members, that also reiterated the original blueprint and called for yet another permanent commission to supervise the war. Since 1989, the National Commission on AIDS has been sponsored by Congress and the President and continues to echo the recommendations of *Confronting AIDS*.

At times these officials of the war on AIDS reveal their intent to manipulate public sentiment with carefully crafted propaganda. In 1993, when a major scientific report concluded that AIDS was remaining strictly in risk groups and not spreading into the general population, AIDS officials could not deny the clear evidence. Instead, they angrily denounced the report *for allowing Americans not to be frightened*. David Rogers, vice chairman of the National Commission on AIDS, told a reporter that "his group had worked hard to try to make AIDS a concern to everyone. 'Now to have someone say, "We can relax,"' Rogers said, 'I would much prefer to have them say, "You should worry about your own son and daughter."'"[19]

The initial *Confronting AIDS* report has indeed served as the unquestioned standard against which the entire AIDS establishment has measured itself. Virtually all of its recommendations have been carried out enthusiastically, in the spirit of true warfare. But when fighting a war, one has little time to ask questions or indulge in scientific skepticism. Such a war requires immediate action, not careful thought. To prevent any mutiny, the AIDS establishment has made sure to control its potential opposition.

A RIGGED DEBATE

To prevent serious criticism or opposition from arising against the war on AIDS, whether among scientists or the general public, the entire nation had to be mobilized to participate. Anyone not enthusiastically joining in would be stigmatized with a label of "apathy," and any person raising questions about the HIV hypothesis could be painted as being in a state of "denial."

The NIH carried out its assignment, lavishing billions of dollars on HIV research. Many scientists quickly learned an easy way to tap into this plentiful grant money, while others knew better than to raise questions, an act that peer reviewers could easily punish. Thus, virtually all scientists marched along without hesitation.

The CDC, on the other hand, tended to occupy the front lines as the federal government's major public health agency. Thinking of themselves more as activists than researchers, CDC officers spent much of their time reaching the lay public with prevention measures. Their AIDS activities consisted of more than just HIV testing or distributing condoms and sterile needles; their largest program dealt with mobilizing the public through "education"— convincing the general public to join in the war on HIV. The job of inducing the nonscientists to march fell upon the CDC.

Traditionally, the CDC has disseminated its views largely through state and local authorities, whether health departments, school systems, or other government structures. But the new mandate provided by *Confronting AIDS* demanded greater action. The crux of its mission was to persuade the public "to see HIV-AIDS as an infectious disease," in the words of one CDC official.[20] With large sums of new money appropriated by Congress for the purpose, the agency launched its new initiative.

Only one major obstacle stood in the way. With its messages always attached to the CDC label, the agency would have limited impact on public opinion. So the CDC chose to expand its existing program for increasing influence with other organizations. The agency started with its traditional partners, the state governments. Ten million dollars began flowing from the CDC to the states in early 1985, allocated for new HIV testing sites. The money was carefully linked to "appropriate" counseling as defined by the CDC, ensuring that anyone being tested would hear the official line on AIDS.[21] The CDC has continued financing these testing programs.

Looking beyond state and local governments, the CDC also recognized the potential for spreading its views through private groups, which have greater credibility as independent voices for their constituencies. The CDC had already targeted some of these "community-based organizations" and quietly developed ties with them before the publication of *Confronting AIDS*.

It began with the United States Conference of Mayors (USCM), which in 1984 received CDC money for distributing AIDS information. Within months, the CDC began sending increased funding to the USCM, which in turn distributed the new money to private AIDS organizations. Under CDC monitoring, the USCM helped AIDS groups organize and expand their efforts, and even used the money to start new AIDS groups. Eventually the CDC added more funding through state health departments, which also dispersed it to AIDS activists. By the early 1990s, about three hundred such groups were funded, directly or indirectly, by the CDC. Under central coordination, these groups became so closely interconnected that they constituted a singular web of activists. Push a button at CDC headquarters in Atlanta, Georgia, and a nationwide network of ostensibly private organizations would act in unison. To the public, the whole thing seemed quite spontaneous.

These organizations, their CDC links invisible to most people, actively spread the ominous message of infectious AIDS. Most of these AIDS activist organizations were homosexual groups, through which the CDC view quickly permeated the entire homosexual community. This influence became so pervasive that some information, even life-saving messages, were blocked from reaching AIDS patients. Michael Callen, a twelve-year AIDS survivor (until 1994) who worked to give hope to patients, described unexpected opposition from an AIDS activist:

> Once, after giving my "hope speech" during a public forum organized by the Gay Men's Health Crisis, I was angrily pulled aside by a gay man who worked in GMHC's Education Department. He begged me to stop saying that AIDS might not be 100 percent fatal. Shocked that a gay man

would make such a request, I asked for reasons. He gave three: (1) efforts to persuade gay men to practice safer sex might be undermined because they would "take AIDS less seriously"; (2) it was bad for fund-raising; and (3) it would make lobbying for increased federal funding more difficult. "After all," he said, "if not everyone who gets it dies, then maybe AIDS isn't really the crisis we're being told it is."[22]

After the publication of *Confronting AIDS* in 1986, the CDC started branching out to new types of organizations. While continuing to fund AIDS activists, the agency now also directed its money and influence toward other civic groups that could influence other segments of American society. Tens of millions of dollars, for example, flowed to the American Red Cross in a cooperative agreement that gave the CDC an immense degree of control over the organization. The Red Cross used the money to create and widely distribute many millions of pamphlets, videos, and guides throughout the country, as well as to sponsor innumerable presentations in local communities. The American branch even pulled influence with international Red Cross and Red Crescent societies to spread the CDC doctrine around the world.

The CDC also infiltrated the National Hemophilia Foundation in New York with its inexhaustible resources, thus selling invisibly the HIV-AIDS hypothesis to the fifteen thousand HIV-positive American hemophiliacs. The consequences of this collaboration are particularly tragic, as thousands of HIV-positive hemophiliacs were now encouraged to take the deadly AZT to prevent AIDS from the hypothetical AIDS virus. Indeed, the mortality of hemophiliacs has increased sharply since 1987, the same year AZT was licensed as an antiviral drug. Moreover, the National Hemophilia Foundation's endorsement of the HIV hypothesis rigged the debate about the benefits of highly purified (foreign-protein free) Factor VIII. In the early 1990s highly purified Factor VIII was shown not only to halt, but even to cure AIDS symptoms in HIV-positive hemophiliacs. The results revealed that foreign proteins contaminating commercial Factor VIII rather than HIV caused hemophilia-AIDS, but the benefits of pure Factor VIII were obscured by the simultaneous treatment of HIV-positive hemophiliacs with the toxic AZT.[23]

Vast sums of CDC money were allocated to dozens of minority and civil rights organizations ranging from the National Urban League and the Southern Christian Leadership Conference to the National Council of La Raza and the Association of Asian/Pacific Community Health Organizations. Special "partnerships" were also formed between the CDC and powerful lobbies. Elected representatives in state governments were influenced through the National Conference of State Legislatures; unionized workers were reached through funding to the AFL-CIO and some of its affiliated unions, including the American Federation of State, County, and Municipal Employees (AFSCME) and the Service Employees International Union (SEIU); and school sex education programs were shaped by grants for the Sex Information and Education Council of the U.S. (SIECUS).

The CDC's influence extended into the schools by funding such organizations as the two major teachers' unions, the National Education Association and the American Federation of Teachers, as well as through the National Parent–Teacher Association, the American Association of School Administrators, the National School Boards Association, the Center for Population Options, and many similar groups.

The CDC even managed to exploit the raging AIDS debate between AIDS activist groups and the religious right. In addition to the funding for AIDS activists mentioned above, the CDC formed a partnership with the National Association for People With AIDS (NAPWA). This group cosponsored annual conferences for AIDS activist groups, and the 1992 meeting "attracted over 1,000 individuals representing 578 community-based agencies in 189 cities and 46 states as well as Puerto Rico, the Dominican Republic, Japan, Kenya, Ireland and Portugal."[24] The CDC simultaneously became a partner of Americans for a Sound AIDS Policy (ASAP), which "serves as a resource to the religious community and disseminates HIV-AIDS information and publications through 23,000 Christian bookstores."[25] ASAP became the central source of AIDS material for the religious right and advised Congressman William Dannemeyer.

Funding these two organizations paid off well for the CDC. The AIDS activist movement championed distribution of condoms and sterile needles, while the religious right advocated mandatory HIV testing, tracing of sexual contacts, and sometimes even quarantines. Both sides called for more funding for AIDS research and CDC programs; the CDC could never lose.

The CDC was not the only part of the AIDS establishment providing funds. Burroughs Wellcome joined the act in 1987, once AZT had been approved and was starting to generate hundreds of millions of dollars in sales, profits that the company certainly wanted to protect. Since the company's market lay disproportionately in the homosexual community, it had to be sure of encountering little criticism from that direction. The homosexual press quickly fell into line as Burroughs Wellcome began placing high-cost advertising for AZT in virtually every such publication, large or small. Chuck Ortleb's *New York Native* remained stalwart as one of the few holdouts; most others embraced the extra revenue.

Next the company turned to financing AIDS activist groups directly. Some sixteen thousand such organizations exist in the United States, ranging from relatively mainstream foundations that support research to the more radical groups like ACT UP. Burroughs Wellcome has given money to most of them, particularly the extremist groups with the strongest reputations for fierce independence. After the money began flowing to AIDS activists, many organizations began blunting their criticisms of AIDS dogma while keeping up the fiery rhetoric. Few constituents of these organizations noticed the change, and HIV-positive people continued relying on their advice.

In the center of the AIDS establishment sits the American Foundation for AIDS Research (AmFAR). It was created in 1985 by Michael Gottlieb, the doctor who reported the first five AIDS cases, and Mathilde Krim, the scientist and socialite who helped launch the War on Cancer and now played a central role in the war on AIDS. AmFAR gained prominence through its Hollywood connections—it recruited such top names as Elizabeth Taylor and Barbra Streisand for fund-raising and publicity—and thus came to dominate public relations for the AIDS establishment. The pharmaceutical companies showed their pleasure with generous donations. Burroughs Wellcome announced a staggering $1 million donation in 1992, and the Bristol-Myers Squibb Foundation, connected with the company producing ddI, has also provided funds.

Toward the more radical end of the spectrum is Project Inform, a San Francisco–based watchdog group founded by activist Martin Delaney. This group gained its fame as a bulldog fighting the FDA, based on Delaney's underground testing network that provided various experimental cures to dying AIDS patients. Delaney started as one of the angriest critics of AZT as a toxic drug improperly rushed through clinical trials. He also co-authored a book in 1987, *Strategies for Survival*, in which he warned active homosexual men against the disastrous health effects of such recreational drugs as poppers, cocaine, heroin, and amphetamines. In addition to emphasizing "immuno-suppression" among the effects, he and co-authors Peter Goldblum and J. Brewer admonished their homosexual readers not to ignore the dangers of drug abuse:

> And don't get bent out of shape if your favorite "recreational" high gets a bad rap. Maybe your use of the drug is so moderate, so cool, that you never bump into the gremlins lurking within. It's also possible that you haven't been a "user" long enough for the effects to occur. The damage from most drugs is long-term and cumulative... But don't discount the information out of hand simply because it doesn't agree with your own experience or because you don't want to hear it.[26]

But aid from the pharmaceutical companies seemed to change all that. Burroughs Wellcome donated $150,000 for an upgrade and expansion of Project Inform's computer system, and Bristol-Myers Squibb pitched in another $200,000.[27] Suddenly well-funded, prestigious, and personally consulted by Anthony Fauci, Delaney changed his mind. When he gave a lecture at Stanford University in 1990, the former nemesis of the FDA now praised that federal agency for its work and sympathetically described it as "overworked and understaffed."[28] In publicly attacking Duesberg for questioning the HIV hypothesis, he has issued furious monographs and letters to newspaper editors.

Delaney has also dropped his former opposition to AZT (see chapter 9). His most recent anti-Duesberg letter, published in *Science* in January 1995, makes that very point. It argued that Duesberg did not deserve funding for his research because he had called AZT prescriptions "genocide."[29] But two months later the director of *Project Inform* confessed to an "inadvertent error":

> Apology: In my letter of 20 January (p. 314), I wrote that Peter H. Duesberg "repeatedly and publicly has accused many of those who disagree with him of... genocide." This term was incorrectly attributed to Duesberg in a newspaper report that was the source for the quote. I apologize for this inadvertent error.[30]

Larry Kramer is another example of a seemingly radical AIDS activist who works closely with the establishment. An angry homosexual-rights activist renowned for his bitter language in denouncing AIDS officials, Kramer founded the Gay Men's Health Crisis (GMHC) in 1982. Under his early leadership, GMHC labored to create AIDS activism in a community that did not want to acknowledge the syndrome's existence. Later, after new leaders replaced Kramer and the AIDS epidemic had become institutionalized, GMHC continued spreading the official CDC view of prevention and treatment. This view included endorsing AZT therapy. A former executive director of GMHC did admit to writer John Lauritsen that the group had been receiving money from Burroughs Wellcome, but declined to say just how much.[31]

Kramer went on to found an even more radical group, ACT UP, in 1987. Intent on pushing more drugs through the FDA approval process, ACT UP protesters gained attention by stopping rush-hour traffic on Wall Street, invading corporate and government offices, and disrupting scientific AIDS conferences. Yet very quickly these activists became an integral part of the AIDS establishment. Anthony Fauci, the *de facto* coordinator of the war on AIDS at NIH, began attending ACT UP meetings in 1989 and brought key members (as well as Martin Delaney) into NIH advisory positions. ACT UP members were soon incorporated into AmFAR and other positions of influence.

Burroughs Wellcome also developed close relations with the group. An editorial note in the *San Francisco Sentinel*, referring to local ACT UP/Golden Gate, stated that "ACT UP chapters elsewhere have received millions in contributions from Burroughs Wellcome over the last few years."[32] Certainly the organization has received thousands of dollars. An offshoot group headed by ACT UP/New York member Pete Staley, the Treatment Action Group (TAG), is funded by the pharmaceutical company and in 1992 arranged for the $1 million grant from Burroughs Wellcome to AmFAR. ACT UP is also sponsored as a participant at the annual International AIDS conferences. According to John Lauritsen, at the Ninth Conference in Berlin in 1993, "Most of the 300 ACT UP members had the 950 DM [deutsche mark] entrance fee waived by the organizers. Many had traveled to Berlin, staying in hotels with swimming pools, with all expenses paid by Wellcome. An ACT UP representative from London admitted that his group had received £50,000 from Wellcome."[33]

The relationship has paid off for Burroughs Wellcome and other pharmaceutical companies. Rather than protesting against AZT, ACT UP has demonstrated for cheaper AZT. But not only was Burroughs Wellcome already intending to lower the price, the demonstration itself helped advertise the drug; some of the signs held by activists read, "What good is a cure if you can't afford it?"[34] Only ACT UP demonstrators could get away with calling AZT a "cure." Recently the group has voiced some criticism of AZT, but nothing that will endanger the drug's status. The group also helped Bristol-Myers Squibb win FDA licensing for the toxic drug ddI. Through colorful street actions and high-pressure negotiations, ACT UP maneuvered the FDA into approving ddI on the fast track, without any controlled animal tests and without any studies at all.

ACT UP has also worked to squelch any criticisms of AZT or the HIV hypothesis. John Lauritsen has described his own experiences being "shouted down and silenced" by members when he attended meetings and tried to raise questions, an experience also relayed to us by other AIDS activists.35 At the 1993 International AIDS Conference in Berlin, ACT UP took direct action against a small contingent of HIV dissenters, whose viewpoint had never previously been represented at such meetings:

> In front of the ICC [the conference center], Christian Joswig and Peter Schmidt were attacked by several dozen members of ACT UP, who destroyed signs, burned leaflets, and attempted to destroy camera equipment. Conference officials witnessed these acts, and then ordered the victims of the assault to stay at least 100 meters from the ICC. Officials took no action against the attackers from ACT UP.
>
> Also on the 10th [of June], 100 ACT UP members destroyed a booth belonging to AIDS-Information Switzerland. They chanted obscenities, smashed panels, destroyed displays and chairs, and tore up literature, before covering the remains of the booth with 30 rolls of toilet paper. The Swiss group's sin had been to criticize condoms.36

The communications media is naturally another target for the war on AIDS. The CDC leads this charge, funding such groups as the National Association of Broadcasters, which is "the broadcasting industry's trade association, representing the major networks and some 6,000 individual radio and television stations."37 More important, the entire network of CDC- and Burroughs Wellcome–funded organizations effectively serves as a powerful lobby, both in government and in the media. Politicians and journalists alike, when not consulting the CDC directly, usually get AIDS information from one or more of these many organizations, on the assumption that all these groups function independently.

These avenues give the CDC an impressive hold on the media, allowing it to promote fear of an AIDS explosion while saving the HIV hypothesis from embarrassing public relations disasters. Thus, the CDC has been able gradually to lengthen the supposed latent period between HIV infection and AIDS from ten months to two years, then to five, then ten, and now approaching twelve or more years. The agency has also created an illusion of the spread of AIDS, pumping up the number of cases by redefining the syndrome. In 1985, the CDC first expanded the list of diseases officially called "AIDS." Again this happened in 1987, artificially increasing the annual caseload by 5 percent or more. The definition was expanded again on January 1, 1993, adding such diseases as bacterial pneumonias, tuberculosis, and even low T-cells—in healthy people—as AIDS-defining conditions (see Table 2, chapter 6). The increase in AIDS cases has resulted more from this statistical trick than from a genuine rise in sickness. A more critical media would have noticed such maneuvers and would have asked questions.

Stories can also be squelched when they become too embarrassing. The January 20, 1990, issue of the medical journal *Lancet* contained two adjoining papers on Kaposi's sarcoma. The first, with CDC official Harold Jaffe as senior author, acknowledged that the disease mostly targeted homosexuals with AIDS, consistently avoiding the other risk groups. The second paper reported six HIV-negative homosexuals with Kaposi's sarcoma. Duesberg had previously raised these data as arguments against the HIV hypothesis, a fact ignored in these papers. The authors therefore recognized that Kaposi's sarcoma might not be caused by HIV, but they, and presumably the CDC, wanted instead to blame some other undiscovered infectious microbe. The news media gave this story some attention, but during the following weeks this news only helped the case against the virus-AIDS hypothesis. As a result, the CDC dropped the issue and the story died completely.

But in December 1994 the story was suddenly resurrected. After ten years of championing the HIV hypothesis, *Science* magazine asked, "Is a New Virus the Cause of KS [Kaposi's sarcoma]?"[38] The journalist in charge of the story apparently did not realize the heresy implicit in such a question. After all, Kaposi's sarcoma has done more for the awareness of AIDS than any other AIDS disease.

Kaposi's sarcoma had risen from pre-AIDS anonymity to become the signal disease of the AIDS epidemic. No other AIDS disease has increased so much over its pre-AIDS background as Kaposi's sarcoma. Therefore, "Kaposi" had become synonymous with "AIDS." It was for this reason that practicing physicians accepted AIDS so readily as a "new disease." Most of them had never seen Kaposi's sarcoma in young men before AIDS.

But if HIV was no longer the cause of Kaposi's sarcoma, was it still the cause of other AIDS diseases? That is the question that Duesberg asked in a letter to *Science* on January 20, 1995.[39] If the "domino theory" applied to AIDS, losing Kaposi's sarcoma to another cause could spell trouble for HIV as the cause of remaining AIDS diseases (see chapter 6). The CDC's carefully crafted empire of thirty AIDS diseases was held together only by their presumed common cause, HIV. The CDC realized the danger and immediately stepped in for damage control.

Their strategy would be first to confine the damage by restricting non-HIV AIDS causes to other viruses and then to cautiously call each new virus into question. The CDC's director for HIV/AIDS, Harold Jaffe, and four other CDC officials jointly wrote back to *Science*:

> We read with interest the report, "Identification of herpes-like DNA sequence in AIDS-associated Kaposi's sarcoma"... We hypothesized that if KS is caused by a herpesvirus, antiviral agents with activity against herpesvirus might also decrease the incidence of KS. To test this hypothesis, we examined [...] persons 13 years old or older with Human Immunodeficiency Virus or with AIDS [...] in 10 metropolitan areas in the United States... of the three antiviral medications evaluated, only foscarnet was associated with a significant reduction in the risk for KS.[40]

The letter put HIV back in the picture by testing an antiviral drug in "persons... with Human Immunodeficiency Virus" rather than in persons with the new Kaposi's virus. The result was as obscure as the design of their experiment: "A recent report indicated that KS improved in three of five patients with the use of foscarnet. While it is known that foscarnet has some antiviral activity against HIV, it is doubtful that the activity against HIV alone could account for the reduced risk of KS."[41] Even if HIV proved to lose out against the new virus as the cause of Kaposi's sarcoma, the CDC's letter reconfirmed the legitimacy of the antiviral therapies and prevention developed in the war on HIV.

However, the CDC's defense of viral AIDS carefully avoided any direct reference to Duesberg's strategy to break all HIV-AIDS links because one HIV-AIDS link had been broken—the legal strategy of *falsus in uno, falsus in omnibus*. This was done in a parallel letter by the discoverers of the new "herpesvirus-like DNA sequences" in Kaposi's sarcoma in no mistaken terms:

> The convoluted logic of Duesberg suggesting that our findings support his hypothesis that HIV is not the cause of AIDS escapes us... If one assumes that KS is caused by a herpesvirus that may be transmitted both sexually and nonsexually, continued safe-sex practices by both HIV-positive and negative individuals may limit the spread of this agent as well as that of HIV.[42]

Although the discoverers of the new "herpesvirus-like DNA sequences" had yet to isolate their virus from a sarcoma, they already assumed it was sexually transmitted.

In sum, the status quo had been restored. To think that HIV was not the cause of AIDS, even if it was no longer the cause of Kaposi's sarcoma, was identified as "convoluted logic." Continued safe-sex practices would now be even more necessary than ever to prevent the spread of two AIDS viruses.

An even more spectacular example briefly stunned the world in July of 1992. HIV dissidents had for some time pointed to the existence of people with AIDS diseases but no HIV infection as the definitive argument against the HIV hypothesis. Then, just before the Eighth International AIDS Conference in Amsterdam, *Newsweek* suddenly published an article by reporter Geoffrey Cowley on several HIV-negative AIDS cases. The article mentioned unpublished research by two laboratories suggesting the discovery of a new retrovirus; rumor had it the scientists had leaked the story to *Newsweek* so they could blame HIV-free AIDS cases on a new virus. In any case, one of them had already submitted a paper reporting a new retrovirus to the *Proceedings of the National Academy of Sciences*, which would not be published for several weeks. Anthony Fauci jumped on the potential bandwagon, calling up the editor of the *Proceedings* to pressure him into publishing the paper immediately.

Researchers at the AIDS conference interpreted the *Newsweek* article as a green light and began unveiling dozens of previously unmentioned HIV-free AIDS cases in the United States and Europe. The situation began reeling out of control. Rather than merely promoting the idea of two AIDS viruses, the media fallout unintentionally started re-opening the question of whether HIV caused AIDS. James Curran of the CDC and Anthony Fauci of the NIH raced to Amsterdam on Air Force Two to take charge of the situation. The best they could do was to listen to all reports of such cases and promise to resolve the situation. In reality, they had decided to drop the whole matter.

Three weeks later, the CDC sponsored a special meeting at its Atlanta headquarters. The scientists reporting HIV-free AIDS cases were invited, as was Cowley, the *Newsweek* reporter who first broke the story. The unexplained AIDS cases were relabeled with a highly forgettable name—*idiopathic CD4 lymphocytopenia*, or ICL—so as to break any connection between these cases and AIDS. The ICL cases were then dismissed as insignificant, and Cowley was apparently persuaded to cooperate more closely with the CDC in the future. His next AIDS article toed the official line perfectly, containing little news, and he never again followed up on the growing list of HIV-free AIDS cases.

In February of 1993, a group of papers was published in the

New England Journal of Medicine, accompanied by an article by Fauci with the title "CD4 T-Lymphocytopenia without HIV Infection—No Lights, No Cameras, Just Facts." Ironically imitating arguments straight from Duesberg's critique of AIDS, he concluded ICL must not be infectious at all. Fauci argued that the number of AIDS-defining diseases now called "ICL" was far too large and the diseases were too heterogenous to be caused by a single virus.43 Fauci also insisted that the epidemiology of ICL cases set ICL apart from AIDS, as about a third of all ICL cases were women, compared to only 10 percent of cases in AIDS. However, Fauci seemed to have forgotten some of his very own "facts": that HIV is said to cause all thirty AIDS diseases by itself, and that HIV is said to cause an epidemic in Africa in which 50 percent of the patients are female. After Fauci's article the issue had died, and so did the media coverage. *Roma locuta, causa finita* (Rome has spoken, the case is closed).

Lawrence Altman, the EIS alumnus who had become the head medical writer for the *New York Times*, meanwhile admitted "he knew of cases for several months but did not break the story because he didn't think it was his paper's place to announce something the CDC was not confident enough of to publish."44 No one bothered asking why a top reporter would feel obligated to follow the CDC line.

Duesberg has personally been informed by at least two scientists in the San Francisco Bay area who work with dozens of HIV-negative AIDS patients who, because of local and national pressures, have been intimidated into concealing these cases.45 This may well be a nationwide problem, with untold numbers of HIV-free AIDS victims remaining unreported.

According to the British magazine *Continuum*, fear of the AIDS establishment is not restricted to the United States: "A doctor at Charing Cross Hospital in London, England, just admitted to *Continuum* that he has a case of [Kaposi's sarcoma] in a 'HIV-negative' gay man. The doctor, who wishes to remain anonymous because he fears the consequences of speaking out, has said that many doctors are aware of major problems with the HIV=AIDS hypothesis but 'no one wants to put their head above the parapet.'"46

By funding AIDS activist groups, the CDC and pharmaceutical companies have created yet another type of influence over the media. These activists can either directly co-opt or intimidate reporters, who often depend on the activists for news stories. Lisa Krieger, medical reporter for the *San Francisco Examiner*, revealed some of the more common tactics:

> I'm often attacked by AIDS activists, whose unspoken assumption is that either I'm on their team or against them. When I write about the need for reform in San Francisco's AIDS infrastructure, I am called "uninformed," "insensitive" and "racist." When I dare criticize an AIDS organization, I am told I am "writing self-serving drivel" or passing on "rumor, gossip and innuendo."47

Though denying she is influenced by such pressure and insisting that her reporting is objective, Krieger nevertheless admitted some of her normally unspoken biases in her attitude toward AIDS reporting:

> How could I resist, in my weaker moments, becoming an AIDS advocate? I want to reach through the newsprint and grab the reader by the collar, as if that would somehow shake the complacency with which the public has come to accept this disease. I want to applaud every new clinical drug trial, elevate AIDS activists to sainthood.48

One startling example of such media bias appeared in September 1993, when a New York City resident who rejects the HIV hypothesis of AIDS wrote a letter to the *New York Times*. Although critical of AZT therapy, the letter was published. However, the newspaper actually *added* words; immediately following a reference to HIV, the *Times* inserted the phrase "which causes AIDS."49

CENSORSHIP IN THE MEDIA

Aside from inviting docile journalists to meetings and conferences and funding AIDS activist groups, the CDC and NIH have one other powerful tool for maintaining media cooperation. Elinor Burkett, a courageous *Miami Herald* reporter who wrote a major article covering the HIV-AIDS debate, explained it best as a question of "access":

> If you have an AIDS beat, you're a beat reporter, your job is everyday to go out there, fill your newspaper with what's new about AIDS. You write a story that questions the truth of the central AIDS hypothesis and what happened to me will happen to you. Nobody's going to talk to you. Now if nobody will talk to you, if nobody at the CDC will ever return your phone call, you lose your competitive edge as an AIDS reporter. So it always keeps you in the mainstream, because you need those guys to be your buddies...
> When you call the CDC on the phone, and I called them certainly on a regular basis when I was writing that piece, they say things to you like "You will be responsible for people in Miami stopping using condoms, if you write that article." Do I want people in Miami to stop using condoms? Of course not!... There's all kinds of blackmail, and I don't mean overt blackmail. It's emotional blackmail of that sort, and it's the fact that exactly what I knew was going to happen, happened, which is, I can't get a phone call returned by any of them.[50]

Faced with growing dissent against the HIV hypothesis since 1987, the generals in the war on AIDS have openly hinted at using such tactics. They have repeatedly made clear their preference that dissidents confine the debate to scientific circles, keeping it out of the public eye. When asked by one reporter to answer Peter Duesberg's challenge, David Baltimore instead condemned the entire viewpoint as "irresponsible and pernicious" and two years later warned in a scientific letter that "Duesberg's continued attempts to persuade the public to doubt the role of HIV in AIDS are not based on facts."[51] Prominent retrovirus hunter Frank Lilly (deceased, October 1995), a member of the original Presidential Commission on the HIV Epidemic, angrily responded to a presentation by Duesberg at the Presidential AIDS Commission in New York in 1988 and declared, "I regret that it has become a public question."[52] Upset by sporadic but growing media coverage of Duesberg, AmFAR sponsored a special scientific meeting in April of 1988. Supposedly the conference would air all views on the cause of AIDS, but it "was really an attempt to put Duesberg's theories to rest," admitted one of the many journalists present.[53]

Fauci stated the point more bluntly in 1989, declaring in an editorial that Duesberg's ideas were nonsense and complaining that his views were receiving too much publicity. "Journalists who make too many mistakes, or who are sloppy," he warned, "are going to find that their access to scientists may diminish."[54] And in a 1993 letter to the journal *Nature*, two of the most powerful virologists in Italy bared their teeth:

> Your subtitle ends: "He should stop." Or, we submit, "should he be stopped?" For example, should he somehow be prevented from appearing on television to misinform individuals who are at risk from the disease? One approach would be to refuse television confrontations with Duesberg, as Tony Fauci and one of us managed to do at the opening day of the VIIth International Conference on AIDS in Florence. One can't spread misinformation without an audience.[55]

Such threats do work when leveled by the well-funded AIDS establishment, frightening the media and overriding their natural fascination with such a newsworthy story. Based on internal documents faxed to Duesberg by an anonymous source, key officials of the United States government specifically engineered a strategy for suppressing the HIV debate in 1987 while Duesberg was still on leave at the NIH. The operation began on April 28, less than a month after Duesberg's first paper on the HIV question appeared in *Cancer Research,* apparently because several journalists and homosexual activists began raising questions.

That day, a memo was sent out from the office of the secretary of Health and Human Services (HHS), headed by the words "MEDIA ALERT." Describing the situation created by Duesberg's paper, the staff member ominously noted that "[t]he article apparently went through the normal pre-publication process and should have been flagged at NIH" (there is no reason any scientific paper should be "flagged" by any government agency). The staffer then pointed out the threat to the government:

> This obviously has the potential to raise a lot of controversy (If this isn't the virus, how do we know the blood supply is safe? How do we know anything about transmission? How could you all be so stupid, and why should we ever believe you again?) and we need to be prepared to respond. I have already asked NIH public affairs to start digging into this.[56]

Copies of the memo were addressed to the secretary, under secretary, and assistant secretary of HHS, as well as to the assistant secretary for public affairs, the chief of staff, the Surgeon General, and the White House.

A parallel memo was issued by the NIH on the same day. Its author was Florence Karlsberg, the public relations officer interviewed at about the same time by John Lauritsen, and the memo was addressed to top NIH officials. "I want to alert you about some incidents that have occurred in the last 24 hours," Karlsberg wrote. She listed several public inquiries about Duesberg and emphasized, "DHHS is quite anxious and is awaiting feedback re NIH/NCI response to, and strategy for, this provocative situation." Commenting that "Bob Gallo and others have tried to educate Peter [Duesberg] re: HTLV-III [HIV] and AIDS—but it's hopeless," Karlsberg recommended creating a response team consisting of NIH epidemiologist William Blattner, Dani Bolognesi,

Anthony Fauci, and Robert Gallo to deal with the controversy. "Perhaps the epidemiologic approach might be more productive in countering Peter's assertions."[57]

Within two days, Blattner drafted a three-page memo. In it he marshaled a list of evasions and pieces of circumstantial evidence that would later become the standard defense of the HIV hypothesis used by all scientists and government agencies.[58] By June, he had reworked a third draft as a potential press release. But the memo was never released to the public. Instead, the NIH and other officials adopted a policy of silence, hoping to discourage further interest by the media.

By December, the strategy was clearly failing. In another internal NIH memo dated December 30, Karlsberg wrote a fellow staffer that the Blattner memo "was not pursued in June because Paul [an NIH staffer] suggested at that time that this project be put aside temporarily—at least until necessary." She continued:

> Alas—in the past few months, inquiries have been mounting... The calls and interest are mounting. Perhaps it's time to review and activate the attached STATEMENT.

The statement, signed "Florence" and entitled "HIV: The Cause of AIDS," contained at the bottom a handwritten response, initialed "PVN," that read "I guess it is time to get off the dime. This isn't going away."[59]

Get off the dime they certainly did. The Blattner memo was apparently revised and expanded and the names of Robert Gallo and Howard Temin were added as co-authors. It was published in July of 1988 as one-half of the debate with Peter Duesberg in *Science* magazine (see chapter 6). Naturally, the piece was not identified as the product of NIH planning. But that was to be the last time the AIDS establishment would publicly engage Duesberg in debate. Heightened controversy, after all, might backfire on the NIH, attracting attention rather than discouraging media interest.

Indeed, the major media were already learning of the controversy over HIV and were becoming curious. So the official war on

AIDS turned to more covert tactics, such as quietly yanking the leash of "access" to pull the media into line.

The *MacNeil-Lehrer News Hour* sent camera crews to interview Duesberg in early 1988, planning to do a major segment on the controversy. But when the February 8 broadcast date arrived, the feature had been pulled. Apparently AIDS officials had heard of its imminent airing and had intercepted it. A few months later, the program aired a short, tepid segment, with half the time now taken up by Fauci debunking Duesberg.

Meanwhile, the ABC daily program *Good Morning, America* also discovered the story and arranged to fly Duesberg to New York for an in-studio interview. He arrived Sunday night, February 20, and was booked into the Barbizon Hotel. But that very evening he received a call from the studio to announce that something had come up, and the interview was canceled. Turning on the television the next morning, he saw Fauci connected by satellite, filling Duesberg's time slot and discussing every aspect of AIDS *except* the controversy over HIV. Pressured by the editor of the *New York Native* and other dissidents, a *Good Morning, America* film crew eventually did fly to Berkeley, and a short segment was broadcast—again "balanced" with Fauci.

The story repeated itself twice with the Cable News Network (CNN). The second time, for example, a film crew flew out to interview Duesberg, planning to broadcast a half-hour special during the 1991 International AIDS Conference in Italy. Once again, the show was killed at the last minute, and a shorter version, only a couple of minutes long, reached the airwaves long after the conference was over.

A similar plan to interview Duesberg on national Italian television was also killed during the conference. One of the "killers" proudly identified himself in a letter to *Nature*: "refuse television confrontations with Duesberg, as Tony Fauci and one of us managed to do at the opening day of the VIIth International Conference on AIDS in Florence."[60]

The *Larry King Live* program, carried on CNN, scheduled a half-hour satellite interview with Duesberg for August 6, 1992.

Suspicious that something might again go awry, Duesberg called the producer a few hours before live broadcast. Sorry, she told him, something urgent has just come up regarding the election. Duesberg turned on the television that evening to discover that he had been replaced, not by an election issue, but by Fauci and the president of AmFAR. Neither surprise guest mentioned the controversy over HIV, nor did Larry King.

Duesberg has appeared on major national television only twice. The first time was on March 28, 1993, on the ABC magazine program *Day One*. Even in this case, according to one producer, Fauci tried get the show canceled days before broadcast. On April 4, 1994, Duesberg got his second chance to make his case on national television. This time it would be on Ted Koppel's *Nightline*, which had promised it would be "Fauci-proof." The interviewer hired by Koppel, Kary Mullis—who had just won the Nobel Prize for the new technique that AIDS scientists relied on so heavily to find traces of the elusive HIV—was not to be intimidated by anybody's AIDS-speak, not even Fauci's. But when the program finally aired, a few months after taping, there was Fauci again. After fifteen taped minutes for the dissidents—Mullis, Duesberg, and others—Fauci took over the balance of time debating Root-Bernstein live on cofactors for HIV. Clearly, Fauci proves to be a faithful stand-in for Duesberg-AIDS television programs, and he certainly can be counted on when it comes to AIDS thought control.

Such influence with the media by AIDS officials extends overseas. An award-winning English producer aired a one-hour documentary on Duesberg and the HIV controversy in June of 1990, timed to coincide with the International AIDS Conference in San Francisco. The program, entitled "The AIDS Catch," leaned in Duesberg's favor, and the British press lavished it with advance praise right up to the day of broadcast. But when the British medical and public health establishment retaliated with stern condemnations, the press turned around and began criticizing the program. After the show aired, the Terrence Higgins Trust, an AIDS organization funded mostly by the British government and partly by Burroughs Wellcome, filed a legal complaint against the program that prevented its rebroadcast or further distribution.

The long arm of the AIDS establishment reaches even the President of the United States. Jim Warner, a Reagan White House advisor critical of AIDS alarmism, heard about Duesberg and arranged to sponsor a debate in January of 1988. This would have forced the HIV issue into the public spotlight, but it was abruptly canceled days ahead of time, on orders from above.

Nor has the print media been exempt from such pressure. The first national publication to show interest was *Newsweek*, where Duesberg met with a senior writer in March 1987. However, the magazine had just arranged a special honorary dinner for Robert Gallo, in its Washington, D.C., office, a few days hence. Maybe a story could be done later, the writer told Duesberg. Four years later, that day seemed to arrive shortly after an editorial in *Nature* that favored Duesberg. Photographers showed up at his laboratory, taking photos for a story to appear immediately. But that article was canceled within days.

Pulitzer Prize-winning journalist John Crewdson, a former *New York Times* writer now on staff with the *Chicago Tribune*, discovered the controversy and became excited at the prospect of breaking a new investigative story. In November of 1987, he took Duesberg to dinner and showed strong interest, and by the following month had written an article on the HIV controversy but then, as he has since admitted to a mutual contact, he ran into editorial roadblocks and ended up writing articles on the Gallo virus-stealing scandal. He has expressed a genuine desire to cover the debate over HIV but feared the political pressures.[61] By early 1993, these pressures finally led him to join forces against Duesberg, and he threatened to publish an article refuting his position for good. Nothing has yet happened. Since Robert Gallo's acquittal on scientific misconduct charges, Crewdson has once again indicated potential interest in the HIV debate.

The *New York Times* has mentioned Duesberg only three times in seven years, every time attacking him. The *Washington Post* has done likewise, with one hostile article and one small, neutral piece. The *San Francisco Chronicle* intended to cover the story, until it encountered opposition from scientists in the local AIDS establishment. In 1989 *Rolling Stone* had commissioned a freelance writer from New York to write a Duesberg article, but then canceled it during the interview with Duesberg at his lab at the University of California at Berkeley. *Harper's* magazine canceled a major article in 1990 after having commissioned it from a freelance reporter who spent three years on the piece.

The *Los Angeles Times*, unlike other major newspapers, has covered the issue a few times. But each article underwent extraordinary editorial review, even when written by veteran staff reporters, ultimately being framed in terms slightly hostile to Duesberg and always accompanied by a piece attacking his position. In June of 1993, its weekly magazine published an article criticizing AZT therapy. The freelance writer had been commissioned for the piece several months earlier, but had been subjected to such a gauntlet of editors that she had to write off nine drafts, with many key facts being deleted in the process. She calculated her final pay for that article as $3 per hour of work. The previous year, she had spent many months writing a specially commissioned article for *Esquire* magazine; that story had been killed altogether.

One chemist who wrote to *Time* magazine discovered that *Time*, too, was consciously refusing to cover the sensational HIV debate. In a noncommittal response letter, the editor wrote, "We appreciated your call for the coverage of the theories of Peter Duesberg, and have brought your comments to the attention of the appropriate editors. We have been aware of Duesberg's challenge to the mainstream concept of AIDS for several years, and continue to monitor the debate he has set in motion."[62] Then the letter referred to unpublished data supposedly refuting Duesberg's position.

Again, this censorship extends to other countries. A star reporter for Germany's *Bild der Wissenschaft* was shocked when her article on Duesberg was canceled without explanation, while *Der Spiegel* went so far as to attack Duesberg in 1993 and again in 1995 without allowing him to respond. In general, smaller or independent regional periodicals have proven much more willing to cover the HIV debate than have national publications. After all, the larger media depend more heavily on access to government scientists and public health officials.

CENSORSHIP IN THE PROFESSIONAL LITERATURE

Having averted serious media publicity, the AIDS establishment directed its power toward isolating and neutralizing Duesberg within scientific circles. A scientist's career depends heavily on peer-reviewed grant money, peer-reviewed opportunities to publish in scientific journals, and invitations to conferences. These vulnerabilities became the targets for sanctions by AIDS officials.

Robert Gallo and some other scientists began refusing, for example, to attend scientific conferences if Duesberg would be allowed to make a presentation. So in 1988, when an old colleague and friend of Duesberg's finally arranged a meeting on retroviruses on the Greek island of Crete, he dropped Duesberg's name from the announcement. Incredulous, Duesberg called back only to find that the apologetic long-term collaborator could not allow him to give a lecture, or the meeting would fall apart. Since that time, Duesberg rarely has been invited to retrovirus meetings and virtually never to AIDS conferences, despite seminal contributions to the field, including the isolation of the retroviral genome, the first analysis of the order of retroviral genes, and the discovery of the first retroviral cancer gene.

Since then, however, Duesberg has received invitations to three major meetings to which Gallo had also been invited. In all three cases, Gallo carried out his threat. These included a retrovirology meeting in New York in 1989, when Gallo excused himself because of disease in the family. The next opportunity for a Duesberg-Gallo match was scheduled for a hematology meeting in Hannover, Germany in 1990. Gallo was already in Hannover for a lecture before the conference. But after a breakfast with the conference organizer and Duesberg at the beginning of the conference, Gallo suddenly disappeared, citing disease in the family for his premature departure. Disease in the family seemed to be a predictable coincidence whenever Gallo was scheduled to meet Duesberg at a public forum. The next opportunity arose at a cancer meeting in Bonn, Germany, in 1993. Gallo was slated to deliver the opening lecture at the Bonn meeting, but canceled a mere three hours ahead of his scheduled appearance, citing disease in the family as his excuse. The notice was sent from Hamburg, only a few hundred miles away, where Gallo had lectured the previous day.

After the 1987 paper in *Cancer Research*, publishing suddenly became unbelievably difficult for Duesberg. Papers, especially on AIDS, would constantly run into obstacles at every turn, from hostile peer reviewers to reluctant editors. Even in the *Proceedings of the National Academy of Sciences*, where Academy members such as Duesberg have an automatic right to publish papers without the standard peer review, he nevertheless encountered serious trouble.

In June of 1988 he submitted a paper to the *Proceedings*, providing new arguments and evidence against the HIV hypothesis. The editor promptly rejected it, citing lack of "originality" in the paper's viewpoint.[63] A new editor meanwhile took the helm, and Duesberg invoked his rights as an Academy member and protested. The new editor took up the issue, nervously pointing out the paper was "controversial" and insisting he could not publish it without peer review.[64] The next several months brought three hostile reviewers, dozens of disputed points, and tense negotiations covering more than sixty pages, but finally the paper appeared in February 1989. The paper hinted at its extraordinary history with only a special disclaimer: "This paper, which reflects the author's views on the causes of AIDS, will be followed in a future issue by a paper presenting a different view of the subject."[65] Robert Gallo was asked to write a rebuttal, but never did.

In August 1990, Duesberg submitted another paper, this time arguing that drug use is more tightly associated with AIDS than is HIV. Again the editor promptly rejected the paper, arguing that it was too long. Forced to split the paper in two, one part documenting that AIDS was not contagious and the other that drugs are the cause, Duesberg submitted the two shorter papers.

Following two peer reviews and several months of protracted haggling, the editor relented and published the less controversial one that questioned infectious AIDS.

Duesberg resubmitted the other half of the paper—the one arguing that drug use causes most AIDS. This time his paper was doomed. Although Duesberg had already taken advice from four scientific colleagues in writing it, the paper was subjected to three anonymous reviewers by the editor. Two of the three voted to block publication, one of them calling any questions of the HIV hypothesis "extreme and highly dubious" and warning that the drug-AIDS hypothesis "has a potential for being harmful to the HIV infected segment of the population." This particular reviewer admitted, "I am no expert in the fields concerned," and none of the three could point to factual errors in the paper.[66] At this point a new editor replaced the previous one, and Duesberg tried again with a modified paper. The new editor added four new reviewers who, though unable to find serious flaws, all voted to kill the paper. One reviewer even suggested the real reason was that if the paper were published, "one is further tempted to blame the victim."[67]

Trying once more, Duesberg had fellow Academy member Harry Rubin submit the paper after running it by four independent reviewers, all of whom recommended changes but favored its publication. The editor completely ignored those opinions, selecting three more anonymous reviewers who again voted down the paper by late 1991. One year after first being sent to the *Proceedings*, the paper was completely dead. This decision made Duesberg the second member in the 128-year history of the Academy to have a paper rejected from its journal; apparently, the other had been Linus Pauling, who had argued vitamin C might prevent cancer.

But the AIDS establishment made its most effective counterattack by going after Duesberg's funding, the lifeblood of any scientist's laboratory. In 1985 the NIH had awarded him an Outstanding Investigator Grant (OIG), a special seven-year grant designed to give accomplished scientists the freedom to explore new ideas and directions without constantly having to apply for new funding. The time for renewal application arrived in 1990, two years before the grant would finish. But that October, Duesberg received the shocking news: His rating by the peer review committee was so low as to guarantee the grant would be discontinued, whereas two-thirds of the competing OIG applications were approved. Though referring to Duesberg as "one of the pioneers of modern retrovirology," the committee betrayed its real motives by complaining that he had ventured off to question the cause of AIDS. According to the reviewers, "Dr. Duesberg has become sidetracked" and "can no longer be considered at the forefront of his field... More recent years have been less productive, perhaps reflecting a dilution of his efforts with nonscientific issues."[68]

The very fact that a group of top researchers would consider the questioning of orthodox views in science as "nonscientific" comments powerfully on how completely science has been turned upside down since it had become totally dependent on the centralization of funding in the NIH. In this case, moreover, the deck had been deliberately stacked against Duesberg. Of the ten specially selected reviewers, two had severe conflicts of interest. Dani Bolognesi was a Burroughs Wellcome consultant who tested AZT for the company, and Flossie Wong-Staal was a former researcher for Robert Gallo. Of the remaining members, Duesberg accidentally discovered that three had never reviewed the grant at all, and a fourth had only given his recommendation by phone—a favorable one. Thus, it would appear that the NIH had rigged the outcome.

Naturally, Duesberg protested vigorously but received only a brush-off. Throughout the next two years, he waged an unceasing battle to save his grant. First, the University of California at Berkeley refused to endorse his appeal to the NIH, without which he could not legally proceed. As with most universities, virtually the largest source of income was from research grants, especially from the NIH, and the university must have feared retaliation. Duesberg also could not get a straight response from the NIH. He then turned to his Congressman, Ron Dellums, whose staff aide began writing inquiry letters. The secretary of Health and Human Services, Louis Sullivan, responded dismissively, admitting familiarity with the Duesberg case but denying any irregularities in procedure. Further correspondence brought equally vague answers from Bernadine Healy, the director of NIH.

This continued for many months but, after an article in a national academic newspaper embarrassing to the university, Duesberg won university endorsement and the NIH agreed to investigate. After stalling yet another nine months, the NIH announced in early 1993 that the grant proposal would be reviewed from scratch. For a short while, the situation seemed to be improving.

Then in March, while the new committee was reviewing the grant proposal, the journal *Nature* suddenly published a string of articles publicized as definitive proofs of the HIV hypothesis. Michael Ascher and a team of epidemiologists, funded on an NIH contract from Anthony Fauci, wrote a commentary asserting that among a group of a thousand San Francisco men, only those with HIV developed AIDS, regardless of drug abuse.[69] Two weeks later, Fauci himself published a paper boasting that he had found large amounts of HIV hiding in the lymph nodes of infected people. A third article backed up Fauci's claim on the virus detection. At the time, *Nature* issued press releases advertising the papers, and the news media excitedly buzzed with the news that Duesberg's AIDS viewpoint had finally been disproved.

Only months later, when the dust began to settle, did the claims begin to unravel: Ascher and colleagues had used improper and misleading statistical methods on poorly collected data.[70] Every one of the AIDS patients in Ascher's study was a homosexual who had used nitrite inhalants in addition to cocaine and amphetamines, and 84 percent had also been on AZT prescriptions.[71]

The definitive argument to refute the drug-hypothesis would have been to find a group of AIDS patients who had never used any drug. Since that was not possible, Ascher and colleagues had to make an arbitrary choice between two independent AIDS correlations, HIV and drugs. Naturally they chose antibodies against HIV as the correlation that was the cause. However, in order to make the HIV correlation 100 percent, Ascher's data had to be "adjusted" in two ways: First, using the AIDS definition—one of thirty diseases plus antibodies against HIV—to their advantage, Ascher and colleagues left out forty-five patients with AIDS-defining diseases but without HIV.[72] This adjusted the HIV antibody-AIDS correlation to 100 percent. Second, Ascher and colleagues drew a curve showing a group of drug-free, HIV-positive patients labeled "seropositive—no drug use," who did not even exist in their article. For emphasis, the curve was even drawn on a blue background, which is unnecessarily expensive and very rare in a scientific journal. But the graph with the nonexistent, drug-free AIDS patients was faithfully reproduced by the *San Francisco Chronicle* and many other newspapers.

Duesberg wrote to *Nature* inquiring about the source of the "drug-free" men and trying to point out the logical holes in the Ascher paper, the biggest of which was Ascher's attempt to refute the drug-hypothesis with AIDS patients who had all used a multiplicity of drugs including nitrites, amphetamines, cocaine, and even AZT.[73] Indeed, Ascher's AIDS patients were nothing short of walking pharmacies.[74] The editor, John Maddox, not only refused to publish the letter, but advertised the censorship in a full-page editorial, boldly entitled "Has Duesberg a Right of Reply?" The answer, according to Maddox, was no. The editor then revealed the hidden reason behind stifling the response: Duesberg had asked "unanswerable rhetorical questions."[75] This was the editorial to which the prominent Italian virologists gleefully responded—openly calling for further censorship, as mentioned above.

But the *Lancet* published the Duesberg letter inquiring about the drug-free AIDS cases,[76] and *Genetica* recently published a re-analysis of Ascher's et al. database that confirmed Duesberg's suspicion that there were no drug-free AIDS patients in Ascher's study.[77] Ascher and colleagues tried to argue their way out of the dilemma making undocumented claims in letters to the *Lancet* and to *Science*. But now, two years later, neither Ascher nor *Nature* ever identified the source of the "drug-free" men with AIDS.[78]

Fauci's own *Nature* paper, boasting large amounts of virus in AIDS patients, actually analyzed just three patients who showed only tiny amounts of dormant HIV genes, even in the lymph nodes and no infectious virus at all.[79] Two patients contained a dormant HIV gene in one thousand T-cells, and one contained a dormant HIV gene in one hundred T-cells.[80] Those were the skimpy data that inspired *Nature* editor Maddox to (1) write his own editorial "Where the AIDS Virus Hides Away";[81] (2) call on two "Bob Club" members, Dani Bolognesi and Howard Temin, to write yet another editorial "Where Has HIV Been Hiding?";[82] and (3) to launch an international press release offering a draft for an article ready for each newspaper to print. All this for a few dormant HIV genes in three AIDS patients. Ironically, even Ascher and his colleagues later turned on Fauci, criticizing his paper in a letter published in *Nature* for its skimpy data on virus in AIDS patients.[83]

The review committee again voted down Duesberg's grant proposal a few months later. This time the rating was low enough to discontinue the grant, but not so startlingly low as to appear abnormal. Nor did any reviewers hold obvious conflicts of interest other than being retrovirologists studying HIV and animal viruses. They did, however, complain about Duesberg's questioning attitude as the major obstacle to funding him and singled out his AIDS debate as an example.

Since then, every one of his seventeen peer-reviewed grant applications to other federal state or private agencies—whether for AIDS research, on AZT and other drugs, or for cancer research—has been turned down. The most spectacular example is the fate of a grant proposal for testing the health hazards of nitrite inhalants, or poppers, in mice. Duesberg had applied with an internationally respected inhalation toxicologist, Professor Otto Raabe from the University of California at Davis. The proposal had actually been inspired by Harry Haverkos, director of the Office on AIDS at the National Institute on Drug Abuse (NIDA), during a visit to Duesberg's lab in 1993. Haverkos had long favored nitrites as a cause of AIDS, particularly of Kaposi's sarcoma. To advance the nitrite-AIDS hypothesis Haverkos had organized a conference on the subject and then edited the conference's proceedings for the NIDA monograph *Health Hazards of Nitrite Inhalants*.[84]

The Duesberg-Raabe proposal followed the classical reductionist approach in which scientists try to reduce a complex problem to a single cause by eliminating competing alternatives. In this case the proposal set out to distinguish between nitrite inhalants and retroviruses as causes of AIDS-defining diseases in experimental mice. However, despite having a high-level ally in the federal agency, the proposal was turned down. The AIDS study section at the NIDA that reviewed the application quickly realized the imminent danger to the virus-AIDS hypothesis: The project could prove that nitrite inhalants are sufficient causes of immunodeficiency, pneumonia, Kaposi's sarcoma, and other AIDS diseases, and would thus discredit the HIV hypothesis. They knew what to do—and did it in 1993, in 1994, and again in 1995. The review committee acknowledged the proposal's strength, but refused to award it any rating at all ("Not recommended for further consideration"). The only consistent argument against the proposal was the lack of "preliminary experiments." But "preliminary experiments" are not a requirement for a grant application: an innovative idea, exhaustive knowledge of the literature, and professional competence are.

Informed about the latest rejection of the popper grant, even Haverkos was dismayed, but he offered a possibly life-saving lesson of "grantsmanship." Haverkos advised to avoid the AIDS issue altogether and to rewrite the application as a response to a specific NIDA *Program Announcement* requesting research on the "medical and health consequences of drug abuse" issued by himself and the director of NIDA. The announcement invites independent investigators to study "a possible link between inhaled nitrite use and Kaposi's sarcoma" and encourages "studies of nitrite inhalant and other drug use... to determine relationships between substance abuse and health outcomes" pointing out that "animal studies are encouraged"—exactly what Duesberg and Raabe had proposed for three years in row. Haverkos even offered to bulletproof the proposal by rewriting it himself; unfortunately, he sighed, it would be hard to rewrite the name "Duesberg." The proposal was resubmitted for a NIDA review in the fall of 1995. On November 21, 1995, even this application, which responded to NIDA's request for research on the medical consequences of nitrite inhalants, was rejected.

Privately, two high-ranking NIDA officials acknowledged to Duesberg that neither NIDA nor any other federal institute was sponsoring even one study on the long-term effects of recreational drug use. Given that we are more than ten years into an epidemic that can't be dissociated from drug use—even with more than $35 billion spent—this is a remarkable situation. The occasions of these private contacts were two recent NIDA conferences on "AIDS and Drug Abuse," one in Gaithersburg, Maryland, in May 1994 and the other in Scottsdale, Arizona, in June 1995. Thus, it would seem that drugs are acceptable topics as possible causes of AIDS at NIDA conferences, but not acceptable study-objects for HIV dissidents.

The chilling effects of silencing tactics extend even onto the campus itself. In March 1993, Duesberg was scheduled to give a keynote speech on AIDS to a Los Angeles meeting of alumni of the University of California at Berkeley. He flew in the evening before, only to learn that three colleagues called up the conference organizers demanding that he be balanced with an opposing speaker or be canceled. The speech was nevertheless delivered as planned and received an enthusiastic response from the audience. Among those who applauded was University of California Berkeley Chancellor Chang-Lin Tien, who is a staunch supporter of academic freedom. In private conversations Tien considered alternative views the only chance to solve the AIDS crisis. Duesberg eventually discovered that one of the three professors who protested his speech was a member of his own department in charge of advising graduate students.

Several fellow professors maneuver against Duesberg in various ways. His promotions in pay are blocked and his teaching assignments are restricted to difficult undergraduate laboratory courses rather than the coveted graduate lecture courses. While other faculty sit on committees governing teaching policies, courses and curricula, speaker invitations, and hiring of faculty, Duesberg is placed in charge of the annual picnic committee. More important, graduate students are discouraged from entering

Duesberg's lab during their decision-making first year, advice that can be psychologically intimidating to such inexperienced students. Under the condition of anonymity, several students have confessed to such pressures more than once.

By 1994 *Nature* editor Maddox had identified himself so much with the HIV hypothesis that he wanted more than the symptomatic treatments of HIV dissent with censorship and the never-ending string of "new studies," Maddox wanted a cure. Maddox had made his wishes perfectly clear in editorials, calling on the dissidents to concede defeat: "When he [Duesberg] offers a text for publication that can be authenticated, it will if possible be published—not least in the hope and expectation that his next offering will be an admission of recent error."[85] "The danger for the Duesbergs of this world is that they will be left high and dry, championing a cause that will have ever fewer adherents as time passes. Now may be the time for them to recant."[86] "Those that have made the running in the long controversy over HIV in AIDS, Dr. Peter Duesberg of Berkeley, California, in particular, have a heavy responsibility that can only be discharged by a public acknowledgment of error, honest or otherwise. And the sooner the better."[87]

Maddox's opportunity to defeat HIV dissent for good offered itself in September 1994. At that time Duesberg got a call from an old friend, who is now a high-ranking geneticist at the NIH, for an urgent personal meeting on a professional matter. An excited Duesberg asked what professional matter could be so private that it required a personal meeting. Was it about AIDS? About cancer? The voice at the other end of the phone said the subject was simply too hot to be discussed over the phone, but he could be in San Francisco in twenty-four hours. The next day the two met at the opera in San Francisco. After some small talk about the old days, the topic quickly shifted to AIDS and suddenly a paper was on the table at the opera café: "HIV Causes AIDS: Koch's Postulates Fulfilled." The paper was signed by three authors: Duesberg's friend, another NIH researcher specializing in epidemiology, and, surprisingly, by Duesberg as well. It had been commissioned by *Nature* editor John Maddox.

The NIH geneticist argued that by now Duesberg could safely sign on because the evidence for HIV had grown so overwhelming that nobody would listen to arguments against it, no matter how reasonable these arguments were. By continuing his opposition to HIV, Duesberg would even risk his credentials for having discovered cancer genes.[88] In his touching appeal, the geneticist deplored that the scientific community had ostracized Duesberg without a fair trial and that the proposed paper would open the doors for Duesberg's reentry into the establishment. The paper would be in press the next Tuesday, when the NIH geneticist would have dinner with Maddox in London—provided the authors would reach consensus on the subject.

With a promise for a carefully considered decision before the Tuesday meeting with Maddox, Duesberg returned his friend to the airport. The decision was to convert the paper in two: one essentially unchanged but without Duesberg's name; the other a rebuttal written by Duesberg. This proposal would have put both sides of the debate on an equal footing—but it proved to be the end of this most unusual invitation to publish in *Nature*.

THE CHANGING TIDE

Although the war on AIDS has achieved a life of its own, its original momentum flowed largely from the power of David Baltimore, the cochairman of the *Confronting AIDS* committee. But despite his many allies, even he ultimately proved to be vulnerable.

Baltimore's reign began quietly unraveling in 1986, but few people noticed at the time. An immunology paper he published that year with several colleagues came under fire when one of the authors stepped forward to charge fraudulent research—that some of the reported experiments were never really performed. Baltimore's clout prevented investigations for several months; then, despite the evidence, both MIT and Tufts University cleared the paper of wrongdoing. The NIH spent a full year probing the matter and also exonerated the authors in January 1989. Disturbed, Michigan Congressman John Dingell held hearings to revive the case. By May of 1989, Baltimore finally ran into serious problems. Dingell had prodded the Secret Service to investigate the experimental notebooks kept by one of Baltimore's fellow authors, Thereza Imanishi-Kari. She was caught having faked her data, using ink not in existence when she supposedly carried out the experiments. Now the NIH reopened its own investigation.[89]

Baltimore's influential friends came to his rescue. Dozens of top scientists campaigned on his behalf, including testifying to Congress. That October, as his reputation was suffering, he even received a career-rescuing offer from the prestigious Rockefeller University in New York. The board of trustees, prodded by wealthy banker and fellow member David Rockefeller, asked Baltimore to serve as president of the university. The faculty opposed the move, embarrassed at the thought of having a fraud-tainted leader:

> In fact, said Richard M. Furlaud, chairman of the board, the opposition was so strong that Baltimore "actually withdrew his candidacy because of it." But the board—and David Rockefeller—weren't giving up. Furlaud and Rockefeller flew to Cambridge to persuade him to change his mind. "Mr. Rockefeller said, look, we still think you're the right person to do the job," recalls Furlaud. "And then he accepted [the role of candidate]."[90]

Over strenuous objections, the trustees pushed the nomination through and handed the presidency to Baltimore in July 1990. Rockefeller himself pulled strings to have Baltimore invited into such exclusive private clubs as the New York–based Council on Foreign Relations. The tensions at the university simmered for another year before the NIH finally released its report on the fraud probe, which after two years of delays concluded that some of the data were indeed faked. Baltimore suddenly had to retract the paper, but dismayed his colleagues by publicly defending it anyway. The controversy erupted into open rebellion as three of the university's top scientists left to take jobs elsewhere. Again Baltimore's friends stepped in, and David Rockefeller donated $20 million to the university as evidence of his "absolute confidence" in Baltimore's presidency.[91] The money briefly held back the opposition, but when yet another leading scientist announced his departure, Baltimore finally had to resign as president. On December 3, 1991, he retreated from the ruins of his former position to continue HIV research in his lab.

Confronting AIDS has not withstood the test of time much better. Even though AIDS officials still refer to its authority, the report has gradually been tarnished by its failures. For example, it predicted a total of more 270,000 American AIDS cases through 1991, including 74,000 new cases during 1991, and a grand total of 179,000 deaths by that time.[92] Using the same 1985 CDC definition of AIDS, only 167,000 AIDS cases had actually been tallied through 1991—a little more than half the predicted level. The CDC filled most of the gap by expanding the AIDS definition, but such tricks cannot work forever.

Ultimately, the war on AIDS has failed to save lives, the only test that really counts. Condoms, sterile needles, and widespread HIV testing have made no measurable impact, except to arouse frustration and despair among the HIV-positives and fear among the HIV-negatives. And such toxic chemotherapies as AZT have been recklessly prescribed to people who might otherwise have lived. The NIH, the CDC, and the virus hunters have been winning this war, but the rest of us have been losing.

To win the war on AIDS one must first know its cause. The proverbial strategy of "First find the cause, then fight the cause," is the only rational strategy to win that war. The next chapter proves that the cause of AIDS is already known, and that the scientific basis for a rational war on AIDS is already at hand, even though the evidence is obstructed by the propaganda of the HIV-AIDS establishment.

CHAPTER ELEVEN

Proving the Drug-AIDS Hypothesis, the Solution to AIDS

MOST AMERICANS FIRST HEARD about the psychedelic drugs in the 1960s when drugs had become chic as symbols of nonconformism. Drugs united nonconformists of all denominations including rock stars, Vietnam war protesters, sex gurus, and intellectuals. Now millions of Americans are daily users of cocaine, nitrite inhalants, amphetamines, heroine, LSD, marijuana, PCP, and other psychoactive drugs. Since the 1960s every administration has paid more than its predecessor into the apparently unwinnable war on drugs. This war is fought to restrict Americas newest vice by "supply control" and education.[1] The costs of this war have escalated just as much as the epidemic it fights—to a current level of $13 billion a year.[2]

But hardly anybody knows that the highest price of the American drug epidemic is the tens of thousands of drug diseases and drug deaths that it generates each year. Indeed, the drug epidemic appears to have generated the first really new disease epidemic in the Western world since World War II. But epidemiologists and medical researchers are quick to explain that exact numbers are hard to come by because the drugs involved are all illegal and because drugs may not be toxic by themselves.

As often in the history of science, the biggest obstacle in finding the truth is not the difficulty in obtaining data but the bias of the investigators on what data to chase and how to interpret them. The wars on drugs and AIDS are perfect examples: The same government spends $13 billion annually to fight the war on drugs and $7.5 billion to fight the war on AIDS, and nothing has been achieved. The drug warriors use unpopular legal and military force to chase drug suppliers, and the AIDS warriors use scientific methods to chase viruses. These strategies are based on the bias of the current medical establishment that recreational drugs are basically not toxic[3] and that all diseases of drug addicts are caused by deadly viruses and microbes. For example, *Science* asserts "heroin is a blessedly untoxic drug,"[4] provided it is injected with a "clean needle." Thus, drugs are fought because they are illegal and microbes because they are thought to be deadly.

But what if drugs caused AIDS? And what if the AIDS epidemic were the product of the drug epidemic? The public would no longer have to fear microbes, but drugs. Thousands of HIV-positive people would no longer have to accept their inescapable and imminent AIDS death. Drug addicts could prevent diseases by stopping drug use. Hundreds of thousands of HIV-positives would be spared the toxicity of AZT. The war on drugs could be won by pointing out that drugs cause AIDS and other diseases, just as the war on tobacco is being won by pointing out that smoking causes emphysema, lung cancer, and heart disease. An unpopular war against drugs would become a popular war against disease.

The answer to these questions depends on unbiased scientists collecting drug-AIDS data and doing drug-AIDS experiments. But since 1984 the AIDS establishment has either ignored,[5] misunderstood,[6] or even misrepresented[7] drug-AIDS connections in favor of its darling HIV. The AIDS establishment has even succeeded in discrediting its very own pre-1984 drug-AIDS hypothesis.[8]

Hardly anybody can remember that only ten years ago AIDS was still considered by many scientists a collection of diseases acquired by the consumption of recreational drugs. Since nearly all early AIDS patients were either male homosexuals who have used nitrite and ethylchloride inhalants, cocaine, heroin, amphetamines, phenylcyclidine, LSD, and other drugs as sexual stimulants (see Table 1, page 418), or were heterosexuals injecting cocaine and heroin intravenously, early AIDS researchers named these drugs as the causes of AIDS (see chapter 8).[9] Drugs seemed to be the most plausible explanation for the near-perfect restriction of AIDS to these risk groups because drug consumption is their most specific, common denominator. This original drug-AIDS hypothesis was called the *lifestyle hypothesis*.[10]

Although official statistics have since replaced drugs by HIV as the common denominator of AIDS, recreational drugs have never left the major AIDS risk groups, i.e., male homosexuals and intravenous drug users, to this date (see chapter 8). In fact, the HIV hypothesis has tightened the drug-AIDS connection, having united all AIDS risk groups, including even hemophiliacs, transfusion recipients, and "other categories" by the prescription of AZT and other anti-HIV drugs. Indeed, the drug hypothesis stated below provides the only consistent explanation for American and European AIDS:

> All AIDS diseases in America and Europe that exceed their long-established, normal backgrounds are caused by the long-term consumption of recreational drugs and by AZT and its analogs. Hemophilia-AIDS, transfusion-AIDS, and the extremely rare AIDS cases of the general population reflect the normal incidence of AIDS-defining diseases in these groups plus the AZT-induced incidence of these diseases under a new name.[11]

The key to the drug hypothesis is that only long-term consumption causes irreversible AIDS-defining diseases. Occasional or short-term recreational drug use causes reversible diseases or no diseases at all. With drugs, *the dose is the poison*. Toxicity of drugs is first a function of how much is taken at any given time. But the untold price of frequent drug use is the cumulative toxicity that builds up over a lifetime, causing irreversible damage. The more drugs are consumed over time, the more toxicity is accumulated. Therefore, it takes twenty years of smoking to acquire irreversible lung cancer or emphysema, and twenty years of drinking to acquire irreversible liver cirrhosis. Therefore, it takes about ten years of nitrites, heroin, amphetamines, or cocaine to develop AIDS.[12] And therefore it takes less than a year of the much more toxic drug AZT to cause AIDS by prescription.

Surprisingly, even the HIV-AIDS orthodoxy acknowledges that drug use is high among American and European AIDS patients.[13] But it insists that "Duesberg's drug hypothesis"[14] must be rejected, and even censored, for ethical reasons: Knowledge of the drug hypothesis would call into question the HIV hypothesis. And questioning the HIV hypothesis would promote unsafe sex and its known and perceived consequences.[15] However, the HIV hypothesis deserves no veto power and no immunity, having achieved no therapy, no prevention, and not even scientific proof. Yet it has vetoed all other alternative AIDS research since 1984!

Despite the escalating drug use epidemic,[16] all drug-AIDS connections have been ignored since 1984:

1. There are three million to eight million American cocaine addicts and about six hundred thousand heroin addicts,[17] and a third of all American AIDS patients are intravenous drug users—but there is not a single experimental study funded by the NIH, the CDC, the National Institute on Drug Abuse (NIDA), or any other division of the Department of Health and Human Services that investigates the long-term effects of cocaine and heroin addiction in animals.

2. Millions of mostly male homosexual Americans, including many with AIDS, are addicted to nitrites (see Table 1)[18]—but there is currently not even one study funded by NIH, the CDC, or even the NIDA to study the health hazards of long-term nitrite consumption in experimental animals.

On the contrary, applications to study the long-term effects of recreational drugs that are consumed by 97 percent of all American AIDS patients[19] (see chapter 8) are rejected by federal institutions with the instruction to the applicant that diseases of drug addicts are caused by HIV. The inside story of how the NIDA

Proving the Drug-AIDS Hypothesis, the Solution to AIDS ▪ 413

has rejected four consecutive applications in 1993, in 1994, and twice in 1995 to study the health hazards of nitrite inhalants in mice by one of the world's leading inhalation toxicologists, Professor Otto Raabe from the University of California at Davis, and the retrovirologist Peter Duesberg from the University of California at Berkeley—each with the score "Not recommended for further consideration"—has been told in chapter 10 and elsewhere.[20] But despite their firm stand against experimental tests of the nitrite hypothesis, both the NIDA and the CDC have just reconfirmed the nitrite-Kaposi's sarcoma "link"[21] and have warned about new increases in nitrite consumption.[22] By contrast to the United States' tight fist on drug-AIDS money, no money seems enough to sponsor HIV-AIDS research. More than one hundred thousand researchers in the United States have studied unsuccessfully for ten years how HIV might cause AIDS—more than one researcher for every one of the seventy-five thousand annual AIDS patients.

3. There are currently at least two hundred thousand HIV-positive people, including many with AIDS, who are prescribed AZT, ddI, ddC, other DNA chain terminators, and other experimental anti-HIV drugs. Since 1986 the AIDS establishment has spent billions of dollars to bring these drugs into human bodies, but it has yet to fund the first study to test the health hazards of the indefinite prescription of such drugs in animals. No animal tests—just human experiments!

Sincere studies of the drug-AIDS hypothesis would measure toxicity over years of recreational or medical use, granting drugs the same "long latent periods" that HIV is granted to cause AIDS. But no such studies are done. This is not an oversight. It reflects the mindset of the current medical orthodoxy—that neither recreational drugs nor even AZT are intrinsically unhealthy.[23]

The complete absence of research on the health hazards of recreational drugs has attracted the attention of Republican Congressman Gil Gutknecht. On March 24, 1995, Gutknecht sent a formal letter to the secretary of Health and Human Services, Donna Shalala, inquiring why the United States does not fund any research on the health hazards of recreational drugs (see chapter 12 for full text). Four months later, on July 10, 1995, the following answer was received from the secretary: "AIDS prevention programs continue to be based on our understanding of scientifically defined HIV transmission modes because prevention of AIDS is prevention of HIV. To deviate funds from scientifically sound findings to those that lack evidence would be unconscionable" (see chapter 12). Thus, direct experimental tests of the drug hypothesis are currently not possible in the United States. Therefore the drug hypothesis stands untried in the courts of experimental science.

However, even a complete prohibition of experimental science by the current orthodoxy cannot suppress scientific truth forever. Proof of drug toxicity already exists in the scientific literature[24] and is provided for open-minded observers on a daily basis in the form of AIDS-diseases in drug users with and without HIV.[25] Despite the current prohibitions on experimental verification, the drug-AIDS hypothesis can be verified by the standard tests of scientific hypotheses, for their ability to identify plausible cause, and, above all, to make valid predictions.

The correct hypothesis of AIDS must (1) explain why an agent is a plausible cause of one or all of the thirty fatal AIDS diseases and (2) predict all clinical and epidemiological aspects of AIDS. The drug hypothesis meets these criteria to the letter, but the HIV hypothesis does not.

DRUGS PLAUSIBLE CAUSES OF AIDS

A plausible cause for immunodeficiency, weight loss, dementia, and muscle atrophy must account for the loss of billions of human blood cells, muscle cells, brain cells, and a plausible cause for Kaposi's sarcoma must be a potent carcinogen. At the doses consumed, recreational drugs can easily provide plausible chemical explanations. A person inhaling 1 milliliter of amylnitrites, takes up about 6×10^{21} nitrite molecules—that is, 6×10^7 nitrite molecules for every one of the 10^{14} cells of the human body. Just a few of these molecules could kill a cell or cause a cancer if they reacted with specific sites of human DNA. A person prescribed

500 mg of AZT per day takes up about 10^{21} AZT molecules—10^6 per cell. And just one of these molecules is sufficient to kill a cell, the task AZT was originally designed to meet in chemotherapy.

Similar numerical ratios of drugs per human cell apply to cocaine, heroin, and amphetamines used at recreational doses of 0.1 to 1 gram. At these concentrations drugs significantly alter the metabolism of neurons and other body cells—the reason why these drugs cause the desired, psychoactive effects. At slightly higher doses, termed *overdoses*, they are directly lethal, accounting for the thousands of "hospital emergencies" and drug deaths recorded annually in the United States[26] (see Figure 2B, chapter 8). At recreational doses, cocaine, heroin, and amphetamines work as catalysts, accelerating and altering normal human functions beyond normal tolerances. They are also indirectly toxic via malnutrition, insomnia, lack of sanitation, and the many economic and social consequences that come with their high price and illegitimacy (see chapter 8). Thus, the drug hypothesis can offer plausible chemical causes and can even offer specific drugs for "risk-group–specific" AIDS diseases (see below and chapter 8).

By contrast, the HIV-AIDS hypothesis cannot offer a plausible scientific cause for AIDS. Even in people dying of AIDS only one in about five hundred T-cells is ever infected by HIV. Moreover, in most of these infected cells, HIV is just a dormant gene making no viral molecules at all[27]. There is not even one authentic precedent in biology of a dormant gene having any effect, let alone causing a fatal disease. André Lwoff's dormant bacterial killer virus (phage) is the classical example of what to expect from a dormant gene—nothing (see chapter 4). As long as the virus' genes remain dormant, they coexist in monotonous harmony with healthy host bacteria over thousands of generations. But once activated by ultraviolet light, the dormant killer virus wakes up, makes plenty of deadly molecules, and kills its bacterial host within twenty minutes. The same rules of gene control apply to humans. Every human cell contains the same genes as every other. But as long as the "nose-gene" is active and the "liver-gene" is not, the nose will remain a nose for the duration of an individual's life.

Thus, the HIV hypothesis cannot provide a plausible chemical cause for any one of the thirty fatal AIDS diseases. This is acknowledged as the Achilles' heel of the HIV hypothesis even by its most pious advocates.[28]

DRUG HYPOTHESIS PREDICTS AIDS—EXACTLY

The correct scientific hypothesis must be able to predict the outcome of an experiment, regardless of whether man or nature is experimenting. The following eight examples show how the drug-AIDS hypothesis meets this condition exactly and how the HIV hypothesis fails each test.

American AIDS is restricted to intravenous drug users and male homosexuals who practice risk behavior.

Drug hypothesis: Since 1981, 94 percent of all American AIDS cases have been from risk groups that have used recreational drugs. About one-third of these were intravenous drug users and two-thirds were male homosexuals[29] who had practiced risk behavior by using oral recreational drugs (see Table 1) and AZT.[30] Thus, recreational drug use and AZT explain the restriction of AIDS to drug users.

HIV hypothesis: Since its beginning in 1981, viral AIDS should have long entered the general population, just like all authentic infectious diseases. The failure to leave specific risk groups in more than a decade discredits the virus hypothesis.

Nine out of ten American and European AIDS patients are males.

Drug hypothesis:
(i) According to the NIDA and the Bureau of Justice Statistics, more than 75 percent of *hard*, recreational drugs are consumed intravenously by males.[31] The CDC reports that women are now

the fastest growing AIDS risk group.[32] This also correlates with drug use statistics. According to the federally supported Drug Strategies program, women now claim an increasing share of hard drugs: "Women account for the fastest-growing population in jails and prisons, in large part because of drug offenses."[33]

(ii) The CDC and independent investigators report that nearly all male homosexuals with AIDS and at risk for AIDS are long-term users of oral drugs such as nitrite inhalants, ethylchloride inhalants, amphetamines, cocaine, and others to facilitate sexual contacts, particularly anal intercourse.[34] The largest study of its kind, which investigated nitrite inhalant use in a cohort of more than three thousand male homosexuals from Chicago, Baltimore, Los Angeles, and Pittsburgh, reports a "consistent and strong cross-sectional association with... anal sex."[35] Table 1 lists examples of drug use by male homosexuals with AIDS or at risk for AIDS reported by the CDC and other investigators.[36]

(iii) Many HIV-positive homosexuals are prescribed AZT as an antiviral drug (see chapter 9).[37]

Since intravenous drug users, who are 75 percent male, make up one-third of all AIDS patients, and male homosexuals make up almost two-thirds of all American AIDS patients, the drug hypothesis explains why nine out of ten American AIDS patients are males. The same applies to European AIDS.[38]

HIV hypothesis: According to the hypothesis that AIDS is a sexually transmitted viral disease, AIDS should have long equilibrated between the sexes—exactly as predicted by the AIDS establishment. All other sexually transmitted diseases are equally distributed between the sexes.[39] Since 1981 the wives of the fifteen thousand HIV-positive hemophiliacs should also have contracted AIDS from their husbands. But none of this has happened to date in the United States and Europe.[40]

418 ■ INVENTING THE AIDS VIRUS

TABLE 1

Drug use by homosexuals with AIDS and at risk for AIDS

Percentage of study participants using drugs

Drugs	(1) Atlanta 1983: 50 AIDS, 120 at risk	(2) San Francisco 1987: 492 at risk	(3) San Francisco 1990: 182 AIDS	(4) Chicago 1990: 3,916 at risk	(5) San Francisco 1993: 215 AIDS	(6) Vancouver 1993: 136 AIDS	(7) Chicago 1995: 76 at risk
nitrite inhalants	96	82	79	most other drugs total: 82	100	98	79
ethylchloride inhalants	35-50				many	many	47
cocaine	50-60	84	69		many	many	
amphetamines	50-70	64	55				
phenylcyclidine	40	22	23				
LSD	40-60		49				
metaqualone	40-60	51	44				
barbiturates	25	41	30				
marijuana	90		85				
heroin	10	20	3				
alcohol			46				
cigarettes			33				16
none reported				18			47
AZT					most	most	

Proving the Drug-AIDS Hypothesis, the Solution to AIDS ▪ 419

Pediatric AIDS in America and Europe is restricted to babies born to drug-addicted mothers.

Drug hypothesis: According to the drug hypothesis, babies acquire AIDS diseases from cocaine and heroin shared with their mothers during pregnancy.[41] Indeed, about 80 percent of pediatric AIDS cases in America and Europe are children born to mothers who were intravenous drug users during pregnancy[42] (see example 8 below). The remainder reflects the normal, low incidence of AIDS-defining diseases among newborns.

HIV hypothesis: All babies born to HIV-positive mothers should have AIDS. However, since HIV is a harmless, perinatally transmitted retrovirus, only babies born to drug-addicted mothers develop AIDS (see chapter 6 and example 8 below). For example, thousands of HIV-positive, healthy recruits are identified (and rejected) each year by the U.S. Army, although they have probably been HIV-positive since their date of birth.[43]

Why AIDS now?

Drug hypothesis: In the United States, recreational drug use has increased over the past decades from statistically undetectable levels to epidemic levels at about the same rate as AIDS.[44] For example, cocaine consumption increased two hundred–fold from 1980 to 1990, based on cocaine seizures that increased from 500 kg in 1980 to 100,000 kg in 1990.[45] During the same time, cocaine-related hospital emergencies increased from 3,296 cases in 1981, to 80,355 cases in 1990, and to 119,843 in 1992.[46]

In the past three years, the increase of cocaine consumption has slowed down at the expense of increases in heroin consumption, which were accompanied by increases in heroin-related hospital emergencies.[47] Heroin-related hospital emergencies doubled, from more than thirty thousand in 1990 to over more than sixty thousand in 1993.[48] Nitrite consumption jumped from a few medical applications to millions of doses annually in the 1980s.[49] According to a recent report from the NIDA and the CDC, "nitrite use has increased in the 1990s in gay men in Chicago and San Francisco" after a decline in the 1980s.[50]

The dosage units of amphetamine confiscated in the war on drugs jumped from 2 million in 1981 to 97 million in 1989.[51] On this basis the Bureau of Justice Statistics estimates that amphetamine consumption has increased one hundred–fold during the same time.[52]

Drug offenders are now the "largest and fastest-growing category in the federal prisons population, accounting for 61 percent of the total, compared with 38 percent in 1986. The number of federal drug offenders increased from about five thousand in 1980 to about fifty-five thousand in 1993. In 1993, between 60 percent and 80 percent of the 1.2 million prisoners in the United States had been on illicit drugs.[53]

The German "Rauschgiftbilanz" reports an 11.2 percent increase in the consumption of illicit recreational drugs in 1994 compared to 1993.[54]

Consider a grace period of about ten years to achieve the dosage needed to cause irreversible disease, and you can date the origin of AIDS in 1981 as a consequence of the drug use epidemic that started in America in the late 1960s during the Vietnam War. Indeed, AIDS increased from a few dozen cases annually in 1981 to about one hundred thousand in 1993 (see chapter 6, Figure 2A).[55] Note that the spread of AIDS and the spread of cocaine and cocaine-related hospital emergencies are parallel since 1981.

Since 1987, AZT and other DNA chain terminators have been added to the list of toxic drugs consumed by AIDS patients and those at risk for AIDS. AZT is now prescribed to about two hundred thousand HIV-positives worldwide.[56]

Thus, the drug hypothesis explains (i) why the AIDS epidemic occurred when it did in America and Europe and (ii) why it spreads steadily according to drug consumption.

HIV hypothesis: Since HIV is an old virus in the United States and is established in a steady population of one million ever since it was detectable in 1984, it cannot explain a new epidemic. Moreover, according to Farr's law (see chapter 6), a new, infectious epidemic

Proving the Drug-AIDS Hypothesis, the Solution to AIDS ■ 421

should have exploded, but AIDS did not. The spread of AIDS and the nonspread of HIV, the hypothetical cause of AIDS, are entirely incompatible with each other since 1984.

Not all drug users get AIDS.

Drug hypothesis: There are currently between 3 million and 8 million cocaine addicts and 0.6 million heroin addicts in the United States.[57] In 1980, 5 million Americans had used nitrite inhalants. In 1989, at least 100 million doses of amphetamines were consumed in the United States.[58] Most of the 401,749 American AIDS cases since 1981[59] have been recruited from this large reservoir of drug users.

According to a 1994 survey of the NIDA, "more than 5 percent (221,000) of the 4 million women who give birth each year use illicit drugs during their pregnancy."[60] These mothers are the reservoir from which most of the 1,017 pediatric AIDS cases reported in the United States in 1994 were recruited.[61]

Unfortunately, scientific documentation of recreational drug use is extremely sporadic and inaccessible, not only because these drugs are illegal, but, more important because the medical-scientific community is totally uninterested in drugs as a cause of AIDS (see above).

In addition, about 150,000 HIV-positive Americans were on AZT in 1992.[62] There are no national statistics available on how many HIV-positive Americans are on anti-HIV drugs that, like AZT, are designed to kill human cells.[63]

The relatively small percentage of AIDS patients among the many American drug users reflects the percentage with the highest lifetime dose of drug use, just like the three hundred thousand annual lung cancer and emphysema patients reflect the highest lifetime tobacco dose of the 50 million smokers in the United States. The long "latent period of HIV" is a euphemism for the time needed to accumulate the drug dosage that is sufficient for AIDS.

Therefore, it takes about ten years of injecting heroin and cocaine to develop weight loss, tuberculosis, bronchitis, pneumonia, and other drug-induced diseases.[64] The time lag from initiating a habit of inhaling nitrites to *acquiring* Kaposi's sarcoma has been determined to be seven to ten years.[65] The different "latent periods of HIV" are simply reflections of the time the human host takes to accumulate sufficient drug dosage for AIDS to occur. Blaming Kaposi's sarcoma on HIV after inhaling carcinogenic nitrites for ten years is like blaming lung cancer and emphysema on a "slow" virus after smoking two packs of cigarettes a day for twenty years.

AZT, at the currently prescribed high doses of 0.5 to 1.5 grams per person per day, causes many of the above-described AZT-specific diseases faster than recreational drugs do, i.e., within weeks or months after administration, because AZT is much more toxic than recreational drugs[66] (see chapter 9). In short, disease from drug use is not an all-or-nothing phenomenon like disease from infection: only high cumulative doses cause irreversible damage and disease.

HIV hypothesis: The virus hypothesis can explain neither why AIDS is linked to drugs nor why the risk of AIDS depends on the lifetime dosage of drugs.

Risk-group–specific AIDS diseases.

Drug hypothesis: Group-specific drug use explains the following risk-group–specific AIDS diseases:

(i) *Kaposi's sarcoma specific for male homosexuals:* Kaposi's sarcoma as an AIDS diagnosis is twenty times more common among homosexuals who use nitrite inhalants than among AIDS patients who are intravenous drug users or hemophiliacs.[67] Due to their carcinogenic potential, nitrites were originally proposed as causes of Kaposi's sarcoma.[68] "Aggressive and life-threatening" Kaposi's sarcoma, particularly pulmonary Kaposi's sarcoma (lung cancer), are exclusively observed in male homosexuals.[69] Up to 32 percent of Kaposi's sarcomas of homosexual men can be diagnosed as pulmonary Kaposi's sarcoma.[70] This lends additional support to the nitrite–Kaposi's sarcoma hypothesis since the lungs are the primary site of exposure to nitrite inhalants. Pulmonary Kaposi's sarcoma has never been observed by Moritz Kaposi, nor was it observed by others prior to the AIDS epidemic.[71]

It appears that the nitrite-induced AIDS Kaposi's sarcoma and the classic Kaposi's sarcomas are entirely different cancers under the same name. The "HIV-associated" Kaposi's sarcomas observed in male homosexuals are "aggressive and life-threatening,"[72] fatal within eight to ten months after diagnosis, and often located in the lung.[73] The classic "indolent and chronic" Kaposi's sarcomas are diagnosed on the skin of the lower extremities and hardly progress over many years.[74] Meduri et al. point out that the "pulmonary involvement by the neoplasma has been an unusual clinical finding" in the Kaposi's sarcomas of male homosexuals compared to all "classic" Kaposi's sarcomas.[75] Nevertheless, the distinction between classic and AIDS Kaposi's sarcoma is hardly ever emphasized. It may have escaped many observers due to the "difficulty in pre-mortem diagnosis" as "pulmonary Kaposi's sarcoma was indistinguishable from opportunistic pneumonia."[76]

The immunotoxicity and cytotoxicity of nitrites also explains the proclivity of male homosexual nitrite users for pneumonia, which is the most common AIDS disease in the United States and Europe[77] (see Table 1, chapter 6). Moreover, the immunotoxins and cytotoxins of cigarette smoke explain why, in two groups of otherwise matched HIV-positive male homosexuals, cigarette smokers developed pneumonia twice as often as nonsmokers over a period of nine months.[78]

(ii) *High mortality of intravenous drug users:* Intravenous drug users suffer from long-term malnutrition and insomnia, which are primary causes of immunodeficiency worldwide.[79] This explains the tuberculosis, pneumonia, and weight loss that are typical of these risk groups.[80] Injection of unsterile drugs combined with immunodeficiency also cause septicemia and endocarditis, which are common in AIDS patients who are intravenous drug users.[81]

424 ■ INVENTING THE AIDS VIRUS

As a result, intravenous drug users have a high mortality. The average age at death is 29.6 years for HIV-free and 31.5 years for HIV-positive addicts, according to a German study,[82] and both HIV-positive and -negative intravenous drug users die from the same disorders, according to an American study.[83]

(iii) *Low birth weight and mental retardation of AIDS babies:* Eighty percent of American/European babies with AIDS are born to mothers who were intravenous drug users during pregnancy (see chapter 8). Their symptoms range from low birth weight and mental retardation to immunodeficiency through maternal drug use.[84] The B-cell deficiencies and certain bacterial infections, which are both considered AIDS-defining only in children, are both consequences of the immunodeficiency "acquired" from the drugs their mothers used during pregnancy.[85]

(iv) *Anemia, wasting, and accelerated death of AZT recipients:* Anemia, leukopenia, pancytopenia, diarrhea, weight loss, hair loss, impotence,[86] hepatitis,[87] and *Pneumocystis* pneumonia[88] are observed in recipients of AZT and other DNA chain terminators. These are predictable consequences of the cytotoxicity of these drugs. In addition, nonrenewal of mitochondrial DNA causes muscle atrophy, hepatitis, and dementia, and carcinogenic activity causes cancers such as lymphoma in AZT recipients.[89] Owing to the carcinogenic activity of AZT, the lymphoma rate of AZT-treated AIDS patients is a staggering 9 percent per year, or 50 percent in three years, according to the National Cancer Institute.[90]

Compared to untreated controls, AZT recipients develop AIDS 4.5 times more often and die 2.4 times more often[91] or 25 percent more often,[92] or live only two years instead of three years with AIDS.[93] In short, specific drugs cause specific diseases.

HIV hypothesis: The virus hypothesis is clueless. On genetic grounds the same virus must cause the same disease (or diseases) in the same host. Just as a specific instrument makes a specific sound, a specific virus causes a specific disease in all risk groups, e.g., hepatitis virus causes the same hepatitis and wart virus causes the same warts in men, women, homosexuals, and heterosexuals.

Noncorrelation between HIV and AIDS.

Drug hypothesis: The drug hypothesis predicts AIDS without HIV, HIV without AIDS, and other noncorrelations. All of these predictions are confirmed:

(i) *Long-term survivors or "nonprogressors":* Persons infected by HIV for more than the ten-year latent period from HIV to AIDS are called long-term survivors and, more recently, *nonprogressors* if they are studied by HIV researchers.[94]

Indeed, the vast majority of HIV-positives are long-term survivors! Worldwide, they number 17 million, including 1 million HIV-positive but healthy Americans and 0.5 million HIV-positive but healthy Europeans.[95] Most of these have been HIV-positive for at least ten years now, because their numbers have not changed since the time between 1984 to 1988, when the HIV-testing epidemic began in the respective countries.[96]

Only about 6 percent (or 1,025,073) of the 18 million HIV-positives (including the 17 million without AIDS) have developed AIDS diseases since AIDS statistics have been kept.[97] Since no more than 6 percent of HIV carriers worldwide have developed AIDS in seven to ten years, the annual AIDS risk of an HIV carrier is less than 1 percent per year. However, even this low figure is not corrected for the normal occurrence of the thirty AIDS-defining diseases (see Table 2, chapter 6) in HIV-free controls. It may well reflect the normal incidence of these diseases in most people. There is no evidence that HIV-positive people who are not drug users have a higher morbidity or mortality than HIV-free controls.[98]

David Ho, director of the Aaron Diamond AIDS Research Center of New York, recently gave the key to long-term survival with HIV: "None had received antiretroviral therapy."[99] Likewise, Alvaro Muñoz from the Johns Hopkins University in Baltimore reported that not one of the long-term survivors of the largest federally funded study of male homosexuals at risk for AIDS, the MACS study, had used AZT.[100] And several survey studies document that, in addition to abstaining from antiviral drugs, long-term survivors are those who have given up or never taken recreational drugs.[101]

(ii) *Intravenous drug users and male homosexuals lose their T-cells prior to HIV infection:* Prospective studies of male homosexuals using psychoactive and sexual stimulants have demonstrated that their T-cells may decline prior to infection with HIV. For example, the T-cells of thirty-seven homosexual men from San Francisco declined steadily prior to HIV infection for 1.5 years, from more than 1,200 to below 800 per microliter.[102] In fact, some had fewer than 500 T-cells 1.5 years before seroconversion.[103] Although recreational drug use was not mentioned in these articles, other studies of the same cohort (technical term for group) of homosexual men from San Francisco described extensive use of recreational drugs, including nitrites.[104] Likewise, thirty-three HIV-free male homosexuals from Vancouver, Canada, had "acquired" immunodeficiency prior to HIV infection.[105] Again this study did not mention drug use, but in other articles the authors reported that all men of this cohort had used nitrites, cocaine, and amphetamines.[106]

About 450 (16 percent of 2,795) homosexual American men of the MACS cohort from Chicago, Baltimore, Pittsburgh, and Los Angeles had acquired immunodeficiency, having fewer than 600 T-cells per microliter, without ever acquiring HIV.[107] Many HIV-positive and -negative men of this cohort had essentially the same degree of lymphadenopathy: "Although seropositive men had a significantly higher mean number of involved lymph node groups than seronegative men (5.7 compared to 4.5 nodes, $p < 0.005$), the numerical difference in the means is not striking."[108] According to previous studies on this cohort, 71 percent of these men had used nitrite inhalants, in addition to other drugs;[109] 83 percent had used one drug, and 60 percent had used two or more drugs during sex in the previous six months (see Table 1).[110]

Proving the Drug-AIDS Hypothesis, the Solution to AIDS ■ 427

Another study of the same cohort observed that the risk of developing AIDS correlated with the frequency of receptive anal intercourse prior to and after HIV infection.[111] And receptive anal intercourse correlates directly with the use of nitrite vasodilators.[112]

Thus, in male homosexuals at risk for AIDS, AIDS often precedes infection by HIV, not vice versa. Since the cause must precede the consequence, drug use remains the only choice to explain "acquired" immunodeficiencies prior to HIV. If male homosexuality were to cause immunodeficiency, about 10 percent of the adult American male population (the estimated percentage of homosexuals) should have AIDS.[113]

Surveys (prospective studies) of intravenous drug users also document T-cell losses prior to infection by HIV. For example, among intravenous drug users in New York, "the relative risk for seroconversion among subjects with one or more CD4 [T-cell] count <500 cells per microliter compared with HIV-negative subjects with all counts >500 cell per microliter was 4.53."[114] In other words, by the time these intravenous drug users were infected by HIV, their T-cells were already below 500. A similar study from Italy showed that a low number of T-cells was the highest risk factor for HIV infection.[115] In other words, the T-cells were dropping before HIV infection. Logic follows that drug consumption dropped the T-cells.

(iii) *HIV-free AIDS:* One summary of the AIDS literature describes more than 4,621 clinically diagnosed AIDS cases who were not infected by HIV.[116] Additional cases are described that were not included in this summary.[117] They include intravenous drug users, male homosexuals using aphrodisiac drugs like nitrite inhalants, and hemophiliacs developing immune suppression from long-term transfusion of foreign proteins contaminating Factor VIII.[118]

Each of these noncorrelations between HIV and AIDS are predicted by the hypothesis that recreational drugs and other noncontagious risk factors cause AIDS.

HIV hypothesis: Since AIDS occurs without HIV, and since T-cells of drug addicts decrease prior to HIV infection, HIV must be discarded as a cause of AIDS. The studies of drug addicts prove instead that HIV is just a marker of drug consumption, rather than the cause of AIDS: The more drugs consumed intravenously or for sex, the higher the risk of HIV infection.[119]

AIDS cured by withdrawal from recreational drugs and by discontinuation of AZT—despite HIV.

Drug hypothesis: If AIDS is caused by drugs, some patients should be able to recover if they abstain from drug use, even if they are HIV-positive. The following examples prove this point:

(i) *AZT:* Ten out of eleven HIV-positive, AZT-treated AIDS patients recovered cellular immunity after discontinuing AZT in favor of an experimental vaccine.[120] Two weeks after discontinuing AZT, four out of five AIDS patients recovered from myopathy.[121] Three of four AIDS patients recovered from severe pancytopenia and bone marrow aplasia four to five weeks after AZT was discontinued.[122]

(ii) *Heroin/cocaine:* The incidence of AIDS diseases among HIV-positive intravenous drug users over sixteen months was 19 percent (23/124) and only 5 percent (5/93) among those who stopped injecting drugs.[123] The T-cell counts of HIV-positive intravenous drug users from New York dropped 35 percent over nine months, compared to HIV-positive controls who had stopped injecting.[124]

(iii) *Recreational drugs and AZT:* The health of male homosexuals is stabilized or even improved by avoiding recreational drugs. For example, in August 1993 there was no mortality during 1.25 years in a group of 918 British HIV-positive homosexuals who had "avoided the experimental medications on offer" and chose to "abstain from or significantly reduce their use of recreational drugs, including alcohol."[125] Assuming an average ten-year latent period from HIV to AIDS, the virus-AIDS hypothesis would have predicted at least 58 (918/10 × 1.25 × 50 percent) AIDS cases among 918 HIV-positives over 1.25 years. Indeed, the absence of mortality in this group over 1.25 years corresponds to a minimal latent period from HIV to AIDS of more than 1,148 (918 × 1.25) years. As of July 1, 1994, there was still not a single AIDS case in this group of 918 HIV-positive homosexuals.[126]

The T-cells of 29 percent of 1,020 HIV-positive male homosexuals and intravenous drug users in a clinical trial even increased over two years.[127] These HIV-positives belonged to the placebo arm of an AZT trial for AIDS prevention and thus were not treated by AZT. It is probable that, under clinical surveillance, the 29 percent whose T-cells increased despite HIV have given up or reduced immunosuppressive recreational drugs in the hope that AZT would prevent AIDS.

(iv) *AIDS babies, born to drug-addicted mothers, recover after birth:* HIV-positive babies, born to mothers who were intravenous drug users during pregnancy, provide the best examples for the prediction that termination of drug use prevents or cures AIDS—despite the presence of HIV. For example, for three years Blanche et al. have observed seventy-one HIV-positive newborns who had shared intravenous drugs with their mothers prior to birth. Ten of these children developed encephalopathy and AIDS-defining diseases, of which nine died during their first eighteen months of life. The study points out that the risk of a newborn to develop AIDS was related "directly with the severity of the disease in the mother at the time of delivery." Based on the severity of their symptoms, 60 percent of the children were treated prophylactically, but apparently briefly, with AZT "for at least one month," and 50 percent were treated with sulfa drugs.[128]

Unexpectedly, sixty-one of the seventy-one HIV-positive children either developed only "intermittent" diseases, from which they recovered during their first eighteen months or developed no disease at all during the three years of observation. The T-cells of these children increased after birth from low to normal levels—despite the presence of HIV.

A very similar picture emerges from a collaborative European study of HIV-positive newborns.[129] The study reports that about 20 percent of the HIV-positive children had died or developed long-term AIDS during the first year after birth, and another 20 percent during the second and third year. About 10 percent of the children were "treated with Zidovudine [AZT]" before six months of age and 40 percent by four years.[130]

More than 60 percent of congenitally infected children proved to be healthy up to six years after birth—despite the presence of HIV. Most of these had experienced transient AIDS diseases— such as pneumonia, bacterial infections, candidiasis, and cryptosporidial infection—during the first year after birth.

Although this study does not even mention the health and health risks of the mothers, previous reports from the European Collaborative Study group have documented that "nearly all children were born to mothers who are intravenous drug users."[131] In 1991, the European Collaborative Study group reported that 80 percent of the children with pediatric AIDS were born to mothers who were intravenous drug users.[132] The 1991 study further points out that "children with drug withdrawal symptoms" were most likely to develop diseases and that children with no withdrawal symptoms but "whose mothers had used recreational drugs in the final six months of pregnancy were intermediate" in their risk to develop diseases, although they were all infected by HIV.[133]

The drug hypothesis explains the fate of the children as a function of the drugs consumed. Those who received the highest doses of drugs before birth would have acquired irreversible diseases, and those who acquired diseases from sublethal thresholds would be able to recover after birth once they were no longer forced to share their mother's drugs. Indeed, both the European Collaborative Study group and Blanche et al. show that the majority of children gained T-cells and recovered from transient diseases after discontinuation of maternal drug input—despite the presence of HIV. The children's risk for AIDS was related "directly with the severity of the disease in the mother,"[134] which is an expression for the extent of drug consumption by the mother.

Moreover, the harm of maternal drug consumption to sick babies was compounded after birth, because "prophylactic treatment [with]... sulfamethoxazale and Zidovudine [AZT] was started earlier and was more frequent among the 16 children born to mothers with class IV disease [AIDS]."[135] The European Collaborative Study group reports that 10 percent to 40 percent of HIV-positive children were treated with AZT.

Although recent American epidemiological studies also avoid revealing the poor correlations between HIV infection and AIDS, the correlation between HIV and pediatric AIDS in the United States appears to be similar to Europe. A recent report from Baltimore confirmed that 67 percent of the mothers of HIV-positive American babies, studied by HIV researchers, are intravenous drug users.[136] And the CDC reported that 12,240 (82 percent) of the 14,920 children born with an HIV diagnosis in the United States from 1978 to 1993 are alive and well—despite the presence of HIV.[137]

It follows that discontinuation of recreational and antiretroviral drug use stabilizes and even cures AIDS in HIV-positives. Likewise, the T-cells of HIV-positive hemophiliacs increase after removal of immunosuppressive foreign proteins from their Factor VIII therapy,[138] and the T-cells of African HIV-positive tuberculosis patients increase after "standard anti-TB treatment" and improved nutrition.[139]

HIV hypothesis: According to the HIV hypothesis, every infected adult and baby should have progressively lost T-cells and developed AIDS. This was not observed. On the contrary, HIV-positive AIDS patients recovered once freed of AZT and recreational drugs—despite the continued presence of the hypothetical T-cell killer, HIV.

In sum, the drug-AIDS hypothesis correctly predicts all aspects of American/European AIDS, while the HIV hypothesis predicts none.[140]

THE SOLUTION TO THE AIDS CRISIS

Testing the drug hypothesis should have a very high priority in AIDS research, because this hypothesis makes verifiable predictions.[141] Drug toxicity must be determined experimentally by exposing animals, such as mice, or humans with carefully monitored doses over appropriate periods of time. Except for experiments measuring immediate effects of psychoactive drugs, no such experiments have ever been done in animals or humans. Alternatively, drug toxicity could be tested epidemiologically in humans, who are addicted to recreational drugs or are prescribed AZT, by comparing their diseases with those (if any) of otherwise matched, drug-free controls.[142] Such tests could be conducted at a fraction of the cost that is now invested in the HIV hypothesis.

But thirty-five billion AIDS dollars have been plowed entirely into studying HIV since 1984, leaving the comparatively tiny field of drug toxicity with virtually no support at all (see chapter 12). Most illegal drugs have been given only to mice or rats in a single dose, looking for the short-term effects (see chapter 6).[143] Until researchers can perform long-term experiments, the role of drugs in AIDS will never be completely understood. Certainly the evidence above strongly proves that drugs can more easily cause AIDS than could any microbe, particularly microbes that are latent and neutralized by antimicrobial immunity.

If the drug hypothesis proves to be correct, AIDS could be prevented entirely by existing technologies and institutions if:

1. AZT use, currently the most toxic, legal threat to public health, were banned immediately.

2. Illicit recreational drugs were reduced or prevented by education that "drugs cause AIDS."

3. AIDS patients were treated for their specific diseases, e.g., for tuberculosis with antibiotics, for Kaposi's sarcoma with conventional

cancer therapy, for weight loss with good nutrition, and were instructed to avoid recreational drugs and AZT.

In addition to saving about seventy-five thousand lives per year from AIDS in the United States alone, the drug hypothesis could save American taxpayers up to $20 billion annually. Currently, the federal government spends annually $7.5 billion on AIDS treatment, research, and education[144] (see chapter 12) and $13 billion on the war on drugs that is mainly concerned with "supply control," interdiction, methadone treatment, and "education."[145]

But neither AIDS education nor drug education ever target the health effects of long-term drug use. However, if programs for AIDS prevention and drug education were based on the health consequences of long-term drug use, AIDS prevention would be as successful as the federal antismoking program. As a result of education that smoking causes lung cancer, emphysema, and heart disease, smoking has dropped in the United States from 42 percent of the adult population in 1965 to 25 percent in 1995.[146]

The solution to AIDS could be as close as a very testable and very affordable alternative hypothesis. The next chapter describes imminent signs of change, showing that the truth is finally emerging and outlines a solution for restoring science to its legitimate roots.

CHAPTER TWELVE

■

The AIDS Debate Breaks the Wall of Silence

ON JUNE 7, 1993, more than fifteen thousand HIV researchers from around the world arrived in Berlin for the Ninth International AIDS Conference, a four-day meeting at which the latest experimental results would be presented. Such a staggering number of scientists naturally brought with them a comparable volume of data, filling eight hundred lectures and forty-five hundred poster displays. The one-paragraph summaries of new papers alone filled "two guides the size of telephone directories." No researcher at the conference could possibly review more than a small fraction of the data, a situation described by one reporter as "information overload."[1] Despite being overwhelmed and accomplishing little of substance, AIDS officials used such meetings as public relations victories. During the previous eight years, the annual conferences had proven to be gala events, generating weeklong sensational media stories on the frightful AIDS epidemic and the heroic efforts of scientists to stop it.

This time, however, things had changed. An atmosphere of pessimism hung over the Berlin conference, the participants widely acknowledging their confusion and the failures of the war on AIDS. "After more than a decade of struggling in frustration as the epidemic gallops on," wrote one correspondent, "researchers are being forced to reexamine assumptions they once held without question."[2] HIV clearly could not be killing T-cells directly, leaving open the question of just how it could cause AIDS. T-cell counts, once thought to represent the ultimate measure of the immune system, no longer seemed to diagnose an AIDS patient's condition accurately. AZT treatment was being discredited by preliminary results from the Concorde study on nearly two thousand patients, showing that the drug did not prolong life. And when veteran polio virologist Jonas Salk presented the results of his new HIV vaccine, the audience concluded it would not work after all. Some listeners even called New York directly on their cellular phones to dump their stock invested in Salk's biotechnology venture. Every belief and expectation based on the HIV hypothesis was proving false in the face of new evidence. Try as they might, AIDS officials could not prevent the general impression that twelve years of research was falling to pieces.

Although no one in the AIDS establishment questioned the HIV hypothesis itself, clearly the confidence of many scientists was weakening. *Science* magazine had anticipated the negative mood the previous week with a special issue entitled "AIDS: The Unanswered Questions," of which more than forty pages were devoted to the cover story.[3] The Berlin conference also marked the first attendance by dissenters against the HIV hypothesis, who were surprised to find serious interest from many conference participants. The mood even affected Robert Gallo, who became touchy with reporters when asked about his conviction on scientific misconduct charges. The conference, in fact, symbolized the changing tide in the AIDS debate.

The next International AIDS Conference was held in Yokohama, Japan, in 1994. Again, more than ten thousand AIDS jet-setters met, and again, nothing was offered to prevent or cure AIDS. An AIDS vaccine, initially promised by Gallo ten years earlier for 1986, was now scheduled for the next century. Once more annual AIDS statistics had doubled in America, after the CDC had once again increased its catalog of AIDS diseases—now to about thirty.

Public health officials could still not demonstrate that they had saved any lives by controlling the blood supply, nor through their programs for promoting and distributing condoms and sterile injection needles. Worst of all, none of the virus-based predictions had been borne out: AIDS has not exploded into the heterosexual population, as do all other sexually transmitted diseases, nor can doctors predict the course of illness in any given patient. And in contrast to the official prediction that HIV would kill virtually all infected people, seventeen million HIV-positives,[4] including more than one million Americans, have remained AIDS-free for nearly a decade. AIDS officials can neither control nor predict the epidemic, leaving AZT therapy as their only consistent answer.

The development of an effective treatment for AIDS has been equally disappointing. The final report of the Concorde study shattered the hope that "antiviral" DNA chain terminators such as AZT might at least prevent AIDS. The chilling news was that instead of preventing AIDS, the drugs helped to bring it on. The mortality of AZT recipients was 25 percent higher than that of those in untreated control groups.[5] This drug, originally developed for cancer chemotherapy, efficiently destroys the immune system and causes symptoms largely indistinguishable from AIDS itself. Even Burroughs Wellcome, the manufacturer of AZT, makes that same assessment, but expresses it in different words: "It was often difficult to distinguish adverse events possibly associated with Zidovudine [AZT] administration from underlying signs of HIV disease."[6]

One can only guess what William Paul, the new American "AIDS czar," was really thinking when he gave his famous "back-to-the-basics" speech to the researchers gathered at the Tenth Annual AIDS conference in Yokohama.[7] Paul, the scientist, called on AIDS researchers to reexamine all of the many assumptions of the HIV-AIDS hypothesis. But Paul, the politician, failed to name the most important one: the central assumption that HIV causes AIDS. Paul even warned against "unreflective allegiance to the status quo" and criticized the funding monopoly of government-sponsored AIDS research: "Research administrators need to remember that breakthroughs would come from insights that cannot be planned. Command science is no more likely to succeed than command economics."[8]

The HIV researchers in Yokohama must have sensed that HIV science was not going anywhere soon. They had grown so pessimistic about their ability to achieve any significant progress in the near future that they voted to hold international AIDS conferences in the future only every other year.

In late July 1995, the Ninth Annual Congress of Immunology met in San Francisco and convened what the *San Francisco Chronicle* called a "high-wattage panel" on AIDS, featuring such HIV-AIDS luminaries as Robert Gallo, Luc Montagnier, and David Baltimore. Repeating the sad refrain of HIV-AIDS research, the high-wattage panelists admitted to their peers that they had no good news and very little news at all. Mocked one national magazine on progress in AIDS research in 1993: "The Good News Is, the Bad News Is the Same."9

Gallo hoped to start antiviral gene therapy for HIV-positives within the next year, but anticipated no breakthroughs in the near future. Baltimore said that despite the testing of many vaccine strategies, "nothing on the horizon at the moment has the potential... of being a good vaccine." Acknowledging that he had been on an HIV vaccine panel a decade before that predicted the development of an AIDS preventive within five to ten years, Baltimore lamented, "Here we are 10 years later, and it is still 10 years away." Panelist Montagnier urged the continuation of the status quo: "What is important is that treatments for HIV start immediately, as soon as infection is known"—even if, as *Chronicle* science writer Charles Petit noted in a charitable understatement, "none [of the available treatments] is terribly good yet." Montagnier's semi-iconoclastic message of the Sixth International AIDS conference in San Francisco five years earlier, that HIV was not able to cause AIDS without a cofactor, was apparently forgotten and forgiven. How else could he have proposed preventive treatment with the available cytotoxic DNA terminators in good faith? To recommend chemotherapy against a virus that is not sufficient to cause AIDS would be irresponsible at the least.

For those who might conclude the outlook for HIV research is hopelessly bleak, Gallo sounded an optimistic note: "There are many, many things to try. The list is almost endless. The crucial thing in this research and others is to find ways to stop the virus from replicating." Clearly, there are "many, many things to try" every year for the $7.5 billion from U.S. taxpayers alone.

For the sake of everyone directly affected by the suffering caused by AIDS, however, the crucial objectives of all research would seem to be how to help sick people get better and prevent people at risk from falling prey to the complex of diseases now called *AIDS*. If the goal of AIDS research is changed from shoring up the never-proven HIV hypothesis to the protection of public health, then a different set of research and prevention objectives emerges—objectives that are likely to bring about real gains in the battle against AIDS much sooner, and at much lower cost, than the current direction.

Taxpayers, and HIV-positives and their relatives, potentially constitute the most explosive opposition to the AIDS establishment. As the failures of the war on AIDS mount up, the size of the imminent backlash grows; the longer AIDS officials resist the inevitable, the harder they will fall. "Command science" cannot forever hide the truth. Time, therefore, has become the most valuable ally of the HIV-AIDS debate.

THE AIDS DEBATE COMES OUT

The Group for the Scientific Reappraisal of the HIV-AIDS Hypothesis has grown from its original two dozen members in 1991 to more than four hundred professionals today, including more than two hundred scientists and medical doctors. The Group's newsletter, *Reappraising AIDS*, now reaches more than fifteen hundred people. Not only do the growing ranks of dissenting scientists serve as a barometer of frustration among researchers and physicians, but these hundreds of skeptics are also beginning to make themselves heard. They are writing books, scientific papers, and popular articles, while giving public lectures and interviews with the media.

Several scientific journals have invited and published dissident papers, including the German *AIDS-Forschung*; the French journals *Research in Immunology* from the Pasteur Institute and

Biomedicine and Pharmacology; the British-based *Pharmacology and Therapeutics*; and the American-based *Perspectives in Biology and Medicine* and *Bio/Technology*. In 1994 the editor of the *International Archives of Allergy and Immunology* commissioned Duesberg's article "Infectious AIDS—Stretching the Germ Theory Beyond Its Limits" under the heading "Controversy: HIV and AIDS."[10] A Swiss virologist defended the orthodoxy. Later that year the editor of the Dutch-based *Genetica*, the oldest genetics journal of its kind, asked Duesberg to edit a special issue on the HIV-AIDS controversy. The issue appeared in the spring of 1995 with a foreword by its editor-in-chief, John McDonald:

> Challenges to the mainstream view that AIDS is caused by HIV have been receiving increasing attention in recent months especially in the popular press. Part of the reason for this attention is no doubt grounded in wide-spread frustration resulting from the fact that after more than a decade of intensive research, there is still no cure for this deadly syndrome. A second issue which seems to be adding fuel to the controversy is the claim that a de facto conspiracy exists within the scientific community to prevent dissenting views and alternative AIDS hypotheses from being presented to the scientific and general public (see, for example, the recent *London Times* article by Neville Hodgkinson entitled "HIV: A Conspiracy of Silence" recently published in the June/July 1994 issue of *The National Times*).
>
> According to the Popperian dictum, a valid scientific hypothesis can ultimately only be strengthened by the challenge of alternative views. On the other hand, ignoring charges of scientific censorship can only work to undermine the public's confidence not only in the prevailing scientific view but also in the entire scientific establishment. In providing this forum for alternative AIDS hypotheses, *Genetica* hopes to dispel the notion that a "conspiracy of silence" exists within the scientific community.[11]

Of course, the virologists are not delighted at this increasing attention given the dissenting viewpoint.

Owing to unexpected interest in the subject, the publisher of *Genetica* has asked Duesberg to edit a book entitled *AIDS: Virus or Drug-Induced?* The book will include all articles of the journal and over a dozen new articles from scientists, mathematicians, a law professor, and journalists from the *London Sunday Times*, the *New York Native*, and *Spin* magazine.

In December 1994 even *Science*, the world's most popular scientific magazine, wrote an eight-page editorial, "'The Duesberg Phenomenon': Duesberg and Other Voices," acknowledging that the "phenomenon has not gone away and may be growing."[12] However, the "phenomenon" was not allowed to describe his "controversial" theory to the readers of *Science* in his own words. The author of the article was instead an experienced "plugged-in" AIDS journalist.[13] In 1995, *The Scientist*, a specialty newspaper for the fast-growing population of American scientists, also took up the issue with several articles. This time Duesberg was granted a full page to explain the drug-AIDS hypothesis in his own words to the fifty thousand readers of that journal.[14] Also in 1995, the *American Journal of Continuing Education in Nursing* published an invited article on the drug-AIDS hypothesis.

In 1994 the NIDA sponsored a conference in Gaithersburg, Maryland, near Washington, D.C., to reconsider the nitrite inhalant (poppers)–AIDS link. Duesberg was invited to attend and to discuss a grant proposal to study the ability of nitrites to cause AIDS diseases in mice. "Gallo... surprised some attendees and panelists by arguing that HIV is not the primary cause of KS [Kaposi's sarcoma is on the list of AIDS diseases], although it may aggravate the condition once KS is caused by 'something else.'" And, "In the true spirit of scientific inquiry, quite different from the rancor of prior discussions of alternative causes of AIDS, Gallo called for funding of Duesberg's nitrite experiments."[15]

Duesberg's speaking invitations have also increased but are typically restricted to smaller universities that are independent of AIDS grants and are more often issued by student groups than by faculty engaged in research. Invitations to debate the HIV-AIDS hypothesis by larger universities and international conferences all come from countries where scientists are less dependent on the approval of the CDC, the NIH, and Burroughs Wellcome than they are in the United States—for example, from Cologne, Dortmund, Berlin, Kiel, Bonn, and Hamburg in Germany; from Vienna in Austria as a featured speaker of the Third Austrian AIDS Conference in 1992; from Bologna and Pavia in Italy for AIDS/cancer conferences in 1993 and 1994; from Barcelona in Spain; from Belo Horizonte in Brazil; and in 1995 from the National Academy of Medicine in Caracas, Venezuela. Such opportunities are growing each year.

The lay public is also beginning to hear more of the HIV debate, despite the general blackout on the issue. The ABC television programs *Day One* and *Nightline* have aired programs featuring Duesberg and other scientists critical of the HIV hypothesis, as well as segments critical of AZT therapy. *Tony Brown's Journal* has featured many HIV dissidents on national television, including Duesberg since 1991.

Favorable attention to HIV dissidents by the American press is rapidly growing in publications ranging from the general-interest magazines *Skeptic, Spin, Omni, Penthouse, Insight, New Republic, Reason, Commentary, New Age,* and *Policy Review*; to newspapers such as the *San Jose Mercury, Philadelphia Inquirer, Miami Herald, Oakland Tribune,* and *Los Angeles Times*; to the gay-interest magazines *New York Native, Genre,* and *Men's Style*; and the drug-interest magazine *High Times.* Outside America, mainstream publications such as the Austrian *Der Standard, Wiener Zeitung,* and *News*; the Canadian *Maclean's* magazine; the Italian *Corriere de la Sera*; the German *Die Woche* and *Hamburger Abendblatt*; the British *London Sunday Times* and the *Continuum*; the Swiss magazine *Der Beobachter*; and the French magazine *Le Lien* have all published favorable reviews of the alternative hypothesis.

England has witnessed the most spectacular crack in the official wall of silence on the HIV debate. It began largely with coverage of the HIV controversy by the *Sunday Times* of London in 1992, spearheaded by medical writer Neville Hodgkinson. Gradually, other major British newspapers were drawn into the fray: the *Independent*, *Financial Times*, *Sunday Express*, *Telegraph*, *Guardian*, and *Daily Mail*. As both sides argued the issue more openly, the London-based scientific journal *Nature* finally decided it had had enough, issuing an editorial blast condemning the *Sunday Times* on December 9, 1993. The *Times* fired back, the fight becoming louder and harder to ignore over the next few months. Charge followed furious countercharge. Even the *New York Times* was forced, for a single day, to break silence on the issue and publish an article on the spectacular British feud. In the meantime, this heated HIV controversy spread to Canada. The debate has slowed down since Hodgkinson took a leave from the *Sunday Times* to write a book on AIDS. However, the open debate in the British press has become a major problem for AIDS officials trying to maintain their war on AIDS.

The absence of an explosive AIDS epidemic has helped create skepticism toward the HIV establishment, causing serious changes in the British AIDS program. Referring to English statistics, the London *Sunday Telegraph* noted in late 1992:

> The initial official estimates that the disease will cut a swathe throughout the nation with an estimated 100,000 new cases a year by the mid-Nineties had to be revised downwards to 30,000 and downwards again to 13,000. Then the Government Actuary looked at the figures and suggested they be reduced downwards yet again to 6,500, but even this has proved to be a six-fold over-estimate of the number of new cases this year...
>
> By this summer when these results were published it had become apparent that with such a low prevalence rate there was no "heterosexual AIDS epidemic," nor was there likely to be one.[16]

As a result, the British government has decided to cut out AIDS "education" programs aimed at the general public, focusing instead on the risk groups.

Even the legal profession has taken an interest in the HIV debate: Duesberg's information and testimony has been used in the courtroom defenses of two HIV-positive men, both accused of "assault with a deadly weapon" for having sexual intercourse with HIV-negative women. Other defense attorneys have sought similar information, always leading to confidential settlements. A group named Project AIDS International has formed in Los Angeles, preparing materials and talking with interested attorneys over the possibility of suing Burroughs Wellcome for the production of AZT.

Such a lawsuit has already become reality in England, the world headquarters of Burroughs Wellcome. Sue Threakall is a schoolteacher whose husband, a hemophiliac, tested positive for HIV in 1985. He remained basically healthy until he began taking AZT in 1989. From that point forward, Bob Threakall's life went downhill. One year later he had to quit his job, suffering "severe weight loss, thrush, stomach upsets, poor sleep patterns, sore mouth, continued sinus infections, weakness, breathlessness, loss of appetite, etc." Early in 1991, he died "confused, delirious, wasted, constant diarrhea, unable to swallow, and with hardly any normal lung tissue left."[17] After contacting Peter Duesberg and absorbing the information refuting the HIV hypothesis, an angry Sue Threakall turned to the courts. In January 1994 she won a government commitment to finance her lawsuit—guaranteeing her case will proceed regardless of her own financial condition—and she filed for damages against Burroughs Wellcome. Several more people, including hemophiliacs, have joined the growing list of plaintiffs.[18]

By 1995, the HIV controversy had even made a convert of noted Wall Street short-seller Michael Murphy of Half Moon Bay, California. *Men's Style* magazine has investigated that story:

> Murphy publishes one of the most influential newsletters for short-sellers, the Overpriced Stock Index. Unlike traditional stockbrokers, short-sellers make money when a

The AIDS Debate Breaks the Wall of Silence ▪ 445

company's stock price falls, not rises. Murphy, who closely follows hi-tech, pharmaceutical and biotech stocks, had become a Duesberg true believer by November and devoted his entire November newsletter to a laymen's explanation of Duesbergian theory and how that would translate into a huge short-selling profit for people who sold Burroughs Wellcome stock soon, before the share price crashed in the wake of impending public realization that HIV did not cause AIDS and, therefore, AZT did not work. The headline read, in all caps: "HIV DOES NOT CAUSE AIDS. AIDS IS NOT CONTAGIOUS. AIDS IS NOT SPREAD BY SEXUAL CONTACT. AZT KILLS PATIENTS." He followed up with an equally hyperbolic December issue.

The double play happened when *Wall Street Journal* columnist William Powers came across the Overpriced Stock Index. Powers, who semi-regularly pens the "Heard on Wall Street" opinion column, was intrigued. "When he first came out with his views, he put out a press release and I called to make sure it wasn't a prank," the affable Powers says. "Murphy does have a following; he isn't the most powerful short-seller but he's one of the few that will put his money where his mouth is. I write about anything that might affect the public stock, so I let him say his piece."

Did he ever. On January 20, 1995 Powers's column—complete with a trademark Journal pen-and-ink sketch of Murphy—gave itself over to Murphy's exhortations to sell Burroughs Wellcome stock, that AZT causes AIDS, and that the disease is not sexually transmitted. The column marked the most credible, if not the highest profile, mainstream depiction of Duesberg to date. "Most people outside of the scientific community don't know anything about HIV," frets NIAID spokesman Greg Folkers, "and when it's in the *Wall Street Journal*, it's given instant credibility."[19]

When money talks, even politicians listen. In March 1995, the new Republican Congressman Gil Gutknecht had grown suspicious of the insatiable appetite of the American AIDS establishment for funding. Seeing an opportunity to save some billions of tax dollars, Gutknecht sent the following critical letter to Secretary of Health and Human Services Donna Shalala, Anthony Fauci, and five other leading American AIDS officials:

GIL GUTKNECHT
1st District, Minnesota

COMMITTEE ON
GOVERNMENT REFORM
AND OVERSIGHT

COMMITTEE ON SCIENCE

Congress of the United States
House of Representatives
March 24, 1995   Washington, DC 20515-2301

Dr. Anthony Fauci
National Institute of Health
9000 Rockville Pike
Bethesda, MD 20892

Dear Dr. Fauci:

As a freshman Representative who sits on the Government Reform and Oversight and Science Committees of the 104th Congress, one of my concerns is the AIDS policy of the U.S. government. Twelve years, $35 billion and 270,000 deaths since the beginning of the AIDS crisis in America there is still no cure, no vaccine, and no effective treatment for the disease. Considering the social and financial costs involved so far, I would like to request your responses to a series of questions:

1. I am told that:
   a) there is not a single documented case of a health care worker (without any other AIDS risk) who contracted AIDS from the over 401,749 American AIDS patients in 10 years;
   b) the partner of AIDS patient Rock Hudson, the wife and 8-year old daughter of late AIDS patient Arthur Ashe, as well as the husband of the late AIDS patient Elizabeth Glaser are HIV and AIDS-free:
   What is the scientific proof that AIDS is contagious?

2. Is there any study showing that HIV-positive American men or women - who are not on recreational drugs, or AZT, or received transfusions - ever got AIDS from HIV? Are there any documented cases of tertiary heterosexual AIDS transmission: AIDS transmitted to a non-risk group heterosexual who in turn transmits AIDS to another non-risk group heterosexual?

3. After more than ten years of intensive research and over 100,000 papers published on HIV/AIDS, is there a study that proves that HIV is the cause of AIDS?

4. How do you explain HIV-free AIDS cases (I am told there are over 4,621 on record) beyond renaming them 'ICL'?

5. If infectious HIV is the cause of AIDS, why is Kaposi's sarcoma - the signal disease of AIDS - exclusively observed in male homosexuals?

HOME OFFICE
MIDWAY OFFICE PLAZA
WASHINGTON OFFICE   3530 Crossview Drive SW, Suite #10
425 CANNON HOUSE OFFICE BUILDING   ROCHESTER, MN 55902
WASHINGTON DC 20515-2301   (507) 252-9841
(202) 225-2472   (507) 252-9915 FAX
(202) 225-3246 FAX   IN MN 1-800-862-8632 TOLL FREE

PRINTED ON RECYCLED PAPER

March 24, 1995

Congress of the United States
House of Representatives
Washington, DC 20515-2301

Dr. Anthony Fauci
National Institutes of Health
9000 Rockville Pike
Bethesda, MD 20892

Dear Dr. Fauci:

As a freshman Representative who sits on the Government Reform and Oversight and Science Committees of the 104th Congress, one of my concerns is the AIDS policy of the U.S. government. Twelve years, $35 billion and 270,000 deaths since the beginning of the AIDS crisis in America there is still no cure, no vaccine, and no effective treatment for the disease. Considering the social and financial costs involved so far, I would like to request your responses to a series of questions:

1. I am told that:
 a) there is not a single documented case of a health care worker (without any other AIDS risk) who contracted AIDS from the over 401,749 American AIDS patients in 10 years;
 b) the partner of AIDS patient Rock Hudson, the wife and 8-year old daughter of late AIDS patient Arthur Ashe, as well as the husband of the late AIDS patient Elizabeth Glaser are HIV and AIDS-free:
 What is the scientific proof that AIDS is contagious?

2. Is there any study showing that HIV-positive American men or women—who are not on recreational drugs, or AZT, or received transfusions—ever got AIDS from HIV? Are there any documented cases of tertiary heterosexual AIDS transmission: AIDS transmitted to a non-risk group heterosexual

who in turn transmits AIDS to another non-risk group heterosexual?

3. After more than ten years of intensive research and over 100,000 papers published on HIV/AIDS, is there a study that proves that HIV is the cause of AIDS?

4. How do you explain HIV-free AIDS cases (I am told there are over 4,621 on record) beyond renaming them 'ICL' [idiopathic CD4 lymphocytopenia]?

5. If infectious HIV is the cause of AIDS, why is Kaposi's sarcoma—the signal disease of AIDS—exclusively observed in male homosexuals?

6. Why are there long-term survivors (12–15 years) of HIV? (Is there medical precedent for a fatal virus with such a long latency period?) Are long-term survivors generally people who do not use recreational drugs and AZT?

7. How does the medical community explain the fact that the median life expectancy of American hemophiliacs has increased from 11 in 1972 to 27 in 1987, although 75 percent were infected by HIV in the decade before 1984?

8. Can federal efforts ignore the theory that recreational drugs and AZT cause AIDS, considering that 30 percent of all American AIDS patients are intravenous drug users and that nearly all others are users of oral recreational drugs and/or AZT, ddI or ddC?

9. Considering that there is little scientific proof of the exact linkage of HIV and AIDS, is it ethical to prescribe AZT, a toxic chain terminator of DNA developed 30 years ago as cancer chemotherapy, to 150,000 Americans—among them pregnant women and newborn babies—as an anti-HIV drug?

The AIDS Debate Breaks the Wall of Silence ■ 449

10. Is there any scientific precedent of a virus causing an autoimmune disease? What do Kaposi's sarcoma, lymphoma, dementia, cervical cancer, and wasting disease have to do with immune deficiency? If HIV never claims more than 1 out of 1,000 cells every other day and the body replaces at least 30 out of 1,000 during the same period, how does HIV damage the immune system?

11. In how many American AIDS cases was HIV actually found? How many presumptive diagnoses of HIV have been recorded? Do HIV antibody tests cross-react with other microbes, viruses, vaccines or other natural or artificial substances?

12. Considering the history of the HIV-AIDS hypothesis and its inability to come up with a cure, vaccine or effective treatment for AIDS in the past ten years, how much money has been spent by government agencies on alternative-hypothesis AIDS research (i.e., Duesberg, Root-Bernstein, Lo)?

Advancement in medicine depends entirely upon experimentation, objectivity, testing all hypotheses, and most importantly, debate, in order to find the truth. Consider this initial inquiry my contribution to this important debate. I eagerly await your response to the above questions.

Sincerely,

Gil Gutknecht
U.S. Representative

cc: Robert Gallo, Harold Jaffe, Bill Paul, Harold Varmus, Patsy Fleming, David Satcher, Donna Shalala

Four months after he had sent out his twelve questions, Gutknecht received, on July 10, a seven-page letter signed by Secretary of Health and Human Services Donna Shalala. According to a footnote on page 1, the letter had been "prepared by [CDC official] P. Drotman." The letter addressed each of Gutknecht's questions with undocumented *ex-cathedra* assertions that are worthless for a scientific debate:

> Although HIV is the underlying cause of AIDS, much remains to be known about exactly how HIV causes immune deficiency. However, this incomplete understanding does not indicate that the virus is harmless. Why some persons exposed to HIV become infected while others do not is also not known. [...] Some individuals effectively combat this viral illness for a longer time than others. [...] The precise mechanism of cell death following HIV infection remains a topic for research. [...] The overwhelming evidence indicates that HIV causes AIDS and that use of contaminated equipment for injection of drugs is one major route of transmission of HIV.

In answer to Gutknecht's question 1, "What is the scientific proof that AIDS is contagious?" the response shifted from evasion to double-talk:

> Regarding scientific proof that AIDS is contagious, before the discovery of HIV, evidence from epidemiologic studies involving tracing of patients' sex partners and cases occurring in blood recipients clearly indicated that the underlying cause of the condition was an infectious agent.

But CDC official Drotman must have known the definitive multi-center study, co-authored and sponsored by the CDC, which showed in 1989 that morbidity and mortality of blood transfusion recipients with and without HIV is exactly the same.[20] Drotman must have even better known two statements that he had published himself in 1995:

1. *Regarding transfusion AIDS:* "But few transfusion recipients

have developed KS [Kaposi's sarcoma, the signal AIDS disease!], even though many of their donors were homosexual and bisexual men who developed KS later."[21]

2. *Regarding sexual transmission of AIDS:* "Few clusters of persons with AIDS-related KS have been reported. Such clusters may be difficult to identify because most persons with AIDS have had contact with many different people. In particular, drug users and homosexual and bisexual men may have had contact with hundreds of partners that they did not know very well."[22]

Clearly, scientific truth was not served by Shalala's letter. The tone of Shalala's answer to Gutknecht is virtually the same as that chosen by Blattner, Gallo, and Temin to answer Duesberg in *Science* in 1988 (see chapter 6). Both answers argue on the basis of authority instead of science. In that spirit Shalala answered Gutknecht's last question about funding "alternative hypothesis AIDS research": "To deviate funds from scientifically sound findings to those that lack evidence would be unconscionable."[23] (The complete Shalala letter is available from the CDC on request.)

Despite these openings, the HIV-AIDS establishment is retaining its full grip on power. The many billions of dollars spent each year by the federal government on biomedical research, and especially those billions devoted to HIV research and control, have purchased enormous influence with every significant interest group involved in AIDS. Scientists, as frustrated or uncertain as they may become over their lack of progress, can never afford to destroy their careers by turning against the peer-enforced dogma. The pharmaceutical industry, particularly Burroughs Wellcome, Bristol-Myers Squibb, and Hoffmann-La Roche—the producers of AZT, ddI, and ddC, respectively—cannot afford to lose their profitable AIDS drugs. AIDS activist groups would hardly want to lose favor with their pharmaceutical patrons or the CDC. The communications media also will not endanger its cozy relationship with the CDC, NIH, and other key agencies. All these groups must continue to support the war on HIV, ignoring or suppressing all genuine debate.

These conflicts of interest between the trusted commitment of scientists to academic freedom and peer pressures to collaborate with the commercial and scientific establishments betray the taxpayers' and the patients' faith. Most important, the great majority of the public at large and the AIDS patients have had little or no opportunity to find out that questions exist about the HIV hypothesis, much less to hear a fair presentation of the arguments. If the public were to discover the facts and how the debate had been hidden from them, they would likely demand an end to the war on AIDS.

HOW COMMAND SCIENCE BETRAYS PUBLIC TRUST

For the public ever to break command science it must first understand the basis of its enormous powers. The medical establishment derives these powers from three sources: (1) enforced consensus through peer review, (2) consensus through commercialization, and (3) the fear of disease, particularly infectious disease.

1. *Enforced consensus through peer review:* The initial power base of the establishment is grant allocation. No medical scientist could even hope to make a career without a research grant from the NIH. Grant allocation selects and rewards conformism with the establishment view. Nonconformists are eliminated by the outwardly democratic "peer review system" advertised by the orthodoxy as an independent jury system that provides checks and balances.

However, a truly independent jury would be fatal for the establishment. Indeed, a grant awarded to test an unorthodox theory of AIDS that proved to be successful would be an end to the orthodoxy itself. Therefore, orthodox scientists (with NIH grants) are carefully selected as the "peers" to review "investigator-initiated" grant applications. The system works because the peers serve the orthodoxy by serving their own vested interests.

Under this review system, a scientist's access to funding, promotions, publication in journals, ability to win prizes, and invitations to conferences are entirely controlled by his peers. This absurd situation puts one's competitors in charge of one's career, a direct conflict of interest. Imagine if every new automobile or new computer had to be approved by competing corporations before being released to the market. Products would cease improving, and the market would experience a steady decline in quality. Innovation and competition would die. Such inherent problems have led to public criticisms of peer review over the past twenty years, with various congressmen and even the Office of Management and Budget voicing objections to the system.

The stifling effect of peer review becomes worse as the number of peers increases, one of the direct effects of over-funded science. The growing number of researchers creates a herd effect, drowning out the voice of the lone scientist who questions official wisdom. Researchers begin spending more time networking and seeking to build coalitions of allies rather than stepping on toes by raising unpopular questions. When the number of scientists in a field is small, they all feel more free to break into conflicting factions with different opinions. A field crowded with peers quickly stifles all such independent thinking, imposing a consensus on the group as a whole. For example, if in a small group an innovative scientist challenges an established opponent, the loser may join the winner or choose another field. But if one innovative scientist takes on one hundred thousand orthodox colleagues and the challenger wins, the consequences for the orthodoxy would be disastrous if it were not for the peer review system. Hundreds of thousands could not practically be converted or easily reemployed, nor would so many be willing to concede defeat to a single challenger. The challenging hypothesis would have to be aborted by the orthodoxy before it is born. The power of peer review!

Former NIH director Bernadine Healy, when questioned during her confirmation hearings before the Senate in 1991, brought up this same problem. She referred to the competition between Mozart and his less competent but more popular rival, Salieri. One journalist summarized her point: "Salieri would probably have fared better than Mozart in the equivalent of today's peer review system, Healy said, but if medicine is to succeed, 'the Mozarts must be allowed to flourish' as well."[24]

As long as a scientist's work is reviewed only by competitors within his own field, peer review will crush genuine science. At a minimum, a scientist should be reviewed only by researchers outside his field, those without such direct conflicts of interest. The further removed from the grant applicant's field, the less biased the reviewers will be. There are numerous well-qualified researchers in related fields who could objectively serve on review committees. Unfortunately, the AIDS establishment, so far from embracing concepts of impartial review, was the first to break even with the standard practice of restricting scientific decisions to scientific peers by appointing faithful "activists" and "risk groups" to regular science policy meetings (e.g., Martin Delaney from Project Inform, San Francisco, serving as advisor to Anthony Fauci at NIH).

Through peer review the federal government has attained a near-monopoly on science. A handful of federal agencies, primarily the NIH, dominate research policies and effectively dictate the official dogma. HIV research provides a case in point. By declaring the virus the cause of AIDS at a press conference sponsored by the Department of Health and Human Services, NIH researcher Robert Gallo swung the entire medical establishment, and even the rest of the world, behind his hypothesis. Once such a definitive statement is made, the difficulty of retracting it only increases with time. As the situation stands now, scientists who sit as peer reviewers for nonfederal granting agencies generally receive NIH funding themselves and thus tend to enforce the federal dogma even on other organizations.

2. *Consensus through commercialization:* The second basis of establishment power is achieved by the commercialization of science. The biotechnology industry arose mainly to supply equipment and reagents to NIH-funded laboratories. As the NIH budget has increased, so has the subsidized market for biotechnology products. The pharmaceutical industry, likewise, has

profited from monopolies granted by the FDA, which bans competing therapies. Both the biotechnology and pharmaceutical industries feed opportunistically off the NIH largesse.

Naturally, some of these federally provided corporate profits find their way back to scientists in the form of patent royalties, consultantships, paid board positions, and stock ownership. These same scientists often sit in judgment of their fellow researchers as peer reviewers, deciding whether a competitor should be funded or allowed to publish. Such commercial conflicts of interest have almost totally permeated biomedical scientific institutions today, whether universities or the NIH or FDA. Researchers have made a regular policy of looking the other way or even rewriting the rules to allow such behavior.

The laws of marketing force consensus even more effectively than peer review. In order for a research product to find a market, the underlying hypothesis for the product must be accepted by a majority of the practitioners in the field. Without such informal support, a product would never stand a chance in the formal review process by peers of the FDA, a division of the Department of Health and Human Services (HHS). The FDA, like its sister HHS divisions, follows the national HIV-AIDS dogma and selects its review committees accordingly.

Thus, commercial success can be achieved only by consensus. For example, an AIDS product that is not based on the HIV-AIDS hypothesis would not be approved unless it miraculously cured AIDS overnight. By contrast, a toxic chemotherapy like AZT, which has yet to cure the first AIDS patient and was even rejected as a cancer drug thirty years ago, proved to be a commercial success overnight after FDA approval.

Moreover, marketing arrests the evolution of research to the status of a successful market product. It would be economic suicide for a scientist to advance research that would render his established commercial products obsolete. This applies to Gallo's NIH patent for the HIV test, for which Gallo's salary is supplemented by the NIH since 1985 by $100,000 annually for the duration of the patent.[25] Many other AIDS researchers are HIV millionaires, since nongovernment scientists are not limited to supplements of $100,000 annually.[26]

Conflicts of interest must inevitably arise when huge sums of money are poured into science. With 7.5 billion federal AIDS dollars available annually, people will always find ways to sell their research. Annually more than 25 million HIV tests at $50 apiece; about two hundred thousand AZT therapies at $10,000 per person per year; innumerable HIV-DNA tests at more than $300 apiece; T-cell counts; blood transfusions; vaccines; and many other commodities are sold. The only workable solution lies in restoring science ethics to their status of twenty years ago, when all commercial applications of government-sponsored biological science were prohibited.

Both peer review and commercialization have killed the evolution of AIDS research by penalizing conceptual innovation and fixing the status quo. And monopolies kill science even faster than economies.

3. *The fear of disease:* Traditionally, the power of medical sciences has been based on the fear of disease, particularly infectious disease. The HIV-AIDS establishment has exploited this instrument of power to its limits. From individual AIDS educators insisting on condoms, to activists calling for clean needles and experimental drugs, to doctors prescribing AZT, to the CDC calling for increased funding of its war against viruses, fear is a nonnegotiable argument.

The bloated science bureaucracy, and particularly the virus-hunting program, have succeeded for decades mostly by using the war motif. Wars are fought most effectively in the name of fear. James Shannon first squeezed extra money out of Congress by declaring war on polio. Richard Nixon mobilized researchers behind the War on Cancer. David Baltimore and his allies engineered the war on AIDS. Now the NIH is declaring a war on breast cancer, which is doomed to finish in the same cemetery as the other War on Cancer; scientific fashion today blames breast cancer on specific genetic mutations, which hard evidence proves harmless and irrelevant to the cause of the tumor.[27] Increased moneys for breast cancer research will lock into place the scientific mistakes and stifle all attempts to discover the real causes of the disease. Progress will be delayed and lives will be lost from too much funding, all in the name of war. But while a state of war can mobilize public fear and support for a time, the tactic eventually backfires, creating public opposition once defeat becomes obvious. The wars on cancer and AIDS have produced nothing but tragedy.

"Germs are back," declared one newspaper article ominously in July of 1993. "Those who track trends in advertising and who sell disinfecting products confirm that today's Americans are acutely attuned to the invisible, microbial world and its potential hazards. Rebounding tuberculosis, undercooked fast-food burgers in Seattle, tainted tap water in Milwaukee and the mystery disease in the Southwest serves [sic] only to encourage our concern about microscopic invaders, it seems." Quoting journalism professor Gail Baker Woods on this rising public fear, the article then noted its biggest cause. "The trend—which Woods says includes not only disinfecting products but also newly introduced clear products with their aura of purity—is in part a response to fear of AIDS and a need to feel immune from such invisible threats."[28]

The fear of microbes is resurging at a most ironic time. Infectious diseases, which once were the leading killers, have stopped killing people in the industrial world. Today, fewer than 1 percent of all deaths in the First World result from contagion; heart disease has become the major cause, followed by cancer, and life expectancies have grown nearly to eighty years.[29] The polio epidemic marked the end of the era of infectious disease for industrial societies. Yet by blaming AIDS, cancer, and other modern, non-contagious diseases on microbes, the virus-hunting research establishment has pointlessly resurrected this old anxiety. That modern science could so effectively terrify the public of a long-vanished threat testifies to its enormous power.

But to some extent, the fear of catching disease has always helped medical authorities, making the lay public more willing to yield money and freedoms for the sake of an answer. Researchers discovered this gold mine of popularity once Robert Koch had proved in 1882 that a bacterium caused tuberculosis. Soon newly graduated medical doctors scurried to find the bacteria causing every conceivable disease. Once blamed, the right germ could open the door to a vaccine or public health measures to control the disease—and to a place in the textbooks, a secure career, and perhaps even a Nobel Prize. Certainly microbes were easier to blame for a disease than spending years of frustrating effort in search of complex or unfamiliar causes. Microbes were tangible, a well-defined target at which to aim.

Enthusiasm turned to fashion, and fashion to a stampede. Bacteria were being found even in noncontagious diseases. Scurvy, pellagra, SMON, and beriberi, among others, were each in turn blamed on a series of microbes, sometimes leading to control measures that only exacerbated these epidemics and always delaying the search for missing vitamins in the diet or toxic medical and recreational drugs. Merely isolating a germ often served to implicate it as being guilty for some disease. Few people stopped to consider the possibility that most germs were simply harmless.

Bacteria hunting temporarily disappeared with the end of contagious epidemics. Meanwhile, virus hunting arrived on the scene, seeing its heyday primarily in the polio epidemic. Once polio disappeared, however, the microbe hunters should have dropped their outmoded specialty, developing new methods and ideas for studying other types of disease. But because of a series of political decisions during the 1950s, they did not.

In 1951, Alexander Langmuir founded the Epidemic Intelligence Service (EIS) of the CDC, intended to act as an early warning detection and control system for contagious epidemics. The EIS, and the CDC, would go on during the next several decades to ring nationwide alarm bells over minor disease outbreaks, while falsely labeling leukemia, Legionnaires' disease, and AIDS as infectious. The NIH, the other federal cornerstone of modern biomedical research, underwent radical restructuring after James

Shannon took over in 1955. Determined to create the largest scientific research establishment in world history, Shannon milked Congress for exponentially increasing budgets by launching massive new programs, most notably the war on polio and the Virus-Cancer Program. The new NIH moneys went to purchase greater quantities of data gathering and to recruit enormous numbers of new people into the swelling ranks of researchers.

The outcome of these two changes was predictable. Microbe hunting reappeared with a vengeance, seizing every available disease to blame on a germ; NIH funding patterns ensured that the virus hunters would predominate over the others. In trying to explain the slow, degenerative diseases of nerve and brain tissues, or of cancer, the virus hunters were forced to improvise. Carleton Gajdusek, who performed questionable research in which he never could isolate a virus from kuru disease, nevertheless became known as the father of the "slow virus" idea and received the 1976 Nobel Prize for the notion. According to this hypothesis, a virus could infect a host one day and, despite being permanently neutralized by the immune system, could somehow cause a fatal disease years later. Although a blatant violation of the logic behind Koch's postulates and the germ theory, Gajdusek's "slow virus" hypothesis captured the imagination of scientists and reshaped their whole approach to medicine. Hilary Koprowski, Robert Gallo, and David Baltimore numbered among the converts who searched avidly for "slow viruses" and eventually led AIDS research.

By the time the AIDS epidemic surfaced, the outcome of research on the new syndrome had been predetermined. AIDS in the United States and Europe fits the pattern of noninfectious diseases and for several reasons is likely to be a product of the drug abuse epidemic of recent years. Nevertheless, virus hunters jumped on the new opportunity, the retrovirologists being in the right position to have a retrovirus, HIV, officially declared the cause. As with so many products of modern virus hunting, HIV completely failed the test of Koch's postulates. But just finding it was all the evidence the virologists needed.

A news article from 1987 shows how "irrational" fear of the AIDS virus affects the lay public:

> Fear of acquired immune deficiency syndrome is sweeping heterosexual society on both coasts...
> Restaurants in gay districts of San Francisco, New York, and the Los Angeles area are shunned by many people, and the Bon Appetit restaurant in suburban Sacramento lost numerous customers after the January AIDS death of a chef who worked there five years earlier...
> At Gay Men's Health Crisis Inc., which helps AIDS patients in New York, "we consistently have television crews who will not come inside our building," said spokeswoman Lori Behrman.
> About half the 4,000 calls received monthly by the center's AIDS hotline are from "worried well" people, half of whom are needlessly concerned about getting AIDS from swimming pools, insects, or nonrisky sex practices, said hotline coordinator Jerry Johnson.[30]

This cynical manipulation of public fear may bring in the money for the AIDS lobby, but it creates human tragedy for everyone else. From Miami, Florida, comes the personal account of some real victims of the war on AIDS—Cesar and Teresa Schmitz and their baby daughter Louise—whose lives were nearly destroyed by the "AIDS virus" propaganda. Teresa herself related the story in December of 1993:

> In January 1992 we found out my husband was HIV+. I will never forget that morning. I will never forget the first three or four days after that test result. It was surely the most devastating experience I ever had in my entire life. Abruptly, it was all gone. No more future. No more nothing. From that moment on life would be waiting for death.
> The [worst] part was to face my beautiful and adorable one-year-old little girl. She was condemned to die.
> Out of my despair I did anything I could to get an answer

about the chances of my baby surviving. The "trained professionals" at the 800 numbers that I called gave me answers like: "Oh my God," after I said that my husband was HIV+ and I had a baby. They even asked me: "Is her hair falling?" "Is she losing weight?"

I could not allow my beautiful and precious baby to go through all that suffering. I could not imagine her going from hospital to hospital, having needles stuck in her little arm, seeing her going skinnier and skinnier. I could not take that...

The only way out of that despair, of that suffering, was to kill ourselves. There was no other solution for us but this one. It would end the pain and the nightmare right at the beginning...

Two weeks later my test result came out—I WAS NEGATIVE!

So, it meant that Louise was negative too... Now Cesar was the only one of us condemned to die...

March 1992 (not even two months after the results) Cesar started with the symptoms of AIDS: diarrhea, nausea, weight loss, and so on. The strange thing was that the symptoms began right after he started taking AZT.

He was feeling so bad, so sick, he decided, against his doctor's will, to stop taking AZT. All of a sudden, like magic, no more symptoms. He was healthy and normal again and remains so, since then [in 1995]. He goes regularly to a clinic for lab tests. The doctor thinks he is doing very well, but insists and pressures him to take AZT or its similar because "it is the only way." The doctor's faith... is so strong that he does not listen to Cesar...

Our marriage was falling apart: no sex life for two years. He did not want to take any chances of contaminating me. The only sure way was abstinence...

About a month ago I decided to write to Dr. Peter Duesberg...

After talking to him my life changed, everything went back to normal. Cesar and I are having a really normal life.

We are planning our second child. We got to the conclusion this whole HIV hypothesis is a mistake, a tragic hoax.³¹

And once the AIDS virus will no longer scare the American public, the CDC stands ready with new superbugs: "After AIDS, Superbugs Give Medicine the Jitters."³² These are the powers of the medical establishment that will continue to suppress the AIDS debate and have been able to obstruct the already-available solution for a long time.

While hundreds of thousands of people die of heavy drug abuse or from their AZT prescriptions, AIDS officials insist on pushing condoms, sterile needles, and HIV testing on a terrified population. "AIDS propaganda is ubiquitous," observes Charles Ortleb, publisher of the homosexual-interest *New York Native*. "Ten percent of every brain in America must be filled with posters, news items, condom warnings, etc., etc. The iconography of 'AIDS' is everywhere. Part of the Big Lie that some activists promote over and over in an Orwellian way is that 'AIDS' is somehow not on the front burner of America. 'AIDS' propaganda has become part of the very air that Americans breathe."³³ All of this is based on a war against a harmless virus waged with deadly "treatments" and misleading public health advice.

This is truly a medical disaster on an unprecedented scale.

Ironically, HIV-positives actually have no reason to fear. As with uninfected people, those who stay off recreational drugs and avoid AZT will never die of "AIDS." Antibody-positive people can live absolutely normal lives. Worldwide, seventeen million of the eighteen million HIV-positives certainly do.³⁴ Those at real risk of AIDS could help their fate if they were only informed that recreational drugs cause AIDS. And those with AIDS could recover if they were informed that AZT and its analogs inevitably terminate DNA synthesis, and thus life.

When the public finally catches on to these deceptive tactics, the HIV hypothesis of AIDS and its proponents will find harsh judgment. AIDS researchers will do their best to control the downfall, accepting the idea of cofactors and gradually relegating HIV to a less important role in the syndrome—a process that has just begun. In December 1994, Kaposi's sarcoma, once the signature disease of AIDS, was taken off the list of HIV diseases. For the time being, another virus was causing that AIDS disease.35 They will probably even try to take credit themselves for discovering the unimportance of HIV, disguising the reversal as further "progress" in AIDS research. But the fact that they are stubbornly fighting such change, and even accelerating the war on AIDS with new, toxic antiviral drugs, indicates they may learn only too late.

This time the public may hold biomedical researchers and public health experts accountable, and misguided microbe hunting will meet its long-overdue judgment.

APPENDIX A

■

Foreign-Protein–Mediated Immunodeficiency in Hemophiliacs With and Without HIV*

Peter H. Duesberg
Dept. of Molecular and Cell Biology
University of California at Berkeley

Abstract

Hemophilia-AIDS has been interpreted in terms of two hypotheses: the foreign-protein–AIDS hypothesis and the Human Immunodeficiency Virus (HIV)-AIDS hypothesis. The foreign-protein–AIDS hypothesis holds that proteins contaminating commercial clotting factor VIII cause immunosuppression. The foreign-protein hypothesis, but not the HIV hypothesis, correctly predicts seven characteristics of hemophilia-AIDS: (1) The increased life span of American hemophiliacs in the two decades before 1987, although 75 percent became infected by HIV—because factor VIII treatment, begun in the 1960s, extended their lives and simultaneously disseminated harmless HIV. After 1987 the life span of hemophiliacs appears to have decreased again, probably because of widespread treatment with the cytotoxic anti-HIV drug AZT. (2) The distinctly low, 1.3 to 2 percent, annual AIDS risk of hemophiliacs, compared to the higher 5 to 6 percent annual risk of intravenous drug users and male homosexual aphrodisiac drug users because transfusion of foreign proteins is less immunosuppressive than recreational drug use. (3) The age bias of hemophilia-AIDS—i.e., that the annual AIDS risk increased two-fold for each ten-year increase in age—because immunosuppression is a function of the lifetime dose of foreign proteins received

---

* Article originally appeared in *Genetica*, 95 (1995): 51–70. Reprinted by permission of the publisher.

from transfusions. (4) The restriction of hemophilia-AIDS to immunodeficiency diseases—because foreign proteins cannot cause nonimmunodeficiency AIDS diseases, like Kaposi's sarcoma. (5) The absence of AIDS diseases above their normal background in sexual partners of hemophiliacs—because transfusion-mediated immunotoxicity is not contagious. (6) The occurrence of immunodeficiency in HIV-free hemophiliacs—because foreign proteins, not HIV, suppress their immune system. (7) Stabilization, even regeneration, of immunity of HIV-positive hemophiliacs by long-term treatment with pure factor VIII. This shows that neither HIV nor factor VIII plus HIV are immunosuppressive by themselves. Therefore, AIDS cannot be prevented by elimination of HIV from the blood supply and cannot be rationally treated with genotoxic antiviral drugs, like AZT. Instead, hemophilia-AIDS can be prevented and has even been reverted by treatment with pure factor VIII.

1. The Drug- and Hemophilia-AIDS Epidemics in America and Europe

About thirty previously known diseases are now called AIDS if they occur in the presence of antibody against human immunodeficiency virus (HIV) (Institute of Medicine, 1988; Centers for Disease Control and Prevention, 1992). These diseases are thought to be consequences for an acquired immunodeficiency syndrome and hence are grouped together as AIDS (Institute of Medicine, 1988). From its beginning in 1981, AIDS has been restricted in America and Europe to specific risk groups (Centers for Disease Control, 1986; World Health Organization, 1992b). Currently, over 96 percent of all American AIDS cases come from AIDS risk groups, rather than from the general population (Centers for Disease Control, 1993). These include over 60 percent male homosexuals who have been long-term oral users of psychoactive and aphrodisiac drugs; 33 percent mostly heterosexual, intravenous drug users and their children; 2 percent transfusion recipients; and about 1 percent hemophiliacs (Duesberg, 1992a; Centers for Disease Control, 1993). Altogether, about 90 percent of all American and European AIDS patients are males (World Health Organization, 1992a; Centers for Disease Control, 1993).

Each risk group has specific AIDS diseases. For example, Kaposi's sarcoma is almost exclusively seen in male homosexuals, tuberculosis is common in intravenous drug users, and pneumonia and candidiasis are virtually the only AIDS diseases seen in hemophiliacs (Duesberg, 1992a).

In view of these epidemiological and clinical criteria, American and European AIDS has been interpreted alternatively as an infectious and a noninfectious epidemic by the following hypotheses:

Appendix A ▪ 467

(1) *The Virus-AIDS hypothesis*. This hypothesis postulates that all AIDS is caused by the retrovirus HIV and thus is an infectious epidemic. The inherent danger of a transmissible disease quickly promoted the HIV hypothesis to the favorite of "responsible" health care workers, scientists, and journalists (Booth, 1988). For example, a columnist of the *New York Times* wrote in July 1994 that all non-HIV–AIDS science is "cruelly irresponsible anti-science" (Lewis, 1994). And the retrovirologist David Baltimore warned in *Nature*, "There is no question at all that HIV is the cause of AIDS. Anyone who gets up publicly and says the opposite is encouraging people to risk their lives" (Macilwain, 1994).

Moreover, the U.S. Centers for Disease Control (CDC) have favored the HIV-AIDS hypothesis from the beginning (Centers for Disease Control, 1982; Shilts, 1985; Centers for Disease Control, 1986; Booth, 1988; Oppenheimer, 1992) because according to Red Cross official Paul Cumming in 1983—"the CDC increasingly needs a major epidemic to justify its existence" (Associated Press, 1994). Indeed, there has been no viral or microbial epidemic in the U.S. and Europe since polio in the 1950s. All infectious diseases combined now account for less than 1 percent of morbidity and mortality in the Western world (Cairns, 1978). And the control of infectious diseases is the primary mission of the CDC.

(2) *The drug-AIDS hypothesis*. This hypothesis holds that AIDS in the major risk groups is caused by group-specific, recreational drugs and by anti-HIV therapy with cytocidal DNA chain terminators, like AZT, and is thus not infectious (Lauritsen & Wilson, 1986; Haverkos & Dougherty, 1988; Duesberg, 1991, 1992a; Oppenheimer, 1992). The drug-AIDS hypothesis was favored by many scientists, including some from the CDC, before the introduction of the HIV-AIDS hypothesis in 1984 (Marmor et al., 1982; Mathur-Wagh et al., 1984; Haverkos et al., 1985; Mathur-Wagh, Mildvan & Senie, 1985; Newell et al., 1985; Haverkos & Dougherty, 1988; Duesberg, 1992a; Oppenheimer, 1992).

(3) *The foreign-protein–hemophilia AIDS hypothesis*. This hypothesis holds that hemophilia-AIDS is caused by the long-term transfusion of foreign proteins contaminating factor VIII and other clotting factors and is thus not infectious. This hypothesis also preceded the virus hypothesis and has coexisted with it, despite the rising popularity of the HIV hypothesis (see section 3).

The infectious and noninfectious AIDS hypotheses indicate entirely different strategies of AIDS prevention and therapy. Here we analyze the cause of hemophilia-AIDS in the lights of the HIV-AIDS hypothesis and the foreign-protein–AIDS hypothesis. The hemophiliacs provide the most accessible group to test AIDS hypotheses of infectious versus noninfectious causation. This is because the time of infection via transfusion can be estimated more accurately than HIV infection from sexual contacts and because the role of treatment-related AIDS risks can be controlled and quantitated much more readily than AIDS risks due to the consumption of illicit, recreational drugs.

2. The HIV-AIDS Hypothesis

The HIV hypothesis claims that AIDS began to appear in hemophiliacs in 1981 (Centers for Disease Control, 1982) because (1) hemophiliacs were accidentally infected via transfusions of factor VIII contaminated with HIV since the 1960s, when widespread prophylactic factor VIII treatment began (but not after 1984 when HIV was eliminated from the blood supply) and because (2) AIDS is currently assumed to follow HIV infection on average only after ten years (Centers for Disease Control, 1986; Institute of Medicine, 1988; Chorba et al., 1994). Indeed, about 15,000 of the 20,000 American hemophiliacs, or 75 percent, are HIV antibody–positive from transfusions of HIV-contaminated clotting factors received before HIV was detectable (Tsoukas et al., 1984; Institute of Medicine and National Academy of Sciences, 1986; Sullivan et al., 1986; McGrady, Jason & Evatt, 1987; Institute of Medicine, 1988; Koerper, 1989). Contamination of factor VIII with HIV reflects the practice, developed in the 1960s and 1970s, of preparing factor VIII and other clotting factors from blood pools collected from large numbers of donors (Aronson, 1983; Koerper, 1989; Chorba et al., 1994).

The HIV hypothesis claims that 2,214 American hemophiliacs developed AIDS-defining diseases between 1982 and the end of 1992 because of HIV (Centers for Disease Control, 1993). However, this corresponds only to a 1.3 percent annual AIDS risk, i.e., 201 cases per 15,000 HIV-positive hemophiliacs per year. (Note that the non-age adjusted annual mortality of an American with a life expectancy of 80 years is 1.2 percent.) Further, the HIV-AIDS hypothesis claims that the mortality of hemophiliacs has increased over two-fold in the three-year period from 1987 to 1989 compared to periods from 1968 to 1986, although infection with HIV via transfusions had already been halted with the HIV-antibody test in 1984 (Chorba et al., 1994).

HIV is thought to cause immunodeficiency by killing T-cells, but paradoxically only after the virus has been neutralized by antiviral immunity and only on average ten years after infection (Institute of Medicine, 1988; Duesberg, 1992a; Weiss, 1993). However, HIV, like all other retroviruses, does not kill T-cells or any other cells in vitro; in fact, it is mass-produced for the HIV antibody test in immortal T-cell lines (Duesberg, 1992a). Moreover, the basis for the ten-year latent period of the virus, which has a generation time of only twenty-four to forty-eight hours, is entirely unknown (Duesberg, 1992a; Weiss, 1993; Fields, 1994). It is particularly paradoxical that the loss of T-cells in hemophiliacs over time does not correspond to viral activity and abundance. No T-cells are lost prior to antiviral immunity, when the virus is most active (Duesberg, 1993a; Piatak et al., 1993). Instead, most T-cells are lost when the virus is least active or latent in hemophiliacs (Phillips et al., 1994a) and other risk groups (Duesberg, 1992a; 1993a, 1994; Piatak et al., 1993; Sheppard, Ascher & Krowka, 1993), namely after it is neutralized by antiviral immunity (a positive HIV-antibody test). Indeed, there are healthy, HIV-antibody–positive persons in which thirty-three to forty-three times more cells are infected by latent HIV than in AIDS patients (Simmonds et al., 1990; Bagasra et al., 1992; Duesberg, 1994). Even Gallo, who claims credit for the HIV-AIDS hypothesis (Gallo et al., 1984), has recently acknowledged: "I think that if HIV is not being expressed and not reforming virus and replicating, the virus is a dud, and won't be causing the disease... nobody is saying that indirect control of the virus is not important..." (Jones, 1994).

There is also no explanation for the profound paradoxes that AIDS occurs only after HIV is neutralized and that antiviral immunity does not protect against AIDS, although this immunity is so effective that free virus is very rarely detectable in AIDS patients (Duesberg, 1990, 1992a, 1993a; Piatak et al., 1993). The high efficiency of this antiviral immunity is the reason that leading AIDS researchers had notorious difficulties in isolating HIV from AIDS patients (Weiss, 1991; Cohen, 1993).

All of the above associations between HIV and AIDS support the hypothesis that HIV is a passenger virus, instead of the cause of AIDS (Duesberg, 1994). A passenger virus differs from one that causes a disease by three criteria:

(1) The time of infection by the passenger virus is unrelated to the initiation of the disease. For example, the passenger may infect ten years prior to, or just immediately before, initiation of the disease—just as HIV does in AIDS.

(2) The passenger virus may be active or passive during the disease, i.e., the primary disease is not influenced by the activity of the passenger virus or the number of virus-infected cells, as is the case for HIV in AIDS.

(3) The disease may occur in the absence of the passenger virus. In the case of AIDS, over 4,621 HIV-free AIDS cases have been clinically diagnosed (Duesberg, 1993b; see also section 4.6).

Therefore, HIV meets each of the classical criteria of a passenger virus—exactly (Duesberg, 1994).

Moreover, since HIV is not active in most AIDS patients and is often more active in healthy carriers than in AIDS patients (Duesberg, 1993a, 1994; Piatak et al., 1993), and since AIDS patients with and without HIV are clinically identical (Duesberg, 1993b), HIV is in fact only a harmless passenger virus. It is harmless, because it does not contribute secondary diseases to AIDS pathogenicity, as for example *Pneumocystis* pneumonia, *Candida*, or herpes virus do. These microbes each cause typical AIDS-defining opportunistic infections. But HIV does not appreciably affect the pathogenicity of AIDS, as HIV-free and HIV-positive AIDS cases are clinically indistinguishable (Duesberg, 1993b, 1994). Likewise, there is no clinical distinction between AIDS cases in which HIV is active and those in which it is totally latent and restricted to very few cells (Duesberg, 1993a; Piatak et al., 1993).

Thus, despite enormous efforts in the past ten years, there is no rational explanation for viral pathogenesis, and the virus-AIDS hypothesis stands unproved (Weiss & Jaffe, 1990; Duesberg, 1992a; Weiss, 1993; Fields, 1994). Above all, the hypothesis has failed to make any verifiable predictions, the acid test of a scientific hypothesis. For example, the predicted explosion of AIDS into the general population, or among female prostitutes via sexual transmission of HIV, or among health care workers treating AIDS patients via parenteral transmission did not occur (Duesberg, 1992a, 1994).

As yet, the hypothesis is supported only by circumstantial evidence, i.e., correlations between the occurrence of AIDS and antibodies against HIV in AIDS patients (Blattner, Gallo & Temin, 1988; Institute of Medicine, 1988; Weiss & Jaffe, 1990; Weiss, 1993). However, because AIDS is defined by correlation between diseases and antibodies against HIV (Institute of Medicine, 1988), the relevance of the correlation argument for AIDS etiology has been challenged (Duesberg, 1992a, 1993b, 1994; Thomas Jr., Mullis & Johnson, 1994). States Mullis, at a *London Sunday Times* Nobel Laureate lecture in 1994, "Any postgraduate student who had written a convincing paper demonstrating that HIV 'causes' AIDS would... have published 'the paper of the century'" (Dickson, 1994).

In view of the circularity of the correlation argument, the apparent transmission of AIDS to hemophiliacs via transfusion of HIV-infected blood or factor VIII has been cited as the most direct support for the virus-AIDS hypothesis (Blattner, Gallo & Temin, 1988; Institute of Medicine, 1988; Weiss & Jaffe, 1990; Weiss, 1993). However, the HIV-hemophilia–AIDS hypothesis is weakened by the extremely long intervals between infection and AIDS, averaging between ten years (Institute of

Medicine, 1988) and thirty-five years, (Duesberg, 1992a; Phillips et al., 1994b), compared to the short generation time of HIV which is only twenty-four to forty-eight hours (see section 4.2). During such long intervals other risk factors could have caused AIDS diseases, particularly in hemophiliacs who depend on regular transfusions of clotting factors for survival. The fact that HIV is typically not more active, and often even less active, in those who develop AIDS than in those who are healthy, further weakens the HIV-hemophilia–AIDS hypothesis (see above).

3. The Foreign-Protein–Hemophilia–AIDS Hypothesis

Before the introduction of the HIV-AIDS hypothesis, but after the introduction of prophylactic long-term treatment of hemophilia with blood-derived clotting factors had begun, numerous hematologists had noticed immunodeficiency and corresponding opportunistic infections in hemophiliacs. Several of these had advanced the foreign-protein–hemophilia–AIDS hypothesis, which holds that the long-term transfusion of foreign proteins contaminating commercial factor VIII, and possibly factor VIII itself, is the cause of immunosuppression in hemophiliacs. Indeed, until recently most commercial preparations of factor VIII contained from 99- to 99.9-percent foreign, non-factor VIII proteins (Brettler & Levine, 1989; Mannucci et al., 1992; Seremetis et al., 1993; Gjerset et al., 1994). According to the foreign-protein hypothesis, immunodeficiency in hemophilia patients is proportional to the lifetime dose of foreign proteins received (Menitove et al., 1983; Madhok et al., 1986; Schulman, 1991).

Long before HIV had been discovered, it was known empirically that "transfusion of patients undergoing renal transplantation is associated with improved graft survival and it has been suggested that transfusion is immunosuppressive in an as yet unidentified way" (Jones et al., 1983). The authors had cited this empirical knowledge to explain immunosuppression in eight British hemophiliacs, and *Pneumocystis* pneumonia in six (Jones et al., 1983). A multicenter study investigating the immune systems of 1,551 hemophiliacs, treated with factor VIII from 1975 to 1979, documented lymphocytopenia in 9.3 percent and thrombocytopenia in 5 percent (Eyster et al., 1985). Further, the CDC reported AIDS-defining opportunistic infections in hemophiliacs between 1968 and 1979, including 60 percent pneumonias and 20 percent tuberculosis (Johnson et al., 1985). An American hematologist commented on such opportunistic infections in hemophiliacs, including two candidiasis and sixty-six pneumonia deaths that had occurred between 1968 and 1979, "...it seems possible that many of the unspecified pneumonias in hemophiliacs in the past would be classified today as AIDS" (Aronson, 1983).

Gordon (1983), from the National Institutes of Health, noted that all hemophiliacs with immunodeficiency identified by the CDC had received factor VIII concentrate. While acknowledging the possibility of a "transmissible agent," Gordon argued that "repeated administration of factor VIII concentrate from many varied donors induces a mild disorder of immune disregulation by purely immunological means, without the intervention of infection." Froebel et al. (1983) also argued against the hypothesis that immunodeficiency in American hemophiliacs was due to a virus and suggested that it was due to treatments with factor VIII because "Scottish patients with hemophilia, most of whom had received no American factor VIII concentrate for over two years, were found to have immunological abnormalities similar to those in their American counterparts..." Menitove et al. (1983) already had described a correlation between immunosuppression of hemophiliacs and the amount of factor VIII received over a lifetime; the more factor a hemophiliac had received, the lower was his T4:T8–cell ratio. Their data were found to be "consistent with the possibility that commercially prepared lyophilized factor VIII concentrates can induce an AIDS-like picture..." In the same year, Kessler et al. (1983) proposed that "Repeated exposure to many blood products can be associated with development of T4/T8 abnormalities" and "significantly reduced mean T4/T8 ratios compared with age- and sex-matched controls."

After the introduction of the HIV-AIDS hypothesis in 1984, Ludlam et al. studied immunodeficiency in HIV-positive and HIV-negative hemophiliacs and proposed "that the abnormalities [low T4:T8-cell ratios] result from transfusion of foreign proteins" (Carr et al., 1984). Likewise, Tsoukas et al. (1984) concluded: "These data suggest that another factor, or factors, instead of, or in addition to, exposure to HTLV-III [old term for HIV] is required for the development of immune dysfunction in hemophiliacs."

In 1985 even the retrovirologist Weiss reported "the abnormal T-lymphocyte subsets are a result of the intravenous infusion of factor VIII concentrates per se, not HTLV-III infection" (Ludlam et al., 1985). Likewise, hematologists Pollack et al. (1985) deduced that "[d]erangement of immune function in hemophiliacs results from transfusion of foreign proteins or a ubiquitous virus rather than contracting AIDS infectious agent." The "AIDS infectious agent" was a reference to HIV, because in 1985 HIV was extremely rare in blood concentrates outside the United States, but immunodeficiency was observed in Israeli, Scottish, and American hemophiliacs (Pollack et al., 1985). A French AIDS-hemophilia group also observed "...allogenic or altered proteins present in factor VIII... seem to play a role of immunocompromising agents." They stated that "A correlation between treatment intensity and immunologic disturbances was found in patients infused with factor VIII preparations, irrespective of their positive or negative LAV [HIV] antibody status" (AIDS-Hemophilia French Study Group, 1985). Likewise, Hollan et al. (1985) reported "an immunodeficiency independent of HTLV-III infection" in Hungarian hemophiliacs.

Madhok et al. (1986) arrived at the conclusion that "clotting factor concentrate impairs the cell mediated immune response to a new antigen in the absence of infection with HIV." Moreover, Jason et al. (1986) from the CDC observed that "Hemophiliacs with immune abnormalities may not necessarily be infected with HTLV-III/LAV, since factor concentrate itself may be immune suppressive even when produced from a population of donors not at risk for AIDS." Sullivan et al. (1986) deduced from a comprehensive study of hemophiliacs that "hemophiliacs receiving commercial factor VIII concentrate experience several stepwise incremental insults to the immune system: alloantigens in factor VIII concentrate [etc.]..."

Sharp et al. (1987) commented that "[f]ive out of 12 such patients had a mild T4 lymphocytopenia, and this may have been related to parenteral administration of large quantities of protein." And Aledort (1988) observed that "chronic recipients... of factor VIII, factor IX and pooled products... demonstrated significant T-cell abnormalities regardless of the presence of HIV antibody." Brettler and Levine proposed in 1989 that "[f]actor concentrate itself, perhaps secondary to the large amount of foreign protein present, may cause alterations in the immune systems of hemophiliac patients." And even Stehr-Green et al. (1989) from the CDC conceded that foreign proteins were at least a cofactor of HIV in immunosuppression: "Repeated exposure to factor concentrate... could also account for more rapid progression of HIV infection with age."

Although Becherer et al. claimed in 1990 that clotting factor does not cause immunodeficiency, they showed that immunodeficiency in hemophiliacs increases with both the age and the cumulative dose of clotting factor received during a lifetime. Likewise, Simmonds et al. observed in 1991 that even among HIV-positive hemophiliacs "[t]he rate of disease progression, as assessed by the appearance or not of AIDS symptoms or signs within five years of seroconversion, was related... to the concentration of total plasma IgM before exposure to infection..." The hematologist Prince noted in a review from 1992 that "[w]hen serum samples from these [immunodeficient hemophilia] patients were tested for antibodies to HIV-1, it was found that a sizable group of hemophilia patients, usually 25 percent to 40 percent, were seronegative for HIV-1," and "...all found marked anergy, lack of response, in HIV seronegative concentrate recipients. Taken together, these findings were interpreted as evidence that clotting factor concentrates suppressed the immunocompetence of recipients..."

In 1991, Schulman concluded that "immunosuppressive components in F VIII concentrates" cause immunodeficiency not only in HIV-positive but also in HIV-negative hemophiliacs. Schulman had observed reversal of immunodeficiency and thrombocytopenia in HIV-positive hemophiliacs treated with purified factor VIII, and that immunity "was inversely correlated with the annual amount of factor VIII, infused" (Schulman, 1991).

At the same time several groups have reported that T-cell counts are stabilized, or even increased in HIV-positive hemophiliacs treated with factor VIII free of foreign proteins (de Biasi et al., 1991; Hilgartner et al., 1993; Seremetis et al., 1993; Goedert et al., 1994) (see also section 4.7). And in 1994, the editor of *AIDS News*, published by the Hemophilia Council of California, granted foreign proteins the role of a cofactor of HIV in hemophilia-AIDS with an editorial "Factor Concentrate Is a Cofactor" (Maynard, 1994).

According to the foreign-protein hypothesis, antibodies against HIV and against other microbes would merely be markers of the multiplicity of transfusions received (Evatt et al., 1984; Pollack et al., 1985; Brettler et al., 1986; Sullivan et al., 1986; Koerper, 1989). Since HIV has been a rare contaminant of blood products, even before 1984, only those who have received many transfusions would become infected. The more immunosuppressive transfusions a person has received, the more likely that person is to become infected by HIV and other microbes that contaminate factor VIII (see section 4.6). For example, only 30 percent of hemophiliacs who had received less than 400 units factor VIII per kg per year were HIV-positive, but 80 percent of those who had received about 1,000 units and 93 percent of those who had received over 2,100 units per kg per year were HIV-positive (Sullivan et al., 1986).

4. Predictions of the Foreign-Protein–AIDS and HIV-AIDS Hypotheses

Here we compare the HIV- and the foreign-protein–AIDS hypotheses in terms of how well their predictions can be reconciled with hemophilia-AIDS:

4.1 Mortality of hemophiliacs with and without HIV. The virus-AIDS hypothesis predicts that the mortality of HIV-positive hemophiliacs will be higher than that of matched HIV-free counterparts. Considering the high, 75 percent rate of infection of American hemophiliacs by HIV since 1984, one would expect that the median age of all American hemophiliacs would have significantly decreased and that their mortality increased. The HIV-AIDS hypothesis predicts that in 1994, at least one ten-year latent period after most American hemophiliacs were infected, over 50 percent of the 15,000 HIV-positive American hemophiliacs would have developed AIDS or died from AIDS (Institute of Medicine, 1988; Duesberg, 1992a). But despite the many claims that HIV causes AIDS in hemophiliacs (Centers for Disease Control, 1986; Institute of Medicine, 1988; Weiss & Jaffe, 1990; Chorba et al., 1994), there is not a single controlled study showing that the morbidity or mortality of HIV-positive hemophiliacs is higher than that of HIV-negative controls matched for the lifetime consumption of factor VIII.

Instead, the mortality of American hemophiliacs has decreased and their median age has increased since 75 percent were infected by HIV. The median age of American hemophiliacs has increased from eleven years in 1972, to twenty years in 1982, to twenty-five years in 1986, and to twenty-seven years in 1987, although 75 percent had become HIV-antibody–positive prior to 1984 (Institute of Medicine and National Academy of Sciences, 1986; Koerper, 1989; Stehr-Green et al., 1989). Likewise, their median age at death has increased from about forty to fifty-five years in the period from 1968 to 1986 (Chorba et al., 1994).

Contrary to the HIV-AIDS hypothesis, one could make a logical argument that HIV, instead of decreasing the life span of hemophiliacs, has in fact increased it. A more plausible argument suggests that the life span of American hemophiliacs has increased as a consequence of the widespread use of factor VIII that started in the late 1960s (see above). As predicted by the foreign-protein hypothesis, the price for the extended life span of hemophiliacs by treatment with commercial factor VIII was immunosuppression due to the long-term parenteral administration of large quantities of foreign protein (see section 4.2). Prior to factor VIII therapy, most hemophiliacs died as adolescents from internal bleeding (Koerper, 1989).

However, a recent CDC study reports that the mortality of American hemophiliacs suddenly increased 2.5-fold in the period from 1987 to 1989, after it had remained almost constant in the period from 1968 to 1986 (Chorba et al., 1994). Since American hemophiliacs became gradually infected via the introduction in the 1960s of pooled factor VIII treatments until 1984, when HIV was eliminated from the blood supply (see above), one would have expected first a gradual increase in hemophilia mortality and then a rather steep decrease. The increase in mortality would have followed the increase of infections with a lag defined by the time that HIV is thought to require to cause AIDS. The presumed lag between HIV and AIDS has been estimated at ten months by the CDC in 1984 (Auerbach et al., 1984) and at ten years by a committee of HIV researchers, including some from the CDC, in 1988 (Institute of Medicine, 1988). Therefore the sudden increase in hemophilia deaths in 1987 is not compatible with HIV-mediated mortality. Hemophilia mortality should have gradually decreased after 1984, when HIV was eliminated from the blood supply, depending on the lag period assumed between infection and AIDS. Even if the lag period from HIV to AIDS were ten years, the mortality of hemophiliacs should have significantly decreased by 1989, five years after new infections had been stopped.

An obvious explanation for the chronological inconsistency between infection of hemophiliacs with HIV since the 1960s and the sudden increase in their mortality twenty years later is the introduction of the cytotoxic DNA chain terminator AZT as an anti-HIV drug in 1987. AZT has been recommended and prescribed to symptomatic HIV carriers since 1987 (Fischl et al., 1987; Richman et al., 1987) and to healthy HIV carriers with lower than 500 T-cells since 1988 (Volberding et al., 1990; Goldsmith et al., 1991; Phillips et al., 1994b). Approximately 200,000 HIV antibody-positives with and without AIDS diseases are currently prescribed AZT worldwide (Duesberg, 1992a). According to a preliminary survey of hemophiliacs from a national group, Concerned Hemophiliacs Acting for Peer Strength (CHAPS), thirty-five out of thirty-five HIV-positive hemophiliacs asked had taken AZT, and twenty out of thirty-five who had taken AZT at some time were currently on AZT (personal communication, Brent Runyon, executive director of CHAPS, Wilmington, North Carolina).

The DNA chain terminator AZT was developed 30 years ago to kill growing human cells for cancer chemotherapy. Because of its intended toxicity, chemotherapy is typically applied for very limited periods of time, i.e., weeks or months, but AZT is now prescribed to healthy HIV-positives indefinitely, despite its known toxicity (Nussbaum, 1990; Volberding et al., 1990). Indeed, AZT has been shown to be toxic in HIV-positives and proposed as a possible cause of AIDS diseases since 1991 (Duesberg, 1991, 1992a, 1992b, 1992c). Recently, the European "Concorde trial" (Seligmann et al., 1994) and several other studies have shown that, contrary to earlier claims, AZT does not prevent AIDS (Oddone et al., 1993; Tokars et al., 1993; Lenderking et al., 1994; Lundgren et al., 1994). The Concorde trial even showed that the mortality of healthy, AZT-treated HIV carriers was 25 percent higher than that of placebo-treated controls (Seligmann et al., 1994). Likewise, an American multicenter study showed that the death risk of hemophiliacs treated with AZT was 2.4 times higher and that their AIDS risk was even 4.5 times higher than that of untreated HIV-positive hemophiliacs (Goedert et al., 1994). Thus, the widespread use of AZT in HIV-positives could be the reason for the sudden increase in hemophilia mortality since 1987.

The AZT-hemophilia–AIDS hypothesis and the foreign-protein–AIDS hypothesis both predict that hemophilia-AIDS would stay constant or

Appendix A ■ 477

increase as long as unpurified factor VIII is used and AZT is prescribed to HIV-positive hemophiliacs. By contrast, the HIV-AIDS hypothesis predicts that hemophilia-AIDS should have decreased with time since 1984 when HIV was eliminated from the blood supply. The HIV hypothesis further predicts that AIDS should have decreased precipitously since 1989 when AZT was prescribed as AIDS prevention to inhibit HIV.

But the decrease in hemophilia-AIDS predicted by the HIV-AIDS hypothesis was not observed. Instead, the data confirm the AZT–foreign-protein–AIDS hypotheses: The CDC reports 300 hemophilia AIDS cases in 1988, 295 in 1989, 320 in 1990, 316 in 1991, 316 in 1992 and, after broadening the AIDS definition as of January 1993 (Centers for Disease Control and Prevention, 1992), 1,096 in 1993 (Centers for Disease Control, 1993, 1994; and prior *HIV/AIDS Surveillance* reports).

4.2 Annual AIDS risk of HIV-positive hemophiliacs compared to other HIV-positive AIDS risk groups. The HIV-AIDS hypothesis predicts that the annual risk of HIV-positive hemophiliacs would be the same as that of other HIV-infected risk groups. One could in fact argue that it should be higher, because the health of hemophiliacs is compromised compared to AIDS risk groups without congenital health deficiencies.

By contrast, the foreign-protein–AIDS hypothesis makes no clear prediction about the annual AIDS risk of hemophiliacs compared to drug-AIDS risk groups, because the relative risks have not been studied and are hard to quantitate.

By the end of 1992, 2,214 American hemophiliacs with AIDS were reported to the CDC (Centers for Disease Control, 1993; Chorba et al., 1994). Since there are about 15,000 HIV-positive American hemophiliacs, an average of only 1.3 percent (201 out of 15,000) have developed AIDS annually between 1981 and 1992 (Tsoukas et al., 1984; Hardy et al., 1985; Institute of Medicine and National Academy of Sciences, 1986; Sullivan et al., 1986; Stehr-Green et al., 1988; Goedert et al., 1989; Koerper, 1989; Morgan, Curran & Berkelman, 1990; Gomperts, De Biasi & De Vreker, 1992). But after the inclusion of further diseases into the AIDS syndrome (Institute of Medicine, 1988) and the introduction of AZT as an anti-HIV drug, both in 1987, the annual AIDS risk of American hemophiliacs appears to have stabilized at 2 percent, i.e., about 300 out of 15,000 per year until 1993 when the AIDS definition was changed again (Centers for Disease Control, 1993) (see section 4.1).

Hemophilia-AIDS statistics from Germany are compatible with American counterparts: about 50 percent of the 6,000 German hemophiliacs are HIV-positive (Koerper, 1989). Only 37, or 1 percent, of these developed AIDS-defining diseases during 1991 (Leonhard, 1992), and

186, or 1.5 percent, annually during the four years from 1988 to 1991 (Schwartlaender et al., 1992).

The 1.3 to 2 percent annual AIDS risk indicates that the average HIV-positive hemophiliac would have to wait for twenty-five to thirty-five years to develop AIDS diseases from HIV. Indeed latent periods of over twenty years have just been calculated for HIV-positive hemophiliacs based on the loss of T-cells over time (Phillips et al., 1994b).

By contrast, the annual AIDS risk of the average, HIV-positive American is currently 6 percent, because there are now about 60,000 annual AIDS cases (Centers for Disease Control, 1993) per 1 million HIV-positive Americans (Curran et al., 1985; Centers for Disease Control, 1992b; Duesberg, 1992a). This reflects the annual AIDS risks of the major risk groups, the male homosexuals and intravenous drug users who make up about 93 percent of all American AIDS patients (Centers for Disease Control, 1993). The annual AIDS risks of intravenous drug users (Lemp et al., 1990) and male homosexuals appear to be the same, as both were estimated at about 5 to 6 percent (Anderson & May, 1988; Lui et al., 1988; Lemp et al., 1990) (Table 1).

In view of the compromised health of hemophiliacs, it is surprising that the annual AIDS risk of HIV-infected hemophiliacs is only 1.3 to 2 percent and thus three to five times lower than that of the average HIV-infected, nonhemophiliac American or European (Table 1). Commenting on the relatively low annual AIDS risk of hemophiliacs compared to that of homosexuals, hematologists Sullivan et al. (1986) noted that "[t]he reasons for this difference remain unclear." Hardy et al. (1985) from the CDC also noted the discrepancy in the latent periods of different risk groups. "The magnitude of some of the differences in rates is so great that even gross errors in denomination estimates can be overcome." And Christine Lee, senior author of the study that had estimated latent periods of over twenty years from infection to hemophilia AIDS (Phillips et al., 1994b), commented on the paradox: "It may be that hemophiliacs have got that cofactor [of foreign blood contaminants], homosexuals have got another cofactor, drug users have got another cofactor, and they all have the same effect, so that at the end of the day you get [approximately] the same progression rate" (Jones, 1994).

Thus, the three- to five-fold difference between the annual AIDS risks of HIV-positive hemophiliacs and the other major risk groups is not compatible with the HIV hypothesis. However, it can be reconciled with the foreign-protein and drug-AIDS hypothesis (Duesberg, 1992a, 1994), because different causes, i.e., drugs and foreign proteins, generate AIDS diseases at different rates.

4.3 *The age bias of hemophilia-AIDS.* The HIV-AIDS hypothesis predicts that the annual AIDS risks of HIV-positive hemophiliacs is independent of their age, because virus replication is independent of the age of the host. Predictions would have to be adjusted, however, by the hypothetical lag period between infection and AIDS. If the average latent period from HIV to AIDS is ten months, as was postulated in 1984 (Auerbach et al., 1984), less-than-ten-month-old HIV-positive hemophiliacs would have a lower probability of having AIDS. If the average latent period from HIV to AIDS is ten years (Institute of Medicine, 1988; Lui et al., 1988; Lemp et al., 1990; Weiss, 1993), HIV-positive hemophiliacs under ten years of age would have a lower probability of having AIDS. In other words, if the time of infection is unknown, the annual AIDS risks of HIV-positive hemophiliacs over ten months or ten years, respectively, would be independent of the age of the HIV-positive hemophiliac.

Table 1. Annual AIDS Risks of HIV-Infected Groups

American/European Risk Group	Annual AIDS (in %)	References
Hemophiliacs	1.3–2	see text
Male homosexuals	5–6	Lui et al., 1988
		Anderson & May, 1988
		Lemp et al., 1990
Intravenous drug users	5–6	Lui et al., 1988
		Anderson & May, 1988
		Lemp et al., 1990

By contrast, the foreign-protein hypothesis predicts that the annual AIDS risk of HIV-positive and -negative hemophiliacs increases with age because immunosuppression is the result of the lifetime dose of proteins transfused (Pollack et al., 1985; Brettler et al., 1986; Sullivan et al., 1986; Koerper, 1989) (see above). The more years a hemophiliac has been treated with unpurified blood products, the more likely he is to develop immunodeficiency. Thus, the foreign-protein hypothesis predicts that the annual AIDS risk of a hemophiliac would increase with age.

Statistics show that the median age of hemophiliacs with AIDS in the United States (Evatt et al., 1984; Koerper, 1989; Stehr-Green et al., 1989) and other countries (Darby et al., 1989; Biggar and the International Registry of Seroconverters, 1990; Blattner, 1991) is about five to fifteen years higher than the average age of hemophiliacs. In the United States, the average age of hemophiliacs was twenty to twenty-seven years from 1980 to 1986, while that of hemophiliacs with AIDS was thirty-two to thirty-five years (Evatt et al., 1984; Koerper, 1989; Stehr-Green et al., 1989).

Likewise, the annual AIDS risk of HIV-positive hemophiliacs shows a strong age bias. An international study estimated the annual AIDS risk of children at 1 percent and that of adult hemophiliacs at 3 percent over a five-year period of HIV infection (Biggar and the International Registry of Seroconverters, 1990). In the United States, Goedert et al. (1989) reported that the annual AIDS risk of one- to seventeen-year-old hemophiliacs was 1.5 percent, that of eighteen- and thirty-four-year-old hemophiliacs was 3 percent, and that of sixty-four-year-old hemophiliacs was 5 percent. Goldsmith et al. (1991) reported that the annual T-cell loss of hemophiliacs under twenty-five years was 9.5 percent and for hemophiliacs over twenty-five years, 17.5 percent.

Lee et al. (1991) reported that the annual AIDS risk of hemophiliacs eleven years after HIV seroconversion was 31 percent under twenty-five years and 56 percent over twenty-five years. They estimated that the relative risk of AIDS increased five-fold over twenty-five years. The same group confirmed in 1994 that the annual AIDS risk of HIV-positive hemophiliacs over thirty years is two times higher than in those under fifteen years of age (Phillips et al., 1994b). Stehr-Green et al. (1989) estimated that "...the risk of AIDS increased two fold for each ten-year increase in age after controlling for year of seroconversion." Likewise, Fletcher et al. (1992) reported a four-fold higher incidence of AIDS in hemophiliacs over twenty-five years of age than in those aged five to thirteen years. Thus, the annual AIDS risk of hemophiliacs increases about two-fold for each ten-year increase in age.

This confirms the foreign-protein hypothesis, which holds that the cumulative dose of transfusions received is the cause of AIDS-defining diseases among hemophiliacs. According to the hematologist Koerper (1989), "this may reflect lifetime exposure to a greater number of units of concentrate...," and to Evatt et al. (1984), "[t]his age bias may be due to differences in duration of exposure to blood products..." A recent study of HIV-free hemophiliacs is directly compatible with the foreign-protein hypothesis. The study showed that despite the absence of HIV "with increasing age, numbers of $CD4^+$ $CD45RA^+$ cells decreased and continued to do so throughout life" (Fletcher et al., 1992).

By contrast, AIDS caused by an autonomous infectious pathogen would be independent of the age of the recipient because the replication cycle of viruses, including HIV, is independent of the age of the host. Thus, the foreign-protein–AIDS hypothesis, rather than the HIV-AIDS hypothesis, correctly predicts the age bias of hemophilia-AIDS.

4.4 *Hemophilia-specific AIDS diseases.* The thirty AIDS diseases fall into two categories, the microbial immunodeficiency diseases and the nonimmunodeficiency diseases, i.e., diseases that are neither caused by nor consistently associated with immunodeficiency (Duesberg, 1992a, 1994). Based on their annual incidence in America in 1992, 61 percent of the AIDS diseases were microbial immunodeficiency diseases, including *Pneumocystis* pneumonia, candidiasis, tuberculosis, etc., and 39 percent were nonimmunodeficiency diseases, including Kaposi's sarcoma, lymphoma, dementia, and wasting disease (Table 2) (Centers for Disease Control, 1993).

The virus-AIDS hypothesis predicts that the probability of all HIV-infected persons to develop a given immunodeficiency or nonimmunodeficiency AIDS disease is the same and independent of the AIDS risk group. By contrast, the hypothesis that AIDS is caused by drugs or by foreign proteins predicts specific diseases for specific causes (Duesberg, 1992a).

In America, 99 percent of the hemophiliacs with AIDS have immunodeficiency diseases, of which 70 percent are fungal and viral pneumonias (Evatt et al., 1984; Koerper, 1989; Papadopulos-Eleopulos et al., 1994). Only one study reports that 1 percent of hemophiliacs with AIDS had Kaposi's sarcoma (Selik, Starcher & Curran, 1987). The small percentage of Kaposi's sarcoma may be due to aphrodisiac nitrite inhalants used by male homosexual hemophiliacs (Haverkos & Dougherty, 1988; Duesberg, 1992a). There are no reports of wasting disease or dementia in American hemophiliacs. An English study also reported predominantly pneumonias and other immunodeficiency diseases among hemophiliacs, and also three cases of wasting syndrome (Lee et al., 1991). It appears that the AIDS diseases of hemophiliacs are virtually all immunodeficiency diseases, whereas 39 percent of the AIDS diseases of intravenous drug users and male homosexuals are nonimmunodeficiency diseases (Table 2). Since AIDS diseases in hemophiliacs and nonhemophiliacs are not the same, their causes can also not be the same.

The almost exclusive occurrence of immunodeficiency AIDS diseases among hemophiliacs is correctly predicted by the foreign-protein–AIDS hypothesis, but not by the HIV-AIDS hypothesis. The prediction of the HIV hypothesis, that the distribution of immunodeficiency and nonimmunodeficiency diseases among hemophiliacs is the same as in the rest of the American AIDS population, is not confirmed.

4.5 *Is hemophilia–AIDS contagious?* The virus-AIDS hypothesis predicts that AIDS is contagious, because HIV is a parenterally and sexually transmitted virus. It predicts that hemophilia-AIDS is sexually transmissible. Indeed, AIDS researchers claim that the wives of hemophiliacs develop AIDS from sexual transmission of HIV (Booth, 1988;

Lawrence et al., 1990; Weiss & Jaffe, 1990; Centers for Disease Control, 1992a, 1993). Further, the HIV-AIDS hypothesis predicts that wives of hemophiliacs will develop the same AIDS diseases as other risk groups.

The foreign-protein hypothesis predicts that AIDS is not contagious and that the wives and sexual partners of hemophiliacs do not contract AIDS from their mates.

To test the hypothesis that immunodeficiency of hemophiliacs is sexually transmissible, the T4:T8–cell ratios of forty-one spouses and female sexual partners of immunodeficient hemophiliacs were analyzed (Kreiss et al., 1984). Twenty-two of the females had relationships with hemophiliacs with T-cell ratios below 1, and nineteen with hemophiliacs with ratios of 1 and greater. The mean duration of relationships was ten years, the mean number of sexual contacts was 111 during the previous year, and only 12 percent had used condoms (Kreiss et al., 1984). Since the T-cell ratios of all spouses were normal, averaging 1.68—exactly like those of fifty-seven normal controls—the authors concluded that "there is no evidence to date for heterosexual or household-contact transmission of T-cell subset abnormalities from hemophiliacs to their spouses..." (Kreiss et al., 1984).

The CDC reports that between 1985 and 1992, 131 wives of American hemophiliacs were diagnosed with unnamed AIDS diseases (Centers for Disease Control, 1993). If one considers that there have been 15,000 HIV-positive hemophiliacs in the United States since 1984 and that one-third are married, then there are 5,000 wives of HIV-positive hemophiliacs. About sixteen of these women have developed AIDS annually during the eight years (131:8) from 1985 to 1992. But these sixteen annual AIDS cases would have to be distinguished from the at least eighty wives of hemophiliacs who are expected to die per year based on natural mortality. Considering the human life span of about eighty years and that on average at least 1.6 percent of all those over twenty years of age die annually, about eighty out of 5,000 wives over twenty would die naturally per year. Thus, until controls show that among 5,000 HIV-positive wives of hemophiliacs, sixteen more than eighty, i.e., ninety-six, die annually, the claim that wives of hemophiliacs die from sexual or other transmission of HIV is unfounded speculation.

Moreover, it has been pointed out that all AIDS-defining diseases of the wives of hemophiliacs are typically age-related opportunistic infections, including 81 percent pneumonia (Lawrence et al., 1990). Kaposi's sarcoma, dementia, lymphoma, and wasting syndrome are not observed in wives of hemophiliacs (Lawrence et al., 1990).

Again, the foreign-protein, but not the HIV hypothesis, correctly predicts the noncontagiousness of hemophilia-AIDS. It also predicts the specific spectrum of AIDS diseases in wives of hemophiliacs. By contrast, the

Table 2. AIDS-defining diseases in the United States in 1992[a]

Immunodeficiencies	Nonimmunodeficiencies
42% pneumonia	20% wasting disease
17% candidiasis	9% Kaposi's sarcoma
12% mycobacterial, including 3% tuberculosis	6% dementia
8% cytomegalovirus	4% lymphoma
5% toxoplasmosis	
5% herpesvirus	
Total = 61% (>61% due to overlap)	Total = 39%

[a] = Centers for Disease Control, 1993

virus-AIDS hypothesis predicts the same spectrum of AIDS diseases among wives of hemophiliacs as among the major risk groups (see Table 2). It appears that the virus-AIDS hypothesis is claiming normal morbidity and mortality of the wives of hemophiliacs for HIV.

4.6 *Immunodeficiency in HIV-positive and -negative hemophiliacs.* The HIV hypothesis predicts that immunodeficiency is observed only in HIV-positive hemophiliacs. By contrast, the foreign-protein hypothesis predicts that immunodeficiency is a function of the lifetime dose of transfusions received and is not dependent on HIV or antibodies against HIV. The foreign-protein hypothesis also predicts that HIV-positive hemophiliacs are more likely to be immunosuppressed than HIV-negatives because HIV is a rare contaminant of blood transfusion and thus is a marker for the number of transfusions received (see section 3 and below) (Tsoukas et al., 1984, Ludlam et al., 1985; Kreiss et al., 1986; Sullivan et al., 1986; Koerper, 1989; Fletcher et al., 1992).

Twenty-one studies, summarized in Table 3, have observed 1,186 immunodeficient hemophiliacs, 416 of whom were HIV-free. Immunodeficiency in these studies was either defined by a T4:T8–cell ratio of about 1 or less than 1, compared to a normal ratio of 2, or by other tests such as immunological anergy. Since immunodeficiency was observed in the absence of HIV, most of the studies listed in Table 3 have concluded that immunodeficiency in hemophiliacs was caused by transfusion of factor VIII and contaminating proteins. According to the first of Koch's postulates (Merriam-Webster, 1965), the absence of a microbe, e.g., HIV, from a disease excludes it as a possible cause of that disease. Thus, transfusion of foreign protein, not the presence of HIV, emerges as the common denominator of all hemophiliacs with immunodeficiency.

Table 3. Immunosuppression in HIV-negative and -positive Hemophiliacs

Study	HIV-negative	HIV-positive
(1) Tsoukas et al., 1984	6/14	9/15
(2) Carr et al., 1984	18/53	
(3) Ludlam et al., 1985	15	
(4) Moffat and Bloom, 1985	23	23
(5) AIDS-Hemophilia French Study Group, 1985	33	55
(6) Hollan et al., 1985	30/104	
(7) Sullivan et al., 1986	28	83
(8) Madhok et al., 1986	9	10
(9) Kreiss et al., 1986	6/17	22/24
(10) Gill et al., 1986	8/24	30/32
(11) Brettler et al., 1986	4	38
(12) Sharp et al., 1987	5/12	
(13) Matheson et al., 1987	5	3
(14) Mahir et al., 1988	6	5
(15) Antonaci et al., 1988	15	10
(16) Aledort, 1988	57	167
(17) Jin et al., 1989	12	7
(18) Lang et al., 1989	24	172
(19) Jason et al., 1990	31	
(20) Becherer et al., 1990	74	136
(21) Smith et al., 1993	7	
Totals	416	770

If two numbers are listed per category, the first reports immunodeficient and the second healthy plus immunodeficient hemophiliacs per study group. In most studies immunodeficiency was expressed by the $T_4:T_8$-cell ratio, in others by anergy. In a normal immune system the $T_4:T_8$-cell ratio is about 2. In immunodeficient persons it is about 1 or below 1. Studies which list both HIV-positive and -negative groups indicate that HIV-positives are more likely to be immunodeficient than -negatives. This is because HIV is a marker for the number of transfusions received, and transfusion of foreign proteins causes immunodeficiency (see sections 3 and 4.6).

Nevertheless, several of the controlled studies listed in Table 3, which compare HIV-negative to HIV-positive hemophiliacs, have shown that immunodeficiency is more often associated with HIV-positives than with -negatives. Although some studies did not report immunodeficiency in HIV-positives, Table 3 lists 770 HIV-positives and 416 HIV-negatives per 1,186 immunodeficient hemophiliacs. In view of this, one could argue that HIV is one of several possible causes of immunodeficiency.

However, some of the investigators listed in Table 3 (Tsoukas et al., 1984; Ludlam et al., 1985; Kreiss et al., 1986; Madhok et al., 1986; Sullivan et al., 1986) and others who have not performed controlled studies (Koerper, 1989) have proposed that HIV is just a marker for the number of transfusions received (section 3). As a rare contaminant of factor VIII, HIV has in fact been a marker for the number of transfusions received before it was eliminated from the blood supply in 1984, just like hepatitis virus infection was a marker of the number of transfusions received until it was eliminated from the blood supply earlier (Anonymous, 1984; Koerper, 1989). According to Kreiss et al. (1986), "[s]eropositive hemophiliac subjects, on average, had been exposed to twice as much concentrate... as seronegative[s]." Sullivan et al. (1986) also reported that "[s]eropositivity to LAV/HTLV-III (HIV) was 70 percent for the hemophiliac population and... varied directly with the amount of factor VIII received" (see section 3). More recently, Schulman (1991) reported that "a high annual consumption" of factor VIII concentrate "predisposed" to HIV-seroconversion, and Fletcher et al. (1992) described a positive "relationship between the amount of concentrate administered and anti-HIV prevalence rate..."

The chronology of studies investigating immunodeficiency in HIV-free hemophiliacs faithfully reflects the popularity of the HIV hypothesis: the more popular the HIV hypothesis became over time, the fewer studies there were that investigated immunodeficiency in HIV-free hemophiliacs. Indeed, most of the controlled studies investigating the role of HIV in immunodeficiency of HIV-positive and matched HIV-negative hemophiliacs were conducted before the virus hypothesis became totally dominant in 1988 (Institute of Medicine, 1988), namely between 1984 and 1988 (Table 3). The studies by Jin, Cleveland, and Kaufman, and Lang et al., both dated 1989, and the studies by Becherer et al. and by Jason et al., both dated 1990, all described data collected before 1988 (Table 3). After 1988 the question whether HIV-free hemophiliacs developed immunodeficiency became increasingly unpopular. As a result, only a few studies have described immunodeficiency in HIV-free hemophiliacs.

For example, Schulman (1991) reported "worrisome evidence of similar immunological disturbances has been observed, albeit to a lesser degree, in anti–HIV-negative hemophiliacs" and that immunodeficiency in hemophiliacs "correlates more strongly with annual consumption of factor concentrates than with HIV status." Fletcher et al. (1992) published a median T4:T8–cell ratio of 1.4, with a low tenth-percentile of 0.8, in a group of 154 HIV-free hemophiliacs, and also showed a steady decline of T-cell counts with treatment years. Likewise, Hassett et al. (1993) reported that "patients with hemophilia A without human immunodeficiency virus type 1 (HIV-1) infection have lower $CD4^+$ counts and $CD4^+$:$CD8^+$ ratios than controls." The study observed an average T4:T8–cell ratio of 1.47 in a group of 307 HIV-free hemophiliacs, differing over 50 years in age, compared to an average of 1.85 in normal controls. Unlike others Hassett et al. attributed the lowered $CD4^+$ counts to a hemophilia-related disorder rather than to foreign proteins, but like others they attributed increased $CD8^+$ counts to treatment with commercial factor VIII. However, Fletcher et al.'s and Hassett et al.'s practice of averaging immunodeficiency markers of large numbers of people, differing over 50 years in age, obscures how far the immunity of the longest, and thus most treated cases, had declined compared to cases which have received minimal treatments.

Since the authors of these studies did not report the lifetime dosage of factor VIII treatments of HIV-free hemophiliacs, a correlation between foreign-protein dosage and immunosuppression cannot be determined. On the contrary, averaging immunodeficiency parameters of newcomers and long-term treatment recipients obscures the relationship between the lifetime dosage of factor VIII and immunosuppression.

Moreover, the CDC reported seven HIV-free hemophiliacs with AIDS (Smith et al., 1993). This study was one of a package that proposed to set apart HIV-free AIDS from HIV-positive AIDS with the new term *idiopathic CD4 lymphocytopenia*. The goal of these studies was to save the virus-AIDS hypothesis, despite the presence of HIV-free AIDS (Duesberg, 1993b, 1994; Fauci, 1993). Nevertheless all of the seven HIV-free hemophiliacs met one or more criteria of the CDC's clinical AIDS definition from 1993 (Centers for Disease Control and Prevention, 1992), i.e., they all had less than 300 T-cells per microliter (range from 88 to 296), and three also had AIDS-defining diseases such as herpes and thrombocytopenia (Smith et al., 1993).

The occurrence of immunodeficiency in HIV-free hemophiliacs demonstrates most directly that long-term transfusion of foreign proteins contaminating factor VIII is sufficient to cause immunodeficiency in hemophiliacs. To prove the foreign-protein hypothesis it would be necessary to show that treatment of HIV-positive hemophiliacs with pure factor VIII does not cause immunodeficiency. It is shown below that this is actually the case.

4.7 *Stabilization, even regeneration of immunity of HIV-positive hemophiliacs by treatment with pure factor VIII.* Commercial preparations of factor VIII contain between 99 and 99.9 percent non-factor VIII proteins (Eyster & Nau, 1978; Brettler & Levine, 1989; Gjerset et al., 1994; Mannucci et al., 1992; Seremetis et al., 1993). The foreign-protein– hemophilia–AIDS hypothesis predicts that long-term transfusion with commercial factor VIII would be immunosuppressive, because of the presence of contaminating proteins. Further, it predicts that pure factor VIII, containing 100 to 1,000 times less foreign protein per functional unit, may not be immunosuppressive.

Several studies have recently tested whether the impurities of factor VIII or factor VIII by itself are immunosuppressive in HIV-positive hemophiliacs. De Biasi et al. (1991) showed that over a period of two years the average T-cell counts of ten HIV-positive hemophiliacs treated with nonpurified, commercial factor VIII declined two-fold, while those of matched HIV-positive controls treated with pure factor VIII remained unchanged. Moreover, four out of six anergic HIV-positive patients treated with purified factor VIII recovered immunological activity. Goldsmith et al. (1991) also found that the T-cell counts of thirteen hemophiliacs treated with purified factor VIII remained stable for 1.5 years. Seremetis et al. (1993) have confirmed and extended de Biasi et al.'s conclusion by establishing that the T-cells of HIV-positive hemophiliacs were not depleted after treatment with pure factor VIII for three years. Indeed, the T-cell counts of fourteen out of thirty-one HIV-positive hemophiliacs increased up to 25 percent over the three-year period of treatment with purified factor VIII—despite infection by HIV. By contrast, in the group treated with unpurified factor VIII, the percentage of those with less than 200 T-cells per µl increased from 7 percent at the beginning of the study to 47 percent at the end.

Likewise Hilgartner et al. (1993) reported individual increases of T-cell counts of up to 50 percent in a group of 36 HIV-positive hemophiliacs treated with purified factor VIII whose average T-cell count had declined 1 percent during six months. Goedert et al. (1994) have also reported that "T-cell counts fell less rapidly with high purity products." Moreover, Schulman (1991) observed that four HIV-positive hemophiliacs recovered from thrombocytopenia upon treatment with pure factor VIII for two to three years, and others from CD8-related immunodeficiency upon treatment for six months.

However, despite the evidence that purified factor VIII is beneficial in maintaining or even increasing T-cell counts, several studies testing purified factor VIII are ambiguous about its effectiveness in preventing or treating AIDS (Goldsmith et al., 1991; Hilgartner et al., 1993; Gjerset et al., 1994; Goedert et al., 1994, Phillips et al., 1994a). Some of these studies have only tested partially purified, i.e., 2–10 units/mg, instead of highly purified, i.e., 2,000–3,000 units/mg, factor VIII (Gjerset et al., 1994). But each of the studies that are ambiguous about the benefits have also treated their patients with toxic antiviral DNA chain terminators like AZT. Indeed, the study by de Biasi et al. was the only one that has tested purified factor VIII in the absence of AZT. The study by Seremetis et al. initially called for no AZT, but later allowed it anyway. Thus, in all but one study, the potential benefits of highly purified factor VIII have been obscured by the toxicity of AZT (see section 5.4).

It is concluded that treatment of HIV-positive hemophiliacs with pure factor VIII provides lasting stabilization of immunity and even allows regeneration of lost immunity. It follows that foreign proteins, rather than factor VIII or HIV, cause immunosuppression in HIV-positive hemophiliacs.

5. Conclusions and Discussion

Four criteria of proof have been applied to distinguish between the virus and the foreign-protein hypothesis of hemophilia-AIDS: (1) correlation, (2) function (Koch's third postulate), (3) predictions, and (4) therapy and prevention. Each of these criteria proved the foreign-protein hypothesis valid and the HIV hypothesis invalid.

5.1 Correlations between hemophilia-AIDS and the long-term administration of foreign proteins or HIV. Although correlation is not sufficient, it is necessary to prove causation in terms of Koch's postulates (Merriam-Webster, 1965). The first of Koch's postulates calls for the presence of the suspected cause in all cases of the disease, i.e., a perfect correlation; the second calls for the isolation of the cause; and the third for causation of the disease with the isolated causative agent.

All hemophiliacs with immunodeficiency described here have been subject to long-term treatment with factor VIII contaminated by foreign protiens. This establishes a perfect correlation between foreign-protein transfusion and hemophilia-AIDS, and fulfills Koch's first postulate.

By contrast, a summary of twenty-one separate studies showed that 416 of 1,186 immunodeficient hemophiliacs were HIV-free (Table 3). Since HIV does not correlate well with hemophilia-AIDS, it fails Koch's first postulate and is thus not even a plausible cause of AIDS.

5.2 Foreign-protein hypothesis, but not HIV hypothesis, meets Koch's third postulate as cause of immunodeficiency. The fact that all hemophiliacs with immunodeficiency had been subject to long-term treatment with foreign proteins and that factor VIII treatment in the absence of foreign proteins does not cause immune suppression and may even revert it provides functional proof for the foreign-protein hypothesis. Thus, the foreign-protein hypothesis meets Koch's third postulate of causation.

Regeneration of immunity of HIV-positives by treatment with pure factor VIII further indicates that HIV by itself or in combination with factor VIII is not sufficient for hemophilia-AIDS. Therefore, HIV fails Koch's third postulate as a cause of AIDS.

5.3 Foreign-protein hypothesis correctly predicts hemophilia-AIDS and resolves paradoxes of HIV hypothesis. The ability to make verifiable predictions is the hallmark of a correct scientific hypothesis. Application of the two competing hypotheses to hemophilia-AIDS proved that the foreign-protein hypothesis, but not the HIV hypothesis, correctly predicts seven characteristics of hemophilia-AIDS (see sections 4.1–4.7):

(1) The increased life span of American hemophiliacs, despite infection of 75 percent by HIV, due to factor VIII treatment, that extended their lives and disseminated harmless HIV.

(2) The three- to five-times lower annual AIDS risk of hemophiliacs, compared to other AIDS risk groups.

(3) The age bias of the annual AIDS risk of hemophiliacs, increasing two-fold for each ten-year increase in age.

(4) The restriction of hemophilia-AIDS to immunodeficiency-related AIDS diseases, setting it apart from the spectrum of AIDS diseases in other risk groups.

(5) The noncontagiousness of hemophilia-AIDS, i.e., the absence of AIDS diseases above their normal background in sexual partners of hemophiliacs.

(6) The occurrence of immunodeficiency in HIV-free, factor VIII–treated hemophiliacs.

(7) The stabilization, even regeneration, of immunity of HIV-positive hemophiliacs upon long-term treatment with pure factor VIII.

It follows that the foreign-protein hypothesis, but not the HIV hypothesis, correctly predicts hemophilia-AIDS. In addition, the foreign-protein hypothesis resolves all remaining paradoxes of the HIV hypothesis (see section 2):

(1) The failure of HIV neutralizing antibody to protect against AIDS—because HIV is not the cause of AIDS.

(2) The noncorrelation between the loss of T-cells and HIV activity—because foreign proteins rather than HIV are immunotoxic.

(3) The failure of HIV to kill T-cells—because T-cell synthesis is suppressed by immunotoxic foreign proteins.

(4) The latent periods of ten to thirty-five years between HIV and hemophilia-AIDS—because the lifetime dosage of foreign proteins, not HIV, causes AIDS.

5.4 Treatment and prevention of AIDS. The prevention or cure of a disease, by eliminating or blocking the suspected cause, provides empirical proof of causation.

(i) *Drug treatment based on HIV hypothesis:* On the basis of the HIV hypothesis, AIDS has been treated since 1987 with anti-HIV drugs, such as the DNA chain terminators AZT, ddI, etc. (Duesberg, 1992a). The rationale of the AZT treatment is to prevent HIV-DNA synthesis at the high cost of inhibiting cellular DNA synthesis, the original target of AZT cancer chemotherapy (see above). However, not a single AIDS patient has ever been cured with AZT. Since 1989, healthy HIV-positive hemophiliacs have also been treated with DNA chain terminators in efforts to prevent AIDS. But the alleged ability of AZT to prevent AIDS has recently been discredited by several large clinical trials (Oddone et al., 1993; Tokars et al., 1993; Goedert et al., 1994; Lenderking et al., 1994; Lundgren et al., 1994; Seligmann et al., 1994). Moreover, all studies of AZT treatments have confirmed the unavoidable cytotoxicity of DNA chain terminators (Duesberg, 1992; Oddone et al., 1993; Tokars et al., 1993; Lenderking et al., 1994; Lundgren et al., 1994; Seligmann et al., 1994). One study observed a 25 percent increased mortality (Seligmann et al., 1994), and another a 4.5-fold higher annual AIDS risk and a 2.4-fold higher annual death risk in AZT-treated HIV-positive hemophiliacs compared to untreated controls (Goedert et al., 1994).

The failure of AZT therapy to cure or prevent AIDS indicates either that the drug is not sufficient to inhibit HIV or that HIV is not the cause of AIDS. The lower mortality and much lower incidence of AIDS-defining diseases among hemophiliacs not treated with AZT compared to those treated indicates that AZT causes AIDS-defining diseases and mortality. Thus, there is currently no rational or empirical justification for AZT treatment of HIV-positives with or without AIDS.

The apparent ability of AZT to cause AIDS-defining and other diseases in hemophiliacs is just one aspect of the many roles that drugs play in the origin of AIDS (see footnote).

(ii) *Treatment based on foreign-protein hypothesis:* In the light of the foreign-protein hypothesis, hemophiliacs have been treated with factor VIII free of foreign proteins. This treatment has provided lasting stabilization of immunity in HIV-positive hemophiliacs. Moreover, the long-term treatment of immunodeficient, HIV-positive hemophiliacs with purified factor VIII has even regenerated lost immunity.[1] Immunological anergy has disappeared and the T-cells in HIV-positive hemophiliacs have increased up to 25 percent in the presence of pure factor VIII (see section 4.7) (de Biasi et al., 1991; Seremetis et al., 1993). Thus, therapeutic benefits including AIDS prevention and even recovery of lost immunity by omission of foreign proteins from factor VIII lend credence to the foreign-protein–AIDS hypothesis.*

(iii) *Two treatment hypotheses—and one treatment dilemma:* The failure to distinguish between two alternative hypothetical AIDS causes, HIV and foreign proteins, has created a dilemma for contemporary hemophilia treatment. For example, Goedert et al. (1994) acknowledge that "CD4 count fell less rapidly with high purity products." But since they are also treating their patients with toxic AZT (see section 4.1), they observe that "F VIII related changes in CD4 concentration may have little relevance to clinical disease" (Goedert et al., 1994). Indeed the group had published a rare comparison between the annual AIDS and death risks of hemophiliacs treated and not treated with AZT which indicated

---

* The drug-AIDS hypothesis, which applies to most American and European AIDS cases other than hemophiliacs (see section 1) (Duesberg, 1992a), also derives support either from the absence of AIDS or from the stabilization of, or spontaneous recovery from, AIDS conditions in HIV-positives who don't use drugs. For example, in August 1993 there was no mortality during 1.25 years in a group of 918 British HIV-positive homosexuals who had "avoided the experimental medications on offer," and chose to "abstain from or significantly reduce their use of recreational drugs, including alcohol" (Wells, 1993). Assuming a ten-year latent period from HIV to AIDS, the virus-AIDS hypothesis would have predicted at least 115 (918/10 × 1.25) AIDS cases among 918 HIV-positives over 1.25 years. Indeed, the absence of mortality in this group over 1.25 years corresponds to a minimal latent period from HIV to AIDS of over 1,148 (918 × 1.25) years. On July 1, 1994, there was still not a single AIDS case in this group of 918 HIV-positive homosexuals (J. Wells, London, personal communication). Further, the T-cell counts of 197 (58% of 326) HIV-positive homosexuals remained constant over three years, despite the presence of HIV (Detels et al., 1988). These were probably those in the cohort who did not use recreational drugs or AZT. Moreover, it has been observed that the T-cells of 29 percent of 1,020 HIV-positive male homosexuals and intravenous drug users even increased up to 22 percent per year over two years (Hughes et al., 1994). These HIV-positives belonged to the placebo arm of an AZT trial for AIDS prevention and thus were not intoxicated by AZT. It is probable that the 29 percent whose T-cells increased despite HIV may have given up or reduced immunosuppressive recreational drug use in the hopes that AZT would work.

that the AIDS risk of AZT-treated hemophiliacs is 4.5 times higher than in untreated controls and the death risk 2.4 times higher.

In order to reconcile the apparent benefits of purified factor VIII on T-cell counts with the apparent toxicity of simultaneous AZT treatment, they try to separate T-cell loss from AIDS diseases. However, despite non-immunodeficiency AIDS diseases (see Table 2, Section 4.4), AIDS is defined as a T-cell deficiency (Institute of Medicine and National Academy of Sciences, 1986; Institute of Medicine, 1988) and dozens of AIDS researchers have observed that "AIDS tends to develop only after patients' CD4 lymphocyte counts have reached low levels..." (Phillips et al., 1994b). Indeed, as of January 1993 the CDC defined less than 200 T-cells per µl as an AIDS disease (Centers for Disease Control and Prevention, 1992), and sequential T-cell counts of hemophiliacs are used as a basis to calculate their long-term survival (Phillips et al., 1994b).

Because of their exclusive faith in the HIV-AIDS hypothesis, readers of the study by Seremetis et al. (1993), which had demonstrated that foreign proteins associated with factor VIII suppress T-cell counts, have even proposed to "consider the use of high-purity factor VIII concentrates in non-hemophiliac–HIV-positive patients" as a treatment for other AIDS patients, i.e., intravenous drug users and homosexuals. Since hemophiliacs treated with pure factor VIII did either not develop immunodeficiency or even recovered lost immunity, they assumed, in view of the HIV hypothesis, that pure factor VIII must inhibit HIV and thus would help all AIDS patients (Schwarz et al., 1994).

The solution to the treatment dilemma can only come from treatments that are each based only on one hemophilia-AIDS hypothesis: To test the foreign-protein hypothesis, two groups of hemophiliacs must be compared that are matched for their lifetime dosage of factor VIII, for their percentage of HIV-positives (for their percentage and dosage of prior AZT treatment, if applicable), and for their age. All AIDS-defining diseases must be diagnosed in each group clinically for the duration of the test. No anti-HIV treatments must be performed. One group would be treated with purified factor VIII, the other with commercial factor VIII contaminated with foreign proteins.

To test the HIV-AIDS hypothesis, two groups of hemophiliacs must be compared that are matched for their lifetime dosage of factor VIII treatment and their age. The two groups must differ only in the presence of antibody against HIV. Both groups would be treated with the same factor VIII preparation. Only the HIV-positive group would receive AZT. All compensatory treatments of AZT recipients, e.g., blood transfusions to treat for AZT-induced anemia, neutropenia, or pancytopenia (Richman et al., 1987; Volberding et al., 1990; Duesberg, 1992), would have to be recorded. During the duration of the test, all AIDS-defining diseases would each be recorded clinically in both groups.

The outcome of each treatment strategy, purified factor VIII or AZT, would be determined based on morbidity and mortality, including AZT morbidity and mortality, and corrected for treatments compensating for AZT toxicity. As yet, no controlled treatment studies based on a single AIDS hypothesis have been performed.

Nevertheless, the study by de Biasi et al. (1991) and, with reservations, that by Seremetis et al. (1993) come close to the stated criteria for a test of the foreign-protein hypothesis (section 4.7). Seremetis et al. initially excluded, but later allowed, AZT treatment. Both studies showed that purified factor VIII improved immunodeficiency (see ii). However, since all subjects in these studies were HIV-positive, one could indeed argue that the improvement of those treated with purified factor VIII was due to a cooperation between HIV and purified factor VIII.

The definitive treatment of immunodeficiency in hemophiliacs, or of hemophilia-AIDS, could be only as far away as the duration of one carefully controlled treatment test.

Acknowledgments

I thank Siggi Sachs, Russell Schoch, and Jody Schwartz (Berkeley) for critical reviews, and Robert Maver (Overland Park, Missouri, USA), Scott Tenenbaum, Robert Garry (Tulane University, New Orleans, Louisiana, USA), Jon Cohen (*Science*, Washington, D.C., USA), and Michael Verney-Elliot (MEDITEL, London) for critical information. This investigation was supported in part by the Council for Tobacco Research, USA, and private donations from Tom Boulger (Redondo Beach, California, USA), Glenn Braswell (Los Angeles, California, USA), Dr. Richard Fischer (Annandale, Virginia, USA), Dr. Fabio Franchi (Trieste, Italy), Dr. Friedrich Luft (Berlin, Germany), and Dr. Peter Paschen (Hamburg, Germany).

References

AIDS-Hemophilia French Study Group. Immunologic and virologic status of multitransfused patients: role of type and origin of blood products. *Blood* 66: 896–901, 1985.

Aledort, L. M. Blood products and immune changes: impacts without HIV infection. *Sem. Hematol.* 25: 14–19, 1988.

Anderson, R. M. & R. M. May. Epidemiological parameters of HIV transmission. *Nature* (London) 333: 514–519, 1988.

Anonymous. The cause of AIDS? *Lancet* i: 1053–1054, 1984.
Antonaci, S., E. Jirillo, D. Stasi, V. De Mitrio, M. R. La Via & L. Bonomo. Immunoresponsiveness in hemophilia: Lymphocyte- and Phagocyte-mediated functions. *Diagn. Clin. Immunol.* 5: 318–325, 1988.
Aronson, D. L. Pneumonia deaths in haemophiliacs. *Lancet* ii: 1023, 1983.
Associated Press. *Red Cross knew of AIDS blood threat.* San Francisco Chronicle, May 16, 1994.
Auerbach, D. M., W. W. Darrow, H. W. Jaffe & J. W. Curran. Cluster of cases of the Acquired Immune Deficiency Syndrome patients linked by sexual contact. *Am. J. Med.* 76: 487–492, 1984.
Bagasra, O., S. P. Hauptman, H. W. Lischner, M. Sachs & R. J. Pomerantz. Detection of human immunodeficiency virus type I provirus in mononuclear cells by in situ polymerase chain reaction. *N. Engl. J. Med.* 326: 1385–1391, 1992.
Becherer, P. R., M. L. Smiley, T. J. Matthews, K. J. Weinhold, C. W. McMillan & G. C. I. White. Human immunodeficiency virus-1 disease progression in hemophiliacs. *Am. J. Hematol.* 34: 204–209, 1990.
Biggar, R. J. & the International Registry of Seroconverters. AIDS incubation in 1891 HIV seroconverters from different exposure groups. *AIDS.* 4: 1059–1066, 1990.
Blattner, W. A. HIV epidemiology: past, present, and future. *FASEB J* 5: 2340–2348, 1991.
Blattner, W. A., R. C. Gallo & H. M. Temin. HIV causes AIDS. *Science* 241: 514–515, 1988.
Booth, W. A rebel without a cause for AIDS. *Science,* 239: 1485–1488, 1988.
Brettler, D. B., A. D. Forsberg, F. Brewster, J. L. Sullivan & P. H. Levine. Delayed cutaneous hypersensitivity reactions in hemophiliac subjects treated with factor concentrate. *Am. J. Med.* 81: 607–611, 1986.
Brettler, D. B. & P. H. Levine. Factor concentrates for treatment of hemophilia: which one to choose? Blood 73: 2067–2073, 1989.
Cairns, J. *Cancer: Science and Society.* (San Francisco: W.H. Freeman and Company, 1978).
Carr, R., E. Edmond, R. J. Prescott, S. E. Veitch, J. E. Peutherer & C. M. Steel. Abnormalities of circulating lymphocyte subsets in haemophiliacs in an AIDS-free population. *Lancet* i: 1431–1434, 1984.
Centers for Disease Control, 1982. Pneumocystis carinii pneumonia among persons with hemophilia A. *Morbid. Mort. Weekly Report* 31: 365–367.
Centers for Disease Control (eds), 1986. Reports on AIDS published in the *Morbidity and Mortality Weekly Report,* June 1981 through February 1986.

Centers for Disease Control and Prevention, 1994. *HIV/AIDS Surveillance Report*, year-end edition 4: 1–33.
Centers for Disease Control, 1992a. *HIV/AIDS Surveillance Report*, January issue.
Centers for Disease Control, 1992b. The second 100,000 cases of Acquired Immunodeficiency Syndrome—United States, June 1981–December 1991. *Morbid. Mort. Weekly Report* 41: 28–29.
Centers for Disease Control, 1993. *HIV/AIDS Surveillance Report*, year-end edition. February: 1–23.
Centers for Disease Control and Prevention, 1992. 1993 revised classification system for HIV infection and expanded surveillance case definition for AIDS among adolescents and adults. *Morb. Mort. Weekly Rep.* 41 (No. RR17) 1–19.
Chorba, T. L., R. C. Holman, T. W. Strine, M. J. Clarke & B. L. Evatt. Changes in longevity and causes of death among persons with hemophilia A. *Am. J. Hematol*, 45: 112–121, 1994.
Cohen, J. Keystone's blunt message: "It's the virus, stupid." *Science* 260: 292–293, 1993.
Curran, J. W., M. W. Morgan, A. M. Hardy, H. W. Jaffe, W. W. Darrow & W. R. Dowdle. The epidemiology of AIDS: current status and future prospects. *Science* 229: 1352–1357, 1985.
Darby, S. C., C. R. Rizza, R. Doll, R. J. D. Spooner, I. M. Stratton & B. Thakrar. Incidence of AIDS and excess mortality associated with HIV in haemophiliacs in the United Kingdom: report on behalf of the directors of haemophilia centers in the United Kingdom. *Br. Med. J.* 298: 1064–1068, 1989.
de Biasi, R., A. Rocino, E. Miraglia, L. Mastrullo & A. A. Quirino. The impact of a very high purity of factor VIII concentrate on the immune system of human immunodeficiency virus-infected hemophiliacs: a randomized, prospective, two-year comparison with an intermediate purity concentrate. *Blood* 78: 1919–1922, 1991.
Detels, R., P. A. English, J. V. Giorgi, B. R. Visscher, J. L. Fahey, J. M. G. Taylor, J. P. Dudley, P. Nishanian, A. Muñoz, J. R. Phair, B. F. Polk & C. R. Rinaldo. Patterns of $CD4^+$ cell changes after HIV-1 infection indicate the existence of a codeterminant of AIDS. *Journal of Acquired Immune Deficiency Syndromes* 1: 390–395, 1988.
Dickson, D. Critic still lays blame for AIDS on lifestyle, not HIV. *Nature* (London) 369: 434, 1994.
Duesberg, P. H. Quantification of human immunodeficiency virus in the blood. *N. Engl. J. Med.* 322: 1466, 1990.
Duesberg, P. H. AIDS epidemiology: inconsistencies with human

immunodeficiency virus and with infectious disease. *Proc. Natl. Acad. Sci.* USA 88: 1575–1579, 1991.

Duesberg, P. H. AIDS acquired by drug consumption and other noncontagious risk factors. *Pharmacology & Therapeutics* 55: 201–277, 1992a.

Duesberg, P. H. HIV as target for zidovudine. *Lancet* 339: 551, 1992b.

Duesberg, P. H. HIV, AIDS, and zidovudine. *Lancet* 339: 805–806, 1992c.

Duesberg, P. H. HIV and AIDS. *Science* 260: 1705, 1993a.

Duesberg, P. H. The HIV gap in national AIDS statistics. *Biotechnology* 11: 955–956, 1993b.

Duesberg, P. H. Infectious AIDS—stretching the germ theory beyond its limits. *Int. Arch. Allergy Immunol.* 103: 131–142, 1994.

Evatt, B. L., R. B. Ramsey, D. N. Lawrence, L. D. Zyla & J. W. Curran. The acquired immunodeficiency syndrome in patients with hemophilia. *Ann. Intern. Med.* 100: 499–505, 1984.

Eyster, M. E. & M. E. Nau. Particulate material in antihemophiliac factor (AHF) concentrates. *Transfusion* September–October: 576–581, 1978.

Eyster, M. E., D. A. Whitehurst, P. M. Catalano, C. W. McMillan, S. H. Goodnight, C. K. Kasper, J. C. Gill, L. M. Aledort, M. W. Hilgartner, P. H. Levine, J. R. Edson, W. E. Hathaway, J. M. Lusher, E. M. Gill, W. K. Poole & S. S. Shapiro. Long-term follow-up of hemophiliacs with lymphocytopenia or thrombocytopenia. *Blood* 66: 1317–1320, 1985.

Fauci, A. S. $CD4^+$ T-lymphocytopenia without HIV infection—no lights, no camera, just facts. *N. Engl. J. Med.* 328: 429–431, 1993.

Fields, B. N. AIDS: time to turn to basic science: *Nature* (London) 369: 95–96, 1994.

Fischl, M. A., D. D. Richman, M. H. Grieco, M. S. Gottlieb, P. A. Volberding, O. L. Laskin, J. M. Leedon, J. E. Groopman, D. Mildvan, R. T. Schooley, G. G. Jackson, D. T. Durack, D. King & the AZT Collaborative Working Group. The efficacy of azidothymidine (AZT) in the treatment of patients with AIDS and AIDS-related complex. *N. Engl. J. Med.* 317: 185–191, 1987.

Fletcher, M. A., J. W. Mosley, J. Hassett, G. E. Gjerset, J. Kaplan, J. W. Parker, E. Donegan, J. M. Lusher, H. Lee & Transfusion Safety Study Group. Effect of Age on Human Immunodeficiency Virus Type 1—induced Changes in Lymphocyte Populations Among Persons with Congenital Cloning Disorders. *Blood* 80: 831–840, 1992.

Froebel, K. S., R. Madhok, C. Forbes, S. E. Lennie, G. D. Lowe & R. D. Sturrock. Immunological abnormalities in haemophilia: are they caused by American factor VIII concentrate? *Br. Med. J.* 287: 1091–1093, 1983.

Gallo, R. C., Salahuddin, S. Z., Papovic, M., Shearer, G. M., Kaplan, M., Haynes, B. F., Palker, T. J., Redfield, R., Oleske, J., Safai, B., White, G., Foster, P. and Markham, P. D. Frequent detection and isolation of cytopathic retrovirus (HTLV-III) from patients with AIDS and at risk for AIDS. *Science* 224: 500-503, 1984.

Gill, J. C., M. D. Menitove, P. R. Anderson, J. T. Casper, S. G. Devare, C. Wood, S. Adair, J. Casey, C. Scheffel & M. D. Montgomery. HTLV-III serology in hemophilia: Relationship with immunologic abnormalities. *J. Pediatr.* 108: 511–516, 1986.

Gjerset, G. F., M. C. Pike, J. W. Mosley, J. Hassett, M. A. Fletcher, E. Donegan, J. W. Parker, R. B. Counts, Y. Zhou, C. K. Kasper, E. A. Operskalski & The Transfusion Safety Study Group. Effect of Low- and Intermediate-Purity Cloning Factor Therapy on Progression of Human Immunodeficiency Virus Infection in Congenital Cloning Disorders. *Blood* 84: 1666–1671, 1994.

Goedert, J. J., A. R. Cohen, C. M. Kessler, S. Eichinger, S. V. Seremetis, C. S. Rabkin, E. J. Yellin, P. S. Rosenberg & L. M. Aledort. Risks of immunodeficiency, AIDS, and death related to purity of factor VIII concentrate. *Lancet* 344: 791–792, 1994.

Goedert, J. J., C. M. Kessler, L. M. Aledort, R. J. Biggar, W. A. Andes, G. C. White II, J. E. Drummond, K. Vaidya, D. L. Mann, M. E. Eyster, M. V. Ragni, M. M. Lederman, A. R. Cohen, G. L. Bray, P. S. Rosenberg, R. M. Friedman, M. W. Hilgartner, W. A. Blanner, B. Kroner & M. H. Gail. A prospective study of human immunodeficiency virus type 1 infection and the development of AIDS in subjects with hemophilia. *N. Engl. J. Med.* 321: 1141–1148, 1989.

Goldsmith, J. M., J. Deutsche, M. Tang & D. Green. CD4 Cells in HIV-1 Infected Hemophiliacs: effect of Factor VIII Concentrates. *Thromb. Haemost.* 66: 415–419, 1991.

Gomperts, E. D., R. De Biasi & R. De Vreker. The Impact of Cloning Factor Concentrates on the Immune System in Individuals with Hemophilia. *Transfus. Med. Rev.* 6: 44–54, 1992.

Gordon, R. S. Factor VIII products and disordered immune regulation. *Lancet* i: 991, 1983.

Hardy, A. M., J. R. Allen, W. M. Morgan & J. W. Curran. The incidence rate of acquired immunodeficiency syndrome in selected populations. *J. Am. Med. Assoc.* 253: 215–220, 1985.

Hassett, J., G. F. Gjerset, J. W. Mosley, M. A. Fletcher, E. Donegan, J. W. Parker, R. B. Counts, L. M. Aledort, H. Lee, M. C. Pike & Transfusion Safety Study Group. Effect on Lymphocyte Subsets of Cloning

Factor Therapy in Human Immunodeficiency Virus-1-Negative Congenital Cloning Disorders. *Blood* 82: 1351–1357, 1993.
Haverkos, H. W. & J. A. Dougherty (eds). Health Hazards of Nitrite Inhalants. NIDA Research Monograph 83, U.S. Dept. Health & Human Services, Washington, D.C., 1988.
Haverkos, H. W., P. E. Pinsky, D. P. Drotman & D. J. Bregman. Disease manifestation among homosexual men with acquired immunodeficiency syndrome: a possible role of nitrites in Kaposi's sarcoma. *J. Sex. Trans. Dis.* 12: 203–208, 1985.
Hilgartner, M. W., J. D. Buckley, E. A. Operskalski, M. C. Pike & J. W. Mosley. Purity of factor VIII concentrates and serial CD4 counts. *Lancet* 341: 1373–1374, 1993.
Hollan, S. R., G. Fuest, K. Nagy, A. Horvath, G. Krall, K. Verebelyi, E. Ujhelyi, L. Varga & V. Mayer. Immunological alterations in anti-HTLV-III negative haemophiliacs and homosexual men in Hungary. *Immunol. Letters* 11: 305–310, 1985.
Hughes, M. D., D. S. Stein, H. M. Gundacker, E. T. Valentine, J. P. Phair & P. A. Volberding. Within-Subject Variation in CD4 Lymphocyte Count in Asymptomatic Human Immunodeficiency Virus Infection: Implications for Patient Monitoring. *The Journal of Infectious Diseases* 169: 28–36, 1994.
Institute of Medicine. *Confronting AIDS—Update 1988*. National Academy Press, Washington, D.C., 1988.
Institute of Medicine and National Academy of Sciences. *Confronting AIDS*. National Academy Press, Washington, D.C., 1986.
Jason, J., R. C. Holman, B. L. Evatt & the Hemophilia–AIDS Collaborative Study Group. Relationship of partially purified factor concentrates to immune tests and AIDS. *Am. J. Hematol.* 34: 262–269, 1990.
Jason, J. M., J. S. McDougal, G. Dixon, D. N. Lawrence, M. S. Kennedy, M. Hilgartner, L. Aledort & B. L. Evatt. HTLV-II/LAV antibody and immune status of household contacts and sexual partners of persons with hemophilia. *J. Am. Med. Assoc.* 255: 212–215, 1986.
Jin, Z., R. P. Cleveland & D. B. Kaufman. Immunodeficiency in patients with hemophilia: an underlying deficiency and lack of correlation with factor replacement therapy or exposure to human immunodeficiency virus. *Allergy Clin. Immunol.* 83: 165–170, 1989.
Johnson, R. E., D. N. Lawrence, B. L. Evatt, D. J. Bregman, L. Zyla, J. W. Curran, L. M. Aledort, M. E. Eyster, A. P. Brownstein & C. J. Carman. Acquired immunodeficiency syndrome among patients attending hemophilia treatment centers and mortality experience of hemophiliacs in the United States. *Am. J. Epidemiol.* 121: 797–810, 1985.
Jones, C. AIDS: Words from the Front. *SPIN*, October 7: 103–104, 1994.

Jones, P., S. Proctor, A. Dickinson & S. George. Altered immunology in haemophilia. *Lancet* i: 120–121, 1983.
Kessler, C. M., R. S. Schulof, A. L. Goldstein, P. H. Naylor, N. L. Luban, J. E. Kelleher & G. H. Reaman. Abnormal T-lymphocyte subpopulations associated with transfusions of blood-derived products. *Lancet* i: 991–992, 1983.
Koerper, M. A. AIDS and Hemophilia. In: *AIDS: Pathogenesis and Treatment*, pp. 79–95, J.A. Levy (ed.) New York: Marcel Dekker, Inc., 1989.
Kreiss, J. K., C. K. Kasper, J. L. Fahey, M. Weaver, B. R. Visscher, J. A. Steward & D. N. Lawrence. Nontransmission of T-cell subset abnormalities from hemophiliacs to their spouses. *J. Am. Med. Assoc.* 251: 1450–1454, 1984.
Kreiss, J. K., L. W. Kitchen, H. E. Prince, C. K. Kasper, A. L. Goldstein, R. H. Naylor, O. Preble, J. A. Stewart & M. Essex. Human T-cell leukemia virus type III antibody, lymphadenopathy, and acquired immune deficiency syndrome in hemophiliac subjects. *Am. J. Med.* 80: 345–350, 1986.
Lang, D. J., A. A. S. Kovacs, J. A. Zaia, G. Doelkin, J. C. Niland, L. Aledort, S. P. Azen, M. A. Fletcher, J. Gauderman, G. J. Gjerst, J. Lusher, E. A. Operskalski, J. W. Parker, C. Pegelow, G. N. Vyas, J. W. Mosley & the Transfusion Safety Group. Seroepidemiologic studies of cytomegalovirus and Epstein-Barr virus infections in relation to human immunodeficiency virus type 1 infection in selected recipient populations. *J. AIDS* 2: 540–549, 1989.
Lauritsen, J. & H. Wilson, 1986. *Death Rush, Poppers and AIDS*. New York: Pagan Press.
Lawrence, D. N., J. M. Jason, R. C. Holman & J. J. Murphy. HIV transmission from hemophilic men to their heterosexual partners. In: *Heterosexual Transmission of AIDS*, pp. 35–53. N. J. Alexander, H. L. Gabelnick & Spieler, J. M. (eds.) New York: Wiley-Liss, 1990.
Lee, C. A., A. N. Phillips, J. Elford, G. Janossy, P. Griffiths & P. Kemoff. Progression of HIV disease in a haemophiliac cohort followed for 11 years and the effect of treatment. *Br. Med. J.* 303: 1093–1096, 1991.
Lemp, G. E., S. E. Payne, G. W. Rutherford, N. A. Hessol, W. Winkelstein, Jr., J. A. Wiley, A. R. Moss, R. E. Chaisson, R. T. Chen, D. W. Feigal, P. A. Thomas & D. Werdegar. Projections of AIDS morbidity and mortality in San Francisco. *J. Am. Med. Assoc.* 263: 1497–1501, 1990.
Lenderking, W. R., R. D. Gelber, D. J. Couon, B. E. Cole, A. Goldhirsch, P. A. Volberding & M. A. Testa. Evaluation of the quality of life associated with Zidovudine treatment in asymptomatic Human Immunodeficiency Virus Infection. *N. Engl. J. Med.* 330: 738–743, 1994.

Leonhard, H.-W. Alles nur ein Irrtum? neue praxis. *Zeitschrift fur Sozialarbeit, Sozialpadogogik und Sozialpolitik* 22: 14–29, 1992.

Lewis, A. Down the tabloid slope. *New York Times*, July, 4, 1994 (Monday).

Ludlam, C. A., J. Tucker, C. M. Steel, R. S . Tedder, R. Cheingsong-Popov, R. Weiss, D. B. L. McClelland, I. Phillip & R. J. Prescott. Human T-lymphotropic virus type III (HTLV-III) inaction in seronegative hemophiliacs after transfusion of factor VIII. *Lancet* ii: 233–236, 1985.

Lui, K.-J., W. W. Darrow & G. W. Rutherford III. A model-based estimate of the mean incubation period for AIDS in homosexual men. *Science* 240: 1333–1335, 1988.

Lundgren, I. D., A. N. Philips, C. Pedersen, N. Clumeck, J. M. Gatell, A. M. Johnson, B. Ledergerber, S. Vella & J. O. Nielsen. Comparison of long-term prognosis of patients with AIDS treated and not treated with Zidovudine. *J. Am. Med. Assoc.* 271: 1088–1092, 1994.

Macilwain, C. AAAS criticized over AIDS sceptics' meeting. *Nature* (London) 369:265, 1994.

Madhok, R., A. Gracie, G. D. O. Lowe, A. Burnett, K. Froebel, E. Follen & C. D. Forbes. Impaired cell mediated immunity in haemophilia in the absence of infection with human immunodeficiency virus. *Br. Med. J.* 293: 978–980, 1986.

Mahir, W. S., R. E. Millard, J. C. Booth & R. T. Flute. Functional studies of cell-mediated immunity in haemophilia and other bleeding disorders. *Br. J. Haematol.* 69: 367–370, 1988.

Mannucci, P. M., A. Gringeri, R. De Biasi, E. Baudo, M. Morfini & N. Ciavarella. Immune status of asymptomatic HIV-infected hemophiliacs: randomised, prospective, two-year comparison of treatment with a high-priority of an intermediate purity factor VIII concentrate. *Thromb. Haemost.* 67: 310–313, 1992.

Marmor, M., A. E. Friedman-Kien, L. Laubenstein, R. D. Byrum, D. C. William, S. D'Onofrio & N. Dubin. Risk factors for Kaposi's sarcoma in homosexual men. *Lancet* i: 1083–1087, 1982.

Matheson, D. S., B. J. Green, M. J. Fritzler, M.-C. Poon, T. J. Bowen & D. I. Hoar. Humoral immune response in patients with hemophilia. *Clin. Immunol. Immunopathol.* 4: 41–50, 1987.

Mathur-Wagh, U., D. Mildvan & R. T. Senie. Follow-up of 4 1/2 years on homosexual men with generalized lymphadenopathy. *N. Engl. J. Med.* 313: 1542–1543, 1985.

Mathur-Wagh, U., R. W. Enlow, I. Spigland, R. J. Winchester, H. S. Sacks, E. Rorat, S. R. Yancovitz, M. J. Klein, D. C. William & D. Mildwan. Longitudinal study of persistent generalized lymphadenopathy in homosexual men: Relation to acquired immunodeficiency syndrome. *Lancet* i: 1033–1038, 1984.

Maynard, T. Factor concentrate is a co-factor. *AIDS News* (Hemophilia Council of California) 8: 1, 1994.

McGrady, G. A., J. M. Jason & B. L. Evatt. The course of the epidemic of acquired immunodeficiency syndrome in the United States hemophilia population. *Am. J. Epidemiol.* 126: 25–80, 1987.

Menitove, J. E., R. H. Aster, J. T. Casper, S. J. Lauer, J. L. Gonschall, J. E. Williams, J. C. Gill, D. V. Wheeler, V. Piaskowski, R. Kirchner & R. R. Montgomery. T-lymphocyte subpopulations in patients with classic hemophilia treated with cryoprecipitate and lyophilized concentrates. *N. Engl. J. Med.* 308: 83–86, 1983.

Merriam-Webster (eds). *Webster's Third International Dictionary* Springfield, Mass.: G. & C. Merriam Co., 1965.

Moffat, E. H. & Bloom, A. L. HTLV-III antibody status and immunological abnormalities in haemophiliac patients. *Lancet* i: 935, 1985.

Morgan, M., J. W. Curran & R. L. Berkelman. The future course of AIDS in the United States. *J. Am. Med. Assoc.* 263: 1539–1540, 1990.

Newell, G. R., P. W. A. Mansell, M. B. Wilson, H. K. Lynch, M. R. Spitz & E. M. Hersh. Risk factor analysis among men referred for possible acquired immune deficiency syndrome. *Preventive Med.* 14: 81–91, 1985.

Nussbaum, B. *Good Intentions: How Big Business, Politics, and Medicine Are Corrupting the Fight Against AIDS.* New York: Atlantic Monthly Press, 1990.

Oddone, E. Z., P. Cowper, J. D. Hamilton, D. B. Matchar, P. Hartigan, G. Samsa, M. Simberkoff & J. R. Feussner. Cost-effectiveness analysis of early zidovudine treatment of HIV infected patients. *Br. Med. J.* 307: 1322–1325, 1993.

Oppenheimer, G. M. Causes, cases, and cohorts: The role of epidemiology in the historical construction of AIDS. In: *AIDS: The Making of a Chronic Disease*, pp. 49 83, F. Fee & D. M. Fox (eds.) Berkeley: University of California Press, 1992.

Papadopulos-Eleopulos, E., V. E. Turner, J. M. Papdimitriou & D. Causer, 1994. Factor VIII, HIV and AIDS in haemophiliacs: an analysis of their relationship. *Genetica* (1994).

Phillips, A. N., C. A. Sabin, J. Elford, M. Bofill, V. Emery, P. D. Griffiths, G. Janossy & C. A. Lee. Viral burden in HIV infection. *Nature* (London) 367: 124, 1994a

Phillips, A. N., C. A. Sabin, J. Elford, M. Bofill, G. Janossy & C. A. Lee. Use of CD4 lymphocyte count to predict long term survival free of AIDS after HIV infection. *Br. Med. J.* 309: 309–313, 1994b.

Piatak, M., L. C. Saag, S. C. Yang, S. J. Clark, J. C. Kappes, K.-C. Luk,

B. H. Hahn, G. M. Shaw & J. D. Lifson. High levels of HIV-1 in plasma during all stages of infection determined by competitive PCR. *Science* 259: 1749–1754, 1993.

Pollack, S., D. Atias, G. Yoffe, R. Katz, Y. Shechter & I. Tatarsky. Impaired immune function in hemophilia patients treated exclusively with cryoprecipitate: relation to duration of treatment. *Am. J. Hematol.* 20: 1–6, 1985.

Prince, H. The significance of T lymphocytes in transfusion medicine. *Transfus. Med. Rev.* 16: 32–43, 1992.

Richman, D. D., M. A. Fischl, M. H. Grieco, M. S. Gonlieb, P. A. Volberding, O. L. Laskin, J. M. Leedom, J. E. Groopman, D. Mildvan, M. S. Hirsch, G. G. Jackson, D. T. Durack, S. Nusinoff-Lehrman & the AZT Collaborative Working Group. The toxicity of azidothymidine (AZT) in the treatment of patients with AIDS and AIDS-related complex. *N. Engl. J. Med.* 317: 192–197, 1987.

Schulman, S. Effects of factor VIII concentrates on the immune system in hemophilic patients. *Annals of Hematology* 63: 145–151, 1991.

Schwartlaender, B., O. Hamouda, M. A. Koch, W. Kiehl & C. Baars. *AIDS/HIV 1991*. AZ Hefte, 1992.

Schwarz, H. P., M. Kunschak, W. Engl & J. Eibl. High-purity factor concentrates in prevention of AIDS. *Lancet* 343: 478–479, 1994.

Seligmann M., D.A. Warrell, J.-P. Aboulker, C. Carbon, J. H. Darbyshire, J. Dormont, E. Eschwege, D. J. Girling, D. R. James, J.-P. Levy, P. T. A. Peto, D. Schwarz, A. B. Stone, I. V. D. Weller, R. Withnall, K. Gelmon, E. Lafon, A. M. Swart, V. R. Aber, A. G. Babiker, S. Lhoro, A. J. Nunn & M. Vray. Concorde: MCR/ANRS randomised double-blind controlled trial of immediate and deferred zidovudine in symptom-free HIV infection. *Lancet* 343: 871–881, 1994.

Selik, R. M., E. T. Starcher & J. W. Curran. Opportunistic diseases reported in AIDS patients: frequencies, associations, and trends. *AIDS* 1: 175–182, 1987.

Seremetis, S. V., L. M. Aledort, G. E. Bergman, R. Bona, G. Bray, D. Brettler, M. E. Eyster, C. Kessler, T.-S. Lau, J. Lusher & E. Rickles. Three-year randomised study of high-purity or intermediate-purity factor VIII concentrates in symptom-free HIV-seropositive haemophiliacs: effects on immune status. *Lancet* 342: 700–703, 1993.

Sharp, R. A., S. M. Morley, J. S. Beck & G. E. D. Urquhart. Unresponsiveness to skin testing with bacterial antigens in patients with haemophilia A not apparently infected with human immunodeficiency virus (HIV) *J. Clin. Pathol.* 40: 849–852, 1987.

Sheppard, H. W., M. S. Ascher & J. E. Krowka. Viral burden and HIV disease. *Nature* (London) 364: 291–292, 1993.

Shilts, R. *And the Band Played On.* New York: St. Martin's Press, 1985.
Simmonds, P., R. Balfe, J. E. Peutherer, C. A. Ludlam, J. P. Bishop & A. J. Leigh-Brown. Human immunodeficiency virus-infected individuals contain provirus in small numbers of peripheral mononuclear cells and at low copy numbers. *J. Virol.* 64: 864–872, 1990.
Simmonds, R., D. Beatson, R. J. G. Cuthbert, H. Watson, B. Reynolds, J. E. Peutherer, J. V. Parry, C. A. Ludlam & C. M. Steel. Determinants of HIV disease progression: six-year longitudinal study in the Edinburgh haemophilia/HIV cohort. *Lancet* 338: 1159–1163, 1991.
Smith, D. K., J. J. Neal, S. D. Holmberg & Centers for Disease Control Idiopathic CD4+ T-lymphocytopenia Task Force. Unexplained opportunistic infections and CD4+ T-lymphocytopenia without HIV infection. *N. Engl. J. Med.* 328: 373–379, 1993.
Stehr-Green, J. K., R. C. Holman, J. M. Jason & B. L. Evatt. Hemophilia-associated AIDS in the United States, 1981 to September 1987. *Am. J. Public Health* 78: 439–442, 1988.
Stehr-Green, J. K., J. M. Jason, B. L. Evatt & the Hemophilia-Associated AIDS Study Group. Geographic variability of hemophilia-associated AIDS in the United States: effect of population characteristics. Am. J. Hematol. 32: 178–183, 1989.
Sullivan, J. L., F. E. Brewster, D. B. Brettler, A. D. Forsberg, S. H. Cheeseman, K. S. Byron, S. M. Baker, D. L. Willitts, R. A. Lew & P. H. Levine. Hemophiliac immunodeficiency: influence of exposure to factor VIII concentrate, LAV/HTLV-III, and herpesviruses. *J. Pediatr.* 108: 504–510, 1986.
Thomas Jr., C. A., K. B. Mullis & P. E. Johnson. What causes AIDS? *Reason* 26, June: 18–23, 1994.
Tokars, J. I., R. Marcus, D. H. Culver, C. A. Schable, P. S. McKibben, C. I. Bandea and D. M. Bell. Surveillance of HIV Infection and Zidovudine Use among Health Care Workers after Occupational Exposure to HIV-infected Blood. *Ann. Intern. Med.* 118: 913–919, 1993.
Tsoukas, C., E. Gervais, J. Shuster, R. Gold, M. O'Shaughnessy & M. Robert-Guroff. Association of HTLV-III antibodies and cellular immune status of hemophiliacs. *N. Engl. J. Med.* 31: 1514–1515, 1984.
U. S. Dept. of Health and Human Services, National Technical Information Service, Springfield, Va.
Volberding, R. A., S. W. Lagakos, M. A. Koch, C. Pettinelli, M. W. Myers, D. K. Booth, H. H. Balfour Jr., R. C. Reichman, I. A. Bartlett, M. S. Hirsch. R. L. Murphy, W. D. Hardy, R. Soeiro, M. A. Fischl, J. G. Bartlett, T. C. Merigan, N. E. Hyslop, D. D. Richman, E. T. Valentine, L. Corey & the AIDS Clinical Trial Group of the National Institute of Allergy and Infectious Disease. Zidovudine in asymptomatic

human immunodeficiency virus infection. A controlled trial in persons with fewer than 500 CD4-positive cells per cubic millimeter. *N. Engl. J. Med.* 322: 941–949, 1990.
Weiss, R. Provenance of HIV strains. *Nature* (London) 349: 374, 1991.
Weiss, R. A. How does HIV cause AIDS? *Science* 260: 1273–1279, 1993.
Weiss, R. & H. Jaffe. Duesberg, HIV and AIDS. *Nature* (London) 345: 659–660, 1990.
Wells, J. We have to question the so-called "facts." *Capital Gay*, August 20: 14–15, 1993.
World Health Organization. Acquired Immunodeficiency Syndrome (AIDS)—Data as of 1 January 1992. World Health Organization, Geneva, 1992a.
World Health Organization. WHO–Report No. 32: AIDS Surveillance in Europe (Situation by 31 December 1991). World Health Organization, Geneva, 1992b.

APPENDIX B

AIDS Acquired by Drug Consumption and Other Noncontagious Risk Factors*

Peter H. Duesberg
Dept. of Molecular and Cell Biology
University of California at Berkeley

Abstract—The hypothesis that human immunodeficiency virus (HIV) is a new, sexually transmitted virus that causes AIDS has been entirely unproductive in terms of public health benefits. Moreover, it fails to predict the epidemiology of AIDS, the annual AIDS risk and the very heterogeneous AIDS diseases of infected persons. The correct hypothesis must explain why (1) AIDS includes 25 previously known diseases and two clinically and epidemiologically very different epidemics, one in America and Europe, the other in Africa; (2) almost all American (90%) and European (86%) AIDS patients are males over the age of 20, while African AIDS affects both sexes equally; (3) the annual AIDS risks of infected babies, intravenous drug users, homosexuals who use aphrodisiacs, hemophiliacs, and Africans vary over 100-fold; (4) many AIDS patients have diseases that do not depend on immunodeficiency, such as Kaposi's sarcoma, lymphoma, dementia, and wasting; and (5) the AIDS diseases of Americans (97%) and Europeans (87%) are predetermined by prior health risk, including long-term consumption of illicit recreational drugs, the antiviral drug AZT, and congenital deficiencies like hemophilia; those of Africans are Africa-specific. Both negative and positive evidence shows that AIDS is not infectious: (1) the virus hypothesis fails all conventional criteria of causation; (2) over 100-fold different AIDS risks in different risk groups show that HIV is not sufficient for AIDS; (3)

---

* Article originally appeared in *Pharmac. Ther.*, 55 (1992): 201–207. Reprinted by permission of the publisher.

Appendix B

AIDS is "acquired," if at all, only years after HIV is neutralized by antibodies; (4) AIDS is new, but HIV is a long-established, perinatally transmitted retrovirus; (5) alternative explanations disprove all assumptions and anecdotal cases cited in support of the virus hypothesis; (6) all AIDS-defining diseases occur in matched risk groups, at the same rate, in the absence of HIV; (7) there is no common, active microbe in all AIDS patients; (8) AIDS manifests in unpredictable and unrelated diseases; and (9) AIDS does not spread randomly between the sexes in America and Europe. Based on numerous data documenting that drugs are necessary for HIV-positives and sufficient for HIV-negatives to develop AIDS diseases, it is proposed that all American/European AIDS diseases that exceed their normal background result from recreational and anti-HIV drugs. African AIDS is proposed to result from protein malnutrition, poor sanitation, and subsequent parasitic infections. This hypothesis resolves all paradoxes of the virus-AIDS hypothesis. It is epidemiologically and experimentally testable and provides a rational basis for AIDS control.

CONTENTS

1. Virus-AIDS Hypothesis Fails to Predict Epidemiology and Pathology of AIDS
2. Definition of AIDS
   2.1. AIDS: 2 epidemics, sub-epidemics and 25 epidemic-specific diseases
      2.1.1. The epidemics by case numbers, gender and age
      2.1.2. AIDS diseases
      2.1.3. AIDS risk groups and risk-group–specific AIDS diseases
   2.2. The HIV-AIDS hypotheses, or the definition of AIDS
   2.3. Alternative infectious theories of AIDS
3. Discrepancies between AIDS and Infectious Disease
   3.1. Criteria of infectious and noninfectious disease
   3.2. AIDS not compatible with infectious disease
   3.3. No proof for the virus-AIDS hypothesis
      3.3.1. Virus hypothesis fails to meet Koch's postulates
      3.3.2. Anti-HIV immunity does not protect against AIDS
      3.3.3. Antiviral drugs do not protect against AIDS
      3.3.4. All AIDS-defining diseases occur in the absence of HIV
   3.4. Noncorrelations between HIV and AIDS

3.4.1. Only about half of American AIDS is confirmed HIV-antibody–positive
3.4.2. Antibody-positive, but virus-negative AIDS
3.4.3. HIV: just one of many harmless microbial markers of behavioral and clinical AIDS risks
3.4.4. Annual AIDS risks of different HIV-infected risk groups, including babies, homosexuals, drug addicts, hemophiliacs, and Africans, differ over 100-fold
3.4.5. Specific AIDS diseases predetermined by prior health risk

3.5. Assumptions and anecdotal cases that appear to support the virus-AIDS hypothesis
3.5.1. HIV is presumed new because AIDS is new
3.5.2. HIV—assumed to be sexually transmitted—depends on perinatal transmission for survival
3.5.3. AIDS assumed to be proportional to HIV infection
3.5.4. AIDS assumed to be homosexually transmitted in the U.S. and Europe
3.5.5. AIDS assumed to be heterosexually transmitted by African lifestyle
3.5.6. HIV claimed to be abundant in AIDS cases
3.5.7. HIV to depend on cofactors for AIDS
3.5.8. All AIDS diseases to result from immunodeficiency
3.5.9. HIV to induce AIDS via autoimmunity and apoptosis
3.5.10. HIV assumed to kill T-cells
3.5.11. Antibodies assumed not to neutralize HIV
3.5.12. HIV claimed to cause AIDS in 50% within 10 years
3.5.13. HIV said to derive pathogenicity from constant mutation
3.5.14. HIV assumed to cause AIDS with genes unique among retroviruses
3.5.15. Simian retroviruses to prove that HIV causes AIDS
3.5.16. Anecdotal AIDS cases from the general population

3.6. Consequences of the virus-AIDS hypothesis

4. The Drug-AIDS Hypothesis
4.1. Chronological coincidence between the drug and AIDS epidemics
4.2. Overlap between drug-use and AIDS statistics
4.3. Drug use in AIDS risk groups
4.3.1. Intravenous drug users generate a third of all AIDS patients

4.3.2. Homosexual users of aphrodisiac drugs generate about 60% of AIDS patients
4.3.3. Asymptomatic AZT users generate an unknown percentage of AIDS patients
4.4. Drug use necessary for AIDS in HIV-positives
4.4.1. AIDS for recreational drugs
4.4.2. AIDS from AZT and AZT plus confounding recreational drug use
4.5. Drug use sufficient for AIDS indicator diseases in the absence of HIV
4.5.1. Drugs used for sexual activities sufficient for AIDS diseases
4.5.2. Long-term intravenous drug use sufficient for AIDS-defining diseases
4.6. Toxic effects of drugs used by AIDS patients
4.6.1. Toxicity of recreational drugs
4.6.2. Toxicity of AZT
4.7. Drug-AIDS hypothesis correctly predicts the epidemiology and heterogeneous pathology of AIDS
4.8. Consequences of the drug-AIDS hypothesis: Risk-specific preventions and therapies, but resentment by the virus-AIDS establishment
5. Drugs and Other Noncontagious Risk Factors Resolve All Paradoxes of the Virus-AIDS Hypothesis
6. Why Did AIDS Science Go Wrong?
6.1. The legacy of the successful germ theory : a bias against noninfectious pathogens
6.2. Big funding and limited expertise paralyze AIDS research
Note Added in Proof
Acknowledgments
References

"It's too late to correct," said the Red Queen. "When you've once said a thing, that fixes it, and you must take the consequences."
Lewis Carroll, *Through the Looking Glass.*

1. VIRUS-AIDS HYPOTHESIS FAILS TO PREDICT EPIDEMIOLOGY AND PATHOLOGY OF AIDS

At a press conference in April 1984, the American Secretary of Health and Human Services announced that the Acquired Immunodeficiency

Syndrome (AIDS) was an infectious disease, caused by a sexually and parenterally transmitted retrovirus, now termed Human Immunodeficiency Virus (HIV). The announcement predicted an antiviral vaccine within two years (Connor, 1987; Adams, 1989; Farber, 1992; Hodgkinson, 1992).

However, the hypothesis has been a complete failure in terms of public health benefits. Despite unprecedented efforts in research and health care, the hypothesis has failed to generate the promised vaccine, and it has failed to develop into a cure (Thompson, 1990; Savitz, 1991; Duesberg, 1992b; Waldholz, 1992). The U.S. government alone spends annually about $1 billion for AIDS research and about $3 billion for AIDS-related health care (National Center for Health Statistics, 1992). The situation has become so desperate that the director for AIDS research at the National Institutes of Health (NIH) promotes via press releases, eight years after HIV was declared the cause of AIDS, an as-yet-unedited paper which has no more to offer than a renewed effort at causing AIDS in monkeys: "The best possible situation would be to have a human virus [HIV] that infects monkeys" (Steinbrook, 1992). This is said nine years after the NIH first started infecting chimpanzees with HIV—over 150 so far at a cost of $40,000 to $50,000 apiece—all of which are still healthy (Hilts, 1992; Steinbrook, 1992) (Section 3.3; Jorg Eichberg, personal communication).

Moreover, the virus-AIDS hypothesis has failed completely to predict the course of the epidemic (Institute of Medicine, 1988; Duesberg, 1989c, 1991a; Duesberg and Ellison, 1990; Thompson, 1990; Savitz, 1991). For example, the NIH and others have predicted that AIDS would "explode" into the general population (Shorter, 1987; Anderson and May, 1992) and the Global AIDS Policy Coalition from Harvard's International AIDS Center declared in June 1992, "The pandemic is dynamic, volatile and unstable... An explosion of HIV has recently occurred in Southeast Asia, in Thailand..." (Mann and the Global AIDS Policy Coalition, 1992). But despite widespread alarm the "general population" has been spared from AIDS, although there is a general increase in unwanted pregnancies and conventional venereal diseases (Institute of Medicine, 1988; Aral and Holmes, 1991). Instead, American and European AIDS has spread, during the last 10 years, steadily but almost exclusively among intravenous drug users and male homosexuals who were heavy users of sexual stimulants and who had hundreds of sexual partners (Sections 2.1.3, 3.3.4, and 4.3.2).

The hypothesis even fails to predict the AIDS diseases that an infected person may develop and whether and when an HIV-infected person is to develop either diarrhea or dementia, Kaposi's sarcoma or pneumonia (Grimshaw, 1987; Albonico, 1991a, b). In addition the hypothesis fails to explain why the annual AIDS risks differ over 100-fold between different HIV-infected risk groups, i.e., recipients of transfusions, babies born to drug-addicted mothers, American/European homosexuals, intravenous drug users, hemophiliacs, and Africans (Section 3.4.4).

Clearly, a correct medical hypothesis might not produce a cure or the prevention of a disease, as for example theories on cancer or sickle-cell anemia. However, a correct medical hypothesis must be able to (1) identify those at risk for a disease, (2) predict the kind of disease a person infected or affected by its putative cause will get, (3) predict how soon disease will follow its putative cause, and (4) lead to a determination of how the putative agent causes the disease. Since this is not true for the virus-AIDS hypothesis, this hypothesis must be fundamentally flawed. Furthermore, it seems particularly odd that an AIDS vaccine cannot be developed, since HIV induces highly effective virus-neutralizing antibodies within weeks after infections (Clark et al., 1991; Daar et al., 1991). These are the same antibodies that are detected by the widely used "AIDS-test" (Institute of Medicine, 1986; Duesberg, 1989c; Rubinstein, 1990).

In view of this, AIDS is subjected here to a critical analysis aimed at identifying a cause that can correctly predict its epidemiology, pathology, and progression.

2. DEFINITION OF AIDS

2.1. AIDS: 2 Epidemics, Sub-epidemics and 25 Epidemic-specific Diseases

AIDS includes 25 previously known diseases and two clinically and epidemiologically very different AIDS epidemics, one in America and Europe, the other in Africa (Table 1) (Centers for Disease Control, 1987; Institute of Medicine, 1988; World Health Organization, 1992a). The American/European epidemic falls into four sub-epidemics: the male homosexual epidemic, the intravenous drug user epidemic, the hemophilia epidemic, and the transfusion recipient epidemic (Table 1).

2.1.1. *The Epidemics by Case Numbers, Gender and Age*

The American/European AIDS epidemics of homosexuals and intravenous drug users are new, starting with drug-using homosexual AIDS patients in Los Angeles and New York in 1981 (Centers for Disease Control, 1981; Gottlieb et al., 1981; Jaffe et al., 1983a). By December 1991, 206,392 AIDS cases had been recorded in the U.S., and 65,979 in Europe (Table 1) (World Health Organization, 1992a; Centers for Disease Control,

Table 1. AIDS Statistics*

Epidemics	American	European	African
AIDS total 1985–1991	206,000	66,000	129,000
AIDS annual since 1990	30–40,000	12–16,000	~20,000
HIV carriers since 1985	1 million	500,000	6 million
Annual AIDS per HIV carrier	3–4%	3%	about 0.3%
AIDS by sex	90% male	86% male	50% male
AIDS by age, over 20 years	98%	96%	?
AIDS by risk group:			
male homosexual	62%	48%	
intravenous drugs	32%	33%	
transfusions	2%	3%	
hemophiliacs	1%	3%	
general population	3%	13%	100%
AIDS by Disease:			
Microbial	50% *Pneumocystis* pneumonia 17% candidiasis 11% mycobacterial disease including 3% tuberculosis 5% toxoplasmosis 8% cytomegalovirus 4% herpes virus	75% opportunistic infections	fever diarrhea tuberculosis slim disease
Microbial total	62% (sum > 62% due to overlap)	75%	about 90%
Nonmicrobial	19% wasting 10% Kaposi's 6% dementia 3% lymphoma	5% wasting 12% Kaposi's 5% dementia 3% lymphoma	
Nonmicrobial total	38%	25%	

*Data from references cited in Section 2. There are small (±1%) discrepancies between some numbers cited here and the most recent surveys cited in the text, because some calculations are based on previous surveys.

1992b). The U.S. has reported about 30,000 to 40,000 new cases annually since 1987, and Europe reports about 12,000 to 16,000 cases annually (World Health Organization, 1992a; Centers for Disease Control 1992b). Remarkably for a presumably infectious disease, 90% of all American and 86% of all European AIDS patients are male. Nearly all American (98%) and European (96%) AIDS patients are over 20 years old; the remaining 2% and 4%, respectively, are mostly infants (Table 1) (World Health Organization, 1992a; Centers for Disease Control, 1992b). There is very little AIDS among teenagers, as only 789 American teenagers have developed AIDS over the last 10 years, including 160 in 1991 and 170 in 1990 (Centers for Disease Control, 1992b).

Since 1985, 129,066 AIDS cases have been recorded in Africa (World Health Organization, 1992b), mainly from the people of Central Africa (Blattner, 1991). Unlike the American and European cases, the African cases are distributed equally between the sexes (Quinn et al., 1986; Blattner et al., 1988; Piot et al., 1988; Goodgame, 1990) and a range "in age from 8 to 85 years" (Widy-Wirski, et al., 1988).

An AIDS crisis that was reported to "loom" in Thailand as of 1990 (Anderson, 1990; Smith, 1990) and was predicted to "explode" now (Mann and the Global AIDS Policy Coalition, 1992) has generated only 123 AIDS patients from 1984 to June 1991 (Weniger et al., 1991).

2.1.2. *AIDS Diseases*

The majority of American (62%) and European (75%) AIDS patients have microbial diseases or opportunistic infections that result from a previously acquired immunodeficiency (World Health Organization, 1992a; Centers for Disease Control, 1992b). In America these include *Pneumocystis* pneumonia (50%), candidiasis (17%), and mycobacterial infections such as herpes virus disease(4%) (Table 1) (Centers for Disease Control, 1992b). *Pneumocystis* pneumonia is often described and perceived as an AIDS-specific pneumonia. However, *Pneumocystis carinii* is a ubiquitous fungal parasite that is present in all humans and that like many others may become active upon immune deficiency (Freeman, 1979; Pifer, 1984; Williford Pifer et al., 1988; Root-Bernstein, 1990a). Since bacterial opportunists of immune deficiency, like tuberculosis bacillus or pneumococcus, are readily defeated with antibiotics, fungal and viral pneumonias predominate in countries where antibiotics are readily available. This is particularly true for risk groups that use antibiotics chronically as AIDS prophylaxis (Callen, 1990; Bardach, 1992). Indeed, young rats treated for several weeks simultaneously with antibiotics and immunosuppressive cortisone all developed *Pneumocystis* pneumonia spontaneously (Weller, 1955).

Contrary to its name, AIDS of many American (38%) and European (25%) patients does not result from immunodeficiency and microbes (Section 3.5.8). Instead, these patients suffer dementia (6%/5%), wasting disease (19%/5%), Kaposi's sarcoma (10%/12%), and lymphoma (3%/3%) (Table 1) (World Health Organization, 1992a; Centers for Disease Control, 1992b).

The African epidemic includes diseases that have been long established in Africa, such as fever, diarrhea, tuberculosis, and "slim disease" (Table 1) (Colebunders et al., 1987; Konotey-Ahulu, 1987; Pallangyo et al., 1987; Berkley et al., 1989; Evans, 1989a; Goodgame, 1990; De Cock et al., 1991; Gilks, 1991). Only about 1% are Kaposi's sarcomas (Widy-Wirski et al., 1988). The African AIDS definition is based primarily on these Africa-specific diseases (Widy-Wirski et al., 1988) "because of limited facilities for diagnosing HIV infections" (De Cock et al., 1991).

2.1.3. AIDS Risk Groups and Risk-group–specific AIDS Diseases

Almost all American (97%) and European (87%) AIDS patients come from abnormal health risk groups whose health had been severely compromised prior to the onset of AIDS: 62% of American (47% of European) AIDS patients are male homosexuals who have frequently used oral aphrodisiac drugs (Section 4), 32% (33%) are intravenous drug users, 2% (3%) are critically ill recipients of transfusions, and 1% (3%) are hemophiliacs (Institute of Medicine, 1988; Brenner et al., 1990; Centers for Disease Control, 1992b; World Health Organization, 1992a). About 38% of the American teenage AIDS cases are hemophiliacs and recipients of transfusions, 25% are intravenous drug users or sexual partners of intravenous drug users, and 25% are male homosexuals (Centers for Disease Control, 1992b). Approximately 70% of American babies with AIDS are born to drug-addicted mothers ("crack babies") and 13% are born with congenital deficiencies like hemophilia (Centers for Disease Control, 1992b). Only 3% of American and 13% of European AIDS patients are from "undetermined exposure categories," i.e., from the general population (Table 1) (World Health Organization, 1992a; Centers for Disease Control, 1992b). Some of the differences between European and American statistics may reflect differences in national AIDS standards between different European countries and the U.S. and differences in reporting between the World Health Organization (WHO) and the American Centers for Disease Control (CDC) (World Health Organization, 1992a). In contrast to the American and European AIDS epidemics, African AIDS does not claim its victims from sexual, behavioral, or clinical risk groups.

Appendix B

The AIDS epidemics of different risk groups present highly characteristic, country-specific, and sub-epidemic–specific AIDS diseases (Table 1 and Table 2):

(1) About 90% of the AIDS diseases from Africa are old African diseases that are very different from those of the American/European epidemic (Section 2.1.2, Table 1). The African diseases do not include *Pneumocystis* and *Candida*, which are ubiquitous in all humans including Africans (Freeman, 1979; Pifer, 1984).

(2) The American/European epidemic falls into several sub-epidemics based on sub-epidemic–specific diseases:

(a) American homosexuals have Kaposi's sarcoma 20 times more often than all other American AIDS patients (Selik et al., 1987; Beral et al., 1990).
(b) Intravenous drug users have a proclivity for tuberculosis (Section 4.5 and Section 4.6).
(c) "Crack" (cocaine) smokers exhibit pneumonia and tuberculosis (Section 3.4.5 and Section 4.6).
(d) Ninety-nine percent of all hemophiliacs with AIDS have opportunistic infections, of which about 70 percent are fungal and viral pneumonias, but less than 1 percent have Kaposi's sarcoma (Evatt et al., 1984; Centers for Disease Control, 1986; Selik et al., 1987; Koerper, 1989).
(e) Nearly all recipients of transfusions have pneumonia (Curran et al., 1984; Selik et al., 1987).
(f) HIV-positive wives of hemophiliacs exhibit only pneumonia and a few other AIDS-defining opportunistic infections (Section 3.4.4.5).
(g) American babies exclusively have bacterial diseases (18%) and a high rate of dementia (14%) compared to adults (6%) (Table 1) (Center for Disease Control, 1992b).
(h) Users of the cytotoxic DNA chain terminator AZT, prescribed to inhibit HIV, develop anemia, leulipenia, and nausea (Section 4.6.2).

(3) The Thai mini-epidemic of 123 is made up of intravenous drug users (20%), heterosexual male and female "sex workers" (50%), and male homosexuals (30%) (Weniger et al., 1991). Among the Thais, 24% have tuberculosis, 22% have pneumonia and other opportunistic infections common in Thailand, and 10% have had septicemia, which is indicative of intravenous drug consumption (Weniger et al., 1991).

2.2. THE HIV-AIDS HYPOTHESIS, OR THE DEFINITION OF AIDS

Based on epidemiological data collected between 1981 and 1983, AIDS researchers from the CDC (Centers for Disease Control, 1986) "found in gay culture—particularly in its perceived 'extreme' and 'non-normative' aspects (that is 'promiscuity' and recreational drugs)—the crucial clue to the cause of the new syndrome" (Oppenheimer, 1992). Accordingly, the CDC had initially favored a "lifestyle" hypothesis for AIDS.

However, by 1983 immunodeficiency was also recorded in hemophiliacs, some women, and intravenous drug users. Therefore, the CDC adopted the "hepatitis B analogy" (Oppenheimer, 1992) and re-interpreted AIDS as a new viral disease, transmitted sexually and parenterally by blood products and the sharing of needles that were used for intravenous drug injection (Francis et al., 1983; Jaffe et al., 1983b; Centers for Disease Control, 1986; Oppenheimer, 1992). In April 1984 the American Secretary of Health and Human Services and virus researcher Robert Gallo announced at a press conference that the new AIDS virus was found. The announcement was made, and a test for antibody against the virus—termed the "AIDS test"—was registered for a patent, before even one American study had been published on this virus (Connor, 1987; Adams, 1989; Crewdson, 1989; Culliton, 1990; Rubinstein, 1990). Since then most medical scientists have believed that AIDS is infectious, spread by the transmission of HIV.

According to the virus-AIDS hypothesis the 25 different AIDS diseases and the very different AIDS epidemics and sub-epidemics are all held together by a single common cause, HIV. There are two strains of HIV that are 50 percent related, HIV-1 and HIV-2. But as yet only one American-born AIDS patient has been infected by HIV-2 (O'Brien et al., 1992). Since nearly all HIV-positive AIDS cases recorded to date are infected by HIV-1, this strain will be referred to as HIV in this article. The HIV-AIDS hypothesis proposes (a) that HIV is a sexually, parenterally, and perinatally transmitted virus; (b) that it causes immunodeficiency by killing T-cells, but on average only 10 years after infection in adults and two years after infection in infants—a period that is described as the "latent period of HIV" because the virus is assumed to become reactivated in AIDS; and (c) that all AIDS diseases are consequences of this immunodeficiency (Coffin et al., 1986; Institute of Medicine, 1986, 1988; Gallo, 1987; Blattner et al., 1988; Gallo and Montagnier, 1988; Lemp et al., 1990; Weiss and Jaffe, 1990; Blattner, 1991; Goudsmit, 1992).

Because of this belief, 25 previously known, and in part entirely unrelated, diseases have been redefined as AIDS, provided they occur in the presence of HIV. HIV is, in practice, detectable only indirectly via antiviral antibodies, because of its chronic inactivity even in AIDS patients (Section 3.3). These antibodies are identified with disrupted HIV, a procedure that is termed the "AIDS test" (Institute of Medicine, 1986; Rubinstein, 1990). Virus isolation is a very inefficient and expensive procedure, designed to activate dormant viruses from leukocytes. It depends on the activation of a single, latent HIV from about 5 million leukocytes from an antibody-positive person. For this purpose the cells must be propagated *in vitro* away from the virus-suppressing immune system of the host. Viruses may then be detected weeks later in the culture medium (Weiss et al., 1988; Duesberg, 1989c).

Antibodies against HIV were originally claimed to be present in most (88%) AIDS patients (Sarngadharan et al., 1984), but have since been confirmed in no more than about 50% of American AIDS patients (Institute of Medicine, 1988; Selik et al., 1990). The rest are presumptively diagnosed base on disease criteria outlined by the CDC (Centers for Disease Control, 1987; Institute of Medicine, 1988). Because of confidentiality laws, more tests are probably done than are reported to the CDC.

Since the "AIDS test" became available in 1985, over 20 million tests have been performed annually in the U.S. alone on blood donors, servicemen and applicants to the Army, AIDS patients, and many others, and millions more are performed in Europe, Russia, Africa, and other countries (Section 3.6). On the basis of such widespread testing, clearly the most comprehensive in the history of virology, about 1 million, or 0.4%, of mostly healthy Americans (Curran et al., 1985; Institute of Medicine, 1988; Duesberg, 1991a; Vermund, 1991; Centers for Disease Control, 1992a); 0.5 million, or 0.2%, of Western Europeans (Mann et al., 1988; Blattner, 1991; World Health Organization, 1992a); 6 million, or 10%, of mostly healthy Central Africans (Curran et al., 1985; Institute of Medicine, 1988; Piot et al., 1988; Goodgame, 1990; Blattner, 1991; Anderson and May, 1992); and 300,000, or 0.5%, of healthy Thais (Weniger et al., 1991) are estimated to carry antibodies to HIV (Table 1). According to the CDC the incidence of HIV-2 is "relatively high" in Western Africa with a record of 9% in one community, but "exceedingly low" in the U.S. where not even one infection was detected among 31,630 blood donors (O'Brien et al., 1992).

2.3. ALTERNATIVE INFECTIOUS THEORIES OF AIDS

In view of the heterogeneity of the AIDS diseases and the difficulties in reducing them to a common, active microbe, several investigators have proposed that AIDS is caused by a multiplicity of infectious agents such as viruses and microbes, or combinations of HIV with other microbes

Appendix B ■ 517

(Sonnabend et al., 1983; Konotey-Ahulu, 1987, 1989; Stewart, 1989; Cotton, 1990; Goldsmith, 1990; Lemaitre et al., 1990; Root-Bernstein, 1990a, c; Balter, 1991; Lo et al., 1991).

However, the proponents of infectious AIDS who rejected HIV as the sole cause or who see it as one of several causes of AIDS have failed to establish a consistent alternative to or cofactor for HIV. Instead, they typically blame AIDS on viruses and microbes that are widespread and either harmless or not life-threatening to a normal immune system, such as *Pneumocystis*, cytomegalovirus, herpes virus, hepatitis virus, tuberculosis bacillus, *Candida*, mycoplasma, treponema, gonococci, toxoplasma, and cryptosporidiae (Section 3.5.7) (Freeman, 1979; Mims and White, 1984; Pifer, 1984; Evans, 1989c; Mills and Masur, 1990; Bardach, 1992). Since such microbes are more commonly active in AIDS patients than in others, they argue that either chronic or repeated infections by such microbes would generate fatal AIDS (Sonnabend et al., 1983; Stewart, 1989; Mills and Masur, 1990; Root-Bernstein, 1990a, c).

Yet all of these microbes also infect people with normal immune systems either chronically or repeatedly without causing AIDS (Freeman, 1979; Mims and White, 1984; Evans, 1989c; Mills and Masur, 1990). It follows that pathogenicity by these microbes in AIDS patients is a consequence of immunodeficiency acquired by other causes (Duesberg, 1990c, 1991a). This is why most of these infections are termed opportunistic.

3. DISCREPANCIES BETWEEN AIDS AND INFECTIOUS DISEASE

3.1. CRITERIA OF INFECTIOUS AND NONINFECTIOUS DISEASE

The correct hypothesis explaining the cause of AIDS must predict the fundamental differences between the two main AIDS epidemics and the bewildering heterogeneity of the 25 AIDS diseases. In addition, the cause of American/European AIDS should make clear why—in an era of ever-improving health parameters, population growth, and decreasing mortality (The Software Toolworks World Atlas^{TM}, 1992; Anderson and May, 1992)—suddenly a subgroup of mostly 20- to 45-year-old males would die from diverse microbial and nonmicrobial diseases. The mortality from all infectious diseases combined has been reduced to less than 1% in the Western world (Cairns, 1978) through advanced sanitation and nutrition (Section 6) (McKeown, 1979; Moberg and Cohn, 1991; Oppenheimer, 1992). Further, 20- to 45-year-olds are the least likely to die from any disease (Mims and White, 1984). Their relative immunity to all diseases is why they are recruited as soldiers. The correct AIDS hypothesis would also have to explain why only a small group of about 20,000 Africans have developed AIDS diseases annually since 1985 (Table 1), during a time in which Central Africa enjoyed the fastest population growth in the world—3% (The Software Toolworks World Atlas,™ 1992).

The sudden appearance of AIDS could signal a new microbe, i.e., infectious AIDS. Yet the suddenness of AIDS could just as well signal one or several new toxins, such as the many new psychoactive drugs that have become popular in America and Europe since the Vietnam War (Section 4).

Based on common characteristics of all orthodox infectious diseases, infectious AIDS would be predicted to:

(1) Spread randomly between the sexes. This is just as true for venereal as for other infectious diseases (Judson et al., 1980; Haverkos, 1990).

(2) Cause primary disease within weeks or months after infection, because infectious agents multiply exponentially in susceptible hosts until stopped by immunity. They are self-replicating, and thus fast-acting, toxins. (Although "slow" viruses are thought to be pathogenic long after neutralization by antiviral immunity (Evans, 1989c), slow pathogenicity by a neutralized virus has never been experimentally proven (Section 6.1).)

(3) Coincide with a common, active, and abundant microbe in all cases of the same disease. (Inactive microbes or microbes at low concentrations are harmless passengers, e.g., lysogenic bacteriophages, endogenous and latent retroviruses (Weiss et al., 1985), latent herpes virus or latent ubiquitous *Pneumocystis* and *Candida* infections (Freeman, 1979; Pifer, 1984; Williford Pifer et al., 1988). Hibernation is a proven microbial strategy of survival, which allows indefinite coexistence with the host without pathogenicity.)

(4) Lyse or render nonfunctional more cells than the host can spare or regenerate.

(5) Generate a predictable pattern of symptoms.

By contrast noninfectious AIDS, caused by toxins, would be predicted to:

(1) Spread nonrandomly, according to exposure to toxins. For example, lung cancer and emphysema were observed much more frequently in men than in women 20 years ago, because men consumed much more tobacco than women 30–40 years ago (Cairns, 1978).

(2) Follow intoxication after variable intervals as determined by lifetime dosage and personal thresholds for disease. These intervals would be considerably longer than those between microbes and disease, because

microbes are self-replicating toxins. For example, lung cancer and emphysema are "acquired" only after 10–20 years of smoking, and liver cirrhosis is "acquired" only after 10–20 years of alcoholism.

(3) Manifest toxin- and intoxication-site-specific diseases, e.g., cigarettes causing lung cancer and alcohol causing liver cirrhosis.

3.2. AIDS NOT COMPATIBLE WITH INFECTIOUS DISEASE

All direct parameters of AIDS are incompatible with classical criteria of infectious disease:

(1) Unlike conventional infectious diseases, including venereal diseases (Judson et al., 1980), American/European AIDS is nonrandomly (90%) restricted to males, although no AIDS disease is male-specific (Table 1).

(2) The long and unpredictable intervals between infection and "acquiring" primary AIDS symptoms—averaging two years in infants and ten years in adults, and termed "latent periods of HIV"—stand in sharp contrast to the short intervals of days or weeks between infection and primary disease observed with all classical viruses, including retroviruses (Duesberg, 1987; Duesberg and Schwartz, 1992). These short intervals reflect the time periods that all exponentially growing microbes with generation times of half hours and viruses including HIV (Clark et al., 1991; Daar et al., 1991) with generation times of 8–48 hours need to reach immunogenic and thus potentially pathogenic concentrations (Fenner et al., 1974; Freeman, 1979; Mims and White, 1984). Once stopped by immunity, conventional viruses and microbes are no longer pathogenic. Thus, long latent periods between immunity against a microbe and a given disease are incompatible with conventional microbial causes, including HIV (Section 3.5.14). The discrepancy of eight years between the hypothetical latent periods is simply a statistical artifact. It is conceived to link HIV with AIDS and to buy time for the real causes of AIDS to generate AIDS-defining diseases.

(3) There is no active microbe common to all AIDS patients, and no common groups of target cells are lysed or rendered nonfunctional (Sections 3.3 and 3.5.10).

(4) There is no common, predictable pattern of AIDS symptoms in patients of different risk groups. Instead, different risk groups have their own characteristic AIDS diseases (Sections 2.1.3, 3.4.4, and 3.4.5).

Thus, AIDS does not meet even one of the classical criteria of infectious disease. In a recent response to these arguments, Goudsmit (1992), a proponent of the HIV-AIDS hypothesis, confirmed that "AIDS does not have the characteristics of an ordinary infectious disease. This view is incontrovertible." Likewise, epidemiologists Eggers and Weyer conclude that "the spread of AIDS does not behave like the spread of a disease that is caused by a single sexually transmitted agent" (Eggers and Weyer, 1991) and hence have "simulated a cofactor [that] cannot be identified with any known infectious agent" (Weyer and Eggers, 1990). Anderson and May (1992) had to invent "assortative scenarios" for different AIDS risk groups to reconcile AIDS with infectious disease. Indeed, AIDS would never have been accepted as infectious without the numerous unique assumptions that have been made to accommodate HIV as its cause (Sections 3.5 and 6.1).

3.3 No Proof for the Virus-AIDS Hypothesis

Despite research efforts that exceed those on all other viruses combined and that have generated over 60,000 papers on HIV (Christensen, 1991), it has not been possible to prove that HIV causes AIDS. These staggering statistics illustrate that the virus-AIDS hypothesis is either not provable or is very difficult to prove.

Proof for pathogenicity of a virus depends either on (1) meeting Koch's classical postulates, (2) preventing pathogenicity through vaccination, (3) curing disease with antiviral drugs, or (4) preventing disease by preventing infection. However, the HIV-AIDS hypothesis fails all of these criteria.

3.3.1. Virus Hypothesis Fails to Meet Koch's Postulates

Koch's postulates may be summarized as follows: (i) the agent occurs in each case of a disease and in amounts sufficient to cause pathological effects; (ii) the agent is not found in other diseases; and (iii) after isolating and propagation in culture, the agent can induce the disease anew (Merriam-Webster, 1965; Weiss and Jaffe, 1990).

But:

(i) HIV is certainly not present in all AIDS patients, and even antibodies against HIV are not found in all patients who have AIDS-defining diseases. HIV is not even present in all persons who die from multiple-indicator diseases plus general immune system failure—the paradigm AIDS cases (Sections 3.4 and 4.5). In addition, HIV is never present "in amounts sufficient to cause pathological effects" based on the following evidence:

(1) On average only 1 in 500 to 3000 T-cells, or 1 in 1500 to 8000 leukocytes of AIDS patients are infected by HIV (Schnittman et al., 1989; Simmonds et al., 1990). (About 35% of leukocytes are T-cells (Walton et

al., 1986). A recent study, relying on *in situ* amplification of a proviral HIV DNA fragment with the polymerase chain reaction, detected HIV DNA in 1 of 10 to 1 of 1000 leukocytes of AIDS patients. However, the authors acknowledge that the in situ method cannot distinguish between intact and defective proviruses and may include false-positives, because it does not characterize the amplified DNA products (Bagasra et al., 1992). Indeed the presence of 1 provirus per 10 or even 100 cells is exceptional in AIDS patients. This is why direct hybridization with viral DNA, a technique that is capable of seeing 1 provirus per 10 to 100 cells, typically fails to detect HIV DNA in AIDS patients (Duesberg, 1989c). According to one study, "The most striking feature... is the extremely low level of HIV provirus present in circulation PBMCs (peripheral blood mononuclear cells) in most cases" (Simmonds et al., 1990).

Since on average only 0.1% (1 out of 500 to 3000) of T-cells are ever infected by HIV in AIDS patients, but at least 3% of all T-cells are regenerated (Sprent, 1977; Guyton, 1987) during the two days it takes a retrovirus to infect a cell (Duesberg, 1989c), HIV could never kill enough T-cells to cause immunodeficiency. Thus, even if HIV killed every infected T-cell (Section 3.5.10), it could deplete T-cells only at 1/30 of their normal rate of regeneration, let alone activated regeneration. The odds of HIV causing T-cells deficiency would be the same as those of a bicycle rider catching up with a jet airplane.

(2) It is also inconsistent with a common pathogenic mechanism that the fraction of HIV-infected leukocytes in patients with the same AIDS diseases varies 30- to 100-fold. One study reports that the fraction of infected cells ranges from 1 in 900 to 1 in 30,000 (Simmonds et al., 1990), and another reports that it ranges from 1 in 10 to 1 in 1000 (Bagasra et al., 1992). In all conventional viral diseases the degree of pathogenicity is directly proportional to the number of infected cells.

(3) It is entirely inconsistent with HIV-mediated pathogenicity that there are over 40 times more HIV-infected leukocytes in many healthy HIV carriers than in AIDS patients with fatal AIDS (Simmonds et al., 1990; Bagasra et al., 1992). Simmonds et al. report that there are from 1 in 700 to 1 in 83,000 HIV-infected leukocytes in healthy HIV carriers and from 1 to 900 to 1 in 30,000 in AIDS patients. Bagasra et al. report that there are from 1 in 30 to 1 in 1000 infected leukocytes in healthy carriers and from 1 in 10 to 1 in 1000 in patients with fatal AIDS. Thus, there are healthy persons with 43 times (30,000:700) and 33 times (1000:30) more HIV-infected cells than in AIDS patients.

(4) In terms of HIV's biological function, it is even more important that the levels of HIV RNA synthesis in AIDS are either extremely low or even nonexistent. Only 1 in 10,000 to 100,000 leukocytes express viral

RNA in 50% of AIDS patients. In the remaining 50% no HIV expression is detectable (Duesberg, 1989c; Simmonds et al., 1990). The very fact that amplification by the polymerase chain reaction must be used to detect HIV DNA or RNA (Semple et al., 1991) in AIDS patients indicates that not enough viral RNA can be made or is made in AIDS patients to explain any, much less fatal, pathogenicity based on conventional precedents (Duesberg and Schwartz, 1992). The amplification method is designed to detect a needle in a haystack, but a needle in a haystack is not sufficient to cause a fatal disease, even if it consists of plutonium or cyanide.

(5) In several AIDS diseases that are not caused by immunodeficiency (Section 3.5.8), HIV is not even present in the diseased tissues, e.g., there is no trace of HIV in any Kaposi's sarcomas (Salahuddin et al., 1988), and there is no HIV in neurons of patients with dementia, because of the generic inability of retroviruses to infect nondividing cells like neurons (Sections 3.5.8 and 3.5.10) (Duesberg, 1989c).

As a result, there is typically no free HIV in AIDS patients (Section 3.5.6). Indeed, the scarcity of infectious HIV in typical AIDS patients is the reason that neutralizing antibodies, rather than viruses, have become the diagnostic basis of AIDS. It is also the reason that on average 5 million leukocytes of HIV-positives must be cultured to activate ("isolate") HIV from AIDS patients. Even under these conditions it may take up to 15 different isolation efforts(!) to get just one infectious virus out of an HIV carrier (Weiss et al., 1988). The scarcity of HIV and HIV-infected cells in AIDS patients is also the very reason for the notorious difficulties experienced by leading American (Hamilton, 1991; Palca, 1991a; Crewdson, 1992) and British (Connor, 1991, 1992; Weiss, 1991) AIDS researchers in isolating, and in attributing credit for isolation HIV from AIDS patients.

(ii) HIV does not meet Koch's second postulate, because it is found not just in one, but in 25 distinct diseases, many as unrelated to each other as dementia and diarrhea, or Kaposi's sarcoma and pneumonia (Table 1, Section 2.1.2).

(iii) HIV also fails Koch's third postulate, because it fails to cause AIDS when experimentally inoculated into chimpanzees which make antibodies against HIV just like their human cousins (Blattner et al., 1988; Institute of Medicine, 1988; Evans, 1989b; Weiss and Jaffe, 1990). Up to 150 chimpanzees have been inoculated since 1983, and all are still healthy (Duesberg, 1989c) (Jorg Eichberg, personal communication, see Section 1). HIV also fails to cause AIDS when accidentally introduced into humans (Duesberg, 1989c, 1991a).

There is, however, a legitimate limitation of Koch's postulates, namely that most microbial pathogens are only conditionally pathogenic

(Stewart, 1968; McKeown, 1979; Moberg and Cohn, 1991). They are pathogenic only if the immune system is low, allowing infection or intoxication of the large numbers of cells that must be killed or altered for pathogenicity. This is true for tuberculosis bacillus, cholera, influenza virus, poliovirus, and many others (Freeman, 1979; Mims and White, 1984; Evans, 1989c).

However, even with such limitations HIV fails the third postulate. The scientific literature has yet to prove that even one health care worker has contracted AIDS from the over 206,000 American AIDS patients during the past 10 years, and that even one of thousands of scientists has developed AIDS from HIV, which they propagate in their laboratories and companies (Section 3.5.16) (Duesberg, 1989c, 1991a). AIDS is likewise not contagious to family members living with AIDS patients for at least 100 days in the same household (Friedland et al., 1986; Sande, 1986; Hearst and Hulley, 1988; Peterman et al., 1988). However, the CDC has recently claimed that seven health care workers have developed AIDS from occupational infection (Centers for Disease Control, 1992c). But the CDC has failed to provide any evidence against nonoccupational causation, such as drug addiction (see Section 4). Indeed, thousands of health care workers, e.g., 2586 by 1988 (Centers for Disease Control, 1988), have developed AIDS from nonprofessional causes. In addition the CDC has failed to report their sex (see next paragraph) and whether these patients developed AIDS only after AZT treatment (see Section 4) (Centers for Disease Control, 1992c). The failure of HIV to meet the third postulate is all the more definitive since there is no antiviral drug or vaccine. Imagine what would happen if there were 206,000 polio or viral hepatitis patients in our hospitals and no health care workers were vaccinated!

Contrary to expectations that health care workers would be the first to be affected by infectious AIDS, the AIDS risk of those health care workers that have treated the 206,000 American AIDS patients is in fact lower than that of the general population, based on the following data. The CDC reports that about 75% of American health care workers are female, but that 92% of AIDS patients among health care workers are male (Centers for Disease Control, 1988). Thus, the AIDS risk of male health care workers is thirty-five times higher than that of females, indicating nonprofessional AIDS causes.

Moreover, the CDC reports that the incidence of AIDS among health care workers is percentage-wise the same as that in the general population, i.e., by 1988, 2586 out of 5 million health care workers, or 1 out of every 2000, had developed AIDS (Centers for Disease Control, 1988), by the same time that 110,000 out of the 250 million Americans, or 1 out of every 2250, had developed AIDS (Centers for Disease Control,

1992b). Since health care workers are nearly all over 20 years old and since there is virtually no AIDS in those under 20 (Table 1), but those under 20 make up about 1/3 of the general population, it can be estimated that the AIDS risk of health care workers is actually 1/3 lower (1/3 x 1/2,000) than that of the general population—hardly an argument for infectious AIDS.

In view of this, leading AIDS researchers have acknowledged that HIV fails Koch's postulates as the cause of AIDS (Blattner et al., 1988; Evans, 1989a, b; Weiss and Jaffe, 1990; Gallo, 1991). Nevertheless, they have argued that the failure of HIV to meet Koch's postulates invalidates these postulates rather than invalidating HIV as the cause of AIDS (Section 6.1) (Evans 1989b, 1992; Weiss and Jaffe, 1990; Gallo, 1991). But the failure of a suspected pathogen to meet Koch's postulates neither invalidates the timeless logic of Koch's postulates nor any claim that a suspect causes a disease (Duesberg, 1989b). It only means that the suspected pathogen cannot be proven responsible for a disease by Koch's postulates—but perhaps by new laws of causation (Section 6).

3.3.2. Anti-HIV Immunity Does Not Protect against AIDS

Natural antiviral antibodies, or vaccination, against HIV—which completely neutralize HIV to virtually undetectable levels—are consistently diagnosed in AIDS patients with the "AIDS test." Yet these antibodies consistently fail to protect against AIDS disease (Section 3.5.11) (Duesberg, 1989b, c, 1991a; Evans, 1989a, b). According to Evans, "The dilemma in HIV is that antibody is not protective" (Evans, 1989a).

By contrast, all other viral disease are prevented or cured by antiviral immunity. Indeed, since Jennerian vaccination in the late 18th century, antiviral immunity has been the only protection against viral disease. In view of this, HIV researchers have argued that antibodies do not neutralize this virus (Section 3.5.11) instead of considering that HIV may not be the cause of AIDS.

3.3.3. Antiviral Drugs Do Not Protect Against AIDS

All anti-HIV drugs fail to prevent or cure AIDS diseases (Section 4).

3.3.4. All AIDS-defining Diseases Occur in the Absence of HIV

The absence of HIV does not prevent AIDS-defining disease from occurring in all AIDS risk groups; it only prevents their diagnosis as AIDS (Sections 3.4.4, 4.5, and 4.7).

Thus, there is no proof for the virus-AIDS hypothesis—not even that AIDS is contagious. Instead, the virus-AIDS hypothesis is based only on circumstantial evidence, including epidemiological correlations and anecdotal cases (Sections 3.4 and 3.5).

3.4. NONCORRELATIONS BETWEEN HIV AND AIDS

Leading AIDS researchers acknowledge that correlations are the only support for the virus-AIDS hypothesis. For example, Blattner et al. state "...overwhelming seroepidemiologic evidence [is] pointing toward HIV as the cause of AIDS... Better methods... show that HIV infection is present in essentially all AIDS patients" (Blattner et al., 1988). According to an editorial in *Science*, Baltimore deduces from studies reporting an 88% correlation between antibodies to HIV and AIDS: "This was the kind of evidence we are looking for. It distinguishes between a virus that was a passenger and one that was the cause" (Booth, 1988). The studies Baltimore relied on are those published by Gallo et al. in *Science* in 1984 that are the basis for the virus-AIDS hypothesis (Gallo et al., 1984; Sarngadharan et al., 1984), but their authenticity has since been questioned on several counts (Beardsley, 1986; Schüpach, 1986; Connor, 1987; Crewdson, 1989, 1992; Hamilton, 1991; Palca, 1991a). Weiss and Jaffe (1990) concur that "the evidence that HIV causes AIDS is epidemiological...," although Gallo (1991) concedes that epidemiology is just "one hell of a good beginning." In view of correlations it is argued that "persons infected with HIV will develop AIDS and those not so infected will not" (Evans, 1989a), or that "HIV... is the *sine qua non* for the epidemic" (Gallo, 1991).

But correlations are only circumstantial evidence for a hypothesis. According to Sherlock Holmes, "Circumstantial evidence is a very tricky thing. It may seem to point very straight to one thing, but if you shift your point of view a little, you may find it pointing in an equally uncompromising manner to something entirely different" (Doyle, 1928). The risk in epidemiological studies is that the cause may be difficult to distinguish from noncausal associations. For example, yellow fingers are noncausally and smoking is causally associated with lung cancer. "In epidemiological parlance, the issue at stake is that of confounding" (Smith and Phillips, 1992). This is true for the "overwhelming seroepidemiologic evidence" claimed to support the virus-AIDS hypothesis on the following grounds.

3.4.1. Only About Half of American AIDS Is Confirmed HIV-antibody-positive

In the United States, antibodies against HIV are confirmed in only about 50% of all AIDS diagnoses; the remainder are presumptively diagnosed (Institute of Medicine, 1988; Selik et al., 1990). Several studies indicate that the natural coincidence between antibodies against HIV and AIDS diseases is not perfect, because all AIDS-defining diseases occur in all AIDS risk groups in the absence of HIV (Section 4). Ironically, the CDC never records the incidence of HIV in its *HIV/AIDS Surveillance Reports* (Centers for Disease Control, 1992b).

It follows that the reportedly perfect correlation between HIV and AIDS is in reality an artifact of the definition of AIDS and of allowances for presumptive diagnoses (Centers for Disease Control, 1987; Institute of Medicine, 1988). Since AIDS has been defined exclusively as diseases occurring in the presence of antibody to HIV (Section 2.2), the diagnosis of AIDS is biased by its definition toward a 100% correlation with HIV. That is why "persons infected by HIV will develop AIDS and... those not so infected will not" (Evans, 1989a), and why HIV is the *"sine qua non"* of AIDS (Gallo, 1991).

3.4.2. Antibody-positive, but Virus-negative AIDS

The correlations between AIDS and HIV are in fact not correlations with HIV, but with antibodies against HIV (Sarngadharan et al., 1984; Blattner et al., 1988; Duesberg, 1989c). But antibodies signal immunity against viruses and neutralization of viruses, and thus, protection against viral disease—not a prognosis for a future disease as is claimed for antibodies against HIV. For example, antibody-positive against poliovirus and measles virus means virus-negative, and thus, protection against the corresponding viral diseases. The same is true for antibodies against HIV: antibody-positive means very much virus-negative. Residual virus or viral molecules are almost undetectable in most antibody-positive persons (Sections 3.3 and 3.5.6). Thus, antibodies against HIV are not evidence for a future or current HIV disease unless additional assumptions are made (Section 3.5.11).

3.4.3. HIV: Just One of Many Harmless Microbial Markers of Behavioral and Clinical AIDS Risks

In addition to antibodies against HIV, there are antibodies against many other passenger viruses and microbes in AIDS risk groups and

AIDS patients (Sections 2.3 and 4.3.2). These include cytomegalovirus, hepatitis virus, Epstein-Barr virus, Human T-cell Leukemia Virus-I (HTLV-I), herpes virus, gonorrhea, syphilis, mycoplasma, amoebae, tuberculosis, toxoplasma, and many others (Gallo et al., 1983; Sonnabend et al., 1983; Blattner et al., 1985; Mathur-Wagh et al., 1985; Darrow et al., 1987; Quinn et al., 1987; Messiah et al., 1988; Stewart, 1989; Goldsmith, 1990; Mills and Masur, 1990; Root-Bernstein, 1990a, c; Duesberg, 1991a; Buimovici-Klein et al., 1988). In addition, there are between 100 and 150 chronically latent retroviruses in the human germ line (Martin et al., 1981; Nakamura et al., 1991). These human retroviruses are in every cell, not just in a few like HIV, and have the same genetic structure and complexity as HIV and all other retroviruses (Duesberg, 1989c). According to Quinn et al. (1987), "Common to African patients with AIDS and output controls and American patients with AIDS and homosexual men was the finding of extremely high prevalence rates of antibody to CMV (range, 92–100%), HSV (range, 90–100%), hepatitis B virus (range, 78–82%), hepatitis A virus (range, 82–95%), EBV capsid antigen (100%), syphilis (11–23%), and *T. gondii* (51–74%). In contrast, the prevalence of antibody to each of these infectious agents was significantly lower among the 100 American heterosexual men..." Thus, the incidence of many human parasites, both rare and common, is high in typical AIDS patients and in typical AIDS risk groups (Sections 2.3 and 5). However, none of these microbes are fatal and nearly all are harmless to a normal immune system (Section 2.3).

Most of these parasites, including HIV, have been accumulated by AIDS risk behavior and by clinical AIDS risks (Blattner et al., 1985; Institute of Medicine, 1988; Stewart, 1989). Such behavior includes the long-term use of unsterile, injected, recreational "street" drugs and a large number of sexual contacts promoted by oral and injected aphrodisiac drugs (Section 4) (Dismukes et al., 1968; Darrow et al., 1987; Des Jarlais et al., 1987, 1988; Espinoza et al., 1987; Moss, 1987; Moss et al., 1987; van Griensven et al., 1987; Messiah et al., 1988; Chaisson et al., 1989; Weiss, S. H., 1989; Deininger et al., 1990; McKegney et al., 1990; Stark et al., 1990; Luca-Moretti, 1992; Seage et al., 1992). Clinical risk groups, such as hemophiliacs, accumulate such viruses and microbes from occasionally contaminated transfusions (Section 3.4.4).

It follows that a high correlation between AIDS and antibodies against one particular virus, such as HIV, does not "distinguish between a virus that was a passenger and one that was a cause" (Baltimore, see above) (Booth, 1988). It is an expected consequence or marker of behavioral and clinical AIDS risks, particularly in countries where the percentage of HIV carriers is low (Duesberg, 1991a). In addition to HIV, many other

Appendix B

Table 2. Annual AIDS Risks of HIV-infected Groups*

HIV-infected group	Annual AIDS in %	Group-specific diseases
American recipients of transfusions	50	pneumonia, opportunistic infections
American babies	25	dementia, bacterial
Male homosexuals using sexual stimulants	4–6	Kaposi's sarcoma
Intravenous drug users	4–6	tuberculosis, wasting
American hemophiliacs	2	pneumonia, opportunistic infections
German hemophiliacs	1	pneumonia, opportunistic infections
American teenagers	0.16–1.7	hemophilia-related
American general population	0.1–1.0	opportunistic infections
Africans	0.3	fever, diarrhea, tuberculosis
Thais	0.05	tuberculosis

*Based on controlled studies, it is proposed that the health risks of all HIV-infected AIDS risk groups are the same as those of matched HIV-free controls (Sections 3.4.4, 4, and 5). The virus hypothesis simply claims the specific morbidity of each of these groups for HIV.

---

microbes and viruses which are rare and inactive or just inactive, such as hepatitis virus, in the general population are "specific" for AIDS patients, and thus markers for AIDS risks (Sections 2.2, 2.3, and 4.3.2). For example, 100% of AIDS patients within certain cohorts, not just 50% as with HIV (Section 2.2.), were shown to have antibodies against, or acute infections of, cytomegalovirus (Gottlieb et al., 1981; Francis, 1983; van Griensven et al., 1987; Buimovici-Klein et al., 1988). A comparison of 481 HIV-positives to 1499 HIV-negative homosexual men in Berlin found that the HIV-positives were "significantly more often carriers of antibodies against hepatitis A virus, hepatitis B virus, cytomegalovirus, Epstein-Barr virus and syphilis" (Deininger et al., 1990). And the frequent occurrence of antibodies against hepatitis B virus in cohorts of homosexual AIDS patients, termed "hepatitis cohorts," was a precedent that helped to convince the CDC to drop the "lifestyle" hypothesis of AIDS in favor of the "hepatitis analogy" (Francis et al., 1983; Centers for Disease Control, 1986; Oppenheimer, 1992) (Section 2.2).

The higher the consumption of unsterile, injected recreational drugs; the more sexual contacts mediated by aphrodisiac drugs and the more transfusions received; the more accidentally contaminating microbes will be accumulated (Sections 3.4.4.5, 4.3.2, and 4.5). In Africa antibodies against HIV and hepatitis virus are poor markers for AIDS risks, because millions carry antibodies against these viruses (Table 1) (Quinn et al., 1987; Evans, 1989c; Blattner, 1991). Thus, it is arbitrary to consider HIV the AIDS "driver" rather than just one of the many innocent microbial passengers of AIDS patients (Francis, 1983), because it is neither distinguished by its unique presence nor by its unique biochemical activity.

3.4.4. *Annual AIDS Risks of Different HIV-infected Risk Groups, Including Babies, Homosexuals, Drug Addicts, Hemophiliacs, and Africans, Differ over 100-fold*

If HIV were the cause of AIDS, the annual AIDS risks of all infected persons should be similar, particularly if they are from the same country. Failure of HIV to meet this prediction would indicate that HIV is not a sufficient cause of AIDS. The occurrence of the same AIDS-defining diseases in HIV-free controls would indicate that HIV is not even necessary for AIDS.

3.4.4.1. *Critically ill recipients of transfusions.* The annual AIDS risk of HIV-infected American recipients of transfusions (other than hemophiliacs) is about 50%, as half of all recipients die within one year after receiving a transfusion (Table 2) (Ward et al., 1989).

Since the AIDS risk of transfusion recipients is much higher than the national 3-4% average, nonviral factors must play a role (Table 1). Indeed, about 50% of American recipients of transfusions without HIV also die within 1 year after receiving a transfusion (Hardy et al., 1985; Ward et al., 1989), and over 60% within 3 years (Bove et al., 1987). Moreover, the AIDS risk of transfusion recipients increases 3-6 times faster with the volume of blood received than does their risk of infection by HIV (Hardy et al., 1985; Ward et al., 1989). This indicates that the illnesses that necessitated the transfusions are responsible for the mortality of the transfusion recipients. Yet the virus hypothesis claims the relatively high morality of American transfusion patients for HIV without considering HIV-free controls. The hypothesis also fails to consider that the effects of HIV on transfusion mortality should be practically undetectable in the face of the high mortality of transfusion recipients and its postulate that HIV causes AIDS on average only 10 years after infection.

3.4.4.2. *HIV-infected babies.* The second highest annual AIDS risk is reported for perinatally infected American babies, whose health has been compromised by maternal drug addiction or by congenital diseases like hemophilia (Section 2.1.3). They develop AIDS diseases on average two years after birth (Anderson and May, 1988; Blattner et al., 1988; Institute of Medicine, 1988; Blattner, 1991). This corresponds to an annual AIDS risk of 25% (Table 2).

Since the AIDS risk of babies is much higher than the national average of 3-4% (Table 1), nonviral factors must play a role in pediatric AIDS. Based on correlations and controlled studies documenting AIDS-defining diseases in HIV-free babies, it is proposed below that maternal drug consumption (Section 4) and congenital diseases, like hemophilia (Section 3.4.4.5), are the causes of pediatric AIDS. Indeed, before AIDS surfaced, many studies had shown that maternal drug addiction was sufficient to cause AIDS-defining diseases in newborns (Section 4.6.1). In accord with this proposal it is shown that HIV is naturally a perinatally transmitted retrovirus—and thus, harmless (Section 3.5.2).

3.4.4.3. *HIV-positive homosexuals.* The annual AIDS risk of HIV-infected male homosexuals who have hundreds of sex partners and who frequently use aphrodisiac drugs (Sections 4) was originally estimated at about 6% (Mathur-Wagh et al., 1985; Anderson and May, 1988; Institute of Medicine, 1988; Lui et al., 1988; Moss et al., 1988; Turner et al., 1989; Lemp et al., 1990; van Griensven et al., 1990; Blattner, 1991). As more HIV-positives became identified, lower estimates of about 4% were reported (Table 2) (Rezza et al., 1990; Biggar and the International Registry of Seroconverters, 1990; Muñoz et al., 1992).

Since the annual AIDS risk of such homosexual men is higher than the national average, group-specific factors must be necessary for their specific AIDS diseases. Based on correlations with drug consumption and studies of HIV-free homosexuals, it is proposed that here the cumulative consumption of sexual stimulants and psychoactive drugs determines the annual AIDS risk of homosexuals (Sections 4.4 and 4.5). Indeed, all AIDS-defining diseases were observed in male homosexuals from behavioral risk groups before HIV was discovered and have since been observed in HIV-free homosexuals from AIDS risk groups (Sections 4.5 and 4.7).

In the spirit of the virus-AIDS hypothesis, many of these HIV-free homosexual AIDS cases have been blamed on various retrovirus-like particles, papilloma viruses, and other viruses and microbes by researchers who have not investigated drug use, particularly not oral drug use. These cases include 153 immunodeficient HIV-free homosexuals with T4:T8-cell ratios below 1 (Drew et al., 1985; Weber et al., 1986; Novick et al., 1986; Collier et al., 1987; Bartholomew et al., 1987; Buimovici-Klein et al., 1988) and 23 HIV-free Kaposi's sarcomas (Afrasiabi et al., 1986; Ho et al., 1989b, Bowden et al., 1991; Safai et al., 1991; Castro et al., 1992; Huang et al., 1992) (see also Note Added in Proof).

3.4.4.4. *HIV-positive intravenous drug users.* Application of the annual AIDS risk of male homosexual risk groups led to valid predictions for the annual AIDS risk of intravenous drug users (Lemp et al., 1990). Therefore the annual AIDS risk of HIV-infected intravenous drug users was originally estimated to be 6% (Table 2) (Lemp et al., 1990; Blattner, 1991; Goudsmit, 1992). More recent studies have concluded that the annual AIDS risk of intravenous drug users is about 4% (Table 2) (Rezza et al., 1990; Muñoz et al., 1992).

These findings argue against a sexually transmitted cause, because sexual transmission predicts a much higher AIDS risk for homosexuals who have hundreds of sexual partners than for intravenous controlled studies have indicated that the morbidity and mortality of intravenous drug users is independent of HIV (Sections 4.4, 4.5, and 4.7). On the basis of such studies it is proposed that the lifetime dose of drug consumption determines the annual AIDS risk of intravenous drug users (Section 4).

3.4.4.5. *HIV-positive hemophiliacs.* Hemophiliacs provide the most accessible group on which to test the virus hypothesis, because the time of infection can be estimated and because the role of other health risks can be controlled by studying HIV-free hemophiliacs.

About 15,000, or 75%, of the 20,000 American hemophiliacs have HIV from transfusions received before the "AIDS test" was developed in 1984 (Tsoukas et al., 1984; Hardy et al., 1985; Institute of Medicine, 1986, 1988; Stehr-Green et al., 1988; Goedert et al., 1989; Koerper, 1989). Based on limited data and antibodies against selected viral antigens, it is generally estimated that most of these infections occurred between 1978 and 1984 (Evatt et al., 1985; Johnson et al., 1985; McGrady et al., 1987; Goedert et al., 1989). This high rate of infection reflects the practice, developed in the 1960s and 1970s, of preparing factor VIII from blood pools collected from large numbers of donors (Johnson et al., 1985; Aronson, 1988; Koerper, 1989). Since only about 300 of the 15,000 HIV-infected American hemophiliacs have developed AIDS annually over the last 5 years (Morgan et al., 1990; Centers for Disease Control, 1992a, b), the annual AIDS risk of HIV-infected American hemophiliacs is about 2% (Table 2). Data from Germany extend these results: about 50% of the 6000 German hemophiliacs are HIV-positive (Koerper, 1989), and only 37 (1%) of these developed AIDS-defining diseases during 1991 and 303 (1% annually) from 1982 until 1991 (Bundesgesundheitsamt, Germany, 1991; Leonhard, 1992). An international study estimated the annual AIDS risk of adult hemophiliacs at 3% and that of children at 1% over a 5-year period of HIV-infection (Biggar and the International Registry of Seroconverters, 1990).

Appendix B

Table 3. Immunosuppression in HIV-negative and -positive Hemophiliacs

Study	Immunosuppression (T4:T8 about or less than 1)	
	HIV-negative	HIV-positive
1. Tsoukas et al. (1984)	6/14	9/15
2. Ludlam et al. (1985)	15	—
3. French Study Group (1985)	33	55
4. Sullivan et al. (1986)	28	83
5. Madhok et al. (1986)	9	10
6. Kreiss et al. (1986)	6/17	22/24
7. Gill et al. (1986)	8/24	30/32
8. Sharp et al. (1987)	5/12	—
9. Matheson et al. (1987)	5	3
10. Mahir et al. (1988)	6	5
11. Antonaci et al. (1988)	15	10
12. Aledort (1988)	57	167
13. Jin et al. (1989)	12	7
14. Lang, D. J. et al. (1989)	24	172
15. Becherer et al. (1990)	74	136
16. Jason et al. (1990)	31	—
17. de Biasi et al. (1991)	—	10/20

In a normal immune system, the T4 to T8 T-cell ratio is about 2, in immunodeficient persons and in many AIDS patients it is about 1 or below 1. Studies which list the fraction of immunodeficient hemophiliacs in HIV-positive and HIV-negative groups indicate, that HIV-positives are more likely to be immunodeficient. This is because HIV is a marker for the number of transfusions received and transfusion of foreign proteins causes immune deficiency. The study by de Biasi et al. (1991) showed that among 20 HIV-positive hemophiliacs only those 10 who received commercially purified factor VIII, but not those who received further purified factor VIII developed immunodeficiency over a period of two years. See text for references.

According to the virus-AIDS hypothesis, one would have expected that by now (about one 10-year HIV-latent period after infection) at least 50% of the 15,000 HIV-positive American hemophiliacs would have developed AIDS or died from AIDS. But the 2% annual AIDS risk indicates that the average HIV-positive hemophiliac would have to wait for 25 years to develop AIDS disease from HIV, which is the same as their current median age. The median age of American hemophiliacs has increased from 11 years in 1972 to 20 years in 1982 and to over 25 years in 1986, despite the infiltration of HIV in 75% (Johnson et al., 1985; Institute of Medicine, 1986; Koerper, 1989). Thus, one could make a logical argument that HIV, instead of decreasing the life span of hemophiliacs, has in fact increased it.

Considering the compromised health of many hemophiliacs compared to the general population, it is also surprising that the 1–2% annual AIDS risk of HIV-infected hemophiliacs is lower than the 3–4% risk of the average HIV-infected, nonhemophilic European or American (Table 1). There is even a bigger discrepancy between the annual AIDS risks of hemophiliacs and those of intravenous drug users and male homosexuals, which are both about 4–6% (Table 2). In an effort to reconcile the relatively low annual AIDS risks of hemophiliacs with that of homosexuals, hematologists Sullivan et al. (1986) noted, "The reasons for this difference remain unclear." And Biggar and colleagues (1990) noted that "AIDS incubation... was significantly faster" for drug users and homosexuals than for hemophiliacs.

In view of the many claims that HIV causes AIDS in hemophiliacs, it is even more surprising that there is not even one controlled study from any country showing that the morbidity or mortality of HIV-positive hemophiliacs is higher than that of HIV-negative controls.

Instead, controlled studies show that immunodeficiency in hemophiliacs is independent of HIV and that the lifetime dosage of transfusions is the cause of AIDS-defining diseases of hemophiliacs. Studies describing immunodeficiency in HIV-free hemophiliacs are summarized in Table 3 (Tsoukas et al., 1984; AIDS-Hemophilia French Study Group, 1985; Ludlam et al., 1985; Gill et al., 1986; Kreiss et al., 1986; Madhok et al., 1986; Sullivan et al., 1986; Sharp et al., 1987; Matheson et al., 1987; Antonaci et al., 1988; Mahir et al., 1988; Aledort, 1988; Jin et al., 1989; Jason et al., 1990; Lang, D. J. et al., 1989; Bercherer et al., 1990). One of these studies even documents an AIDS-defining disease in an HIV-free hemophiliac (Kreiss et al., 1986). Immunodeficiency in these studies is typically defined by a T_4 to $T8$-cell ratio of about 1 or less than 1, compared to a normal ratio of 2.

Most of the studies listed in Table 3 and additional ones conducted before HIV was discovered have concluded or noted that immunodeficiency is directly proportional to the number of transfusions received over a lifetime (Menitove et al., 1983; Kreiss et al., 1984; Johnson et al., 1985; Hardy et al., 1985; Pollack et al., 1985; Prince, 1992; Ludlam et al., 1985; Gill et al., 1986). According to the hematologists Pollack et al. (1985), "derangement of immune function in hemophiliacs results from transfusion of foreign proteins or a ubiquitous virus rather than contracting AIDS infectious agent." The "ubiquitous virus" was a reference to the virus-AIDS hypothesis but a rejection of HIV, because in 1985 HIV was extremely rare in blood concentrates outside the U.S., but immunodeficiency was observed in Israeli, Scottish, and American hemophiliacs (Pollack et al., 1985). Madhok et al. also arrived at the conclusion that "clotting factor concentrate impairs the cell mediated immune response to a new antigen in the absence of infection with HIV" (Madhok et al., 1986). Aledort (1988) observed that "chronic recipients... of factor VIII, factor IX and pooled products... demonstrated significant T-cell abnormalities regardless of the presence of HIV antibody" (Aledort, 1988). Even those who claim that clotting factor does not cause immunodeficiency show that immunodeficiency in hemophiliacs increases with both the age and the cumulative dose of clotting factor received during a lifetime (Becherer et al., 1990).

One controlled study showed directly that protein impurities of commercial factor VIII, rather than factor VIII or HIV, were immunosuppressive among factor VIII-treated, HIV-positive hemophiliacs. Over a period of two years the T-cells of HIV-positive hemophiliacs treated with commercial factor VIII declined two-fold, while those of matched HIV-positive controls treated with purified factor VIII remained unchanged (Table 3) (de Biasi et al., 1991).

Before AIDS, a multicenter study investigating the immune systems of 1551 hemophiliacs treated with factor VIII from 1975 to 1979 documented lymphocytopenia in 9.3% and thrombocytopenia in 5% (Eyster et al., 1985). Accordingly, AIDS-defining opportunistic infections, including 6% pneumonias and 20% tuberculosis, have been recorded in hemophiliacs between 1968 and 1979 (Johnson et al., 1985). These transfusion-acquired immunodeficiencies could more than account for the 2% annual incidence of AIDS-defining diseases in HIV-positive hemophiliacs recorded now (Centers for Disease Control, 1992b). An American hematologist who recorded opportunistic infections in hemophiliacs occurring between 1968 and 1979, including 2 candidiasis and 66 pneumonia deaths, commented in 1983, "...it seems possible that many of the unspecified pneumonias in hemophiliacs in the past would be classified today as AIDS" (Aronson, 1983).

It follows that long-term transfusion of foreign proteins causes immunodeficiency in hemophiliacs with or without HIV. The virus hypothesis has simply claimed normal morbidity and mortality of hemophiliacs for HIV, by ignoring HIV-free controls.

Nevertheless, several investigators have compared HIV-negatives to HIV-positives (Table 3) and have observed that HIV correlates with the number of transfusions received (Tsoukas et al., 1984; Kreiss et al., 1986; Sullivan et al., 1986; Koerper, 1989; Becherer et al., 1990). According to Kreiss et al., "seropositive hemophiliac subjects, on average, had been exposed to twice as much concentrate... as seronegative[s]" (Kreiss et al.,

1986). And according to Goedert et al., "the prevalence of HIV-1 antibodies was directly associated with the degree of severity [of hemophilia]" (Goedert et al., 1989). Thus, HIV appears just to be a marker of the multiplicity of transfusions, rather than a cause of immunodeficiency.

The conclusion that long-term transfusion of foreign proteins causes immunodeficiency makes three testable predictions:

(1) It predicts that hemophiliacs with "AIDS" would be older than the average hemophiliac. Indeed, the median age of hemophiliacs with AIDS in the U.S. (Evatt et al., 1984; Koerper, 1989; Stehr-Green et al., 1989), England (Darby et al., 1989), and other countries (Biggar and the International Registry of Seroconverters, 1990; Blattner 1991) is significantly higher (about 34 years in the U.S.; Johnson et al., 1985; Koerper, 1989; Becherer et al., 1990) than the average age of hemophiliacs (20–25 years in the U.S.; see above). Goedert et al. reported that the annual AIDS risk of 1- to 17-year-old hemophiliacs was 1.5%, that of 18- to 34-year-old hemophiliacs was 3% and that of 64-year-old hemophiliacs was 5% (Goedert et al., 1989). This confirms that the cumulative dose of transfusions received is the cause of AIDS-defining diseases among hemophiliacs. According to hematologist Koerper, "this may reflect lifetime exposure to a greater number of units of concentrate...," and to Evatt et al., "[t]his age bias may be due to differences in duration of exposure to blood products..." (Evatt et al., 1984; Koerper, 1989).

By contrast, AIDS caused by an autonomous infectious pathogen would be largely independent of the age of the recipient. Even if HIV were that pathogen, the hemophilic population with AIDS should have the same age distribution as the hemophilic population over 10 years, because HIV is thought to take 10 years to cause AIDS and nearly all hemophiliacs were infected about 10 years ago (Johnson et al., 1985; McGrady et al., 1987; Koerper, 1989).

(2) Foreign-protein–mediated immunodeficiency further predicts that all AIDS diseases of hemophiliacs are opportunistic infections. If hemophilia AIDS were due to HIV, only 62% of hemophiliacs' AIDS diseases would be opportunistic infections, because 38% of all American AIDS patients have diseases that are not dependent on, and not consistently associated with, immunodeficiency (Table 1, Section 3.5.8). These include wasting disease (19%), Kaposi's sarcoma (10%), dementia (6%), and lymphoma (3%) (Table 1).

The AIDS pathology of hemophiliacs confirms the prediction of the foreign-protein–hypothesis exactly. In America 99% of the hemophiliacs with AIDS have opportunistic infections, of which about 70% are fungal and viral pneumonias, and less than 1% have Kaposi's sarcoma (Evatt et al., 1984; Selik et al., 1987; Stehr-Green et al., 1988; Goedert et al.,

1989; Koerper, 1989; Becherer et al., 1990). The small percentage of Kaposi's sarcoma is due to the nitrite inhalants used as sexual stimulants by male homosexual hemophiliacs (Section 4). There are no reports of wasting disease and dementia in hemophiliacs.

(3) If hemophilia AIDS is due to transfusion of foreign proteins, the wives of hemophiliacs should not contract AIDS from their mates. But if it were due to a parenterally or sexually transmitted virus, hemophilia AIDS would be sexually transmissible. Indeed, AIDS researchers claim that the wives of hemophiliacs develop AIDS from sexual transmission of HIV (Lawrence et al., 1990; Weiss and Jaffe, 1990; Centers for Diseases Control, 1992b). For example, AIDS researcher Fauci asks, "How about the 60-year-old wife of a hemophiliac who gets infected? Is she cruising, too?" (Booth, 1988).

However, (a) statistical scrutiny and (b) a controlled study unconfirm the hypothesis that hemophilia AIDS is sexually transmissible: (a) The CDC reports that 94 wives of hemophiliacs have been diagnosed with unnamed AIDS diseases since 1985 (Centers for Disease Control, 1992b). If one considers that there have been 15,000 HIV-positive hemophiliacs in the U.S. since 1985 and assumes that a third are married, then there are 5,000 wives of HIV-positive hemophiliacs. During the seven years (94:7) from 1985 to 1991, about 13 of these women developed AIDS each year (Centers for Disease Control, 1992b). By contrast, at least 80 of these women would be expected to die per year, considering the human lifespan of about 80 years and that on average at least 1.6% of all those over 20 years of age die annually. Thus, until controls show that among 5000 HIV-negative wives of hemophiliacs only 67 (80–13) die annually, the claim that wives of hemophiliacs die from sexual transmission of HIV is unfounded speculation.

Moreover, it has been pointed out that all AIDS-defining diseases of the wives of hemophiliacs are typically age-related opportunistic infections, including 81% pneumonia (Lawrence et al., 1990). Kaposi's sarcoma, dementia, lymphoma, and wasting disease are not observed in wives of hemophiliacs (Lawrence et al., 1990). Thus, the virus-AIDS hypothesis seems to claim, once more, normal morbidity and mortality of the wives of hemophiliacs for HIV.

(b) To test the hypothesis that immunodeficiency of hemophiliacs is sexually transmissible, the T4 to T8–cell ratio of forty-one spouses and female sexual partners of immunodeficient hemophiliacs were analyzed (Kreiss et al., 1984). Twenty-two of the females had relationships with hemophiliacs with T-cell ratios below 1, and 19 had relationships with hemophiliacs with ratios of 1 and greater. The mean duration of relationships was 10 years, the mean number of sexual contacts was 111 during the previous year, and only 12% had used condoms (Kreiss et al., 1984). Since the T-cell ratios of all spouses were normal, averaging 1.68—exactly like those of 57 normal controls—the authors concluded that "there is no evidence to date for heterosexual or household-contact transmission of T-cell subset abnormalities from hemophiliacs to their spouses..." (Kreiss et al., 1984).

It follows that the foreign-protein hypothesis, but not the HIV hypothesis, correctly predicts (1) the pathology, (2) the age bias, (3) the non-contagiousness of hemophilia AIDS, and (4) HIV-free immunodeficiency in hemophiliacs. It also explains the discrepancies between the annual AIDS risks of hemophiliacs and other risk groups (Table 2).

Since the virus hypothesis has become totally dominant in 1988, no new studies have described HIV-free immunodeficient hemophiliacs (Table 3), and the question of whether HIV-free immunodeficient hemophiliacs ever developed AIDS-defining diseases became taboo. The study by Jason et al. described data collected in the mid-1980s; the studies by Jin et al. and Becherer et al. collected data before 1988; and the one by de Biasi et al. compared the effects of purified to unpurified factor VIII only in HIV-positive hemophiliacs (Table 3).

In response to the argument that hemophiliacs began to develop AIDS diseases only when HIV appeared (Centers for Disease Control, 1986; Oppenheimer, 1992), it is proposed that "new" AIDS-defining diseases among hemophiliacs are an indirect consequence of extending their life with factor VIII treatment. Long-term treatment with factor VIII has prolonged the median life of hemophiliacs from 11 years in 1972 to 25 years in 1986. But contaminating foreign proteins received over periods of 10 years of treatment have also caused immunodeficiencies, and various viral and microbial contaminants have caused infections in some and HIV infections in 75%. HIV has been a marker for the number of transfusion and factor VIII treatments received, just like hepatitis virus infection was a marker of the number of transfusions received until it was eliminated from the blood supplies (Anonymous, 1984; Koerper, 1989). Prior to factor VIII therapy most hemophiliacs died as adolescents from internal bleeding (Koerper, 1989).

3.4.4.6. *HIV-positive teenagers.* The annual AIDS risk of HIV-infected American teenagers can be calculated as follows: There are about 30 million American teenagers, of which 0.03% (10,000) (Burke et al., 1990) to 0.3% (100,000) (St. Louis et al., 1991) are HIV-positive. Since only 160 developed AIDS in 1991 and only 179 in 1990 (Centers for Disease Control, 1992b), their annual AIDS risk is between 0.16% and 1.7% (Table 2).

Thus, the AIDS risk of teenagers with HIV is less than the national average of 3–4%. There are no statistics to indicate that the annual risk for AIDS-defining diseases of the HIV-infected teenage population is higher than that of HIV-free controls (Section 3.5.2). Since most American teenagers with AIDS are hemophiliacs (38%), intravenous drug users (25%), or male homosexuals (25%) (Section 2.1.3), it is proposed that the associated risk factors, rather than HIV, are the cause of teenage AIDS (Sections 3.4.4.5 and 4).

3.4.4.7. *HIV-positive general U.S. population.* The CDC reports that 3% of all American AIDS cases are from the general population, corresponding to 900 to 1200 of the 30,000 to 40,000 annual AIDS cases (Table 1) (Centers for Diseases Control, 1992b). Since at least 0.03% to 0.3%, or 80,000 to 800,000, of the general American population of 250 million are infected (Section 3.5.4) (U.S. Department of Health and Human Services, 1990; Burke et al., 1990; Morgan et al., 1990; St. Louis et al., 1991), the annual AIDS risk of HIV-infected Americans of the general population is similar to that of teenagers.

There are no statistics to indicate that the annual AIDS risk of the general HIV-infected population is higher than the annual risk for AIDS-defining diseases in HIV-free controls. Because the incidence of AIDS in the general population is exceedingly low, it is proposed again that it reflects the normal, low incidence of AIDS-defining diseases, rather than HIV-mediated diseases.

3.4.4.8. *HIV-positive Africans.* The annual AIDS risks of HIV-infected Africans is only 0.3% (Tables 1 and 2), because 6 million HIV carriers generated 120,000 AIDS cases from 1985 to the end of 1991 (Table 1). There are no controlled studies indicating that the risk from AIDS-defining diseases of HIV-infected Africans differs from that of HIV-negative controls.

Since the annual AIDS risk of HIV-infected Africans is (1) 10 times lower than the average American and European risk, (2) up to 100-fold less than that of American/European risk groups, (3) the same for both sexes, unlike that in America and Europe, and (4) very low considering that the annual mortality in Africa is around 2% and that AIDS includes the most common African diseases, it is proposed that African AIDS is just a new name for indigenous African diseases (Section 2.1.2).

Instead of a new virus, malnutrition, parasitic infections, and poor sanitary conditions have all been proposed as causes of African AIDS-defining diseases (Editorial, 1987; Konotey-Ahulu, 1987, 1989; Rappoport, 1988; Adams, 1989). Further, it has been proposed that the incidence of tuberculosis, diarrhea, fever, and other African AIDS-defining diseases may be the same in Africans with and without HIV (Editorial, 1987). And prior to the discovery of HIV, protein malnutrition was identified by AIDS researchers Fauci et al., as the world's leading cause of immunodeficiency, particularly in underdeveloped countries (Seligmann et al., 1984).

Indeed, recent studies document that only 2168 out of 4383 (49.5%) African AIDS patients with slim disease, tuberculosis, and other Africa-specific diseases, who all met the WHO definition of AIDS, were infected by HIV. These patients were from Abidjan, Ivory Coast (De Cock et al., 1991; Taelman et al., 1991); Lusaka, Zambia; and Kinshasa, Zaire (Taelman et al., 1991). Another study reports 135 (59%) HIV-free patients from Ghana out of 227 diagnosed by clinical criteria of the WHO. These patients meet the weight loss, diarrhea, chronic fever, tuberculosis, and neurological diseases of the WHO definition of AIDS (Hishida et al., 1992). An earlier study documents 116 HIV-negatives among 424 African patients that meet the WHO definition of AIDS (Widy-Wirski et al., 1988). According to an African AIDS doctor, "Today, because of AIDS, it seems that Africans are not allowed to die from these conditions any longer" (Konotey-Ahulu, 1987). Another asks, "What use is a clinical case definition for AIDS in Africa?" (Gilks, 1991).

The 10-fold difference between the average annual AIDS risks of Africans and Americans/Europeans (Table 1) can thus be resolved as follows: (1) The high AIDS risk of HIV-positive Americans and Europeans is the product of the low absolute numbers of HIV in AIDS risks groups, e.g., consumers of recreational drugs and the antiviral drug AZT (Section 4) and recipients of transfusions (Section 3.4.3). (2) The low AIDS risk of Africans is a product of large absolute numbers of HIV carriers and their relatively low, spontaneous, and malnutrition-mediated AIDS risks.

3.4.4.9. *HIV-positive Thais.* Given that there have been only 123 Thai AIDS cases in the past one to two years and that there are an estimated 300,000 HIV carriers in Thailand (Weniger et al., 1991), the annual AIDS risk of HIV-infected Thais can be calculated to be less than 0.05% (Table 2). Since most of these 123 were either intravenous drug users or "sex workers" (Section 2.1.3), it is proposed that these specific health risks are their cause of AIDS (Section 4), rather than the HIV that they share, unspecifically, with 300,000 healthy Thais.

The over 100-fold range in the annual AIDS risks of different AIDS risks groups, summarized in Table 2, clearly indicates that HIV is not sufficient to cause AIDS. It confirms and extends an earlier CDC conclusion:

"The magnitude of some of the differences in rates is so great that even gross errors in denominator estimates can be overcome" (Hardy et al., 1985). Moreover, analysis of the specific health risks of each risk group has identified nonviral health risks that are necessary and sufficient causes of AIDS (Section 4.5; Table 3).

3.4.5. *Specific AIDS Diseases Predetermined by Prior Health Risks*

If HIV were the cause of AIDS, every AIDS case should have the same risk of having one or more of the 25 AIDS diseases. However, the data listed above (Section 2.1) and in Table 2 indicate that, per AIDS cases, different risk groups have very specific AIDS diseases:

(1) Male homosexuals have 20 times more Kaposi's sarcoma than all other American and European AIDS risk groups.

(2) Hemophiliacs and other recipients of transfusions have fungal and viral pneumonia and other opportunistic infections, and practically no Kaposi's sarcoma or dementia.

(3) The AIDS diseases of the "general population" are either spontaneous, hemophilia- or age-related opportunistic infections. Typical examples are cited below (Section 3.5.16).

(4) Babies exclusively have bacterial infections (18%) and a high rate of dementia (14%), compared to adults (6%) (Table 1).

(5) Africans develop Africa-specific AIDS diseases 10 times more and Kaposi's sarcoma 10 times less often than Americans or Europeans.

The epidemiological data summarized in Section 3.4 indicated that HIV is sufficient to determine neither the annual AIDS risk nor the type of AIDS disease an infected person may develop. Instead, prior health risks including drug consumption, malnutrition, and congenital diseases like hemophilia, and their treatments and even the country of residence predetermine AIDS diseases. The correlations between HIV and AIDS that are claimed to support the virus-AIDS hypothesis are not direct, not complete, not distinctive, and, above all, not controlled. Controlled studies indicate that the incidence of AIDS-defining diseases in intravenous drug users and in male homosexuals engaging in high-risk behavior and hemophiliacs is independent of HIV.

Therefore, it is proposed that various group-specific health risk factors, including recreational and antiviral drugs (Section 4) and malnutrition, are necessary and sufficient causes of AIDS. The existence of risk-group–specific AIDS-defining diseases in the absence of HIV confirms this conclusion (Section 3.4.4 and 4.5).

3.5. ASSUMPTIONS AND ANECDOTAL CASES THAT APPEAR TO SUPPORT THE VIRUS-AIDS HYPOTHESIS

The following assumptions and anecdotal cases are frequently claimed to prove the virus-AIDS hypothesis. Despite the popularity of these claims they are either uncontrolled for alternative explanations or they are natural coincidences between HIV infection and naturally occurring diseases.

Figure 1. Distribution over Time of a Hypothetical Flu Epidemic

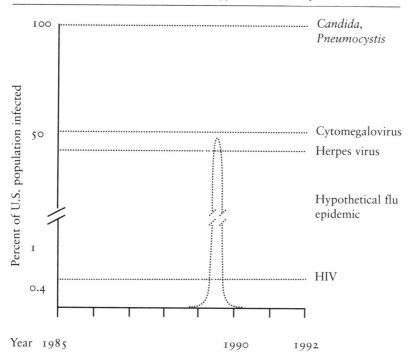

Determination of the age of a microbe in a population based on Farr's law. Farr's law holds that a microbe entering a population spreads exponentially until a susceptible pool is saturated. Subsequently, those microbes that are incompatible with long-term survival of the host are eliminated exponentially, to generate a bell-shaped curve. The rise and fall of a hypothetical flu epidemic caused by a new strain of influenza virus is an example. But microbes that can coexist with their host become established. Examples are *Candida*, *Pneumocystis* (Freeman, 1979), cytomegalovirus, herpes virus (Evans 1989c), and HIV (see text); these are shown at the percentages at which they are established in the American population.

3.5.1. HIV Is Presumed New Because AIDS Is New

HIV is presumed new in all countries with AIDS, because AIDS is new (Blattner et al., 1988; Gallo and Montagnier, 1988; Weiss and Jaffe, 1990). The presumed newness of HIV is used as a primary argument for the virus-AIDS hypothesis: "...the time of occurrence of AIDS in each country is correlated with the time of introduction of HIV into that country; first HIV is introduced, then AIDS appears" (Blattner et al., 1988). Or: "In every country and city where AIDS has appeared, HIV infection preceded it just by a few years" (Weiss and Jaffe, 1990).

However, according to Farr's law, the age of a microbe in a population is determined by changes in its incidence over time (Bregman and Langmuir, 1990). If a microbe is spreading from a low to a high incidence, it is new; however, if its incidence in a population is constant, it is old (Figure 1) (Freeman, 1979; Duesberg, 1991a). Figure 1 shows the incidence of long-established microbes in the U.S. population, i.e., *Candida* and *Pneumocystis* each at about 100% (Freeman, 1979; Pifer, 1984; Williford Pifer et al., 1988), and cytomegalovirus and herpes virus at about 50% and 40%, respectively (Evans, 1989c). In addition, it shows the typical exponential rise and subsequent fall of a hypothetical epidemic by a new influenza virus strain (Freeman, 1979).

Ever since antibodies against HIV were first detected by the "AIDS test" in 1985, the number of antibody-positive Americans has been fixed at a constant population of 1 million, or 0.4% (Section 2.2; Table 1). The U.S. Army also reports that from 1985 to 1990 an unchanging 0.03% of male and female applicants have been HIV-positive (Burke et al., 1990). This is the predicted distribution of a long-established virus (Figure 1). Since there are over 250 million uninfected Americans, and since there is no antiviral vaccine or drug to stop the spread of HIV, the nonspread of HIV in the U.S. in the past seven years is an infallible indication that the American "HIV epidemic" is old. The Central Africa HIV epidemic has also remained fixed at about 10% of the population since 1985 (Section 2.2). Likewise, HIV has remained fixed at 500,000 Europeans since 1988 (World Health Organization, 1992a). The nonspread of HIV confirms exactly the conclusion reached below that HIV behaves in a population as a quasi-genetic marker (Section 3.5.2). Hence, the assumption that HIV is new in the U.S. or in Africa is erroneous.

Indeed, HIV existed in the U.S. long before its fictitious origin in Africa (Gallo, 1987; Gallo and Montagnier, 1988; Anderson and May, 1992) and its fictitious entry into this country in the 1970s (Shilts, 1985). For example, in the U.S. in 1968 an HIV-positive, male homosexual prostitute died from Kaposi's sarcoma and immunodeficiency (Garry et al.,

1988), and 45 out of 1,129 American intravenous drug users were found to be HIV-positive in 1971 and 1972 (Moore et al., 1986).

The putative novelty of HIV is an anthropocentric interpretation of new technology that made it possible to discover HIV and many other latent retroviruses like HTLV-I (Duesberg and Schwartz, 1992). Indeed, the technology to detect a latent virus like HIV became available only around the time AIDS appeared. Given a new virus-scope, the assertion that HIV is new is just like claiming the appearance of "new" stars with a new telescope. Thus, the claims that "...first HIV is introduced, then AIDS appears" (Blattner et al., 1988) and that "HIV... preceded it [AIDS]" (Weiss and Jaffe, 1990) are ironically more true than the proponents of the virus hypothesis had anticipated. HIV preceded AIDS by many, perhaps millions, of years.

3.5.2. HIV—Assumed to Be Sexually Transmitted—Depends on Perinatal Transmission for Survival

AIDS is said to be a sexually transmitted disease, because HIV is thought to be a sexually transmitted virus (Section 2.2). However, HIV is not by nature a sexually transmitted virus. Sexual transmission of HIV is extremely inefficient. Based on studies measuring heterosexual and homosexual transmission, transmission depends on an average of 1000 heterosexual contacts and 100 to 500 homosexual contacts with antibody-positive people (Rosenberg and Weiner, 1988; Lawrence et al., 1990; Blattner, 1991; Hearst and Hulley, 1988; Peterman et al., 1988). According to Rosenberg and Weiner, "HIV infection in non–drug-using prostitutes tends to be low or absent, implying that sexual activity alone does not place them at high risk" (Rosenberg and Weiner, 1988). Moreover, unwanted pregnancies and venereal diseases, but not HIV infections, have increased significantly in the U.S. since HIV has been known (Institute of Medicine, 1988; Aral and Holmes, 1991). This argues directly against sexual transmission of HIV.

Sexual transmission is so inefficient because there is no free, non-neutralized HIV anywhere in antibody-positive persons, particularly not in semen (Section 3.3). In a group of 25 antibody-positive men, only one single provirus of HIV could be found in over 1 million cells of semen in one of the men and no HIV at all was found in the semen of the other 24 (Van Voorhis et al., 1991). Likewise, HIV could be isolated or reactivated only from ejaculates of 9 out of 95 antibody-positive men by cocultivation with 2 million phytohemagglutinin-activated leukocytes (Anderson et al., 1992). No virus or microbe could survive if it depended on a transmission strategy that is as inefficient as 1 in 1000 contacts.

Indeed, HIV depends on perinatal, instead of sexual, transmission for survival—just like other animal and human retroviruses. Therefore, the efficiency of perinatal transmission must be high. This appears to be the case. Based on HIV-tracking via the "AIDS test," perinatal transmission from the mother is estimated to be 13–50% efficient (Blattner et al., 1988; Blattner, 1991; Duesberg, 1991a; Institute of Medicine, 1988; European Collaborative Study, 1991). This number does not include paternal HIV transmission to the baby via semen, for which there are currently no data. The real efficiency of perinatal transmission must be higher than the antibody tests suggest, because in a fraction of recipients HIV becomes immunogenic only when its hosts are of an advanced age (Quinn et al., 1986; St. Louis et al., 1991). During the antibody-negative phase, latent HIV can be detected by the polymerase chain reaction (Rogers et al., 1989; European Collaborative Study, 1991). This is also true for other perinatally transmitted human (Blattner, 1990; Duesberg, 1991a) and animal retroviruses (Rowe, 1973; Duesberg, 1987).

HIV survival via perinatal transmission leads to two predictions: (1) HIV cannot be inherently pathogenic—just like all other perinatally transmitted viruses and microbes (Freeman, 1979; Mims and White, 1984). No microbe-host system could survive if the microbe were perinatally transmitted and at once fatal. (2) HIV must function as a quasi-genetic marker, because it is quasi-nontransmissible by sex, or other natural horizontal modes of transmission, just like known murine retrovirus prototypes (Rowe, 1973; Duesberg, 1987).

Both predictions are confirmed:

(1) Overwhelming statistical evidence from the U.S. and Africa documents that the risk for AIDS-defining diseases for HIV-positive babies, in the absence of other risk factors (Sectors 3.4.4 and 4), is the same as that of HIV-free controls:

(a) "AIDS tests" from applicants to the U.S. Army and the U.S. Job Corps indicate that between 0.03% (Burke et al., 1990) and 0.3% (St. Louis et al., 1991) of the 17- to 19-year-old applicants are HIV-infected but healthy. Since there are about 90 million Americans under the age of 20, there must be between 27,000 and 270,000 (0.03% and 0.3% of 90 million, respectively) HIV carriers. In Central Africa there are even more, since 1% to 2% of healthy children are HIV-positive (Quinn et al., 1986).

Most, if not all, of these adolescents must have acquired HIV from perinatal infection for the following reasons: sexual transmission of HIV depends on an average of 1000 sexual contacts and only 1 in 250 Americans carries HIV (Table 1). Thus, all positive teenagers would have had to achieve an absurd 1000 contacts with a positive partner, or an even more absurd 250,000 sexual contacts with random Americans to acquire HIV by sexual transmission. It follows that probably all of the healthy adolescent HIV carriers were perinatally infected, as, for example, was the 22-year-old Kimberly Bergalis (Section 3.5.16).

The AIDS risk of perinatally infected babies of the general population can be estimated as follows. Between 27,000 and 270,000 Americans under the age of 20 carry HIV. But only about 4260 AIDS cases have been recorded in this age group in the past 10 years (Centers for Disease Control, 1992b). Therefore, between 85% and 98% of HIV-infected youths do not develop AIDS up to 20 years after perinatal infection (Section 2.1). Since the above number includes the AIDS babies from drug-addicted mothers (Sections 3.4.2 and 4), the AIDS risk of HIV-infected babies from mothers who don't use drugs probably reflects normal infant mortality.

(b) A controlled study from Africa compared 218 newborns from HIV-positive mothers to 218 from HIV-negative mothers, and the "rates of prematurity, low birth weight, congenital malformations and neonatal mortality were comparable in the two groups" (Lepage et al., 1991). The mothers were matched for age and parity, and the "frequency of signs and symptoms was not statistically different in the two groups."

(2) The incidence of HIV in American teenagers of different ethnic backgrounds is predictable on genetic grounds. It is about 10-fold higher in blacks than in whites, i.e., 0.3% compared to 0.03% (U.S. Department of Health and Human Services, 1990; Burke et al., 1990; Blattner, 1991; Palca, 1991b; St. Louis et al., 1991; Vermund, 1991). HIV was even 50-fold more common in black mothers in inner-city hospitals in New York (36%) than in whites (0.7%) (Landesmann et al., 1987). This reflects the 25- to 50-fold higher incidence of HIV in the blacks' African ancestors (10%), compared to the whites' European ancestors (0.2 to 0.4%) (Section 2.2, Table 1). Likewise, the different ethnic groups of the Caribbean reflect the distinct HTLV-I incidences of their ancestors in Africa, Europe, and Japan, despite generations of coexistence on the Caribbean islands (Blattner, 1990). The unchanging incidence of HIV in the American population (Figure 1) also confirms the view that HIV is a quasi-genetic marker. Since there is virtually no horizontal transmission of retroviruses, murine retroviruses have functioned as classical genetic markers of mice that could be distinguished from cellular genes only by fastidious genetic crosses (Rowe, 1973).

Thus, the assumption that AIDS is sexually transmitted by HIV is not consistent with the natural perinatal mode of HIV transmission. If natural transmission of HIV caused a disease, AIDS would be a pediatric disease. Instead, HIV is merely a marker of either an average of 1000 sexual contacts and thus of many other possible AIDS risks associated with very high sexual activity or of long-term intravenous drug use (Sections 3.4.3 and 5).

3.5.3. AIDS Assumed to Be Proportional to HIV Infection

The incidence of AIDS is assumed to be proportional to the incidence of HIV via a constant factor. For example, a 10-fold higher incidence of AIDS in American and European males compared to females is assumed to reflect a 10-fold higher incidence of HIV in men (Blattner et al., 1988; Blattner, 1991; Goudsmit, 1992).

However, there is no evidence that the incidence of HIV is 10 times higher in males than in females of the general American and European population, although this is the case for AIDS (Table 1). Indeed, the most recent claim for a 90% bias of HIV for males of the general population (Blattner, 1991) is supported only by a reference to an editorial (Palca, 1991b), which itself provides nothing more than an unreferenced cartoon showing global patterns of HIV infection. According to a CDC epidemiologist, estimates of how HIV is distributed between the sexes of the general population are "approximations" based on the distribution of AIDS (Tim Dondero, personal communication; see also Anderson and May, 1992)—a tautology.

Proportionality between HIV and AIDS via a constant is also incompatible with the following statistics. The U.S. Army (Burke et al., 1990) and the U.S. Job Corps (St. Louis et al., 1991) report, based on millions of tests, that HIV has been equally distributed between the sexes among 17- to 21-year-olds of the general population over the past five years for which data were available (Sections 3.5.1 and 3.5.2). Since testing 17- to 19-year-olds annually for 5 years is equivalent to testing 17- to 24-year-olds, the U.S Army data predict that among 17- to 24-year-olds, AIDS risks should be distributed equally between the sexes. However, the CDC documents that 85% of the AIDS cases among 17- to 24-year-olds were males (Centers for Disease Control, 1992b).

In response to this, some proponents of the virus-AIDS hypothesis have speculated that teenage homosexuals exclude themselves from the Army. However, Randy Shilts, a homosexual writer, reports that just the opposite is true (Shilts, 1991). Moreover, most teenagers are not as yet aware of a definite homosexual persuasion and are not likely to understand the implications nor to fear the consequences of a positive "AIDS test."

The over 100-fold discrepancies between the AIDS risks of different HIV-infected risk groups also disprove the claim that the incidence of AIDS is proportional via a constant to the incidence of HIV (Table 2).

Appendix B ▪ 547

The proportionality between HIV and AIDS holds only if the analysis is restricted to groups with the same AIDS risks. In groups with the same percentage of HIV but with different AIDS risks, AIDS segregates specifically with nonviral AIDS risks, i.e., illicit recreational drugs, the antiviral drug AZT (Section 4), and frequent transfusions (Section 3.4.4).

3.5.4. *AIDS Assumed to Be Homosexually Transmitted in the U.S. and Europe*

In view of a sexually transmitted AIDS virus, it is paradoxical that AIDS is 90% male in America and 86% male in Europe (Sections 3.1 and 3.2). Therefore, it is assumed that "the virus first got its footing in the U.S." in male homosexuals (Booth, 1988) and has remained with homosexuals because it is transmitted preferentially by anal intercourse and because homosexuals have no sex with heterosexuals (Shilts, 1985; Centers for Disease Control, 1986; Blattner et al., 1988; Institute of Medicine, 1988; Blattner, 1991; Bardach, 1992; Project Inform, 1992).

However, this assumption is inconsistent with the fact that about 10% of all males and females prefer anal intercourse (Bolling and Voeller, 1987; Turner et al., 1989) and that American and European heterosexuals have sufficient access to HIV. The females would be infected by HIV-positive, heterosexual intravenous drug users, hemophiliacs, and bisexual males. Thus, if HIV were transmitted by anal intercourse, about the same percentage of women as men should develop AIDS, particularly since the efficiencies of transmission of anal and vaginal intercourse are approximately the same, i.e., between 1 to 100 and 1 to 500 for anal and 1 to 1000 for vaginal intercourse (Blattner, 1991) (see also Section 3.5.2). Yet, despite widespread alarm, this has not occurred in the past 10 years in the U.S. (Table 1), although the first women with AIDS had been diagnosed as early as in 1981 (Centers for Disease Control, 1986; Guinan and Hardy, 1987). The risk of women for both HIV infection and AIDS is the same for those who practice anal intercourse as it is for those who practice other types of intercourse (Guinan and Hardy, 1987).

The preferred anal-transmission hypothesis is also incompatible with the sexually equal distribution of HIV and AIDS in Africa. Since it is postulated that HIV appeared in America and Africa at about the same time 10–20 years ago (Institute of Medicine, 1986; Blattner et al., 1988; Gallo and Montagnier, 1988), HIV should have reached the same equilibria between the sexes in all countries.

Instead, it is shown below that the male bias for AIDS in America and Europe reflects male-specific behavior, including the facts that over

75% of all intravenous drug users are males and that long-term consumption of sexual stimulants, like amyl nitrite and ethyl chloride inhalants, is almost entirely restricted to male homosexuals (Section 4). HIV is just a marker of the many sexual stimulants used to achieve 500–1000 sexual contacts (Section 4). The difference between the AIDS risks of men in America and Europe, namely drugs, and those of Africans, namely country-specific but not sex-specific, risk factors (Section 3.4.4.8) resolves the paradox between the different sexual distributions of AIDS in these countries.

3.5.5. AIDS Assumed to Be Heterosexually Transmitted by African "Lifestyle"

AIDS in Africa is assumed to affect both genders equally, because HIV is distributed equally between the sexes by "prostitution" (Institute of Medicine, 1988), lack of "circumcision" (Klein, 1988; Marx, 1989; Blattner, 1991), African "lifestyle" (Quinn et al., 1987; Blattner et al., 1988; Goodgame, 1990), and "voodoo rituals" (Gallo, 1991). These assumptions are compatible with the sexually equal distributions of HIV and AIDS in Africa.

However, AIDS in Africa is hard to reconcile with the known efficiency of sexual transmission of HIV. Since it takes 1000 HIV-positive sexual contacts to transmit HIV and about 10% of all Central Africans, or 6 million, are HIV-positive (Section 2.2), 6 million Africans would have had to achieve on average at least 10,000 sexual contacts with random Africans to pick up HIV. Since this is highly improbable, it is also highly improbable that sexual transmission of HIV is the cause of AIDS in Africa. The true reason for the sexually equal distribution of HIV in Africa is perinatal transmission of HIV (Section 3.5.2). Nonsexual, country-specific risk factors are the reason for the "sexually" equal distribution of AIDS in Africa (Section 3.4.4.8).

3.5.6. HIV Claimed to Be Abundant in AIDS Cases

HIV is said to be abundant or viremic in AIDS patients (Baltimore and Feinberg, 1989; Coombs et al., 1989; Ho et al., 1989a; Semple et al., 1991) and thus compatible with orthodox viruses which cause disease only at high titers (Duesberg and Schwartz, 1992). In other words, HIV is assumed to meet Koch's first postulate (Section 3.3). The assumption is based on two papers which reported HIV titers of 10^2 to 10^3 infectious units per mL of blood in 75% of AIDS patients and in 25% to 50% of asymptomatic HIV carriers (Coombs et al., 1989; Ho et al.,

1989a). The authors and an accompanying editorial, *HIV Revealed, Toward a Natural History of the Infection* (Baltimore and Feinberg, 1989), concluded that these findings established HIV viremia as an orthodox criterion of viral pathogenicity. Viremia of similar titers was recently also implied in some AIDS patients and asymptomatic carriers based on an indirect assay that amplifies HIV RNA *in vitro* (Semple et al., 1991).

However, several arguments cast doubt on the claim that HIV viremia is relevant for AIDS:

(a) Since viremia was observed in 25% to 50% of asymptomatic HIV carriers (Coombs et al., 1989; Ho et al., 1989a; Semple et al., 1991), it cannot be sufficient for AIDS.

(b) Since no viremia was observed in 25% of the AIDS cases studied by two groups (Coombs et al., 1989; Ho et al., 1989a), it is not necessary for AIDS.

(c) Viremia initiated from a previously suppressed virus and observed years after infection is a classical consequence, rather than the cause of immunodeficiency. Indeed, many normally latent parasites become activated and may cause chronic "opportunistic infections" in immunodeficient persons, as for example *Candida*, *Pneumocystis*, herpes virus, cytomegalovirus, hepatitis virus, tuberculosis bacillus, toxoplasma (Sections 2.3 and 3.4.3)—and sometimes even HIV. It is consistent with this view that HIV viremia is observed more often in AIDS patients than in asymptomatic carriers (Duesberg, 1990c).

(d) The HIVs that make up the "viremias" are apparently not infectious *in vivo*, because only a negligible fraction of leukocytes, on average only 1 in 1500 to 8000, of AIDS patients are infected (Section 3.3). The probable reason is that the "viremias" consist of viruses that are neutralized by the antiviral antibodies of "seropositive" AIDS patients (Duesberg, 1992d). Since viruses, as obligatory cellular parasites, can only be pathogenic by infecting cells, these noninfectious viremias cannot be relevant to the cause of AIDS. If assayed *in vitro*, in the absence of free antiviral antibodies, antibodies may dissociate from neutralized viruses and thus render the virus infectious for cells in culture. This explains the discrepancy between the noninfectious "viremias" *in vitro* and the relatively high infectivity recorded in vitro (Coombs et al., 1989; Ho et al., 1989a).

Thus, HIV viremia is a rare, predictable consequence of immunodeficiency rather than its cause.

3.5.7. HIV to Depend on Cofactors for AIDS

Conceding that HIV is not sufficient to cause AIDS, it is assumed to depend on cofactors. Montagnier (Goldsmith, 1990; Lemaitre et al., 1990; Balter, 1991) and Lo et al. (1991) have proposed mycoplasmas that were discovered in their laboratories; Gallo has proposed two viruses, herpes virus-6 and HTLV-I, which were both discovered in his laboratory (Cotton, 1990; Gallo, 1990, 1991; Lusso et al., 1991). Others have proposed cytomegalovirus, Epstein-Barr virus (Quinn et al., 1987; Evans, 1989a; Root-Bernstein, 1990c), "age" (Evans, 1989a; Goedert et al., 1989; Weiss and Jaffe, 1990; Biggar and the International Registry of Seroconverters, 1990), unidentified "coagents" (Weyer and Eggers, 1990; Eggers and Weyer, 1991), "clinical illness promotion factors" (Evans, 1989b, 1992), and even "pre-existing immune abnormalities" (Ludlam et al., 1985; Marion et al., 1989; Ludlam, 1992) as cofactors of HIV.

However, cofactor hypotheses only replace HIV-specific AIDS problems with the following HIV-plus-cofactor–specific AIDS problems:

(a) Since HIV is extremely rare and dormant in most antibody-positive AIDS patients (Sections 2.2 and 3.3), it is hard to imagine how its various AIDS allies could benefit from their dormant "cofactor" HIV.

(b) Since HTLV-I is just as dormant and unable to kill cells as HIV (Duesberg, 1987; Blattner, 1990; Duesberg and Schwartz, 1992), it is even harder to imagine how one dormant virus could help another dormant virus to generate the biochemical activity that would be necessary to cause a fatal disease.

(c) Since mycoplasma (Freeman, 1979; Cotton, 1990; Goldsmith, 1990; Balter, 1991), herpes virus-6 (Cotton, 1990; Lusso et al., 1991), cytomegalovirus, and Epstein-Barr virus (Mims and White, 1984; Evans, 1989c) are each very common, if not ubiquitous, parasites (Freeman, 1979; Froesner, 1991), AIDS should develop in most people as soon as they are infected by HIV. Likewise, "aged" people should develop AIDS as soon as they are infected by HIV. Yet not more than 3–4% of HIV-antibody–positive Americans or Europeans and 0.3% of antibody-positive Africans develop AIDS each year (Tables 1 and 2).

Moreover, if infectious cofactors helped HIV to cause AIDS, the AIDS risk of Africans would be expected to be higher than that of Americans. This is because the incidence of hypothetical, microbial cofactors in Africans without AIDS was found to be the same as in those with AIDS, while the incidence of microbial cofactors in Americans without AIDS risks was significantly lower than in those with AIDS (Section 3.4.3) (Quinn et al., 1987). Even the cofactor HIV was present in 6% of African

AIDS-free controls (Quinn et al., 1987). Yet the annual AIDS risk of HIV-infected Africans is 10 times lower than that of Americans (Table 1).

(d) Contrary to the claims that "age" is an AIDS cofactor of HIV, the virus-AIDS hypothesis postulates that the latent period for HIV is longer in adults (10 years) than in children (2 years) (Section 2.2). However, the proposal that "age" is a cofactor for HIV becomes more compelling the more the hypothetical "latent period" of HIV grows. Clearly, if a 70-year-old will be infected by a virus with a "latent period" of 10 years, "age" will be a predictable cofactor (see, for example, hemophiliacs, Section 3.4.4.5, and Paul Gann, Section 3.5.16).

(e) The claims that HIV depends on "clinical illness promotion factors" (Evans, 1992) or on a "pre-existing immune abnormality" (Marion et al., 1989; Ludlam, 1992) for AIDS are euphemisms for saying that HIV cannot cause AIDS until something else does (Duesberg, 1989b). The additional hypothesis that a "pre-existing immune abnormality" (Ludlam, 1992) or a "prior immune dysfunction" (Marion et al., 1989) makes a subject more susceptible to HIV is erroneous, because a pre-existing immune deficiency affects only the progression of an infection, but not the risk of infection.

In view of this, I share Gallo's concerns about cofactors of HIV, which he expresses with a quotation from Lewis Thomas: "Multifactorial is multi-ignorance. Most factors go away when we learn the real cause of a disease" (Gallo, 1991). The "cofactor" HIV may be no exception. Until any one of these hypothetical cofactors is actually shown to depend on HIV to cause AIDS, HIV must be considered just one of many innocent bystanders found in AIDS patients (Section 3.4.3).

3.5.8. All AIDS Diseases to Result from Immunodeficiency

All AIDS diseases are said to reflect a primary immunodeficiency (Coffin et al., 1986; Institute of Medicine, 1986; Blattner et al., 1988).

However, immunodeficiency is not a common denominator of all AIDS diseases. About 38% of all AIDS diseases, i.e., dementia, wasting disease, Kaposi's sarcoma, and lymphoma (Table 1), are neither caused by, nor necessarily associated with, immunodeficiency. Cancer is not a consequence of immunodeficiency (Stutman, 1975; Duesberg, 1989c). Indeed, Kaposi's sarcoma frequently has been diagnosed in male homosexuals in the absence of immunodeficiency. For example, the immune systems of 20 out of 37 HIV-positive homosexuals with Kaposi's sarcoma were normal when their disease was first diagnosed (Spornraft et al., 1988). Another study also describes 19 male homosexual Kaposi's sarcoma patients with normal immune systems

(Murray et al., 1988). Likewise, Kaposi's sarcomas have been diagnosed in HIV-free male homosexuals with normal immune systems (Afrasiabi et al., 1986; Archer et al., 1989; Friedman-Kien et al., 1990; Marquart et al., 1991).

Dementia and wasting disease also are not consequences of immunodeficiency (Duesberg, 1989c, 1991a). Thus, the assumption that all AIDS diseases are caused by immunodeficiency is erroneous.

3.5.9. HIV to Induce AIDS via Autoimmunity and Apoptosis

In view of the extremely low number of HIV-infected cells in AIDS patients (Section 3.3), HIV has recently been proposed to cause AIDS by inducing autoimmunity (Hoffmann, 1990; Maddox, 1991a; Mathe, 1992) or apoptosis (Laurent-Crawford et al., 1991; Goudsmit, 1992). According to these new ideas HIV is assumed either to confuse the immune system into attacking itself or to persuade the immune cells to commit suicide, termed apoptosis. The autoimmune hypothesis postulates homology between HIV and human cells, and currently relies only on mouse and monkey models (Hoffmann, 1990; Maddox, 1991a) and on precedents for autoimmunity induced in humans as a consequence of graft rejection and blood transfusions (Root-Bernstein, 1990a, b; Mathe, 1992). One autoimmunologist claims that "each of Duesberg's paradoxes might be understood in the context of the model without sacrificing the idea that HIV is usually involved in pathogenesis" (Hoffmann, 1990). This strategy of crediting me rather than the virus-AIDS hypothesis for its paradoxes shifts the discussion from a problem with science to a problem with a scientist (Booth, 1988; Weiss and Jaffe, 1990).

However, both the autoimmune and the apoptosis hypotheses are incompatible with human AIDS on several grounds:

(a) Autoimmunity or apoptosis cannot account for all those AIDS diseases that are not caused by immunodeficiency, e.g., Kaposi's sarcoma, dementia, wasting disease, and lymphoma (Section 3.5.8).

(b) Autoimmunity or apoptosis fails to explain risk-group–specific AIDS diseases (Section 2.1.3, Tables 1 and 2).

(c) Autoimmunity and apoptosis fail to explain the long average intervals, "latent periods," from conventional immunity against HIV, detected by the "AIDS test," to hypothetical autoimmunity 10 years later (Section 3.2).

(d) Autoimmunity and apoptosis fail to explain the over 100-fold discrepancies between the annual AIDS risks of different HIV-infected groups (Table 2).

Appendix B ■ 553

(e) HIV-induced autoimmunity or apoptosis fails to explain the consistent 90% bias of American/European AIDS for males (Section 2.1, Table 1).

(f) In view of the autoimmunity or apoptosis hypothesis, it is paradoxical that 80% of antibody-positive Americans (1 million minus the 206,000 who have developed AIDS) and 98% of antibody-positive Africans (6 million minus the 129,000 who have developed AIDS) have not developed AIDS since 1984 (Table 1). Obviously, these figures are not even corrected for the normal and drug-induced incidence of AIDS-defining diseases in those groups (Section 3.4.4, Table 2).

(g) There is no sequence homology between HIV and human DNA detectable by hybridization to predict autoimmunity (Shaw et al., 1984). Therefore, autoimmunologists argue that antibodies against those antibodies, which are directed at the viral proteins that bind to cellular receptors, would also react with cellular receptors and thus cause AIDS (Hoffmann, 1990). However, if this were true, all viruses should cause AIDS.

Thus, the HIV-autoimmunity and apoptosis hypotheses of AIDS are (a) not compatible with essential parameters of human AIDS and (b) arbitrary, because they are not based on an autoimmunogenic or apoptogenic property of HIV that is distinct from all other viruses.

3.5.10. *HIV Assumed to Kill T-cells*

Based on an early observation by Gallo et al., HIV is assumed to cause immunodeficiency by specifically killing T-cells (Gallo et al., 1984; Weiss and Jaffe, 1990). Gallo's observation was restricted to primary T-cells (Gallo et al., 1984) but not to established T-cell lines (Rubinstein, 1990). However, according to Montagnier, the discoverer of HIV, "In a search for a direct cytopathic effect of the virus on (primary) T-lymphocytes, no gross changes could be seen in virus-producing cultures, with regard to cell lysis or impairment of cell growth" (Montagnier et al., 1984). Others have confirmed that HIV does not kill infected, primary T-cells *in vitro* (Hoxie et al., 1985; Anand et al., 1987; Langhoff et al., 1989; Duesberg, 1989c). Moreover, HIV-infected primary T-cells are considered the natural "reservoir" of HIV *in vivo* (Schnittman et al., 1989).

Thus, Gallo's controversial observation probably reflects the notorious difficulties experienced by his laboratory in maintaining primary blood cells alive in culture instead of a genuine cytocidal function of HIV (Crewdson, 1989; Culliton, 1990; Rubinstein, 1990; Hamilton, 1991). Gallo showed in a later study from his laboratory that about 50% of uninfected T-cells died within 12 days in culture (Gallo, 1990).

Indeed, the assumption that HIV is cytocidal is incompatible with generic properties of retroviruses and with specific properties of HIV:

(1) The hallmark of retrovirus replication is to convert the viral RNA into DNA and to deliberately integrate this DNA as a parasitic gene into the cellular DNA (Weiss et al., 1985). This process of integration depends on mitosis to succeed, rather than on cell death (Rubin and Temin, 1958; Duesberg, 1989c). The resulting genetic parasite can then be either active or passive, just like other cellular genes (Duesberg, 1987). Transcription of viral RNA from chromosomally integrated proviral DNA also works only if the cell survives infection, because dying cells are not transcriptionally active. Thus, this strategy of replication depends entirely on the survival of the infected cell.

Noncytocidal replication is the reason that retroviruses were all considered potential carcinogens before AIDS (Weiss et al., 1985; Duesberg, 1987). For example, Gallo's first candidate for an AIDS virus is called Human T-cell Leukemia Virus-I (Gallo et al., 1983), and Gallo's second candidate for an AIDS virus was originally described at a press conference in April 1984 by Gallo and the Secretary of Health and Human Services as "a variant of a known human cancer virus called HTLV III" (Crewdson, 1989; Rubinstein, 1990). It used to be called Human T-cell Leukemia Virus-III by Gallo (Gallo et al., 1984; Shaw et al., 1984) before it was renamed HIV in 1986 (Coffin et al., 1986).

(2) Limited cytotoxicity of HIV has been observed soon after infection of cells *in vitro* (Duesberg, 1989c; Bergeron and Sodroski, 1992). Therefore, it has been proposed that multiple copies of unintegrated proviral DNA, generated by multiple infections before all cellular receptors are blocked by newly replicated viruses, could kill T-cells (Bergeron and Sodroski, 1992). However, cells infected by every retrovirus, including HIV (Bergeron and Sodroski, 1992), survive multiple unintegrated proviral DNAs during the early phase of the infection (Weiss et al., 1985). Rare cell death during this phase of infection is a consequence of cell fusion, which is mediated by viruses on the surface of infected cells binding to receptors of uninfected cells. In some conditions retrovirus-mediated fusion occurs so reliably that it has been used to quantitate retroviruses in tissue culture. However, virus-mediated fusion is blocked by antiviral antibodies and thus not relevant to the loss of T-cells in persons with antibodies against HIV (Duesberg, 1989c).

Alternatively, it has been proposed that HIV proteins are directly toxic because of structural similarities with scorpion and snake poisons (Gallo, 1991; Garry et al., 1991; Garry and Koch, 1992). However, no such toxicity is observed in millions of asymptomatic HIV carriers, and there is no reason that it should occur, if it did, only after latent periods of 10 years.

(3) The propagation of HIV in indefinitely growing human T-cells for the "AIDS test" was patented by Gallo et al. in 1984 (Rubinstein, 1990)

and was recently confirmed by Montagnier (Lemaitre et al., 1990). It is totally incompatible with Gallo's claim that HIV kills T-cells. Such HIV-producing T-cells have been growing in many laboratories and companies since 1984 producing viruses at titers of up to 10^6 virus particles per mL, which is many orders of magnitude more than is ever observed in humans with or without AIDS (Duesberg, 1989c, 1991a).

In view of this, Gallo postulates that T-cell lines in culture have all acquired resistance to HIV killing (Gallo, 1991). However, there is no precedent for this *ad hoc* hypothesis, as no other cytocidal virus has ever been observed that is cytocidal *in vivo* and in primary cells *in vitro*, but is noncytocidal in cell lines in culture. It is also implausible that a potentially life-saving cellular mutation, such as resistance to the hypothetical "AIDS virus," would be restricted just to cells in culture, particularly if these mutations occur so readily that they are found in all T-cell lines. There is not even one T-cell line that is consistently killed by HIV.

(4) HIV, like all other retroviruses, does not specifically infect T-cells. It also infects monocytes, epithelial cells, B-cells, glial cells, and macrophages, etc., and none of these are killed by HIV (Levy, 1988; Duesberg, 1991a). Most other retroviruses also infect T-cells, which is why so many of them are suspected "T-cell leukemia" viruses (Weiss et al., 1985; Duesberg, 1987; Blattner, 1990).

Thus, the assumption that HIV causes AIDS by killing T-cells is not tenable.

3.5.11. *Antibodies Assumed Not to Neutralize HIV*

Antibodies against HIV, detected by a positive "AIDS test," are claimed not to protect against AIDS because they do not neutralize HIV (Institute of Medicine, 1988; Evans, 1989a; Weiss and Jaffe, 1990; Gallo, 1991). "It is a test for anti-HIV antibodies and not, as Duesberg states, 'neutralizing antibodies'" (Baltimore and Feinberg, 1990).

However, antiviral immunity completely neutralizes HIV and restricts it to undetectable levels in healthy HIV carriers as well as in AIDS patients (Section 3.3.1) (Duesberg, 1989b, c). Indeed, two recent studies have just confirmed that HIV activity is "rapidly and effectively limited" by antiviral immunity (Clark et al., 1991; Daar et al., 1991) to less than 1 in 1000 T-cells (Section 3.3). By contrast, HIV replicates in the absence of antiviral immunity in human T-cells in culture to titers of 10^6 virus particles per mL (Section 3.5.10). Thus, the assumption that HIV causes AIDS because of inadequate antiviral immunity is unconfirmed. Baltimore's, Feinberg's, and Evans's paradox "that antibody is not protective" (Evans, 1989a) is their failure to recognize the nonrole of HIV in AIDS (Section 3.3.2).

3.5.12. HIV Claimed to Cause AIDS in 50% Within 10 Years

All HIV-infected persons are said to die from AIDS after a medium latent period of 10 years (Anderson and May, 1988; Institute of Medicine, 1987; Moss et al., 1988; Lemp et al., 1990; Blattner, 1991; Duesberg, 1991a).

However, according to statistics from the CDC, only about 30,000–40,000, or 3–4%, of a reservoir of 1 million HIV-infected Americans develop AIDS annually (Table 1). Likewise, 3% of infected Europeans develop AIDS per year (Table 1). Accordingly, 50% of HIV-infected Americans and Europeans would have to wait 12–16 years, and 100% would have to wait 24–33 years to develop AIDS. During this time, many would die from other causes. Since only 0.3% of infected Africans develop AIDS diseases annually (Tables 1 and 2), 50% of Africans would have to wait about 150 years and 100% would have to wait 300 years to develop AIDS.

Thus, it is presumptuous to claim that HIV causes AIDS in 50% of infected persons after median latent periods of 10 years, particularly since the virus has been known for only nine years.

3.5.13. HIV Said to Derive Pathogenicity from Constant Mutation

During its long latent periods, HIV is claimed to acquire pathogenicity by mutation, for example by generating variants that escape immunity (Hahn et al., 1986; Levy, 1988; Eigen, 1989; Gallo, 1990; Weiss and Jaffe, 1990; Anonymous, 1992; Anderson and May, 1992) or by generating defective variants (Eigen, 1989; Haas, 1989; Weiss, R. A., 1989).

However, a recent study just demonstrated that the replicative and functional properties of HIVs from AIDS patients are the same as those from asymptomatic carriers (Lu and Andrieu, 1992). Indeed, most essential structural and replicative proteins of a virus cannot be mutated without eliminating its viability. Functionally relevant mutations of any virus are also severely restricted by the necessity to remain compatible with the host (Duesberg, 1990b). Moreover, there is no precedent for an immune system that has been able to neutralize a virus completely and is then unable to catch up with an occasional subsequent mutation. If viruses in general could evade the immune system by mutation, the immune system would be a useless burden to the host.

Likewise, the proposals that defective HIVs could generate pathogenicity are untenable. Defective viruses are viable only in the presence of nondefective helper viruses and thus unlikely to survive in natural transmission from host to host at low multiplicity of infection, particularly with helper viruses that never achieve high titers like HIV (Duesberg, 1989a).

There are, however, examples of new antigenic variants of retroviruses (Beemon et al., 1974) or influenza viruses (Duesberg, 1968) that have arisen upon rare double infection by two antigenically distinct virus strains via genetic recombination. Yet antigenically new variants of HIV have never been observed in American and European AIDS patients, as all HIV strains diagnosed to date cross-react with the very same standard HIV-1 strain that is patented in America and Europe for the "AIDS test" (Connor, 1991, 1992; Palca, 1991a; Weiss, 1991).

Moreover, if recombination or spontaneous mutation could generate pathogenic HIV mutants from nonpathogenic strains, one would expect all those who are infected by HIV from AIDS patients to develop AIDS within weeks after infection. Such HIV mutants should be pathogenic just as soon as conventional, nonpathogenic HIV strains are immunogenic. But this is not observed.

Thus, the assumption that HIV acquires pathogenicity by mutation during the course of the infection is not tenable.

3.5.14. *HIV Assumed to Cause AIDS with Genes Unique Among Retroviruses*

AIDS researchers assert that HIV causes AIDS with unique genetic information that all other animal and human retroviruses lack and that these unique genes would regulate HIV down during the "latent period" and up during AIDS (Gallo and Montagnier, 1988; Haseltine and Wong-Staal, 1988; Institute of Medicine, 1988; Eigen, 1989; Temin, 1990; Fauci, 1991; Gallo, 1991). Further, it is claimed that HIV-infected cells export factors encoded by these genes that promote neoplastic growth of uninfected cells to cause, for example, Kaposi's sarcoma (Salahuddin et al., 1988; Ensoli et al., 1990; Gallo, 1990); at the same time such genes are said to export "scorpion-poison"-related toxins that kill uninfected neurons to cause dementia (Gallo, 1991; Garry et al., 1991; Garry and Koch, 1992). By contrast, all other known bacterial, animal, and human viruses, including retroviruses, are only able to kill or alter those cells they infect, because viruses are manufactured inside cells and would not benefit from proteins released to uninfected cells.

However, the claims of unique retroviral HIV genes with unique control functions raises several unresolvable problems:

(1) Despite its presumed unique properties HIV has the same genetic complexity, i.e., 9000 nucleotides, and the same genetic structure as all

other retroviruses (Beemon et al., 1974; Wang et al., 1976; Institute of Medicine, 1988). It shares with other retroviruses the three major genes *gag-pol-env*, which are linked in this order in all animal and human retroviruses (Wang et al., 1976). Although "novel" genes that overlap with the major retroviral genes have been discovered in HIV by computerized sequence analysis and by new protein detection technology (Varmus, 1988), such genes have also been found with the same technology in other retroviruses that do not cause AIDS, such as HTLV-I, other human retroviruses, bovine retroviruses, simian retroviruses, and sheep retroviruses (Varmus, 1988; Weiss, 1988; Duesberg, 1989c; Palca, 1990). Thus, there are no unique genetic material and no uncommon genetic structure in HIV RNA that could indicate where this unique AIDS-specific information of HIV is hiding.

(2) Since all retroviral genes share just one common promoter, it would be impossible to differentially activate one HIV gene while the others are latent. Thus, the idea that different viral genes would regulate latency and virulence, as with lambda phage, is not compatible with HIV (Haseltine and Wong-Staal, 1988; Eigen, 1989; Temin, 1990; Fauci, 1991). Since all HIV genes share the same promoter, latent HIV can be activated only by the host—just like all other latent retroviruses. In addition HIV cannot make specific AIDS factors, while its major genes are dormant. Since viral RNA synthesis *in vivo* is detectable in only 1 out of 10,000 to 100,000 leukocytes and then only in half of all AIDS patients (Section 3.3), HIV cannot make Kaposi's sarcomagenic and neurotoxic factors in amounts sufficient to cause fatal tumors and dementias. This is why such factors have not been detectable *in vivo* (Weiss and Jaffe, 1990; Gallo, 1991).

Thus, based on the structure, information, and function of its RNA, HIV is a profoundly conventional retrovirus. It does not contain unique genes that distinguish it from other retroviruses, nor can its genes be differentially regulated at the transcriptional level.

3.5.15. Simian Retroviruses to Prove That HIV Causes AIDS

Animal retroviruses may cause diseases in experimental animals that overlap with the wide spectrum of AIDS diseases. Such systems are now studied for analogies to gain experimental support for the virus-AIDS hypothesis (Blattner et al., 1988; Weiss and Jaffe, 1990; Goudsmit, 1992). For example, a retrovirus isolated from macaques (Fultz et al., 1990), termed Simian Immunodeficiency Virus (SIV), that is 40% related to HIV, is said to cause AIDS-like diseases in rhesus monkeys (Kestler et al., 1990; Temin, 1990). According to an editorial in *Science*, "if SIV infection is all that is needed to cause simian AIDS, that's one more indication that HIV is all that is needed to cause human AIDS" (Palca, 1990).

However, the presumed role of SIV in the diseases of infected monkeys is very different from that of HIV in human AIDS:

(a) According to one study, about half of the infected monkeys developed diseases within several months to one year after infection (Kestler et al., 1990). By contrast only 3–4% of HIV-infected Americans or Europeans and 0.3% of infected Africans develop AIDS annually (Table 1).

(b) In the same study, the absence of antiviral antibodies predicted the incidence of diseases in monkeys, while the opposite is claimed for humans infected with HIV. Another study has confirmed that the monkey's risk of disease is directly proportional to the titer of SIV (Fultz et al., 1990).

(c) The simian retroviruses barely reduce the T-cell levels of ill monkeys (Kestler et al., 1991), while HIV is claimed to deplete T-cells in humans.

(d) The spectrum of diseases observed in the SIV-infected monkeys is different from AIDS, including bacteremia and lacking, among others, Kaposi's sarcoma and dementia (Kestler et al., 1990; Fultz et al., 1990).

(e) In follow-up studies, SIV failed to cause disease in rhesus and mangabey monkeys despite extensive sequence variation of the virus which is thought to enhance pathogenicity (Fultz et al., 1990; Burns and Desrosiers, 1991; Villinger et al., 1991).

(f) Since SIV has never caused any disease in wild monkeys, although about 50% are naturally infected (Duesberg, 1987, 1989c; Blattner et al., 1988; Fultz et al., 1990; Burns and Desrosiers, 1991; Villinger et al., 1991), it is not an appropriate model for the hypothesis that HIV causes AIDS in naturally infected humans.

It would appear that SIV causes disease in monkeys like all viruses cause disease soon after infection and in the absence of effective immunity. This is not a model for the hypothesis that HIV causes AIDS 10 years after it is neutralized by antibodies. Indeed, in the vast literature on retroviruses there is not even one proven example of a latent retrovirus that, in the presence of antiviral immunity, has ever caused a disease in any animal, including chickens, mice, cattle, and monkeys (Weiss et al., 1985; Duesberg, 1987, 1989c).

Moreover, the observation that a retrovirus that is 60% unrelated to HIV causes disease in monkeys cannot prove that HIV causes AIDS in humans, even if all parameters of infection were completely analogous. It can only prove that under analogous conditions other retroviruses may also cause disease, which has been demonstrated with numerous avian and murine retroviruses long ago (Weiss et al., 1985).

3.5.16. *Anecdotal AIDS Cases from the General Population*

Rare AIDS cases occurring outside the major risk groups are claimed to prove that HIV alone is sufficient to cause AIDS in persons with no other AIDS risks (Blattner et al., 1988; Booth, 1988; Baltimore and Feinberg, 1989; Weiss and Jaffe, 1990). Four examples illustrate this point:

(1) Ryan White, an 18-year-old hemophiliac, was said to have died from AIDS in April 1990. However, information from the National Hemophilia Foundation revealed that White had died from unstoppable internal bleeding and had also been treated for an extended period with the cytotoxic DNA chain terminator AZT prior to his death (Duesberg and Ellison, 1990). It appears that hemophilia and AZT (Section 4) would each be sufficient causes of death, and certainly a combination of both would be more than adequate to explain the death of Ryan White. Thus, there is no convincing evidence that White died from HIV.

To prove that HIV played a role in White's death, it would be necessary to compare mortality of matched hemophiliacs with and without HIV. To prove that AZT contributed to his death, matched HIV-positive hemophiliacs with and without AZT must be compared. Without such evidence the HIV death of White is just a hypothesis. Yet White was generally described as an innocent victim of HIV (practicing no-risk behavior), which is why the U.S. Senate approved the Ryan White Comprehensive AIDS Resources Act for over $550 million in aid to hospitals for AIDS emergencies and treatment of children (Anonymous, 1990).

(2) In 1989 the California tax-reformer Paul Gann was reported to have died from AIDS at the age of 77 after receiving HIV from a blood transfusion. However, a close examination of Gann's case reveals that he had quintuple bypass heart surgery for blocked arteries in 1982, when he may have received the blood transfusion with HIV. In 1983 he needed further bypass surgery for blocked intestinal arteries. In 1989, at the age of 77, he was hospitalized again for a broken hip. While recovering from the hip fracture, Gann was immobilized for weeks and developed a pneumonia from which he died (Folkart, 1989). This is a rather typical death for a 77-year-old man in poor health.

To determine whether HIV played any role at all in his death, a controlled study would be necessary showing that the mortality of HIV-positive 77-year-old bypass patients with broken hips is higher than that of HIV-negative counterparts. No such study exists.

Appendix B ■ 561

(3) Kimberly Bergalis, a 22-year-old woman, developed candidiasis and a transient pneumonia 17 and 24 months, respectively, after the extraction of two molars (Centers for Disease Control, 1990). After her dentist had publicly disclosed that "he had AIDS," she was tested for HIV, although Bergalis was a virgin and did not belong to an AIDS risk group (Breo and Bergalis, 1990). Since she was HIV-antibody–positive the CDC concluded that she had contracted AIDS from her dentist (Centers for Disease Control, 1990), who was a homosexual with Kaposi's sarcoma (Ou et al., 1992).

Clearly, prior to the virus-AIDS hypothesis, the story of a doctor transmitting his Kaposi's sarcoma in the form of a yeast infection to his client via a common infectious cause would have hardly made *The New York Times* and certainly not the scientific literature (Lambert, 1991). But since the two entirely unrelated diseases are both labeled AIDS and because of the tremendous popularity of the virus-AIDS hypothesis, the paradoxical story became a cause célèbre for AIDS in the general population.

Once diagnosed for AIDS Bergalis was treated with the cytotoxic DNA chain terminator AZT, which is prescribed to inhibit HIV, until she died in December 1991, with weight loss (15 kg), hair loss, uncontrollable candidiasis, anemia, and muscle atrophy (requiring a wheelchair) (Breo and Bergalis, 1990; Anonymous, 1991; Lauritsen, 1991)—the symptoms of chronic AZT toxicity (Section 4). It is not clear whether her AZT therapy started before or after her pneumonia, since it was only mentioned in an edited interview conducted for the American Medical Association (Breo and Bergalis, 1990) and in some newspapers (Anonymous, 1991), but not in a single one of several scientific reports (Centers for Disease Control, 1990; Witte and Wilcox, 1991; Ou et al., 1992; Palca, 1992a, b) and not in *The New York Times* (Lambert, 1991). Since her fatal condition was attributed to HIV, she received $1 million in compensation from her dentist's, rather than from her AZT doctor's (Section 4) malpractice insurance (Palca, 1992a).

In view of the celebrity of the case and the fear it inspired among patients, 1100 further patients of the dentist came forward to be tested for HIV (Ou et al., 1992; Palca, 1992a). Seven of these, including Bergalis, tested positive. Four or 5 of these, including Bergalis and another woman, did not belong to an AIDS risk group, but 2 or 3 did. At least 3 of those who did not belong to a risk group received $1 million settlements from the dentist's malpractice insurance (Palca, 1992b). However, a plausible mechanism of HIV transmission from the dentist to his 4–5 positive clients without AIDS risks was never identified, and there is no consensus as to whether the viruses of the 3 carriers studied by the

CDC and the insurance companies were sufficiently related to claim a common source (Palca, 1992a, b).

Statistically, it can be shown that the incidence of HIV infections among the dentist's clients reflects, almost to the decimal point, the national incidence of the virus in the U.S. The national incidence of HIV-positives among all Americans is 0.4% (1 out of 250) (Table 1), the incidence of HIV-positives among 1100 patients of the Florida dentist was 0.4% (4 to 5 out of 1100), and the incidence among 15,795 patients from 32 HIV-positive doctors, determined by the CDC for the Bergalis case, was 0.5% (84 out of 15,795). Thus, the incidence of HIV in patients from HIV-positive doctors reflects the national incidence of HIV. This suggests noniatrogenic and, most likely, perinatal infection as the source of HIV in these patients, particularly in the case of the virgin Bergalis (Section 3.5.2). In addition, it identifies a rich source of insurance income for 0.4% of American patients of HIV-positive doctors!

To determine whether HIV had contributed to Bergalis's death, a controlled study would be necessary comparing the mortality of women with yeast infections, with and without antibodies against HIV, and with and without AZT therapy. Since such a study is not available, the assumption that Bergalis died from HIV is pure speculation.

(4) A doctor, presumably infected with HIV from a needle stick in 1983 (Aoun, 1992), described himself in a letter to the *New England Journal of Medicine* as an AIDS patient (Aoun, 1989). He was diagnosed HIV-positive in 1986 (Aoun, 1992). His only AIDS symptom at that time was a weight loss of 4.5 kg (Aoun, 1989). In 1991, then 8 years after the presumed date of the infection, the doctor described his case again in a speech "From the eye of the storm..." published in the *Annals of Internal Medicine* (Aoun, 1992). The speech did not describe any current AIDS symptoms. This case has been cited as an example that HIV is sufficient to cause AIDS (Baltimore and Feinberg, 1989).

However, the weight loss diagnosed in 1986 could have been the result of the anxiety that HIV infection causes in believers of the HIV-AIDS hypothesis, rather than the work of HIV. This interpretation is consistent with the fact that since 1985 at least 800,000 Americans (1 million minus the 206,000 AIDS cases recorded by the end of 1991; see Table 1) have not lost weight or developed other AIDS diseases (Duesberg, 1991a). Likewise, 6 million Central Africans (minus the 129,000 with AIDS) have been healthy HIV carriers since at least 1985 (Table 1).

Thus, there are no convincing anecdotal cases to prove that HIV causes AIDS in persons outside the major risk groups. The use of the above assumptions and anecdotal cases as proof for the virus-AIDS hypothesis is misleading, although they may provide valuable clues for future research.

3.6. CONSEQUENCES OF THE VIRUS-AIDS HYPOTHESIS

Despite the lack of proof and numerous discrepancies with orthodox criteria of infectious disease, the virus-AIDS hypothesis has remained since 1984 the only basis for all efforts in predicting, preventing, investigating, and even treating AIDS. AIDS prevention is based entirely on preventing the spread of HIV. This includes promotion of safe sex (Booth, 1988; Institute of Medicine, 1988; Weiss and Jaffe, 1990; Mann and the Global AIDS Policy Coalition, 1992; Anderson and May, 1992), clean injection equipment for intravenous drugs (National Commission on AIDS, 1991), and the exclusion of HIV-antibody–positive blood donations from transfusions (Vermund, 1991; Duesberg and Schwartz, 1992).

The Food and Drug Administration mandated in 1985 that the 12-million plus annual blood donations in the U.S. (Williams et al., 1990) are tested for HIV-1, and as of 1992 also for HIV-2, although there is as yet only one single American AIDS patient infected by HIV-2 (O'Brien et al., 1992). Since 1985 over 2 million tests have also been performed annually by the U.S. Army (Burke et al., 1990). By 1986 already over 20 million "AIDS tests" were performed in the U.S. (Institute of Medicine, 1986), at a minimum cost to the client of $12 to $70 (Irwin Memorial Blood Bank, San Francisco, personal communication) or $45 (U.S. Immigration Service). The former U.S.S.R. conducted 20.2 million "AIDS tests" in 1990 and 29.4 million in 1991 to detect 112 and 66 antibody-positives, respectively (Voevodin, 1992).

The detection of antibodies in healthy persons is interpreted as a 50% certain prognosis for AIDS within 10 years (Section 3.5.12). Therefore, a positive "AIDS test" is psychologically toxic (Grimshaw, 1987; Albonico, 1991b) and often the basis for the physiologically toxic antiviral therapy with AZT (Section 4) (Duesberg, 1992b, d). A negative test for HIV is a condition for admission to the U.S. Army (Burke et al., 1990), for admission to health insurance programs, for residence in many countries, and even for travel into the U.S. and China. Currently, over 50 countries restrict one or more classes of entrants based on positive-antibody tests for HIV (Duckett and Orkin, 1989). Antibody-positive Americans who had sex with antibody-negatives have been convicted of "assault with a deadly weapon" (Duesberg, 1991c; McKee, 1992). In communist Cuba about 600 antibody-positive persons are quarantined in the name of the virus-AIDS hypothesis (Scheper-Hughes and Herrick, 1992; Treichler, 1992).

Based on the assumption that HIV had either originated recently or spread recently from isolation to its current levels, at the same rates as AIDS had spread in the risk groups in the U.S. and Europe, and on the assumption that AIDS would follow the presumed spread of HIV with a hiatus of 10 years, epidemiologists have made apocalyptic predictions about an AIDS epidemic that has raised fears and funding to unprecedented levels (Heyward and Curran, 1988; Mann et al., 1988; Mann and the Global AIDS Policy Coalition, 1992; Anderson and May, 1992).

Above all, over 180,000 antibody-positives, with and without AIDS, are currently treated indefinitely with the cytotoxic DNA chain terminator AZT in an effort to inhibit HIV (Section 4.4).

4. THE DRUG-AIDS HYPOTHESIS

After the global acceptance of the virus-AIDS hypothesis, several investigators have recently revived the original hypothesis that AIDS is not infectious (Section 2.2). In view of (1) the almost complete restriction (97%) of American AIDS to groups with severely compromised health, (2) the predetermination for certain AIDS diseases by prior health risks, and (3) the many links between AIDS and drug consumption (Sections 2.1.3 and 3.4; Table 2), it has been proposed that recreational drugs and AZT may cause AIDS (Lauritsen and Wilson, 1986; Haverkos, 1988a, 1990; Holub, 1988; Papadopulos-Eleopulos, 1988; Rappoport, 1988; Duesberg, 1990a, 1991a, 1992c, f; Lauritsen, 1990; Albonico, 1991a, b; Pillai et al., 1991; Cramer, 1992; Leonhard, 1992). Here the hypothesis is investigated that all American and European AIDS diseases, above the normal background of hemophilia and transfusion-related diseases, are the result of the long-term consumption of recreational and anti-HIV drugs.

4.1. Chronological Coincidence between the Drug and AIDS Epidemics

The appearance of AIDS in America in 1981 followed a massive escalation in the consumption of psychoactive drugs that started after the Vietnam War (Newell et al., 1985b; Kozel and Adams, 1986; National Institute on Drug Abuse, 1987; Bureau of Justice Statistics, 1988; Haverkos, 1988b; Office of National Drug Control Policy, 1988; Flanagan and Maguire, 1989; Lerner, 1989; Shannon et al., 1990). The Bureau of Justice Statistics reports that the number of drug arrests in the U.S. has increased from about 450,000 in 1980 to 1.4 million in 1989 (Bureau of Justice Statistics, 1988; Shannon et al., 1990). About 500 kg of cocaine were confiscated by the Drug Enforcement Administration in 1980, about 9000 kg in 1983, 80,000 kg in 1989, and 100,000 kg in 1990 (Bureau of Justice Statistics, 1988, 1991; Flanagan and Maguire, 1989). In 1974, 5.4 million Americans had used cocaine at some point in their lives and in 1985 that number had gone up to 22.2 million (Kozel and

Adams, 1986). Currently, about 8 million Americans are estimated to use cocaine regularly (Weiss, S. H., 1989; Finnegan et al., 1992). The number of dosage units of domestic stimulants confiscated, such as amphetamines, increased from 2 million in 1981 to 97 million in 1989 (Flanagan and Maguire, 1989).

Several arguments indicate that these increases reflect increased drug consumption rather than just improved drug control, as has been suggested (Maddox, 1992a):

(1) The Bureau of Justice Statistics estimates that, at most, 20% of the cocaine smuggled into the U.S. was confiscated each year (Anderson, 1987).

(2) The National Institute on Drug Abuse reports that between 1981 and 1990 cocaine-related hospital emergencies increased 24-fold from 3296 to 80,355 and deaths increased from 195 to 2483 (Kozel and Adams, 1986; National Institute on Drug Abuse, 1990a, b). Thus, cocaine-related hospital emergencies had increased 24-fold during 9 of the 10 years in which cocaine seizures had increased 100-fold.

(3) It is highly improbable that, before the jet age, the U.S. would have imported annually as much cocaine as it did in 1990 plus the 100,000 kg that were confiscated in that year.

Further, the recreational use of psychoactive and aphrodisiac nitrite inhalants began in the 1960s and reached epidemic proportions in the mid-1970s, a few years before AIDS appeared (Newell et al., 1985b, 1988). The National Institute on Drug Abuse reports that in 1979–1980 over 5 million people used nitrite inhalants in the U.S. at least once a week (Newell et al., 1988), a total of 250 million doses per year (Wood, 1988). In 1976 the sales of nitrite inhalants in one American city alone amounted to $50 million annually (Newell et al., 1985b, 1988), at $5 per 12 mL dose (Schwartz, 1988).

Since 1987 the cytocidal DNA chain terminator AZT has been prescribed as an anti-HIV drug to AIDS patients (Kolata, 1987; Yarchoan and Broder, 1987b) and to asymptomatic carriers of HIV since 1990 (Editorial, 1990). Currently, about 120,000 Americans and 180,000 HIV-positive persons worldwide, with and without AIDS, take AZT in efforts to inhibit HIV. This estimate is based on the annual AZT sales of $364 million and a wholesale price of $2,000 per year for a daily dose of 500 mg AZT per person (Burroughs-Wellcome Public Relations, 3 April 1992). In addition, an unknown number take other DNA chain terminators like ddI and ddC (Smothers, 1991; Yarchoan et al., 1991).

4.2. OVERLAP BETWEEN DRUG-USE AND AIDS STATISTICS

Drugs and AIDS appear to claim their victims from the same risk groups. For instance, the CDC reports that the annual mortality of 25- to 44-year-old American males increased from 0.21% in 1983 to 0.23% in 1987, corresponding to about 10,000 deaths among about 50 million in this group (Buehler et al., 1990). Since the annual AIDS deaths had also reached 10,000 by 1987, HIV was assumed to be the cause (Institute of Medicine, 1986; Centers for Disease Control, 1987, 1992b). Further, HIV infection was blamed for a new epidemic of immunological and neurological deficiencies, including mental retardation, in American children (Blattner et al., 1988; Institute of Medicine, 1988; Centers for Disease Control, 1992b).

However, mortality in 25- to 44-year-old males from septicemia, considered an indicator of intravenous drug use, rose almost 4-fold from 0.46 per 100,000 in 1980 to 1.65 in 1987, and direct mortality from drug use doubled (National Center for Health Statistics, 1989; Buehler et al., 1990), indicating that drugs played a significant role in the increased mortality of this group (Buehler et al., 1990). In addition, deaths from AIDS diseases and non-AIDS pneumonia and septicemia per 1000 intravenous drug users in New York increased at exactly the same rates, from 3.6 in 1984 to 14.7 and 13.6, respectively, in 1987 (Selwyn et al., 1989). Indeed, the cocaine-related hospital emergencies alone could more than account for the 32% of American AIDS patients that are intravenous drug users (Section 2.1.3). The emergencies had increased from "a negligible number of people" in 1973 to 9946 nonfatal and 580 fatal cases in 1985 (Kozel and Adams, 1986), when a total of 10,489 AIDS cases were recorded, and to 80,355 nonfatal and 2483 fatal cases in 1990 (National Institute on Drug Abuse, 1990a, b), when a total of 41,416 AIDS cases were recorded by the CDC (Centers for Disease Control, 1992a). Moreover 82% of the cocaine-related and 75% of the morphine-related hospital emergency patients were 20–39 years old (National Institute on Drug Abuse, 1990a), the age distribution typical of AIDS patients (Section 2.1.1).

Another striking coincidence is that over 72% of all American AIDS patients (Centers for Disease Control, 1992b) and about 75% of all Americans who consume "hard" psychoactive drugs such as cocaine, amphetamines, and inhalants (National Institute on Drug Abuse, 1987, 1990a, b; Ginzburg, 1988) or get arrested for possession of such drugs (Bureau of Justice Statistics, 1988) or are treated for such drugs (National Institute on Drug Abuse, 1990a) are 20- to 44-year-old males. Thus, there is substantial epidemiological overlap between the two epidemics (Lerner, 1989), reported as *The Twin Epidemics of Substance Use and HIV* by the National Commission on AIDS (National Commission on AIDS, 1991).

Moreover, maternal drug consumption was blamed by some for the new epidemic of immunological and neurological deficiencies, including dementias, of American children (Toufexis, 1991). In view of this, the CDC acknowledges, "We cannot discern, however, to what extent the upward trend in death rates from drug abuse reflects trends in illicit drug use independent of the HIV epidemic" (Buehler et al., 1990).

4.3. Drug Use in AIDS Risk Groups

4.3.1. *Intravenous Drug Users Generate a Third of All AIDS Patients*

Currently, 32% of American (National Commission on AIDS, 1991; Centers for Disease Control, 1992b) and 33% of European (Brenner et al., 1990; World Health Organization, 1992a) AIDS patients are intravenous or intrauterine users of heroin, cocaine, and other drugs (Section 2.1.3). These include:

(1) 75% of all heterosexual AIDS cases in America and about 70% of those in Europe,
(2) 71% of American and 57% of European females with AIDS,
(3) over 10% of American and 5% of European male homosexuals,
(4) 10% of American hemophiliacs with AIDS,
(5) 70% of American children with AIDS, including 50% born to mothers who are confirmed intravenous drug users and another 20% to mothers who had "sex with intravenous drug users" and are thus likely users themselves (Amaro et al., 1989), and
(6) 80–85% of European children with AIDS who were born to drug-addicted mothers (Mok et al., 1987; European Collaborative Study, 1991).

In an article entitled "AIDS and Intravenous Drug Use: The Real Heterosexual Epidemic" AIDS researcher Moss (1987) points out that "90% of infected prostitutes reported in Florida, Seattle, New York and San Francisco have been intravenous drug users... Drug use is also the source of most neonatal AIDS, with 70% of cases occurring in children of intravenous drug users..." Indeed, all studies of American and European prostitutes indicate that HIV infection is almost exclusively restricted to drug users (Rosenberg and Weiner, 1988), although all prostitutes should have the same risks of HIV infection, if HIV were sexually transmitted. Surprisingly, all of these studies only mention the incidence of HIV, rather than of AIDS, in prostitutes.

4.3.2. Homosexual Users of Aphrodisiac Drugs Generate about 60% of AIDS Patients

Approximately 60% of American AIDS patients are male homosexuals over the age of 20 (Table 1). They are generated by risk groups that have sex with large numbers of partners (Centers for Disease Control, 1982; Jaffe et al., 1983b; Darrow et al., 1987; Oppenheimer, 1992) that often average over 100 per year and have exceeded 1000 over a period of several years (Mathur-Wagh et al., 1984; Newell et al., 1985a; Turner et al., 1989; Callen, 1990). The following evidence indicates that these sexual activities and the corresponding conventional venereal diseases are directly proportional to the consumption of toxic sexual stimulants, which include nitrite and ethylchloride inhalants, cocaine, amphetamines, methaqualone, lysergic acid, phenylcyclidine, and more (Blattner et al., 1985; Shilts, 1985; Lauritsen and Wilson, 1986; Darrow et al., 1987; Haverkos, 1988a; Rappoport, 1988; Raymond, 1988; Adams, 1989; Turner et al., 1989; Weiss, S. H., 1989; Ostrow et al., 1990; Lesbian and Gay Substance Abuse Planning Group, 1991a).

An early CDC study of 420 homosexual men attending clinics for sexually transmitted diseases in New York, Atlanta, and San Francisco reported that 86.4% had frequently used amyl and butyl nitrites as sexual stimulants. The frequency of nitrite use was proportionate to the number of sexual partners (Centers for Disease Control, 1982).

In 1983 Jaffe et al. investigated AIDS risk factors of 170 male homosexuals from sexual disease clinics, including 50 with Kaposi's sarcoma and pneumonia and 120 without AIDS. In this group, 96% were regular users of nitrite inhalants and 35–50% were users of ethyl chloride inhalants. In addition, 50–60% had used cocaine, 50–70% amphetamines, 40% phenylcyclidine, 40–60% lysergic acid, 40–60% methaqualone, 25% barbiturates, 90% marijuana, and 10% heroin (Jaffe et al., 1983b). Over 50% had also used prescription drugs. About 80% of these men had past or current gonorrhea, 40–70% had syphilis, 15% mononucleosis, 50% hepatitis, and 30% parasitic diarrhea. Those with Kaposi's sarcoma had a median of 61 sex partners per year and those without AIDS about 26. The study points out that "lifetime exposure to nitrites... (and) use of various 'street' drugs... was greater for cases than controls." The lifetime drug dose of "cases" was reported to be two times higher than of asymptomatic HIV carriers (Jaffe et al., 1983b).

A study of a group of 359 homosexual men in San Francisco reported in 1987 that 84% had used cocaine, 82% alkyl nitrites, 64% amphetamines, 51% methaqualone, 41% barbiturates, 20% injected drugs, an 13% shared needles (Darrow et al., 1987). About 74% had past or current infection by gonococcus, 73% by hepatitis B virus, 67% by HIV, 30% by amoebae, and 20% by treponema (Darrow et al., 1987). This group had been randomly selected from a list of homosexuals who had volunteered to be investigated for hepatitis B virus infection and to donate antisera to hepatitis B virus between 1978 and 1980. For the same group the 50% "progression rate" from HIV to AIDS was calculated to be 8–11 years (Table 2) (Moss et al., 1988; Lemp et al., 1990) and was reported to be relevant for "the [HIV-infected] population as a whole" (Moss et al., 1988)!

A study investigating AIDS risk factors among French homosexuals reported that 31% of those with AIDS, but only 12% of those without AIDS, had achieved "over 100 nitrite inhalations" (Messiah et al., 1988). The study included 53, or 45%, of all homosexual AIDS patients recorded in France by 1987.

The staggering oral drug use among male homosexuals at risk for AIDS was confirmed in 1990 by the largest survey if its kind. It reports that 83% of 3916 self-identified American homosexual men had used one, and about 60% two or more, drugs with sexual activities during the previous six months (Ostrow et al., 1990). Similar drug use has been reported for European homosexuals at risk for AIDS (van Griensven et al., 1987).

A survey of homosexual men from Boston, conducted between 1985 and 1988, documented that among 206 HIV-positives, 92% had used nitrite inhalants, 73% cocaine, 39% amphetamines, and 29% lysergic acid in addition to six other psychoactive drugs as sexual stimulants; among 275 HIV-negative controls, 71% had used nitrites, 57% cocaine, 21% amphetamines, and 17% lysergic acid, again in addition to six other psychoactive drugs (Seage et al., 1992). A similar survey of 364 HIV-positive homosexual men in Berlin conducted between 1983 and 1987 stated that 194 (53.3%) had used nitrite inhalants (Deininger et al., 1990).

According to Newell et al. (1985b), volatile nitrites had penetrated "every corner of gay life" by 1976. Surveys studying the use of nitrite inhalants found that in San Francisco 58% of homosexual men were users in 1984 and 27% in 1991, compared to less than 1% of heterosexuals and lesbians of the same age group (Lesbian and Gay Substance Abuse Planning Group, 1991b).

Several investigators have pointed out that nitrite inhalants, and possibly other drugs, are preferred by male homosexuals as aphrodisiacs because they facilitate anal intercourse by relaxing smooth muscles (Section 4.4.1) (Mirvish and Haverkos, 1987; Newell et al., 1985b; Ostrow et al., 1990; Lesbian and Gay Substance Abuse Planning Group, 1991a; Seage et al., 1992). "Nitrites were used primarily for heightened sexual stimulation during sexual activity by reducing social and sexual inhibitions, prolonging

570 ■ *Appendix B*

duration, heightening sexual arousal, relaxing the anal sphincter during anal intercourse, and prolonging orgasm" (Newell et al., 1985b).

4.3.3. *Asymptomatic AZT Users Generate an Unknown Percentage of AIDS Patients*

The DNA chain terminator AZT has been licensed in the U.S. since 1987 as a treatment for AIDS patients (Chernov, 1986; Kolata, 1987; Lauritsen, 1990; Yarchoan et al., 1991) based on a placebo-controlled study sponsored by Burroughs-Wellcome, the manufacturer of AZT (Section 4.4.2) (Fischl et al., 1987; Richman et al., 1987). In 1990 AZT was also licensed as an AIDS prophylaxis for healthy HIV carriers (Section 4.4.2) (Volberding et al., 1990; Yarchoan et al., 1991).

The choice of this drug as anti-AIDS treatment is based entirely on the virus-AIDS hypothesis. According to Broder et al., "[t]he rationale for anti-retroviral therapy for AIDS is... that HIV is the etiologic agent of AIDS" and that HIV RNA-dependent DNA synthesis is inhibited by AZT (Yarchoan et al., 1991). In view of this and their faith in the virus-AIDS hypothesis, about 120,000 American HIV carriers, with and without AIDS, and 180,000 worldwide currently take AZT every day (Section 4.1). It follows that probably a high percentage of the 40,000 Americans and 15,000 Europeans that currently develop AIDS per year (Table 1) have used AZT and other DNA chain terminators prior to AIDS.

The drug is now recommended as AIDS prophylaxis for all AIDS-free persons with less than 500 T-cells per microliter by the director of AIDS research at the NIH (Kolata, 1992) and with some reservations also by the National Hemophilia Association of New York (personal communication), despite recent doubts about its usefulness (Kolata, 1992). For instance, AZT has been used indefinitely by over 1200 AIDS-free, but presumably HIV-infected, homosexual men from the Multicenter AIDS Cohort Study referenced above (Ostrow et al., 1990), including 7% of 3670 with over 500 T-cells per microliter, 16% of 1921 with 350–499 T-cells, 26% of 1374 with 200–349 T-cells, and 51% of 685 with fewer than 200 T-cells (Graham et al., 1991). Yet the large study acknowledges finding "...no effects [of AZT] on rates of progression to lower CD4+ lymphocyte counts in any of the transition intervals" (Graham et al., 1991). In San Francisco 3.3% of 151 AIDS-free male homosexuals with over 500 T-cells, 11% of 128 with 200–500 T-cells, and 36% of 42 with less than 200 T-cells were on AZT in 1989 (Lang et al., 1991). Another study reports that, in 1989, 26 out of 322 HIV-positive but AIDS-free homosexuals from San Francisco, Chicago, and Denver had taken AZT for less than six months and 101 for over six months (Holmberg et al., 1992).

Appendix B ■ 571

To distinguish between HIV and drugs as causes of AIDS, it is necessary to identify either HIV-carriers that develop AIDS only when they use drugs (Section 4.4) or to identify HIV-free drug users that develop AIDS indicator diseases (Section 4.5) and to demonstrate drug toxicity (Section 4.6).

4.4. Drug Use Necessary for AIDS in HIV-Positives

Studies demonstrating that drugs are necessary for AIDS among HIV-positives fall into two subgroups: (1) those demonstrating that AIDS among HIV-positives depends on the long-term use of recreational drugs and (2) those demonstrating that HIV-positive AIDS-free persons and AIDS patients on the antiviral drug AZT develop new AIDS diseases or AZT-specific diseases. Since the health of AIDS-free persons selected for AZT prophylaxis is compromised by prior AIDS risks, e.g., less than 500 T-cells, and since nearly all American and European AIDS patients have used recreational drugs or have been immunosuppressed by long-term transfusions, evaluating the role of AZT in the progression of AIDS is complicated by these confounding risk factors (Sections 3.4.4 and 4.3.3).

4.4.1. AIDS from Recreational Drugs

(1) A study of 65 HIV-infected drug users from New York showed that their T-cell count dropped over nine months proportionately with drug injection, on average 35%, compared to controls who had stopped (Des Jarlais et al., 1987).

(2) The incidence of AIDS diseases and death among HIV-positive, asymptomatic intravenous drug users over 16 months were 19% (23/124) among those who persisted in injecting psychoactive drugs, 5% (5/93) among those who had stopped injecting drugs, and 6% (5/80) among those on methadone treatment (Weber et al., 1990).

(3) Among male homosexuals, receptive anal intercourse carries a 2.75 times (Warren Winkelstein, personal communication) to 4.4 times (Haverkos, 1988b) higher AIDS risk than insertive intercourse, presumably reflecting a higher risk of infection by HIV (Moss et al., 1987; van Griensven et al., 1987; Winkelstein et al., 1987; Seage et al., 1992). However, if HIV were the cause of AIDS, the donors should have the same AIDS risk as the recipients, because recipients can be infected only by HIV donors. No microbe can survive that is only unidirectionally transmitted. All venereal microbes are therefore bitransitive. Indeed, Haverkos found no differences in sexually transmitted diseases between those practicing receptive and insertive intercourse (Haverkos, 1988b). The

probable reason for the higher AIDS risk associated with receptive anal intercourse is that this sexual practice directly correlates with a 2-fold (van Griensven et al., 1987; Seage et al., 1992) to an 8-fold (Moss et al., 1987; Haverkos, 1988b) enhanced use of nitrite inhalants and other aphrodisiac drugs that facilitate anal intercourse (Sections 4.3.2 and 4.6).

(4) A Canadian study reports that every one of 87 HIV-positive male homosexual AIDS patients had used nitrite inhalants. Those who had used over 20 "hits" per month were more likely to have Kaposi's sarcoma and sarcoma plus pneumonia than those who had used less than 20 hits per month. HIV-free controls, described in a previous report of the same cohort (Section 4.5) (Marion et al., 1989), were not mentioned in this study (Archibald et al., 1992). The authors concluded that a "sexually transmitted agent," which is even more difficult to transmit than HIV(!) (Section 3.5.1), would explain the Kaposi's sarcomas among the AIDS patients. The nitrites were proposed to be a cofactor of this cofactor of HIV (Archibald et al., 1992). Thus, nitrites were necessary for AIDS in HIV-positives.

To determine whether HIV was indeed necessary for these AIDS cases, the incidence of AIDS-defining diseases in HIV-positive and -negative homosexuals who are matched for the duration and extent of drug consumption must be compared. This is what the Canadian team has recently attempted to do in a study termed "HIV Causes AIDS: A Controlled Study" (Craib et al., 1992). The study asserts to meet the challenge of "Duesberg [who] wrote in 1988 (*Science*, 1988; 242: 997–998) and repeated in public addresses in 1991 that the necessary comparisons in controlled cohorts were not available..."

However, the study failed to match the HIV-free control group with the HIV-positives for the extent and duration of drug consumption. It mentions that 49% of the HIV-negatives had used "psychoactive drugs," but fails to mention the percentage of drug users among the HIV-positives. In their previous study 100% of the HIV-positive AIDS patients had used such drugs (Archibald et al., 1992). In addition the authors failed to recognize that HIV-infection is a marker for the duration of drug consumption. Since an average of 1000 sexual contacts is required for sexual transmission of HIV (Section 3.5.2), HIV is a marker for the dosage of sexual stimulants that is used for 1000 contacts. Thus, HIV-positives would have used more sexual stimulants, the equivalent for 1000 contacts, than HIV-negatives. Indeed, the authors acknowledge problems with "claims that AIDS is caused by other exposures and not by HIV... the problem may be semantics. No one has ever disputed that cofactors play a very important role..." (Craib et al., 1992). Moreover, the authors failed to mention whether AZT was prescribed to the HIV-positives.

(5) A survey of 99, including 92 "gay or bisexual," AIDS patients from an "HIV clinic" at St. Mary's Hospital in London reports that 78% used "poppers" (nitrite inhalants), 78% cannabis, 76% cigarettes, 68% alcohol, and 48% "ecstasy" (amphetamines). In addition, the patients received an average of three unspecified medications, probably including AZT (Valentine et al., 1992). HIV-tests were not reported but are assumed to be positive because the patients were in an "HIV clinic." Clearly, the multiplicity of drugs consumed by these patients could be relevant to their pathogenesis.

(6) A European survey of HIV-positive infants with AIDS found that "nearly all children were born to intravenous-drug–abusing mothers" and that AIDS was 9.4 times more likely in children whose mothers had AIDS symptoms before delivery than in those who had no symptoms (Mok et al., 1987). "Children with drug withdrawal symptoms" were most likely to develop diseases; those with no withdrawal symptoms but "whose mothers had used recreational drugs in the final six months of pregnancy were intermediate on all indices, whereas children of former drug users did not significantly differ from those born to women who had no history of i.v. drug use" (European Collaborative Study, 1991). An American survey reported that 63 of 68 infants "with symptomatic HIV infections" had "at least one parent who had AIDS or was in an AIDS high-risk group" (Belman et al., 1988). Since the risk of infants to develop AIDS increased with maternal drug consumption and increased 10-fold with maternal AIDS symptoms, it would appear that disease or subclinical deficiencies during pregnancy rather than perinatal infection by HIV are responsible for pediatric AIDS.

4.4.2. AIDS from AZT and AZT plus Confounding Recreational Drug Use

(1) A placebo-controlled study, sponsored by Burroughs-Wellcome, the manufacturer of AZT, for the licensing of the drug as AIDS therapy in the United States investigated 289 patients with "unexplained" weight loss, fever, oral candidiasis, night sweats, herpes zoster, and diarrhea. (Fischl et al., 1987; Richman et al., 1987). All but 13 of these patients were males. The study was planned for 6 months, but it was interrupted after 4 months, because by then the therapeutic benefits of AZT seemed too obvious to continue the placebo control:

(a) After 4 months on AZT 1 out of 145 in the AZT group but 19 out of 137 in the placebo group had died. Therefore, the study claimed that AZT can "decrease mortality."

(b) T-cell counts first increased from 4–8 weeks and then declined to pretreatment levels within 4 months.

(c) The lymphocyte count decreased over 50% in 34% of the AZT recipients but in only 6% of the control group.

(d) Sixty-six in the AZT group suffered from severe nausea, compared to only 25 in the control group.

(e) Muscle atrophy was observed in 11 AZT recipients but observed in only 3 from the control group.

Yet, the primary claim of the study, "decreased mortality" from AZT is not realistic if one considers that 30 out of the 145 in the AZT group depended on multiple transfusions to survive anemia, compared to only 5 out of the 137 in the placebo group. Thus, the number of subjects in the AZT group who would have died from severe anemia if untreated was larger (i.e., 30) than the AIDS deaths and anemias of the control group combined, namely 19 + 5. The "decreased mortality" claim is further compromised by numerous "concomitant medications" other than transfusions for AZT-specific diseases and failure to match the AZT and placebo groups for the cumulative effects of prior and parallel recreational drug use. In addition some of the AZT-specific AIDS diseases observed in the placebo group appear to be due to patient-initiated "drug sharing" between AZT and placebo recipients (Lauritsen, 1990; Duesberg, 1992d; Freestone, 1992) and falsification of the case report forms (Lauritsen, 1992).

Moreover, the low mortality of 0.7% (1/145) claimed by the licensing study for the first four months on AZT could not be extended in a follow-up study which found the "survival benefits" of AZT rapidly declining after the original 4-month period. By 18 months, 32% of the original AZT group had died, as had 35% of the former control group, which by then had also received AZT for 12 months (Fischl et al., 1989).

Since the original study considered AZT effective in decreasing AIDS mortality, subsequent placebo-controlled studies were deemed unethical. But the low mortality claimed by the licensing study has not been confirmed by later studies, which observed mortalities of 12–72% within 9–18 months (see items (3) to (6) below). In addition, a CDC study recently reported a mortality of 82% in a cohort of 55 AIDS patients that had been on AZT for up to 4 years (Centers for Disease Control, 1991), hardly recommending AZT as an AIDS therapy.

The brief transient gains of T-cells observed upon AZT treatment by the licensing study may reflect compensatory hemopoiesis, random killing of pathogenic parasites (Elwell et al., 1987), and the influence of concomitant medication, including multiple transfusions (Richman et al., 1987). Indeed, the study concluded, based on the "hematological toxicity" described above, that "...the initial beneficial immunological effects of AZT may not be sustained" (Richman et al., 1987). A French study confirms "...the decrease of cell counts below the initial value after a few months of AZT suggests that this drug might be toxic to cells" (see item (3) below) (Dournon et al., 1988). And a recent American study also confirms "... no effects on rates of progression to lower CD4+ Lymphocyte counts in [6-month] transition intervals" (Section 4.3.3) (Graham et al., 1991). Moreover, the manufacturer states, "A modest increase in mean CD4 (T4) counts was seen in the zidovudine group but the significance of this finding is unclear as the CD4 (T4) counts declined again in some patients" (Medical Economics Data, 1992).

(2) In view of the reported success of AZT as AIDS therapy, the drug was also tested for licensing as an AIDS prophylaxis by much of the same team, including Fischl, Richman, and Volberding, and again with support from the manufacturer Burroughs-Wellcome (Volberding et al., 1990). The study treated AIDS-free, HIV-positive 25- to 45-year-old male homosexuals and intravenous drug users who had "fewer than 500 T-cells" for one year either with AZT or with a placebo. The expected annual AIDS risk for intravenous drug users and male homosexual risk groups is about 4–6% per year without AZT (Section 3.4.4.4).

The study reports AIDS diseases in (1) 11 out of 453 on 500 mg AZT per day, (2) 14 out of 457 on 1500 mg AZT per day, and (3) 33 out of 428 on a placebo (Volberding et al., 1990). Thus, the AZT groups appeared to do better than expected and the placebo groups did as expected. Therefore, it was claimed that AZT prevents AIDS.

However, the price for the presumed savings of 22 (33–11) and 19 (33 –14) AIDS cases with AZT, compared to the placebo group, was high because 19 AZT-specific cases of potentially fatal anemia, neutropenia and severe nausea appeared in the 500 mg AZT group, and 72 such cases, including 29 anemias requiring life-saving blood transfusions, appeared in the 1500 mg AZT group. This indicates cytocidal effects of AZT on hemopoiesis and on the intestines. Although the AZT-specific diseases were not diagnosed as AIDS, neutropenia generates immunodeficiency (Walton et al., 1986) and thus AIDS. If these AZT-specific cases were included in the calculation of benefits from AZT compared to the placebo group, the 500 mg group no longer benefited and the 1500 mg group tripled its disease risk.

The study was further compromised by its failure to match the treatment groups for their cumulative recreational drug use prior to and during the study and for the many compensatory treatments for the AZT-specific diseases of the subjects analyzed. The fact that 8 cases in the control group but only 3 and 1 in the 500 mg and 1500 mg-AZT groups, respectively, developed AIDS cancers suggests that the control group could have been exposed to higher recreational drug doses.

Since the licensing study considered AZT effective in preventing AIDS, subsequent controlled trials were deemed unethical. However, several subsequent studies cast further doubt on the claim that AZT is a useful AIDS prophylactic. One study reported that persons with "early" AIDS, i.e., AIDS-free persons at risk for AIDS, died at the same rate of 12–14% as AIDS controls and that 82% developed leukopenia within less than a year (see item 6 below) (Hamilton et al., 1992). Another study described "no effects on rates of progression to lower CD4+ lymphocytes..." recorded within 6-month periods in over 1200 AIDS-free men on AZT (Section 4.3.3) (Graham et al., 1991). A third study reported that 26 out of 127 HIV-positive, AIDS-free homosexuals had discontinued an unreported dose of AZT within less than 6 months, most because of severe toxicity (Section 4.3.3) (Holmberg et al., 1992). In view of these and other data, it is surprising that a loss of T-cells was not noted in the licensing study (Kolata, 1987).

(3) A French study investigated the effects of AZT on 365 AIDS patients. The patients included 72% male homosexuals and 11% intravenous drug users with a median age of 36 years and with opportunistic infections and Kaposi's sarcoma. The study, the largest of its kind, observed new AIDS diseases, including leukopenia, in over 40% and death in 20% within 9 months on AZT (Dournon et al., 1988). The AIDS diseases of 30% worsened during AZT treatment. The study reported no therapeutic benefits 6 months after initiating AZT therapy. The authors concluded: "...the rationale for adhering to high-dose regimens of AZT, which in many instances leads to, toxicity and interruption of treatment, seems questionable."

(4) A Dutch study treating 91 male AIDS patients, averaging 39 years, after 67 weeks on AZT, observed mortality in 72% and AZT-specific myelotoxicity, requiring on average 5 blood transfusions, in 57%. About 34% of the myelotoxicity manifested in anemia and 20% in leukopenia. The authors concluded that "the majority of patients... cannot be maintained on these (AZT) regimens, most commonly due to the development of hematological toxicity" (van Leeuwen et al., 1990).

(5) An Australian study involving 308 homosexual and bisexual men with Kaposi's sarcoma, lymphoma, and opportunistic infections and a median age of 36 years, reported 30% mortality within 1–1.5 years on AZT. In addition one or more new AIDS diseases, including pneumonia, candidiasis, fever, night sweats, and diarrhea were observed in 172 (56%) within one year (Swanson et al., 1990). Moreover, 50% needed at least one blood transfusion, and 29% needed multiple blood transfusions to survive AZT treatment. Yet the authors concluded that the "risk:benefit ratio [is] advantageous to AIDS patients" (Swanson et al., 1990).

(6) A comparison of the effects of indefinite AZT treatment on 170 HIV-positive AIDS-free persons with "early" AIDS to 168 with "late" AIDS indicated that the mortality was the same in both groups, i.e., 12–14% per 1–1.5 years (Hamilton et al., 1992). The median age of the AZT recipients was 40 years; 63% were male homosexuals and 25% were intravenous drug users. AZT-specific diseases were observed in most "early AIDS" cases, i.e., leukopenia in 82%, severe leukopenia in 14%, anemia in 20%, severe anemia requiring transfusions in 5%, nausea in 40%, and skin rashes in 47%. This indicates directly that AZT is toxic for AIDS-free HIV carriers, and that AZT toxicity is sufficiently dominant over other AIDS causes that it accelerates the progression to death of AIDS-free HIV carriers to the same rate that is observed in late AIDS patients (Duesberg, 1992d). The authors concluded that AZT, contrary to the Wellcome–sponsored study from 1987 conducted for licensing AZT, does not extend life.

(7) The annual lymphoma incidence of AZT-treated AIDS patients with Kaposi's sarcoma, pneumonia, and wasting disease was reported to be 9% by the National Cancer Institute and was calculated to be 50% over three years (Pluda et al., 1990). The estimate of the 3-year incidence of lymphoma from this study was recently revised down to 31% (Yarchoan et al., 1991). An independent study observed in a group of 346 AIDS patients in London, most of whom were on AZT, "during the past three years a progressive increase in the number of patients dying from lymphoma...," to a current total of 16% in 1991 (Peters et al., 1991). And a CDC study reported a 15% lymphoma incidence during 24 months on AZT (Centers for Disease Control, 1991).

The lymphoma incidence of untreated, HIV-positive AIDS risk groups is 0.3% per year, derived from the putative average progression rate of 10 years from HIV to AIDS (Moss et al., 1988; Lemp et al., 1990; Duesberg, 1991a) and the 3% incidence of lymphoma in AIDS patients (Centers for Disease Control, 1992b). Therefore, the annual lymphoma risk of AZT recipients is about 30 times higher than that of untreated HIV-positive counterparts. It appears that the chronic levels of the mutagenic AZT, at 20–60 μm (500–1500 mg/person/day), were responsible for the lymphomas (Section 4.6.2).

An alternative interpretation suggests that AZT had prolonged life sufficiently to allow HIV to induce the lymphomas directly or via immunodeficiency (Pluda et al., 1990; Centers for Disease Control, 1991). However, this interpretation is flawed for several reasons: (i) Cancers, including malignant lymphomas, are not consequences of a defective immune system (Section 3.5.8). (ii) There is as yet only a model for how HIV, the presumed killer of T-lymphocytes, could also cause cancer (Section 3.5.14) (Gallo, 1990). (iii) AZT-induced lymphomas lack HIV-

578 ■ *Appendix B*

specific markers (McDunn et al., 1991). (IV) Several studies indicate that AZT does not prolong life (see above) (Dournon et al., 1988; van Leeuwen et al., 1990; Hamilton et al., 1992; Kolata, 1992).

(8) Ten out of 11 HIV antibody-positive, AZT-treated AIDS patients recovered cellular immunity after discontinuing AZT in favor of an experimental HIV vaccine (Scolaro et al., 1991). The vaccine consisted of an HIV strain that was presumed to be harmless, because it had been isolated from a healthy carrier who had been infected by the virus for at least 10 years. Since there was no evidence that the hypothetical vaccine strain differed from that by which the patients were already naturally vaccinated, the only relevant difference between the patients before and during the vaccine trial was the termination of their AZT treatment. It follows that AZT treatment is at least a necessary, if not a sufficient, cause of immunodeficiency in HIV-positives.

(9) Four out of 5 AZT-treated AIDS patients recovered from myopathy two weeks after discontinuing AZT; two redeveloped myopathy on renewed AZT treatment (Till and MacDonnell, 1990), indicating that AZT is at least necessary for myopathy in HIV-positives.

(10) Four patients with pneumonia developed severe pancytopenia and bone marrow aplasia 12 weeks after the initiation of AZT therapy. Three out of 4 recovered within 4–5 weeks after AZT was discontinued (Gill et al., 1987), indicating that AZT is necessary for pancytopenia in HIV-positives.

4.5. Drug Use Sufficient for AIDS Indicator Diseases in the Absence of HIV

Studies demonstrating AIDS-defining diseases in drug users in the absence of HIV are chronologically and geographically censored by the virus-AIDS hypothesis. Before the general acceptance of this hypothesis in the U.S., there were numerous American studies blaming AIDS on recreational drugs, but afterward there was but one American report describing HIV-free Kaposi's sarcomas in homosexuals who had used such drugs, and only a few American and some European studies describing AIDS-defining diseases in HIV-free intravenous drug users (see below).

If HIV were necessary for AIDS among drug users, only HIV-positive drug users should develop AIDS. However, there is not even one controlled study showing that among matched drug users only HIV-positives get AIDS. On the contrary, such studies all indicate that drugs are sufficient to cause AIDS.

4.5.1. Drugs Used for Sexual Activities Sufficient for AIDS Diseases

(1) The first five AIDS cases, diagnosed in 1981 before HIV was known, were male homosexuals who had all consumed nitrite inhalants and presented with *Pneumocystis* pneumonia and cytomegalovirus infection (Gottlieb et al., 1981).

(2) In 1985 and again in 1988 Haverkos analyzed the AIDS risks of 87 male homosexual AIDS patients with Kaposi's sarcoma (47), Kaposi's sarcoma plus pneumonia (20), and pneumonia only (20) (Haverkos et al., 1985; Haverkos, 1988b). All men had used several sexual stimulants; 98% had used nitrites. Those with Kaposi's sarcomas reported double the amount of sexual partners and 4.4 times more receptive anal intercourse than those with only pneumonia. The median number of sexual partners in the year prior to the illness was 120 for those with Kaposi's and 22 for those with pneumonia only. The Kaposi's cases reported 6 times more amyl nitrite and ethyl chloride use, 4 times more barbiturate use, and twice the methaqualone, lysergic acid, and cocaine use than those with pneumonia only. Since no statistically significant differences were found for sexually transmitted diseases among the patients, the authors concluded that the drugs had caused Kaposi's sarcoma.

Although the data for Haverkos's analysis had been collected before HIV was declared the cause of AIDS, Haverkos's conclusion is valid. This is because (1) all patients had AIDS but only the heavy drug users had Kaposi's sarcoma in addition to immunodeficiency and because (2) not all can be assumed to be infected by HIV because transmission depends on an average of 1000 contacts (Section 3.5.2). Indeed, HIV was found in only 24% (Deininger et al., 1990), 31% (van Griensven et al., 1990), 43% (Graham et al., 1991; Seage et al., 1992), 48% (Winkelstein et al., 1987), 49% (Lemp et al., 1990), 56% (Marion et al., 1989), and 67% (Darrow et al., 1987) of cohorts of homosexuals at risk for AIDS in Berlin, Amsterdam, Chicago–Washington, D.C.–Los Angeles–Pittsburgh, Boston, San Francisco, and Canada that were similar to those described by Haverkos.

(3) A 4.5-year tracking study of 42 homosexual men with lymphadenopathy but not AIDS reported that 8 had developed AIDS within 2.5 years (Mathur-Wagh et al., 1984) and 12 within 4.5 years of observation (Mathur-Wagh et al., 1985). All of these men had used nitrite inhalants and other recreational drugs including amphetamines and cocaine, but they were not tested for HIV. The authors concluded that "a history of heavy or moderate use of nitrite inhalant before study entry was predictive of ultimate progression to AIDS" (Mathur-Wagh et al., 1984).

(4) Before HIV was known, three controlled studies compared 20 homosexual AIDS patients to 40 AIDS-free controls (Marmor et al.,

1982), 50 patients to 120 controls (Jaffe et al., 1983b), and 31 patients to 29 controls (Newell et al., 1985a) to determine AIDS risk factors. Each study reported that multiple "street drugs" were used as sexual stimulants. And each study concluded that the "lifetime use of nitrites" (Jaffe et al., 1983b) was a 94% to 100% consistent risk factor for AIDS (Newell et al., 1985a).

(5) Early CDC data indicate that 86% of male homosexuals with AIDS had used oral drugs at least once a week and 97% occasionally (Centers for Disease Control, 1982; Haverkos, 1988b). The National Institute on Drug Abuse reports correlations from 69% (Lange et al., 1988) to virtually 100% (Haverkos, 1988a; Newell et al., 1988) between nitrite inhalants and other drugs and subsequent Kaposi's sarcoma and pneumonia.

(6) A 27- to 58-fold higher consumption of nitrites by male homosexuals compared to heterosexuals and lesbians (Lesbian and Gay Substance Abuse Planning Group, 1991a, b) correlates with a 20-fold higher incidence of Kaposi's sarcoma (Selik et al., 1987; Beral et al., 1990) and a higher incidence of all other AIDS diseases in male homosexuals compared to most other risk groups (Tables 1 and 2).

(7) During the last 6–8 years the use of nitrite inhalants among male homosexuals decreased, e.g., from 58% in 1984 to 27% in 1991 in San Francisco (Lesbian and Gay Substance Abuse Planning Group, 1991b). In parallel, the incidence of Kaposi's sarcoma among American AIDS patients decreased from a high of 50% in 1981 (Haverkos, 1988b), to 37% in 1983 (Jaffe et al., 1983a), to a low of 10% in 1991 (Centers for Disease Control, 1992b). It follows that the incidence of Kaposi's sarcoma is proportional to the number of nitrite users.

(8) After the discovery of HIV, 5 out of 6 HIV-free male homosexuals from New York who have Kaposi's sarcoma have reported the use of nitrite inhalants (Friedman-Kien et al., 1990). Some of these men had no immunodeficiency. Soon after, another 6 cases of HIV-free Kaposi's sarcoma were reported in a "high risk population" from New York (Safai et al., 1991). This indicates directly that HIV is not necessary and suggests that drugs are sufficient for AIDS.

(9) A 44-year-old, HIV-free homosexual man from Germany developed Kaposi's sarcoma and had a T4 to T8-cell ratio of only 1.2. The man "had used nitrite inhalants for about 10 years," but had no apparent immunodeficiency (Marquart et al., 1991). Likewise, Kaposi's sarcoma was diagnosed in a 40-year-old, promiscuous HIV-free homosexual from England who admitted "frequent use of amyl nitrite." The patient was otherwise symptom-free, with a normal T4:T8–cell ratio (Archer et al., 1989). In 1981 an English male homosexual with a "history of amylnitrite inhalation," hepatitis B, gonorrhea, and syphilis was also diagnosed with

Kaposi's sarcoma. In 1984 he was found to be free of HIV, but in 1986 he became antibody-positive (Lowdell and Glaser, 1989).

(10) A prospective study from Canada identified immunodeficiency in 33 out of 166 HIV-free homosexual men (Marion et al., 1989). The study did not mention drug consumption, but a later report on homosexual men with AIDS from the same cohort documented that all had been using either more or less than 20 "hits" of nitrites per month (Section 4.4) (Archibald et al., 1992). Thus, nitrites and possibly other drugs were sufficient for immunodeficiency.

Likewise, W. Lang et al. (1989) described a steady decline of T4-cells in 37 homosexual men in San Francisco from 1200 per µL prior to HIV infection to 600 or less at the time of infection. Although recreational drug use and AZT were not mentioned, other studies of the same cohort of homosexual men from San Francisco described extensive use of recreational drugs (Section 4.3.2) (Darrow et al., 1987; Moss, 1987) and AZT (Lang et al., 1991).

4.5.2. Long-term Intravenous Drug Use Sufficient for AIDS-defining Diseases

(1) Among intravenous drug users in New York representing a "spectrum of HIV-related diseases," HIV was observed in only 22 out of 50 pneumonia deaths, 7 out of 22 endocarditis deaths, and 11 out of 16 tuberculosis deaths (Stoneburner et al., 1988).

(2) Pneumonia was diagnosed in 6 out of 289 HIV-free and in 14 out of 144 HIV-positive intravenous drug users in New York (Selwyn et al., 1988).

(3) Among 54 prisoners with tuberculosis in New York state, 47 were street-drug users, but only 24 were infected with HIV (Braun et al., 1989).

(4) In a group of 21 long-term heroin addicts, the ratio of helper to suppressor T cells declined during 13 years from a normal of 2 to less than 1, which is typical of AIDS (Centers for Disease Control, 1987; Institute of Medicine, 1988), but only 2 of the 21 were infected by HIV (Donahoe et al., 1987).

(5) Thrombocytopenia and immunodeficiency were diagnosed in 15 intravenous drug users on average 10 years after they became addicted, but 2 were not infected with HIV (Savona et al., 1985).

(6) The annual mortality of 108 HIV-free Swedish heroin addicts was similar to that of 39 HIV-positive addicts, i.e., 3–5%, over several years (Annell et al., 1991).

(7) A survey of over a thousand intravenous drug addicts from Germany reported that the percentage of HIV-positives among drug deaths

(10%) was exactly the same as that of HIV-positives among living intravenous drug users (Püschel and Mohsenian, 1991). Another study from Berlin also reported that the percentage of HIV-positives among intravenous drug deaths was essentially the same as that among living intravenous drug users, i.e., 20–30% (Bschor et al., 1991). This indicates that drugs are sufficient for and that HIV does not contribute to AIDS-defining diseases and deaths of drug addicts.

(8) In 1989 the annual mortality of 197 HIV-positive, parenteral drug users from Amsterdam with an average age of 29 years was 4% and that of 193 age-matched HIV-negatives was 3% (Mientjes et al., 1992). The annual incidence of pneumonia was 29% in the HIV-positives and 9% in the negatives. Clearly, a 3-fold higher morbidity is intrinsically inconsistent with a near identical mortality. However, the slightly higher mortality of HIV-positives is compatible with the fact that the positives had injected more drugs for a longer time, e.g., 84% of the positives versus 64% of the negatives had injected over the past 5 years, 85% versus 72% over the past 6 months, and 59% versus 50% had injected heroin and cocaine.

(9) Lymphocyte reactivity and abundance were depressed by the absolute number of injections of drugs not only in 111 HIV-positive, but also in 210 HIV-free drug users from Holland (Mientjes et al., 1991).

(10) The same lymphadenopathy, weight loss, fever, night sweats, diarrhea, and mouth infections were observed in 49 out of 82 HIV-free, and in 89 out of 136 HIV-positive, long-term intravenous drug users in New York (Des Jarlais et al., 1988).

(11) Among intravenous drug users in France, lymphadenopathy in 41 and an over 10% weight loss in 15 out of 69 HIV-positives was observed, and in 12 and 8, respectively, out of 44 HIV-negatives (Espinoza et al., 1987). The French group had used drugs for an average of 5 years, but the HIV-positives had injected drugs about 50% longer than the negatives.

(12) In a group of 510 HIV-positive intravenous drug users in Baltimore, 29% reported one and 19% reported two or more AIDS-defining diseases. In a control group of 160 HIV-negative intravenous drug users matched with the HIV-positives for "current drug use," again 29% reported one and 13% reported two or more AIDS-defining diseases (Muñoz et al., 1992).

Nevertheless, the average T-cell count of HIV-negatives was about 2 times higher than that of HIV-positives (Muñoz et al., 1992). As in the above French study (Espinoza et al., 1987), this appears to reflect a higher lifetime dose of drugs, because HIV is a marker for the duration and extent of drug consumption (Sections 3.4.3, 4.4, and 5).

(13) Among 97 intravenous drug users in New York with active tuberculosis, 88 were HIV-positive and 9 were HIV-negative; among 6 "crack" (cocaine) smokers with tuberculosis, 3 were HIV-negative and 3 were positive (Brudney and Dobkin, 1991).

(14) The mental development and psychomotor indices of 8 HIV-infected and 6 uninfected infants were observed from 6–21 months of age. The mothers of each group were HIV-positive and had used intravenous drugs and alcohol during pregnancy (Koch, 1990; Koch et al., 1990; T. Koch, R. Jeremy, E. Lewis, P. Weintrub, C. Rumsey, and M. Cowan, unpublished data). The median indices of both groups were significantly below average, e.g., 80/100 mental development and 85/100 psychomotor units. The uninfected infants remained on average about 5/100 units higher. A control group of 5 infants, born to HIV-negative mothers who had also used intravenous drugs and alcohol during pregnancy, also had subnormal indices averaging about 95/100 for both criteria.

The degree of neurological retardation of the infants correlated directly with maternal drug consumption: 80% of the mothers of infected infants were "heavy" and 10% occasional parenteral cocaine users, and 33% were "heavy" and 33% occasional alcohol users during pregnancy; 45% of the mothers of uninfected infants were "heavy" and 30% occasional parenteral cocaine users, and 35% were "heavy" and 30% occasional alcohol users; and 21% of the HIV-free mothers were "heavy" and 58% occasional parenteral cocaine users, and 12% were "heavy" and 44% occasional alcohol users. In addition 66% of the HIV-positive and 63% of the negative mothers reported the use of opiates during pregnancy (T. Koch, R. Jeremy, E. Lewis, P. Weintrub, C. Rumsey, and M. Cowan, unpublished data).

(15) The psychomotor indices of infants "exposed to substance abuse *in utero*" were "significantly" lower than those of controls, "independent of HIV status." Their mothers were all drug users but differed with regard to drug use during pregnancy. The mean indices of seventy children exposed during pregnancy were 99 and those of 25 controls were 109. Thus, maternal drug use during pregnancy impairs children independent of HIV (Aylward et al., 1992).

The same study also reports a "significant difference" based on the HIV status of these children. The mean scores of 12 HIV-positives was 88 and that of 75 HIV-negatives was 102. But the study did not break down the scores of the HIV-positive infants based on "exposure to substance abuse *in utero*." Indeed, the scores of 4 of the 12 HIV-infected infants were "above average," i.e., 100 to 114, and 4 of the 12 mothers did not inject drugs during pregnancy.

584 ▪ *Appendix B*

(16) Ten HIV-free infants born to intravenous drug-addicted mothers had the following AIDS-defining diseases, "failure to thrive, persistent generalized lymphadenopathy, persistent oral candidiasis, and developmental delay..." (Rogers et al., 1989).

(17) One HIV-positive and 18 HIV-free infants born to intravenous drug-addicted mothers had only half as many leukocytes at birth than normal controls. At 12 months after birth, the capacity of their lymphocytes to proliferate was 50–70% lower than that of lymphocytes from normal controls (Culver et al., 1987).

(18) Two studies to test the role of HIV on neurological function confirm the drug-AIDS hypothesis indirectly and directly. The first of these, which excluded users of psychoactive drugs, found that neuropsychometric functions of 50 HIV-negative homosexuals were the same as those of 33 HIV-positives (Clifford et al., 1990). Another study of intravenous drug users on methadone found that neither the drug-impaired neuropsychological functions of 137 HIV-negatives nor those of 83 HIV-positives were deteriorating over 7.4 months (McKegney et al., 1990). However, the study notes that the functions of HIV-positives were lower than those of HIV-negatives because "a greater number of injections per month, more frequent use of cocaine... were strongly associated with HIV seropositivity."

Thus, a critical lifetime dosage of drugs appears necessary in HIV-positives and sufficient in HIV-negatives to induce an AIDS indicator and other diseases.

4.6. TOXIC EFFECTS OF DRUGS USED BY AIDS PATIENTS

4.6.1. *Toxicity of Recreational Drugs*

From as early as 1909 (Achard et al., 1909) evidence has accumulated that long-term consumption of psychoactive drugs leads to immune suppression and clinical abnormalities similar to AIDS, including lymphopenia, lymphadenopathy, fever, weight loss, septicemia, increased susceptibility to infections, and profound neurological disorders (Terry and Pellens, 1928; Briggs et al., 1967; Dismukes et al., 1968; Sapira, 1968; Harris and Garret, 1972; Geller and Stimmel, 1973; Brown et al., 1974; Louria, 1974; McDonough et al., 1980; Cox et al., 1983; Kozel and Adams, 1986; Selwyn et al., 1989; Turner et al., 1989; Kreek, 1991; Pillai et al., 1991; Bryant et al., 1992). Since the early 1980s, when T-cell ratios became measurable, low T4 to T8–cell ratios averaging 1 or less were reported in addicts who had injected drugs for an average of 10 years (Layon et al., 1984)

Intravenous drugs can be toxic directly and indirectly. Indirect toxicity can be due to malnutrition because of the enormous expense of illicit drugs or to septicemia because most illicit drugs are not sterile (Cox et al., 1983; Stoneburner et al., 1988; Lerner, 1989; Buehler et al., 1990; Pillai et al., 1991; Luca-Moretti, 1992). Typically, intravenous drug users develop pneumonia, tuberculosis, endocarditis, and wasting disease (Layon et al., 1984; Stoneburner et al., 1988; Braun et al., 1989; Brudney and Dobkin, 1991). Oral consumption of cocaine and other psychoactive drugs has been reported to cause pneumonitis, bronchitis, edema (Ettinger and Albin, 1989), and tuberculosis (Brudney and Dobkin, 1991). Physiological and neurological deficiencies, including mental retardation, are observed in children born to mothers addicted to cocaine and other narcotic drugs (Fricker and Segal, 1978; Lifschitz et al., 1983; Alroomi et al., 1988; Blanche et al., 1989; Root-Bernstein, 1990a; Toufexis, 1991; Finnegan et al., 1992; Luca-Moretti, 1992). According to the National Institute on Drug Abuse, "[c]ocaine is currently the drug of greatest national concern, from a public health point of view..." (Schuster, 1984).

Because inhalation of alkyl nitrites relaxes smooth muscles, it has been prescribed since 1867 against angina pectoris and heart pain at doses of 0.2 mL (Cox et al., 1983; Newell et al., 1985b; Shorter, 1987; Seage et al., 1992). No AIDS-defining diseases have been reported at these doses in patients with those relatively severe, terminal cardiovascular diseases (Cox et al., 1983; Shorter, 1987), possibly because they did not live long enough to develop them. However, immediate and late toxicities have been observed in recreational users who have inhaled milliliters of nitrite inhalants (Newell et al., 1985b; Schwartz, 1988). Alkyl nitrites are directly toxic as they are rapidly hydrolyzed *in vivo* to yield nitrite ions, which react with all biological macromolecules (Osterloh and Olson, 1986; Maikel, 1988). Addicts with 0.5 mm nitrite derivatives and 70% methemoglobin in blood have been recorded (Osterloh and Olson, 1986). Toxicity for the immune system, the central nervous system, the hematologic system, and pulmonary organs has been observed after short exposure to nitrites in humans and in animals (Newell et al., 1985b, 1988; Wood, 1988). In 1982 Goedert et al. found that the helper to suppressor T-cell ratio was lower in homosexual men who had used volatile nitrite inhalants than among nonusers. Further, alkyl nitrites were shown to be both mutagenic and carcinogenic in animals (Jorgensen and Lawesson, 1982; Hersh et al., 1983; Mirvish et al., 1988; Newell et al., 1985b, 1988).

By comparing the AIDS risk factors of 31 homosexual men with AIDS to 29 without, Newell et al. and others determined a direct "dose-response gradient" that the higher the nitrite usage the greater the risk for

AIDS (Marmor et al., 1982; Newell et al., 1985a; Haverkos and Dougherty, 1988) and deduced a 7–10-year lag time between chronic consumption and Kaposi's sarcoma (Newell et al., 1985b). Likewise, a French study of homosexual men with and without AIDS who had inhaled nitrites documents that "cases were significantly older (approximately 10 years) than controls" (Section 4.3.2) (Messiah et al., 1988). Also, a German study observed Kaposi's sarcoma in an HIV-free man after he had inhaled nitrites for 10 years (Section 4.5.1) (Marquart et al., 1991). These studies indicate that about 10 years of nitrite inhalation are necessary to convert "controls" to "cases."

In view of this several investigators have proposed that nitrite inhalants cause pulmonary and skin Kaposi's sarcoma and pneumonia by direct toxicity on the skin and oral mucosa (Centers for Disease Control, 1982; Marmor et al., 1982; Haverkos et al., 1985; Mathur-Wagh et al., 1985; Newell et al., 1985a; Lauritsen and Wilson, 1986; Haverkos, 1990). Because of their toxicity a prescription requirement was instated for the sale of nitrite inhalants by the Food and Drug Administration in 1969 (Newell et al., 1985b) and because of an "AIDS link" (Cox, 1986) the sale of nitrites was banned by the U.S. Congress in 1988 (Public Law 100–690) (Haverkos, 1990) and by the Crime Control Act of 1990 (January 23, 1990).

Although a necessary role of HIV in HIV-positive AIDS patients cannot be excluded, this role would be stoichiometrically insignificant compared to that of the drugs. This is because drug molecules exceed HIV molecules by over 13 orders of magnitude. Given about 10^{10} leukocytes per human, of which at most 1 in 10^4 are actively infected (Section 3.5.1), and that each actively infected cell makes about 100 viral RNAs per day, there are only 10^6 T-cells with 10^2 HIV RNAs in an HIV-positive person. By contrast, 1 mL (or 0.01 mol) of amyl nitrite with a molecular weight of 120 contains 6×10^{21} molecules, or 6×10^7 nitrite molecules, for every one of the 10^{14} cells in the human body. Thus, based on molecular representation, HIV's role in AIDS, if it existed, would have to be catalytic in comparison to that of drugs.

Pillai, Nair, and Watson conclude from a recent review on the role of recreational drugs in AIDS: "Circumstantial and direct evidence suggesting a possible role for drug... induced immunosuppression appears overwhelming. What is required now is better and more accurate detection of substance abuse, a direct elucidation of the immune and related mechanisms involved, and appropriate techniques to analyze it" (Pillai et al., 1991).

4.6.2. Toxicity of AZT

Since 1987 AZT has been used as an anti-HIV agent (Section 4.3.3) based on two placebo-controlled studies reporting therapeutic and prophylactic benefits (Section 4.4.2). However, AZT was originally developed in the 1960s for cancer chemotherapy to kill human cells via termination of DNA synthesis (Cohen, 1987; Yarchoan and Broder, 1987a; Yarchoan et al., 1991). The primary AZT metabolites are 3′-termini of DNA which are cell-killing, 3′-amino-dT which is more toxic than AZT, and 5′-O-glucuronide which is excreted (Cretton et al., 1991). As a chain terminator of DNA synthesis, AZT is toxic to all cells engaged in DNA synthesis. AZT toxicity varies a great deal with the subject treated due to differences in its uptake and in its cellular metabolism (Chernov, 1986; Elwell et al., 1987; Yarchoan and Broder, 1987b; Smothers, 1991; Yarchoan et al., 1991).

AZT is prescribed as an AIDS prophylaxis or therapy at 500–1500 mg per day, corresponding to a concentration of 20–60 µmol/L in the patient. Prior to the licensing of AZT, Burroughs-Wellcome (the manufacturer of the drug) and the NIH have jointly claimed selective inhibition of HIV by AZT *in vitro* because human lymphoblasts and fibroblasts appeared over 1000-fold more resistant to AZT (inhibited only at 1–3 mM) than was replication of HIV (inhibited at 50–500 mM) (Furman et al., 1986). On this basis they calculated an *in vitro* antiviral therapeutic index of 10^4. This "selective" sensitivity of HIV to AZT was explained in terms of a "selective interaction of AZT with HIV reverse transcriptase" (Furman et al., 1986). Accordingly the manufacturer informs AZT recipients: "The cytotoxicity of zidovudine [AZT] for various cell lines was determined using a cell growth assay… ID_{50} values for several human cell lines showed little growth inhibition by zidovudine except at concentrations >50 µg/mL ($\geq$200 µM) or less" (Medical Economics Data, 1992). Further, it informs them that enterobacteria including *E. coli* are inhibited "by low concentrations of zidovudine [AZT]," between 0.02 and 2 µM AZT, just like HIV (Medical Economics Data, 1992).

However, an independent study showed in 1989 that AZT is about 1000-times(!) more toxic for human T-cells in culture, i.e., at about 1µM than the study conducted by its manufacturer and the NIH (Avramis et al., 1989). Other studies have also found that AZT inhibits T-cells and other hemopoietic cells *in vitro* at 1–8 µM (Balzarini et al., 1989; Mansuri et al., 1990; Hitchcock, 1991). Since normal deoxynucleotide triphosphates are present in the cell at micromolar concentrations, toxicity of AZT should be expected in the micromolar range. Indeed, when AZT is added at a micromolar concentration to the culture medium, it and its phosphorylated derivatives quickly reach an equivalent or higher concentration in the cell and thus effectively compete with their natural thymidine counterparts (Avramis et al., 1989; Balzarini et al., 1989; Ho and Hitchcock, 1989; Hitchcock, 1991).

Thus, the low cellular toxicity reported by the manufacturer and the NIH for human cells appears erroneous—possibly because "the clinical development of AZT was exceedingly rapid; it was approved for clinical use in the U.S. about two years after the first *in vitro* observation of its activity against HIV" (Yarchoan et al., 1991). It follows that AZT does not selectively inhibit viral DNA synthesis and is prescribed at concentrations that exceed 20- to 60-fold the lethal dose for human cells in culture.

In view of its inevitable toxicity, the rationale of using AZT as an anti-HIV drug must be reconsidered and its potential antiviral effect must be weighed against its toxicity.

4.6.2.1. *AZT not a rational anti-HIV drug.* A rational antiviral therapy depends on proof that the targeted virus is the cause of the disease to be treated and that toxicity for the virus outweighs that for the host cell. Such proof cannot be supplied for AZT for the following reasons:

(1) There is no proof that HIV causes AIDS (Section 3.3).

(2) Even if the hypothesis that HIV causes AIDS by killing T-cells were correct, it would be irrational to kill the same infected cells twice, once presumably with HIV and once more with AZT.

(3) Since many healthy persons with antibodies against HIV have equal or even higher percentages of infected T-cells than do AIDS patients (Section 3.3), there is no reverse transcription of HIV during progression to AIDS that could be targeted with AZT. Even if some reverse transcription occurred in antibody-positive persons, AZT could not differentially inhibit viral DNA, because HIV DNA comprises only 9 kb, but cell DNA comprises 10^6 kb. Thus, cell DNA is a 100,000-fold bigger target for AZT than HIV. And even if AZT showed a 100-fold preference for reverse transcriptase of HIV over cellular DNA polymerase, as has been claimed by the study conducted by Burroughs-Wellcome and the NIH (Furman et al., 1986), cell DNA would still be a 1000-fold bigger target for AZT than viral DNA. It follows that cell DNA is the only realistic target of AZT in antibody-positive persons.

(4) Since AZT cannot distinguish infected from uninfected leukocytes and on average less than 1 in 1000 is infected (Section 3.3), AZT must kill at least 1000 leukocytes in AIDS patients and in asymptomatic HIV

carriers to kill just one infected cell—a very high toxicity index, even if HIV were the cause of AIDS.

It follows that there is no rational basis for AZT therapy or prophylaxis for AIDS (Duesberg, 1992d).

4.6.2.2. *Toxicity of AZT in AIDS patients and AIDS-free persons.* The following AZT-specific diseases or dysfunctions have been recorded in AIDS patients, in AIDS-free persons and animals treated with AZT, based on studies listed here (Section 4.4.2) and reviewed elsewhere (Smothers, 1991; Medical Economics Data, 1992):

(1) anemia, neutropenia, and leukopenia in 20–80%, with about 30–57% requiring transfusions within several weeks (Gill et al., 1987; Kolata, 1987; Richman et al., 1987; Dournon et al., 1988; Walker et al., 1988; Swanson et al., 1990; van Leeuwen et al., 1990; Smothers, 1991; Hamilton et al., 1992);

(2) severe nausea from intestinal intoxication in up to 45% (Richman et al., 1987; Volberding et al., 1990; Smothers, 1991);

(3) muscle atrophy and polymyositis, due to inhibition of mitochondrial DNA synthesis in 6–8% (Richman et al., 1987; Bessen et al., 1988; Gorard and Guilodd, 1988; Helbert et al., 1988; Dalakas et al., 1990; Till and MacDonnell, 1990; Yarchoan et al., 1991; Hitchcock, 1991);

(4) lymphomas in about 9% within one year on AZT (Section 4.4.2);

(5) acute (nonviral) hepatitis (Dubin and Braffman, 1989; Smothers, 1991);

(6) nail dyschromia (Don et al., 1990; Smothers, 1991);

(7) neurological diseases including insomnia, headaches, dementia, mania, Wernicke's encephalopathy, ataxia, and seizures (Smothers, 1991), probably due to inhibition of mitochondrial DNA (Hitchcock, 1991);

(8) 12 out of 12 men reported impotence after one year on AZT (Callen, 1990); and

(9) AZT is carcinogenic in mice, causing vaginal squamous carcinomas (Cohen, 1987; Yarchoan and Broder, 1987a), and it transforms mouse cells *in vitro* as effectively as methylcholanthrene (Chernov, 1986).

Overall AZT is not a rational prophylaxis or a therapy for AIDS and is capable of causing potentially fatal diseases, such as anemia, leukopenia, and muscle atrophy. Yet, despite its predictable toxicity, AZT is thought to have serendipitous therapeutic and prophylactic benefits according to those investigators who have studied its effects together with the manufacturer for licensing of the drug (Section 4.4.2) (Fischl et al., 1987; Richman et al., 1987; Volberding et al., 1990). Confronted with the difficulties in rationalizing anti-HIV prophylaxis and therapy with AZT, the Wellcome researcher Freestone cites the Burroughs-Wellcome study analyzed above (Section 4.4.2, item 1): "the primary end-point for the study was death (1 in 145 zidovudine recipients, 19 in 137 placebo recipients...)—an end-point little subject to observer error or bias" (Freestone, 1992).

The popularity of AZT as an anti-HIV drug can be explained only by the widespread acceptance of the virus-AIDS hypothesis, the failure to consider the enormous difference between the viral and cellular DNA targets, and a general disregard for the long-term toxicity of drugs (Section 6). In the words of retrovirologist Temin, "but the drug generally becomes less effective after six months to a year..." (Nelson et al., 1991)—a euphemism for its fatal toxicity by that time. This is a probable reason that AZT was licensed without long-term studies in animals compatible with human applications and that the need for such studies is neither mentioned nor called for in reviews of its toxic effects in humans (Chernov, 1986; Yarchoan and Broder, 1987b; Smothers, 1991; Yarchoan et al., 1991), although AZT must be the most toxic drug ever approved for indefinite therapy in America. Even the manufacturer acknowledges that "...the drug has been studied for limited periods of time and long term safety and efficacy are not known" (Shenton, 1992) and recommends that "patients should be informed... that the long-term effects of zidovudine are unknown at this time" (Medical Economics Data, 1992). And after prescribing it for five years, even AIDS "experts" have recently expressed doubts about the "survival benefit" of AZT (Kolata, 1992).

4.7. DRUG-AIDS HYPOTHESIS CORRECTLY PREDICTS THE EPIDEMIOLOGY AND HETEROGENEOUS PATHOLOGY OF AIDS

(1) The long-term consumption of drugs, but not the hosting of a latent virus, predicts drug-specific pathogenicity after "long latent periods." These long latent periods of HIV are in reality the lag periods that recreational drugs (Schuster, 1984; Newell et al., 1985b) and frequent transfusions of foreign proteins take to cause AIDS-defining diseases (Section 3.4.4.5). Drugs are molecularly abundant (Section 4.6.1) and biochemically active as long as they are administered and thus are cumulatively toxic over time. It is for this reason that it typically takes 5–10 years for recreational drugs, and months for AZT, to cause AIDS-defining and other diseases (Sections 3.1 and 5). But HIV, after a brief period of immunogenicity (Clark et al., 1991; Daar et al., 1991), is chronically dormant and thus molecularly and biochemically irrelevant for the rest of the host's life.

Appendix B ■ 591

(2) Drugs and other noninfectious agents also exactly predict the epidemiology of AIDS. About 32% of American AIDS patients are confirmed intravenous drug users, 60% appear to use recreational drugs orally, and an unknown but large percentage of people in both behavioral and clinical AIDS risk groups use AZT. Moreover, the consumption of recreational drugs by AIDS patients is probably underreported because the drugs are illicit and because medical scientists and support for research are currently heavily biased in favor of viral AIDS (Section 6) (Ettinger and Albin, 1989; Lerner, 1989; Duesberg, 1991b). In sum, more than 90% of American AIDS is correlated with drugs. The remainder would reflect the natural background of AIDS-defining diseases in the U.S. (Duesberg, 1992f). Indeed, only drug users do not benefit from the ever-improving health parameters and increasing life spans of the Western world (Hoffman, 1992; The Software Toolworks Atlas^{TM}, 1992). Unfortunately, the widespread use of AZT in hemophiliacs (Section 4.3.3) predicts a new increase in their mortality.

The dramatic increase in America in the consumption of all sorts of recreational drugs since the Vietnam War also explains the simultaneous increase of AIDS in intravenous drug users and male homosexuals (Centers for Disease Control, 1992b). AIDS of both risk groups followed closely the above-listed drug-use statistics during the past 15 years, with increases in 1987 that corresponded to the expanded AIDS definition (Centers for Disease Control, 1987) and the introduction of AZT treatment. By contrast a sexually transmitted AIDS would have soared much faster among homosexuals than among intravenous drug users (Weyer and Eggers, 1990; Eggers and Weyer, 1991). The apparent exponential spread of AIDS during the period from 1984 to 1987 (Heyword and Curran, 1988; Mann et al., 1988; Weyer and Eggers, 1990) probably reflected an exponential spread of "AIDS testing," which resulted in the spread of AIDS diagnoses for drug diseases (Section 4.2). AIDS testing had increased from 0 tests per year in 1984 to 20 million in 1986 in the U.S. alone (Section 3.6).

(3) The drug hypothesis further predicts that the 50–70% of American and the 50–80% of European intravenous drug users who are HIV-free (Stoneburner et al., 1988; Turner et al., 1989; Brenner et al., 1990; U.S. Department of Health and Human Services, 1990; National Commission on AIDS, 1991) and the HIV-free male homosexuals who use sexual stimulants will develop the same diseases as their HIV-positive counterparts—except that the HIV-free diseases will be diagnosed by their old names. This has been amply confirmed for intravenous drug users (Section 4.5, Note added in proof). Yet, more such cases must exist because the CDC allows "presumptive diagnosis" of HIV infection and only about 50% of

all American AIDS cases are confirmed positives (Sections 2.2 and 3.4.1) and because only about 50% of homosexuals from many different cohorts at risk for AIDS are confirmed HIV-positive (Section 4.5.1).

(4) The drug hypothesis also correctly predicts drug-specific AIDS diseases in distinct risk groups due to distinct drugs (Sections 2.1.3, 3.4.5, and 5, Table 2).

4.8. CONSEQUENCES OF THE DRUG-AIDS HYPOTHESIS: RISK-SPECIFIC PREVENTIONS AND THERAPIES, BUT RESENTMENT BY THE VIRUS-AIDS ESTABLISHMENT

The drug-AIDS hypothesis predicts that the AIDS diseases of the behavioral AIDS risk groups in the U.S. and Europe can be prevented by stopping the consumption of recreational and anti-HIV drugs, but not by "safe sex" (Institute of Medicine, 1988; Weiss and Jaffe, 1990; Maddox, 1991b) and "clean" needles, i.e., sterile injection equipment (National Commission on AIDS, 1991) for toxic and unsterile street drugs. Indeed AIDS has continued to increase in all countries that have promoted safe sex to prevent AIDS for over five years now (Centers for Disease Control, 1992b; World Health Organization, 1992a; Anderson and May, 1992). Further, the hypothesis raises the hopes for risk-specific groups.

According to the drug-AIDS hypothesis, AZT is AIDS by prescription. Screening of blood for antibodies to HIV is superfluous, if not harmful, in view of the anxiety that a positive test generates among believers in the virus-HIV hypothesis (Grimshaw, 1987) and the toxic AZT prophylaxis prescribed to many who test "positive." Eliminating the test would also reduce the cost of the approximately 12 million annual blood donations in the U.S. (Williams et al., 1990) and of examining annually 200,000 recruits and 2 million servicemen for the U.S. Army (Burke et al., 1990) by $12 to $70 each (Irwin Memorial Blood Bank, San Francisco, personal communication). Further, it would lift travel restrictions for antibody-positives to many countries including the U.S. and China, it would lift quarantine for HIV-positive Cubans, it would acquit all those antibody-positive Americans who are currently imprisoned for having had sex with antibody-negatives and would grant to HIV "antibody-positives" the same chances to be admitted to a health insurance program as to those who have antibodies only to other viruses.

Despite its many potential blessings, the drug hypothesis is currently highly unpopular—not because it would be difficult to verify, but because of its consequences for the virus-AIDS establishment (Section 6). The drug hypothesis is very testable epidemiologically and experimentally by studying the effects of the drugs consumed by AIDS patients in animals. Indeed, most tests have already been done (Section 4). To disprove this hypothesis it would be necessary to document that an infectious agent exists which—in the absence of AZT(!)—causes AIDS diseases above their normal background in the non–drug-using population. The medical, ethical, and legal consequences of the drug-AIDS hypothesis, should it prevail, have recently been summarized under the title "Duesberg: An Enemy of the People?" (Ratner, 1992). Ratner points out that "[t]he loss of confidence of Americans in their scientists and perhaps, by extension, their physicians, could rival their current disillusionment with politicians" and wonders, "What would happen to the reservoir of good will painstakingly built up for the victims of AIDS?"

5. DRUGS AND OTHER NONCONTAGIOUS RISK FACTORS RESOLVE ALL PARADOXES OF THE VIRUS-AIDS HYPOTHESIS

A direct application of the hypothesis that drugs and other noncontagious risk factors cause AIDS proves that it can resolve all paradoxes of the virus-AIDS hypothesis:

(1) It is paradoxical to assume that AIDS is new because HIV is new. HIV is a long-established, perinatally transmitted retrovirus. It just appears new because, being a chronically latent virus, it became detectable only with recently developed technology (Section 3.5.1). Instead, drugs are the only new health risks in this era of ever-improving health parameters. Thus, AIDS is new because the drug epidemic is new.

(2) According to the virus-AIDS hypothesis it is paradoxical that AIDS did not "explode" into the general population as predicted (Institute of Medicine, 1986; Shorter, 1987; Fineberg, 1988; Heyward and Curran, 1988; Blattner, 1991; Mann and the Global AIDS Policy Coalition, 1992). AIDS has remained restricted for over 10 years to only 15,000 annual cases (0.015%) of the over 100 million sexually active heterosexual Americans, and to only 25,000 (0.3%) of the 8 million homosexuals (Centers for Disease Control, 1992b), although venereal diseases (Aral and Holmes, 1991), unwanted pregnancies and births (Hoffman, 1992; The Software Toolworks World Atlas^{TM}, 1992) are on the increase in America. (Homosexuals represent about 10% of the adult male population (Turner et al., 1989; Lesbian and Gay Substance Abuse Planning Group, 1991a).) This is because psychoactive drugs and AZT, not HIV, are the causes of AIDS.

(3) The paradox of a virus causing risk-group–specific and country-specific AIDS diseases is resolved by distinct nonviral AIDS causes, including drugs and other noncontagious pathogens like long-term transfusions and malnutrition (Sections 2.1.3 and 3.4.5, Tables 1 and 2).

(4) The paradox of a male-specific AIDS virus (i.e., 90% of all American and 86% of all European AIDS cases are males), although no AIDS disease is male-specific, is resolved by male-specific behavior and by male genetic disorders. In America and Europe males consume over 75% of all "hard" injected psychoactive drugs (Section 4.3.1), homosexual males are almost exclusive users of oral aphrodisiacs like nitrites (Section 4.3.2), and nearly all hemophiliacs are males.

(5) The paradox of a 10-year slow AIDS virus—i.e., AIDS occurs only after "latent(!) periods" of HIV that average 10 years in adults and 2 years in babies (Section 2.2)—is resolved by the cumulative toxicity of long-term drug use. According to the CDC the "lifetime use" of drugs determines the AIDS risk (Jaffe et al., 1983b). On average 5–10 years elapse in adult drug addicts between the first use of drugs and "acquiring" drug-induced AIDS diseases (Layon et al., 1984; Schuster, 1984; Savona et al., 1985; Donahoe et al., 1987; Espinoza et al., 1987; Weber et al., 1990). The time lag from a nitrite habit to Kaposi's sarcoma has been determined to be 7–10 years (Newell et al., 1985b). Severe T-cell depletion and immunodeficiency are also "acquired" by hemophiliacs on average only after 14–15 years of treatment with blood concentrates (Section 3.4.4.5).

In babies of drug-addicted mothers, AIDS appears much sooner than in adults because of a much lower threshold of the fetus for drug-pathogenicity. This also resolves the secondary paradox of a discrepancy of 8 years between the "latent periods" of HIV in babies and in adults.

(6) It is paradoxical that American teenagers do not get AIDS, although over 70% are sexually active, about 50% are promiscuous (Turner et al., 1989; Burke et al., 1990; Congressional Panel, 1992), and 0.03% to 0.3% carry HIV (Section 3.5.2). The paradox that a sexually transmitted "AIDS virus" would spare American and European teenagers is resolved by the fact that only years of drug consumption and years of transfusions for hemophilia (Section 3.4.4.5) will cause AIDS—by which time these teenagers will be in their twenties.

(7) The apparent paradox that the same virus would at the same time cause two entirely different AIDS epidemics, one in Africa and the other in America and Europe, is an artifact of the AIDS definition. Because of the HIV-based AIDS definition, a new drug epidemic in America and Europe and an epidemic of old Africa-specific diseases like fever, diarrhea, and tuberculosis (Section 3.4.4.4) were both called AIDS when HIV

became detectable. Since HIV is endemic in over 10% of Central Africans, over 10% of their AIDS-defining diseases are now called AIDS (Section 2.2).

6. WHY DID AIDS SCIENCE GO WRONG?

6.1. THE LEGACY OF THE SUCCESSFUL GERM THEORY: A BIAS AGAINST NONINFECTIOUS PATHOGENS

Unlike any other scientific hypothesis, the virus-AIDS hypothesis became national American dogma before it could be reviewed by the scientific community. It had been announced by the Secretary of Health and Human Services in 1984 before it had been published in the scientific literature. Unlike any other medical hypothesis it captured the world without ever bearing any fruits in terms of public health benefits. From the beginning the hypothesis has absorbed the critical potential of its many followers with the question of whether Montagnier from France or Gallo from the U.S. had won the race in isolating the "AIDS virus" and who owned the lucrative patent rights for the "AIDS test." This question was so consuming that the presidents of the two countries were called to sign a settlement, and a revisionist paper was published by the opponents describing their fierce controversy as an entente cordiale against the real enemy, the "deadly" AIDS virus (Gallo and Montagnier, 1987). During the 1980s press accounts consistently called HIV 'the deadly virus' (Duesberg, 1989c).

Clearly, the enthusiastic acceptance of the virus-AIDS hypothesis was not based on its scientific rigor or its fruits. It was instead grounded on the universal admiration and respect for the germ theory. The germ theory of the late 19th century ended the era of infectious diseases, which now account for less than 1% of all mortality in the Western world (Cairns, 1978). It celebrated its last triumph in the 1950s with the elimination of the polio epidemic by antiviral vaccines.

But the germ theory continues to inspire both scientists and the public to believe that a "good" body can be protected against "evil" microbes. Accordingly, even the greatly feared and highly stigmatizing "AIDS test" for a presumably new, sexually transmitted "AIDS virus" was readily sold to all governments, medical associations, and even to the AIDS risk groups (Section 6.2), despite the absence of convincing evidence for transmissibility. In the words of one observer, "The rationale for such programs is often the historical precedent of syphilis screening," which "never proved to be effective" and led to "toxic treatments with arsenical drugs, assuming the tests were correct..." and "deep stigma and disrupted relationships..." "Patients required a painful regimen of injections, sometimes for as long as two years" (Brandt, 1988). Even epidemiologists failed to recognize that AIDS and HIV were spreading only in newly established behavioral and clinical risk groups and that HIV was a long-established virus in the general populations of many countries (Section 3.5.1). Instead of considering noninfectious causes, they simulated "coagents" (Eggers and Weyer, 1991) and "assortative scenarios" (Anderson and May, 1992) to hide the growing discrepancies between HIV and AIDS, and they intimidated skeptics with apocalyptic predictions of AIDS pandemics in the general populations of many countries that have raised fears and funds to unprecedented levels (Section 1) (Heyward and Curran, 1988; Mann et al., 1988; Mann and the Global AIDS Policy Coalition, 1992; Anderson and May, 1992).

Even now, in an era free of infectious diseases but full of man-made chemicals, scientists and the public share an unthinking preference for infectious over noninfectious pathogens. Both groups share an obsolete microbophobia but tolerate the use or even indulge in the consumption of numerous recreational and medical drugs. Moreover, progressive scientists and policymakers are not interested in recreational and medical drugs and man-made environmental toxins as causes of diseases, because the mechanisms of pathogenesis are predictable. Further, prevention of drug diseases is scientifically trivial and commercially unattractive.

By contrast, microbial and particularly viral pathogens are scientifically and commercially attractive to scientists. Beginning with Peyton Rous, at least 10 Nobel Prizes have been given to virologists in the past 25 years. And many virologists have become successful biotechnologists. For example, a blood test for a virus is good business if the test becomes mandatory for the 12 million annual blood donations in the U.S., e.g., the "AIDS test." The same is true for a vaccine or an antiviral drug that is approved by the Food and Drug Administration.

Thousands of lives have been sacrificed to this bias for infectious theories of disease, even before AIDS appeared. For example, the U.S. Public Health Service insisted for over 10 years in the 1920s that pellagra was infectious, rather than a vitamin B deficiency as had been proposed by Joseph Goldberger (Bailey, 1968). Tertiary syphilis is commonly blamed on treponemes but is probably due to a combination of treponemes and long-term mercury and arsenic treatments used prior to penicillin or merely to these treatments alone (Brandt, 1988; Fry, 1989). "Unconventional" viruses were blamed for neurological diseases like Kreutzfeld-Jacob's disease, Alzheimer's disease, and kuru (Gajdusek, 1977). The now-extinct kuru was probably a genetic disorder that affected just one tribe of natives from New Guinea (Duesberg and Schwartz, 1992). Although a Nobel Prize was given for this theory, the viruses never materialized, and an unconventional protein, termed "prion," is now blamed for some of these diseases (Evans, 1989c; Duesberg and Schwartz, 1992). Shortly after this incident, a virus was also blamed for a fatal epidemic of neuropathy, including blinding, that started in the 1960s in Japan, but it turned out later to be caused by the prescription drug clioquinol (Enterovioform, Ciba-Geigy) (Kono, 1975; Shigematsu et al., 1975). In 1976 the CDC blamed an outbreak of pneumonia at a convention of Legionnaires on a "new" microbe, without giving consideration to toxins. Since the "Legionnaires' disease" did not spread after the convention and the "Legionnaires' bacillus" proved to be ubiquitous, it was later concluded that "CDC epidemiologists must in the future take toxins into account from the start" (Culliton, 1976). The Legionnaires' disease fiasco is in fact the probable reason that the CDC initially took toxins into account as the cause of AIDS (Oppenheimer, 1992).

The pursuit of harmless viruses as causes of human cancer, supported since 1971 by the Virus-Cancer Program of the National Cancer Institute's War on Cancer, was also inspired by indiscouragable faith in the germ theory (Greenberg, 1986; Duesberg, 1987; Shorter, 1987; Anderson, 1991; Editorial, 1991; Duesberg and Schwartz, 1992). For example, it was claimed in the 1960s that the rare Burkitt's lymphoma was caused by the ubiquitous Epstein-Barr virus, 15 years after infection (Evans, 1989c). But the lymphoma is now accepted to be nonviral and attributed to a chromosome rearrangement (Duesberg and Schwartz, 1992). Further, it was claimed that noncontagious cervical cancer is caused by the widespread herpes virus in the 1970s and by the widespread papilloma virus in the 1980s—but in each case cancer would occur only 30-40 years after infection (Evans, 1989c). Noninfectious causes like chromosome abnormalities, possibly induced by smoking, have since been considered or reconsidered (Duesberg and Schwartz, 1992). Further, ubiquitous hepatitis virus was proposed in the 1960s to cause regional adult hepatomas fifty years(!) after infection (Evans, 1989c). In the 1980s the rare, but widely distributed, human retrovirus HTLV-I was claimed to cause regional adult T-cell leukemias (Blattner, 1990). Yet the leukemias would appear only at advanced age, after "latent periods" of up to 55 years, the age when these "adult" leukemias appear spontaneously (Evans, 1989c; Blattner, 1990; Duesberg and Schwartz, 1992). Although the Virus-Cancer Program has generated such academic triumphs as retroviral oncogenes (Duesberg and Vogt, 1970) and reverse transcriptase (Temin and Mitzutani, 1970), it has been a total failure in terms of clinical relevance. Indeed, the pride of retrovirologists in retrovirus-specific reverse transcription is the probable reason that inhibition of DNA synthesis with AZT is perceived, even now, as a "specific" antiretroviral therapy (Section 4.3.3).

The wishful thinking that viruses cause "slow" diseases and cancers faces four common problems: (1) the diseases or tumors occur on average only decades after infection; (2) the viruses are all inactive, if not defective, during fatal disease or cancer; (3) the "viral" tumors are all clonal, derived from a single cell (with a tumor-specific chromosome abnormality) that had emerged out of billions of identically infected cells of a given carrier; and (4) above all, no human cancers and none of the "slow viral diseases" are contagious (Rowe, 1973; Duesberg and Schwartz, 1992).

Therefore, these viruses all fail Koch's postulates, the acid test of the germ theory. And therefore these viruses are all assumed to be very "slow," causing diseases only after long "latent periods" that exceed by decades the short periods of days or weeks that these viruses need to replicate and to become immunogenic. Because of their consistent scarcity, defectiveness, and even complete absence from some tumors and slow diseases (Duesberg and Schwartz, 1992), the search for the presumably pathogenic latent viruses has been directed either at antiviral antibodies, i.e., "seroepidemiological evidence" (Blattner et al., 1988), or at artificially amplified viral DNA and RNA (Section 3.3), or at the "activation" of latent viruses, euphemistically called "virus isolation" (Section 2.2).

Accordingly cancer-, AIDS-, and other slow-virologists try to discredit Koch's postulates in favor of "modern concepts of causation." For example, Evans states that "...Koch's postulates, great as they were for years, should be replaced with criteria reflecting modern concepts of causation, epidemiology, and pathogenesis and technical advances" (Evans, 1992). And Blattner, Gallo, and Temin point out that Koch's postulates are just a "useful historical reference point" (Blattner et al., 1988), and Weiss and Jaffe find it "bizarre that anyone should demand strict adherence to these unreconstructed postulates 100 years after their proposition" (Weiss and Jaffe, 1990)—but they all fail to identify a statute of limitation for adherence to the virus-AIDS hypothesis. In addition, "cofactors" are assumed (a) to make up for the typical inertia of the viral pathogens or carcinogens, (b) to account for the clonality of the cancers via a clonal cellular cofactor, and (c) to help to close the enormous gaps between the very common infections and the very rare incidences of "slow" disease or cancer, that even the long "latent periods" could not close (Duesberg and Schwartz, 1992). The tumor virologist Rowe "recognized that the latent period may cover much of the life span of the animal and that the virus did not act alone but that the tumor response might require... treatment with a chemical carcinogen" (Rowe, 1973).

Despite the total lack of public health benefits and even negative consequences of these theories, such as the psychologically toxic prognoses that antibodies against HTLV-I or against papilloma virus signal future cancers (Duesberg and Schwartz, 1992) or that antibodies against HIV signal future AIDS and the need for AZT prophylaxis, the public and the majority of scientists have held on to them much longer than was justified in terms of scientific evidence. The irresistible appeal of the germ theory was the basis for each of these unproductive theories of the past, as it is the basis now for the universal and enthusiastic approval of the virus-AIDS hypothesis.

But unlike the mistaken germ theories of the past, the virus-AIDS hypothesis was a windfall not only for (1) the virologists and epidemiologists, but also for (2) the biotechnology companies who could develop virus tests and antiviral drugs, (3) the AIDS patients who were relieved that a God-given, egalitarian virus rather than behavioral factors were to blame for their diseases, and (4) the politicians who had to confront the public and the gay (homosexual) lobby requesting action against AIDS. Indeed, a thoroughly intimidated public was happy, once more, to be offered protection by its scientists against another "deadly" virus, albeit for the highest price tag ever.

6.2. BIG FUNDING AND LIMITED EXPERTISE PARALYZE AIDS RESEARCH

Ironically, AIDS research suffers not only from being tied to an unproductive hypothesis, it also suffers from the staggering funds it receives from governments (Section 1) and from conceptually matched private sources. Intended to buy a fast solution for AIDS, these funds have instead paralyzed AIDS research by creating an instant orthodoxy of retrovirologists that fiercely protects its narrowly focused scientific expertise and global commercial interests (Booth, 1988; Rappoport, 1988; Nussbaum, 1990; Duesberg, 1991b, 1992b; Savitz, 1991; Connor, 1991, 1992).

The leaders of the AIDS orthodoxy are all veterans from the wars on "slow" and cancer viruses. Naturally, they were highly qualified to fill the growing gaps in the virus-AIDS hypothesis with their "modern concepts of causation" (Evans, 1992), including long "latent periods," "cofactors," and "seroepidemiological" arguments of causation (Sections 3.3, 3.4, and 3.5). When it became apparent that the first order mechanism of viral pathogenesis, postulating direct killing of T-cells, failed to explain immunodeficiency, the bewildering diversity of AIDS diseases, the many asymptomatic HIV infections, and HIV-free AIDS cases, the scientific method would have called for a new hypothesis. Instead, the virus hunters have shifted the virus-AIDS hypothesis from a failed first order mechanism to a multiplicity of hypothetical second order mechanisms, including cofactors and latent periods, to fill the ever growing discrepancies between HIV and AIDS. By conjugating these second order mechanisms with a multiplicity of unrelated diseases, the virus-AIDS hypothesis has become by far the most mercurial hypothesis in biology. It predicts either diarrhea or dementia or Kaposi's sarcoma or no disease 1, 5, 10, or 20 years after 1 or 2000 sexual contacts with an HIV-antibody–positive person with or without an AIDS disease.

But the coup to rename dozens of unrelated diseases with the common name AIDS, proved to be the most effective weapon of the AIDS establishment in winning unsuspecting followers from all constituencies. By making AIDS a synonym for Kaposi's sarcoma and candidiasis and dementia and diarrhea and lymphoma and lymphadenopathy, the road was paved for a common cause. Who would have accepted, prior to AIDS, that a dental patient caught candidiasis from her doctor's Kaposi's sarcoma? Or which scientist would accept it even now knowing the original data rather than just the corresponding press release? According to sociologist David Phillips "researchers use newspapers as a 'filter' to help them decide which scientific article is worth reading" (Briefings, 1991) or more often which article is worth knowing about.

The control of AIDS research by the nationally and internationally funded AIDS orthodoxy via the popular and scientific press is almost total. It instructs science writers that faithfully report every "breakthrough" in HIV research and every "explosion" of the epidemic. It feeds scientific journals with over 10,000 HIV-AIDS papers annually and with advertisements for HIV tests and antiviral drugs (Schwitzer, 1992). The AIDS doctors are controlled by the companies created, consulted, or owned by the AIDS establishment (Baringa, 1992; Schwitzer, 1992). For example, the *Physician's Desk Reference 1992* instructs AIDS doctors about AZT with an exact copy of Burroughs-Wellcome's instructions. Science writers are warned against reporting minority views. For example, Fauci states: "Journalists who make too many mistakes, or who are sloppy, are going to find that their access to scientists may diminish" (Fauci, 1989). And Ludlam points out, "Whilst I support, and encourage the reporting of, minority views... If the belief that AIDS is not due to HIV becomes prevalent... (it) could lead directly to the deaths of countless misinformed individuals" (Ludlam, 1992). Any challengers are automatically outnumbered and readily marginalized by the sheer volume of the AIDS establishment. For example, the 12,000 scientists attending the annual international AIDS conference held in San Francisco in 1990 were only a fraction of the many who study the information encoded in the 9000 nucleotides of HIV. Says the HIV virologist Gallo when asked about a dissenter: "Why does the Institute of Medicine, WHO, CDC, National

Academy of Sciences, NIH, Pasteur Institute and the whole body of world science 100% agree that HIV is the cause of AIDS?" (Liversidge, 1989).

Consequently, there is no "peer-reviewed" funding for researchers who challenge the virus-AIDS hypothesis (Duesberg, 1991b; Maddox, 1991a; Bethell, 1992; Farber, 1992; Hodgkinson, 1992). Since HIV became the dominant focus of the billion-dollar AIDS research (Coffin et al., 1986; Institute of Medicine, 1988), there has not been even one follow-up of the many previous studies blaming sexual stimulants and psychoactive drugs for homosexual AIDS (Sections 4.4 and 4.5). None of the former "lifestyle" advocates (Section 2.2) have investigated whether drugs might cause AIDS without HIV. Instead drugs, if mentioned at all, were since described as risk factors for infection by HIV (Darrow et al., 1987; Moss et al., 1987; van Griensven et al., 1987; Chaisson et al., 1989; Weiss, S. H., 1989; Goudsmit, 1992; Seage et al., 1992)—as if HIV could discriminate between hosts on the basis of their drug habits (Duesberg, 1992a). For example, Friedman-Kien concluded in 1982 and 1983 with Marmor et al. (1982) and Jaffe et al. (1983b) that the "lifetime exposure to nitrites..." was responsible for AIDS (Section 4.3.2). In 1990 he and his collaborators just mentioned nitrite use in HIV-free Kaposi's sarcoma cases (Friedman-Kien et al., 1990) and in 1992 they blamed viruses other than HIV for HIV-free AIDS cases, and drug use was no longer mentioned (Huang et al., 1992).

Likewise, all studies investigating transfusion-mediated immunodeficiency in hemophiliacs were frozen around 1987 (Table 3), once the virus-AIDS hypothesis had monopolized AIDS research. The question whether immunodeficient(!) HIV-free hemophiliacs would ever develop AIDS-defining diseases was left unanswered and even became unaskable.

Fascinated by the past triumphs of the germ theory, the public, science journalists, and even scientists from other fields never question the authority of their medical experts, even if they fail to produce useful results (Adams, 1989; Schwitzer, 1992). Medical scientists are typically credited for the virtual elimination of infectious diseases with vaccines and antibiotics, although most of the credit for eliminating infectious diseases is actually owed to vastly improved nutrition and sanitation (Stewart, 1968; McKeown, 1979; Moberg and Cohn, 1991; Oppenheimer, 1992). Indeed, the belief in the infallibility of modern science is the only ideology that unifies the 20th century. For example, in the name of the virus-AIDS hypothesis of the American government and the American researcher Gallo, antibody-positive Americans have been convicted for "assault with a deadly weapon" because they had sex with antibody-negatives, Central Africa dedicates its limited resources to "AIDS testing," the former U.S.S.R. conducted 20.2 million AIDS tests in 1990 and

29.4 million in 1991 to identify a total of 178 antibody-positive Soviets, and communist Cuba even quarantines its own citizens if they are antibody-positive (Section 3.6).

Predictably, the AIDS virus hunters, on their last crusade for the germ theory, have no regard for the current drug-use epidemic and its many overlaps with American and European AIDS. Even direct evidence for the role of drugs in AIDS is fiercely rejected by the virus-AIDS orthodoxy (Booth, 1988; Moss et al., 1987; Kaslow et al., 1989; Baltimore and Feinberg, 1990; Ostrow et al., 1990). Merely questioning the therapeutic or prophylactic benefits of AZT is protested by the AIDS establishment (Baltimore and Feinberg, 1990; Weiss and Jaffe, 1990; Anonymous, 1992; Freestone, 1992; Tedder et al., 1992). The prejudice against noninfectious pathogens is so popular that the virus-AIDS establishment uses it regularly to intimidate those who propose noninfectious alternatives, to censor their papers (Duesberg, 1992e), and even to question their integrity.

For example, an editorial in *Science* called me a "rebel without a cause for AIDS," because denying HIV was to deny a cause altogether. The editorial quoted Baltimore as saying I was "irresponsible and pernicious" (Booth, 1988). An article in *Nature* called my drug hypothesis a "perilous message" that would "belittle 'safe sex', would have us abandon AIDS screening... and curtail research into anti-HIV drugs." "Arguments that AIDS [is] the result of evil vapors [poppers (!)], malaria... [are from] the last century." "We... regard the critics as 'flat-earthers' bogged down in molecular minutiae and miasmal theories of disease, while HIV continues to spread" (Weiss and Jaffe, 1990). This is said even though the article agrees that, "Duesberg is right to draw attention to our ignorance of *how* HIV causes disease..." (Weiss and Jaffe, 1990). Others declare "All attempts by epidemiologists to link AIDS to the use of amyl nitrite or other drugs as a direct cause of disease have failed... Duesberg's continued attempts to persuade the public to doubt the role of HIV in AIDS are not based on facts" (Baltimore and Feinberg, 1990). Gallo called the author of the article, "Experts mount startling challenge to AIDS orthodoxy" in *The Sunday Times* (London) (Hodgkinson, 1992), "irresponsible both to myself [Gallo] and to HIV as the cause of AIDS" (Gallo, 1992). Further, Vandenbrouke and Pardoel (1989) argue, "If one is allowed to compare the evolution of scientific theories with the evolution of biologic nature in general, the poppers (nitrite inhalants) episode is the Neanderthal of modern epidemiology."

As a consequence there are no studies that investigate the long-term effects of psychoactive drugs (Lerner, 1989; Pillai et al., 1991; Bryant et al., 1992). The toxicologist Lerner points out that "fewer than 60 are currently enrolled in fellowship programs on alcoholism and drug abuse in the entire country" (Lerner, 1989), although about 8 million Americans alone are estimated to use cocaine (Weiss, S. H., 1989; Finnegan et al., 1992) and many more use other psychoactive drugs regularly (Section 4). This stands in contrast to the 40,000 annual AIDS cases that are studied by at least 40,000 AIDS researchers of which just 12,000 attended the annual International AIDS Conference in San Francisco in 1990.

Instead of warning against drugs, the AIDS establishment "educates" the public with its "clean needle" campaigns that drugs (albeit illegal) are safe, but bugs are not. For example, AIDS researcher Moss, citing Napoleon's line "On s'engage et puis on voit," recommends "clean needles" for "harm reduction" (Moss, 1987). Mindful of its educators, the public is unaware and even disinformed about the health risks of recreational drugs. A popular joke in point is the response of two "junkies" (drug addicts) sharing a syringe filled with an intravenous drug to a concerned colleague: "We are safe, because we use a clean needle and condoms." The long "latent periods" between the gratification from recreational drugs, such as tobacco, alcohol, cocaine, and nitrite inhalants, to their irreversible health effects unfortunately give credence to the "perilous message" that drugs are safe but bugs are not.

Particularly the victims of drug consumption prefer egalitarian infectious causes over noninfectious behavioral ones that imply personal responsibility (Shilts, 1985; Lauritsen and Wilson, 1986; Rappoport, 1988; Callen, 1990). For example, the executive director of the San Francisco–based national Project Inform, an organization operated mainly for and by male homosexuals, Martin Delaney, informs its clients about a study documenting a "level of sexual contact and drug use which was shocking to the general public" as follows: "It (the study) might just as well have noted that most wore Levi's (jeans) for all this told us about the cause of AIDS" (Project Inform, 1992). The organization collaborates with the NIH and is supported by grants from pharmaceutical companies including Burroughs-Wellcome, the manufacturer of AZT (Project Inform, 1992).

In 1987, before AZT, Delaney advised gay men in his book *Strategies for Survival: A Gay Men's Health Manual for the Age of AIDS* about the health effects of nitrite inhalants: "Possible heart damage; fibrillation (compulsive, erratic heart rhythms); possible stroke and resulting brain damage. Conducive to high-risk sexual behavior; distortion of judgment and senses. Statistical link to Kaposi's sarcoma (KS, an AIDS-related cancer); suspected immuno-suppression" (Delaney and Goldblum, 1987). Delaney's advice about amphetamines reads as follows: "Liver and heart damage; neuropathy (nerve damage); possible brain damage; weight loss;

nutritional and vitamin depletion; adrenal depletion (uses up the body's energy reserves). Distorted judgment, values, senses, delusions of strength, anxiety, paranoia, rebound depression, financial strain, powerful addiction, conducive to high-risk sexual activity. Likely immunosuppression (not currently measured), potential for unknown and risky drug interactions, complication in treatment of brain disorders." Delaney also warns about the effects of cocaine: "Heart and lung damage, stroke, cardiovascular irregularities, possible physical addiction. Distortion of judgment, values, and senses, dangerous delusions of grandeur and strength, intense anxiety, paranoia, financial strain, leads to poor judgment about high-risk sexual activity. Likely immunosuppression (not currently measured); increased stress, if smoked, complicates treatment of pneumonia." The book also gives the basis for Delaney's intimate knowledge of drug toxicity: "He... has done work for the National Institute on Drug Abuse" (Delaney and Goldblum, 1987).

Clearly, big science is not always good science, particularly if it is conceptually paralyzed by an unproductive hypothesis. I hope that the scientific evidence collected for this article will focus attention on the noninfectious causes of AIDS and prove that it is not "too late to correct" (quoted material as said by the Red Queen in *Through the Looking Glass*, written by Lewis Carroll) the spell of the virus-AIDS hypothesis by the scientific method. Considering noninfectious causes may prove to be as beneficial to the challenge of AIDS as it was, for example, to the challenge of pellagra. Indeed, a few investigators have recently smuggled recreational drugs as "cofactors" of HIV (Haverkos and Dougherty, 1988; Haverkos, 1990) or even more cautiously as cofactors of cofactors of HIV (Archibald et al., 1992) into the highly fundable virus-AIDS hypothesis. One investigator even dared to document that drugs are sufficient for pediatric AIDS, if only in preliminary reports (Koch, 1990; Koch et al., 1990). A complete report of the data (Section 4.5) was not published for political reasons (Thomas Koch, personal communication). And the "100 percent" consensus on HIV claimed by Gallo in 1989 (Liversidge, 1989) is eroding just a bit in the face of a growing group of dissenters, some of which united in the Group for the Scientific Reappraisal of the HIV/AIDS Hypothesis (DeLoughry, 1991; Bethell, 1992; Bialy and Farber, 1992; Farber, 1992; Hodgkinson, 1992; Project Inform, 1992; Nicholson, 1992; Ratner, 1992; Schoch, 1992).

NOTE ADDED IN PROOF

Sparked by an article in *Newsweek* (Cowley, 1992), numerous HIV-free AIDS cases were unexpectedly reported by many independent(!)

investigators at the VIII International Conference on AIDS/III STD World Congress in Amsterdam in July 1992 (now a joint meeting with sexually transmitted diseases, STDs). Surprisingly, some of the recently announced HIV-free AIDS cases had been studied for years (Altman, 1992a; Cohen, 1992a, b; Laurence et al., 1992), even by the CDC (Spira and Jones, 1992). As a result the CDC had to alter its long-held position that HIV causes all AIDS to "HIV causes the vast majority of AIDS cases..." (Nicholson, 1992). In its monthly *HIV/AIDS Surveillance Reports* the CDC still states that "AIDS is a specific group of diseases which are indicative of severe immunosuppression related to infection with the Human Immunodeficiency Virus (HIV)" (Centers for Disease Control, 1992b). The AIDS risk factors of most of these HIV-free "AIDS patients" were reported to be "intravenous drugs, unprotected sex and transfusions" and the corresponding diseases were Kaposi's sarcoma and pneumonia (Cowley, 1992).

AIDS-virus matchmakers soon reached the consensus that an as-yet-undiscovered "new AIDS virus," which "doesn't appear any more contagious than HIV" (Cowley, 1992), was to blame for HIV-negative AIDS (Bowden et al., 1991; Castro et al., 1992; Huang et al., 1992; Altman, 1992a, b; Cohen, 1992a, b; Laurence et al., 1992). And the director for AIDS research at the NIH reassured the public, "If there is something, scientists will find it" (News Report, 1992). States *The New York Times*, "Arguably, the greatest thrills for a scientist are in discovering a new microbe, a new disease, cure and prevention... Many... know how quickly the exhilaration that comes from believing they are on the verge of making such a discovery vanishes when the initial findings cannot be confirmed" (Altman, 1992b).

However, the new HIV-free AIDS cases are entirely consistent with those listed above that were caused by drug consumption and other non-contagious risk factors (Section 4.5). Although public recognition of HIV-free AIDS cases is new, the new cases just complement the over 1200 cases of "acquired" immunodeficiency and AIDS-defining diseases described above including 334 hemophiliacs (Section 3.4.4.5; Table 3), 265 male homosexuals (Sections 3.4.4.3 and 4.5), 44 intravenous drug users (Section 4.5), and 183 mostly male tuberculosis patients from Florida (Pitchenik et al., 1987, 1990). If the 2466 HIV-free AIDS cases from Africa were included (Section 3.4.4.8), the number of documented HIV-free AIDS cases would exceed 3000!

Moreover, healthy HIV carriers who have been infected for over 10 years and have transmitted their HIV to at least 5 healthy persons via blood transfusions over 7–10 years ago have now received public recognition (Altman, 1992c; Learmont et al., 1992). These cases supplement the 1 million Americans, 0.5 million Europeans, 0.3 million Thais, and 6 million Africans who are healthy, although most had been infected by 1985 (Section 3.5.1).

Thus, both predictions of the hypothesis that AIDS is noncontagious are now generally accepted: (1) HIV-free AIDS and (2) AIDS-free transmission of HIV. Asks John Maddox, editor of *Nature*, "Does that mean Duesberg has been right all along, and that HIV plays no part in the causation of AIDS?" (Maddox, 1992b). Indeed, it would be an evolutionary miracle if the past decade had generated three different AIDS viruses, HIV-1, HIV-2 and the "new AIDS virus," when no such virus has ever emerged before in the history of medicine.

Acknowledgments—Dedicated to (1) all intravenous drug users, oral users of recreational drugs, and AZT recipients who were never told that drugs cause AIDS diseases, and (2) all "antibody-positives" who were never told that the virus-AIDS hypothesis is unproven.

I thank Janie Stone (Berkeley) for numerous corrections of the manuscript, David Shugar (Associate Editor of *Pharmacology and Therapeutics*, Warsaw, Poland) for his courage to take up the AIDS controversy and for playing the devil's advocate of HIV, Annette Gwardyak (Managing Editor) for accommodating many "final" revisions, Hansueli Albonico (Langrau, Switzerland), Harvey Bialy (New York), Julie Castiglia (San Diego), Robert Cramer (Montlhery/Paris), Bryan Ellison (Berkeley), Celia Farber (New York), Harry Haverkos (Rockville, Maryland), Robert Hoffman (San Diego), Geoff Hoffmann (Vancouver, Canada), Phillip E. Johnson (Berkeley), Bill Jordan (Los Angeles), John Lauritsen (New York), Nathaniel Lehrman (New York), Anthony Liversidge (New York), Claus Pierach (Minneapolis), Paul Rabinow (Berkeley), Robert Root-Bernstein (East Lansing, Michigan), Harry Rubin (Berkeley), Frank Rothschild (Berkeley), Russell Schoch (Berkeley), Craig Schoonmaker (New York), Jody Schwartz (Berkeley), Joan Shenton (London), Gordon Stewart (Bristol, U.K.), Richard Strohman (Berkeley), Charles Thomas, Jr. (San Diego), Fritz Ulmer (Wuppertal, Germany), Michael Verny-Elliott (London), Warren Winkelstein (Berkeley), and Yue Wu (Berkeley) for critical and catalytic information, Gedge Rosson (Berkeley) for Figure 1, Brian Davis for fact-checking and typing, Osias Stutman for suggesting the Lewis Carroll quotation, Ted Gardner (Santa Barbara) for a generous donation and encouragement, and the librarians of UC-Berkeley, particularly Chris Campbell, Ingrid Radkey, Pat Stewart, and Norma Kobzina, for AIDS references that were not even cited in the daily American newspapers. I am still supported by Outstanding Investigator Grant #5-R35CA39915-07 from the National Cancer Institute.

REFERENCES

Achard, C., H. Bernard, and C. Gagneux. Action de la morphine sur les proprietes leucocytaires; leuco-diagnostic du morphinisme. *Bull. Memoires Societe Med. Hopitaux Paris* 28 (1909): 958–966.

Adams, J. *AIDS: The HIV Myth*. St Martin's Press, New York (1989).

Afrasiabi, R., R. T. Mitsuyasu, K. Schwartz, and J. L. Fahey. Characterization of a distinct subgroup of high-risk persons with Kaposi's sarcoma and good prognosis who present with normal T4 cell number and T4:T8 ratio and negative HTLV-III/LAV serologic results. *Am. J. Med.* 81 (1986): 969–973.

AIDS-Hemophilia French Study Group. Immunologic and virologic status of multitransfused patients: role of type and origin of blood products. *Blood* 66 (1985): 896–901.

Albonico, H. Lichtblicke zum zweiten Jahrzehnt in der AIDS-Forschung. *Schweizerische Arztezeitung* 72 (1991a): 379–380.

Albonico, H. *Relativierung des HIV-Dogmas—Ein Beitrag zur Erweiterten Sicht von AIDS*. Padagogische Arbeitsstelle, Dortmund (1991b).

Aledort, L. M. Blood products and immune changes: impacts without HIV infection. *Sem. Hematol.* 2S (1988): 14–19.

Alroomi, L. G., J. Davidson, T. J. Evans, P. Galea, and R. Howat. Maternal narcotic abuse and the newborn. *Arch. Dis. Child.* 63 (1988): 81–83.

Altman, L. K. New virus said to cause a condition like AIDS. *New York Times*, July 23 (1992a).

Altman, L. K. Working in public to explain AIDS-like ills. *New York Times*, August 18 (1992b).

Altman, L. K. Group with HIV has no symptoms. *New York Times*, October 9 (1992c).

Amaro, H., B. Zuckerman, and H. Cabral. Drug use among adolescent mothers: profile of risk. *Pediatrics* 84 (1989): 144–151.

Anand, R., C. Reed, S. Forlenza, F. Siegal, T. Cheung, and J. Moore. Non-natural variants of Human Immunodeficiency Virus isolated from AIDS patients with neurological disorders. *Lancet* ii (1987): 234–238.

Anderson, D. J., T. R. O'Brien, J. A. Politch, A. Martinez, G. R. Seage, III, N. Padian, R. Horsburgh, Jr., and K. H. Mayer. Effects of disease stage and zidovudine therapy on the detection of Human Immunodeficiency Virus type I in semen. *J. Am. Med. Ass.* 267 (1992): 2769–2774.

Anderson, J. AIDS in Thailand. *Br. Med. J.* 300 (1990): 415–416.

Anderson, L. F. Cancer Act anniversary encourages reflections, new visions. *J. Natn. Cancer Inst.* 83 (1991): 1795–1796.

Anderson, R. M. and R. M. May. Epidemiological parameters of HIV transmission. *Nature (Lond)* 333 (1988): 514–519.

———. Understanding the AIDS pandemic. *Sci. Am.* 266 (1992): 20–26.
Anderson, W. *Drug Smuggling*. (1987) U.S. General Accounting Office, Washington, D.C.
Annell, A., A. Fugelstad, and G. Agren. HIV-prevalence and mortality in relation to type of drug abuse among drug addicts in Stockholm 1981–1988. In: *Drug Addiction and AIDS*, pp. 16–22, Loimer, N., Schmid, R. and Springer, A. (ed.) (1991) Springer-Verlag, New York.
Anonymous. The cause of AIDS? *Lancet* i (1984): 1053–1054.
———. Senate approves measures on transit and AIDS. *New York Times*, August 6, 1990.
———. Dentist-infected AIDS patient dies. *Dallas Morning News*, December 9, 1991.
———. Doubts about zidovudine. *Lancet* 339 (1992): 421.
Antonaci, S., E. Jirillo, D. Stasi, V. De Mitrio, M. F. La Via, and L. Bonomo. Immunoresponsiveness in hemophilia: lymphocyte- and phagocyte-mediated functions. *Diagn. Clin. Immun.* 5 (1988): 318–325.
Aoun, H. (1989) When a house officer gets AIDS. *New Engl. J. Med.* 321: 693–696.
———. From the eye of the storm, with the eyes of a physician. *Ann. Intern. Med.* 116 (1992): 335–338.
Aral, S. O. and K. K. Holmes. Sexually transmitted diseases in the AIDS era. *Sci. Am.* 264 (1991): 62–69.
Archer, C. B., M. F. Spittle, and N. P. Smith. Kaposi's sarcoma in a homosexual—10 years on. *Clin. Exp. Derm.* 14 (1989): 233–236.
Archibald, C. P., M. T. Schechter, T. N. Le, K. J. P. Craib, J. S. G. Montaner, and M. V. O'Shaughnessy. Evidence for a sexually transmitted cofactor for AIDS-related Kaposi's sarcoma in a cohort of homosexual men. *Epidemiology* 3 (1992): 203–209.
Aronson, D. L. Pneumonia deaths in haemophiliacs. *Lancet* ii (1983): 1023.
———. Cause of death in hemophilia patients in the United States from 1968–1979. *Am. J. Hematol.* 27 (1988): 7–12.
Avramis, V. I., W. Markson, R. L. Jackson, and E. Gomperts. Biochemical pharmacology of zidovudine in human T-lymphoblastoid cells (CEM). *AIDS* 3 (1989): 417–422.
Aylward, E. H., A. M. Butz, N. Hutton, M. L. Joyner, and J. W. Vogelhut. Cognitive and motor development in infants at risk for Human Immunodeficiency Virus. *Am. J. Dis. Children* 146 (1992): 218–222.
Bagasra, O., S. P. Hauptman, H. W. Lischner, M. Sachs, and R. J. Pomerantz. Detection of Human Immunodeficiency Virus type 1 provirus in

mononuclear cells by *in situ* polymerase chain reaction. *New Engl. J. Med.* 326 (1992): 1385–1391.
Bailey, H. *The Vitamin Pioneers.* (1968) Rodale, Emmaus, Pa.
Balter, M. Montagnier pursues the mycoplasma-AIDS link. *Science* 251 (1991): 27B.
Baltimore, D. and M. B. Feinberg. HIV revealed, toward a natural history of the infection. *New Engl. J. Med.* 321 (1989): 1673–1675.
———. Quantification of Human Immunodeficiency Virus in the blood. *New Engl. J. Med.* 322 (1990): 1468–1469.
Balzarini, J., P. Herdewijn, and E. De Clercq. Differential patterns of intracellular metabolism of 2′,3′-didehydro-2′,3′-dideoxythymidine and 3′-azido-2′,3′-dideoxythymidine, two potent anti-Human Immunodeficiency Virus compounds. *J. Biol. Chem.* 264 (1989): 6127–6133.
Bardach, A. L. The heretic. *Buzz*, January/February (1992): 68–73, 90, 92.
Baringa, M. Confusion on the cutting edge. *Science* 257 (1992): 616–619.
Bartholomew, C., W. C. Saxinger, J. W. Clark, M. Gail, A. Dudgeon, B. Mahabir, B. Hull-Drysdal, F. Cleghorn, R. C. Gallo, and W. A. Blattner. Transmission of HTLV-I and HIV among homosexual men in Trinidad. *J. Am. Med. Ass.* 257 (1987): 2604–2608.
Beardsley, T. French virus in the picture. *Nature (London)* 320 (1986): 563.
Becherer, P. R., M. L. Smiley, T. J. Matthews, K. J. Weinhold, C. W. McMillan, and G. C. White, II. Human immunodeficiency virus-1 disease progression in hemophiliacs. *Am J. Hematol.* 34 (1990): 204–209.
Beemon, K., P. Duesberg, and P. Vogt. Evidence for crossing-over between avian tumor viruses based on analysis of viral RNAs. *Proc. Natn. Acad. Sci. U.S.A.* 71 (1974): 4254–4258.
Belman, A. L., G. Diamond, D. Dickson, D. Horoupian, J. Llena, G. Lantos, and A. Rubinstein. Pediatric acquired immunodeficiency syndrome. *Am. J. Dis. Child.* 142 (1988): 29–35.
Beral, V., T. A. Peterman, R. L. Berkelman, and H. W. Jaffe. Kaposi's sarcoma among persons with AIDS: a sexually transmitted infection? *Lancet* 335 (1990): 123–128.
Bergeron, L. and J. Sodroski. Dissociation of unintegrated viral DNA accumulation from single-cell lysis induced by human immunodeficiency virus type 1. *J. Virol.* 66 (1992): 5777–5787.
Berkley, S., S. Okware, and W. Naamara. Surveillance for AIDS in Uganda. *AIDS* 3: (1989) 79–85.

Bessen, L. J., J. B. Greene, E. Louie, L. E. Seitzman, and H. Weinberg. Severe polymyositis-like syndrome associated with zidovudine therapy of AIDS and ARC. *New Engl. J . Med.* 318 (1988): 708.
Bethell, T. Heretic. *The American Spectator*, May 1992: 18–19.
Bialy, H. and C. Farber. It's time to re-evaluate the HIV–AIDS hypothesis. *Rethinking AIDS*, 1 (1992): 1–2.
Biggar, R. J. and the International Registry of Seroconverters. AIDS incubation in 1891 HIV seroconverters from different exposure groups. *AIDS* 4 (1990): 1059–1066.
Blanche, S., C. Rouzloux, M. L. Moscato, F. Veber, M. J. Mayauo, C. Jacomet, J. Tricoire, A. Deville, M. Vial, G. Firtion, and HIV Infection In Newborns French Collaborative Study Group. A prospective study of infants born to women seropositive for human immunodeficiency virus type B. *New Engl. J. Med.* 320 (1989): 1643–1648.
Blattner, W. A. (ed.) *Human Retrovirology: HTLV.* (1990) Raven Press, New York.
———. HIV epidemiology: past, present, and future. *FASEB J.* 5 (1991): 2340–2348.
Blattner, W. A., R. J. Biggar, S. H. Weiss, J. W. Clark, and J. J. Goedert. Epidemiology of human lymphotropic retroviruses: an overview. *Cancer Res.* 45 (1985) (Suppl.): 4598–4601.
Blattner, W. A., R. C. Gallo, and H. M. Temin. HIV causes AIDS. *Science* 241 (1988): 514–515.
Bolling, D. R. and B. Voeller. AIDS and heterosexual anal intercourse. *J. Am. Med. Ass.* 258 (1987): 474.
Booth, W. A rebel without a cause for AIDS. *Science* 239 (1988): 1485–1488.
Bove, J. R., P. R. Rigney, P. M. Kehoe, and J. Campbell. Look back: preliminary experience of AABB members. *Transfusion* 27 (1987): 201–202.
Bowden, E. J., D. A. McPhee, N. J. Deacon, S. A. Cumming, R. R. Doherty, S. Sonza, C. R. Lucas, and S. M. Crowe. Antibodies to gp41 and nef in otherwise HIV-negative homosexual man with Kaposi's sarcoma. *Lancet* 337 (1991): 1313–1314.
Brandt, A. M. AIDS in historical perspective: four lessons from the history of sexually transmitted diseases. *Am. J. Pub. Health* 78 (1988): 367–371.
Braun, M. M., B. I. Truman, B. Maguire, G. T. Di Ferdinando, Jr., G. Wormser, R. Broaddus, and D. L. Morse. Increasing incidence of tuberculosis in a prison inmate population, associated with HIV-infection. *J. Am. Med. Ass.* 261 (1989): 393–397.
Bregman, D. J. and A. D. Langmuir. Farr's law applied to AIDS projections. *J. Am. Med. Ass.* 263 (1990): 50–57.

Brenner, H., P. Hernando-Briongos, and C. Goos. AIDS among drug users in Europe. *Drug Alcohol Depend.* 29 (1990): 171–181.
Breo, D. L. and K. Bergalis. Meet Kimberly Bergalis—the patient in the "dental AIDS case." *J. Am. Med. Ass.* 264 (1990): 2018–2019.
Briefings. All the [Science] that's fit to print. *Science* 254 (1991): 649.
Briggs, L. H., C. G. McKerron, R. L. Souhami, D. J. E. Taylor, and H. Andrews. Severe systemic infections complicating "mainline" heroin addiction. *Lancet* ii (1967): 1227–1231.
Brown, S. M., B. Stimmel, R. N. Taub, S. Kochwa, and R. E. Rosenfield. Immunologic dysfunction in heroin addicts. *Arch. Intern. Med.* 134 (1974): 1001–1006.
Brudney, K. and L. Dobkin. Resurgent tuberculosis in New York City. *Am. Rev. Respir. Dis.* 144 (1991): 744–749.
Bryant, H. U., K. A. Cunningham, and T. R. Jerrells. Effects of cocaine and other drugs of abuse on immune responses. In: *Cocaine: Pharmacology, Physiology and Clinical Strategies*, pp. 353–369, Lakoski, J. M., Galloway, M. P. and White, F. J. (eds) (1992) CRC Press, Boca Raton, Fl.
Bschor F., R. Bornemann, C. Borowski, and V. Schneider. Monitoring of HIV-spread in regional populations of injecting drug users—the Berlin experience. In: *Drug Addiction and AIDS*, pp. 102–109, Loimer, N., Schmid, R. and Springer, A. (eds) (1991) Springer-Verlag, New York.
Buehler, J. W., O. J. Devine, R. L. Berkelman, and F. M. Chevarley. Impact of the Human Immunodeficiency Virus epidemic on mortality trends in young men, United States. *Am. J. Publ. Health* 80 (1990): 1080–1086.
Buimovici-Klein, E., M. Lange, K. R. Ong, M. H. Grieco, and L. Z. Cooper. Virus isolation and immune studies in a cohort of homosexual men. *J. Med. Virol.* 25 (1988): 371–385.
Bundesgesundheitsamt (Germany). Bericht des AIDS-Zentrums des Bundesgesundheitsamtes uber Aktuelle Epidemiologische Daten (bis zum 31.7.1991). *AIDS-Forschung* 6 (1991): 509–512.
Bureau of Justice Statistics. (1988) Special Report—Drug Law Violators, 1980–1986. U.S. Department of Justice, Washington, D.C.
———. (1991) *Catalog of Federal Publications on Illegal Drug and Alcohol Abuse.* U.S. Department of Justice, Washington, D.C.
Burke, D. S., J. F. Brundage, M. Goldensaum, M. Gardner, M. Peterson, R. Visintine, R. Redfield, and the Walter Reed Retrovirus Research Group. Human immunodeficiency virus infections in teenagers; seroprevalence among applicants for the U.S. military service. *J. Am. Med. Ass.* 263 (1990): 2074–2077.
Burns, D. P. W. and R. C. Desrosiers. Selection of genetic variants of

simian immunodeficiency virus in persistently infected rhesus monkeys. *J. Virol.* 65 (1991): 1843–1854.

Cairns, J. *Cancer Science and Society.* (1978) W. H. Freeman and Company, San Francisco.

Callen, M. *Surviving AIDS.* (1990) Harper Perennial, New York.

Castro, A., J. Pedreira, V. Soriano, I. Hewlett, B. Jhosi, J. Epstein, and J. Gonzalez-Lahoz. Kaposi's sarcoma and disseminated tuberculosis in HIV-negative individual. *Lancet* 339 (1992): 868.

Centers for Disease Control. Kaposi's sarcoma and *Pneumocystis* pneumonia among homosexual men—New York City and California. *Morb. Mort. Weekly Rep.* 30 (1981): 305–308.

———. Epidemiologic aspects of the current outbreak of Kaposi's sarcoma and opportunistic infections. *New Engl. J. Med.* 306 (1982): 248–252.

———. Reports on AIDS published in the *Morbidity and Mortality Weekly Report,* June 1981 through May 1986. U.S. Dept. of Health and Human Services, National Technical Information Service, Springfield, Va. (1986).

———. Revision of the CDC surveillance case definition for acquired immunodeficiency syndrome. *J. Am. Med. Ass.* 258 (1987): 1143–1154.

———. Update: acquired immunodeficiency syndrome and human immunodeficiency virus infection among health-care workers. *Morb. Mort. Weekly Rep.* 37 (1988): 229–239.

———. Possible transmission of human immunodeficiency virus to a patient during an invasive dental procedure. *Morb. Mort. Weekly Rep.* 39 (1990): 489–492.

———. Opportunistic non-Hodgkin's lymphomas among severely immunocompromised HIV-infected patients surviving for prolonged periods on antiretroviral therapy—United States. *Morb. Mort. Weekly Rep.* 40 (1991): 591–601.

———. The second 100,000 cases of acquired immunodeficiency syndrome—United States, June 1981–December 1991. *Morb. Mort. Weekly Rep.* 41 (1992a): 28–29.

———. *HIV/AIDS Surveillance, Year-end Edition.* U.S. Department of Health and Human Services, Atlanta, GA (1992b).

———. Surveillance for occupationally acquired HIV infection—United States, 1981–1992. *Morb. Mort. Weekly Rep.* 41 (1992c): 823–825.

Chaisson, R. E., P. Bacchetti, D. Osmond, B. Brodie, M. A. Sande, and A. R. Moss. Cocaine use and HIV infection in intravenous drug users in San Francisco. *J. Am. Med. Ass.* 261 (1989): 561–565.

Chernov, H. I. Document on New Drug Application 19-655. Food and Drug Administration, Washington, D.C. (1986)

Christensen, A. C. Novel reading. *Nature (Lond)* 351 (1991): 600.
Clark, S. J., M. S. Saag, W. D. Decker, S. Campbell–Hill, J. L. Roberson, P. J. Veldkamp, J. C. Knappes, B. H. Hahn, and G. M. Shaw. High titers of cytopathic virus in plasma of patients with symptomatic primary HIV-infection. *New Engl. J. Med* 324 (1991): 954–960.
Clifford, D. B., R. G. Jacoby, J. P. Miller, W. R. Seyfred, and M. Glicksman. Neuropsychometric performance of asymptomatic HIV-infected subjects. *AIDS* 4 (1990): 767–774.
Coffin, J., A. Haase, J. A. Levy, L. Montagnier, S. Oroszlan, N. Teich, H. Temin, H.Varmus, P. Vogt, and R. Weiss. Human immunodeficiency viruses. *Science* 232 (1986): 697.
Cohen, J. Mystery virus meets the sceptics. *Science* 257 (1992a): 1032–1034.
———. New virus reports roil AIDS meeting. *Science* 257 (1992b): 604–605.
Cohen, S. S. Antiretroviral therapy for AIDS. *New Engl. J. Med* 317 (1987): 629.
Colebunders, R., J. Mann, H. Francis, K. Bila, L. Izaley, N. Kakonde, K. Kabasele, L. Ifoto, N. Nzilambi, T. Quinn, G. Van Der Groen, J. Curran, B. Vercauten, and P. Piot. Evaluation of a clinical case definition of Acquired Immunodeficiency Syndrome in Africa. *Lancet* i (1987): 492–494.
Collier, A. C., J. D. Meyers, L. Corey, V. L. Murphy, P. L. Roberts, and H. H. Hansfield. Cytomegalovirus infection in homosexual men. *Am. J. Med.* 82 (1987): 593–601.
Congressional Panel. Federal response to teen AIDS called "National Disgrace." *AIDS Weekly*, April 20, 1992: 12–13.
Connor, S. One year in pursuit of the wrong virus. *New Scientist* 113 (1987): 49–58.
———. Million pound row over AIDS test. *The Independent on Sunday*, January 20, 1991.
———. Scientists dispute royalties on HIV blood-test patent. *The Independent*, August 14, 1992.
Coombs, R. W., A. C. Collier, J.-P. Allain, B. Nikora, M. Leuther, G. E. Gjerset, and L. Corey. Plasma viremia in human immunodeficiency virus infection. *New Engl. J. Med.* 321 (1989): 1626–1631.
Cotton, P. Cofactor question divides codiscoverers of HIV. *J. Am. Med Ass.* 264 (1990): 3111–3112.
Cowley, G. Is a new AIDS virus emerging? *Newsweek*, July 27, 4 (1992): 41.
Cox, G. D. County Health Panel urges "poppers" ban, cites AIDS link. *Los Angeles Daily Journal*, March 24, 1986.

Cox, T. C., M. R. Jacobs, A. E. Leblanc, and J. A. Marshman. *Drugs and Drug Abuse*. Addiction Research Foundation, Toronto, Canada. (1983).

Craib, K. J. P., M. T. Schechter, T. N. Le, M. V. O'Shaughnessy, and J. S. G. Montaner. HIV causes AIDS: a controlled study. VIII *International conference on AIDS/III STD* World Congress, Amsterdam. (1992)

Cramer, R. Aids-Forschung—eine Besinnung. *Neue Zurcher Zeitung*, January 3, 1992.

Cretton, E. M., M. Y. Xie, R. J. Bevan, N. M. Goudgaon, R. E. Schinazi, and J. P. Sommadossi. Catabolism of 3'- azido-3'-deoxythymidine in hepatocytes and liver microsomes, with evidence of formation of 3'-amino-3'-deoxythymidine, a highly toxic catabolite for human bone marrow cells. *Molec. Pharmac.* 39 (1991): 258–266.

Crewdson, J. The great AIDS quest. *Chicago Tribune*, November 19, 1989.

———. House critique triggers another Gallo inquiry. *Chicago Tribune*, June 14, 1992.

Culliton, B. J. Legion fever: postmortem on an investigation that failed. *Science* 194 (1976): 1025–1027.

———. Inside the Gallo probe. *Science* 248 (1990): 1494–1498.

Culver, K. W., A. J. Ammann, J. C. Partridge, D. E. Wong, D. W. Wara, and M. J. Cowan. Lymphocyte abnormalities in infants born to drug abusing mothers. *J. Pediat.* 111 (1987): 230–235.

Curran, J., D. N. Lawrence, H. Jaffe, J. E. Kaplan, L. D. Zyla, M. Chamberland, R. Weinstein, K.-J. Lui, L. B. Schonberger, T. J. Spira, W. J. Alexander, G. Swinger, A. Ammann, S. Solomon, D. Auerbach, D. Mildvan, R. Stoneburner, J. M. Jason, H. W. Haverkos, and B. L. Evatt. Acquired immunodeficiency syndrome (AIDS) associated with transfusions. *New Engl. J. Med.* 310 (1984): 69–75.

Curran, J. W., M. W. Morgan, A. M. Hardy, H. W. Jaffe, W. W. Darrow, and W. R. Dowdle. The epidemiology of AIDS: current status and future prospects. *Science* 229 (1985): 1352–1357.

Daar, E. S., T. Moudcil, R. D. Meyer, D. D. and Ho. Transient high levels of viremia in patients with primary human immunodeficiency virus type I infection. *New Engl. J. Med.* 324 (1991): 961–964.

Dalakas, M. C., I . Illa, G. H. Pezeshkpour, J. P. Laukaitis, B. Cohen, and J. L. Griffin. Mitochondrial myopathy caused by long-term zidovudine therapy. *New Engl. J. Med.* 322 (1990): 1098–1105.

Darby, S. C., C. R. Rizza, R. Doll, R. J. D. Spooner, I. M. Stratton, and B. Thakrar. Incidence of AIDS and excess mortality associated with HIV in haemophiliacs in the United Kingdom: report on behalf of the directors of haemophilia centers in the United Kingdom. *Br. Med. J.* 298 (1989): 1064–1068.

Darrow, W. W., D. E. Echenberg, H. W. Jaffe, P. M. O'Malley, R. H. Byers, J. P. Getchell, and J. W. Curran. Risk factors for Human Immunodeficiency Virus (HIV) infections in homosexual men. *Am. J. Publ. Health* 77 (1987): 479–483.
be Biasi, R., A. Rocino, E. Miraglia, L. Mastrullo, and A. A. Quirino. The impact of a very high purity of factor VIII concentrate on the immune system of human immunodeficiency virus-infected hemophiliacs: a randomized, prospective, two-year comparison with an intermediate purity concentrate. *Blood* 78 (1991): 1919–1922.
De Cock, K. M., R. M. Selik, B. Soro, H. Gayle, and R. L. Colebunders. AIDS surveillance in Africa: a reappraisal of case definitions. *Br. Med. J.* 303 (1991): 1185–1188.
Deininger, S., R. Muller, I. Guggenmoos-Holzmann, U. Laukamm Josten, and U. Bienzle. Behavioral characteristics and laboratory parameters in homo- and bisexual men in West Berlin: an evaluation of five years of testing and counselling on AIDS. *Klin. Wochenschr.* 68 (1990): 906–913.
Delaney, M. and P. Goldblum. *Strategies for Survival: A Gay Men's Health Manual for the Age of AIDS.* (1987) St. Martin's Press, New York.
DeLoughry, T. J. 40 scientists call on colleagues to re-evaluate AIDS theory. *The Chronicle of Higher Education*, December 4, 1991.
Des Jarlais, D., S. Friedman, M. Marmor, H. Cohen, D. Mildvan, S. Yancovitz, U. Mathur, W. El-Sadr, T. J. Spira, and J. Garber. Development of AIDS, HIV seroconversion, and potential cofactors for T4 cell loss in a cohort of intravenous drug users. *AIDS* 1 (1987): 105–111.
Des Jarlais, D. C., S. R. Friedman, and W. Hopkins. Risk reduction of the acquired immunodeficiency syndrome among intravenous drug users. In: *AIDS and IV Drug Abusers: Current Perspectives*, pp. 97–109, Galea, R. P., B. F. Lewis, and L. Baker, (eds) National Health Publishing, Owings Mills, MD (1988).
Dismukes, W. E., A. W. Karchmer, R. F. Johnson, and W. J. Dougherty. Viral hepatitis associated with illicit parenteral use of drugs. *J. Am. Med. Ass.* 206 (1968): 1048–1052.
Don, P. C., F. Fusco, P. Fried, A. Batterman, F. P. Duncanson, T. H. Lenox, and N. C. Klein. Nail dyschromia associated with zidovudine. *Ann. Intern. Med.* 112 (1990): 145–146.
Donahoe, R. M., C. Bueso-Ramos, F. Donahoe, J. J. Madden, A. Falek, J. K. A. Nicholson, and P. Bokos. Mechanistic implications of the findings that opiates and other drugs of abuse moderate T-cell surface receptors and antigenic markers. *Ann. N.Y. Acad. Sci.* 496 (1987): 711–721.

Dournon, E., S. Matheron, W. Rozenbaum, S. Gharakhanian, C. Michon, P. M. Girard, C. Perrone, D. Salmon, P. Detruchis, C. Leport, and the Claude Bernard Hospital AZT Study Group. Effects of zidovudine in 365 consecutive patients with AIDS or AIDS-related complex. *Lancet* ii (1988): 1297–1302.

Doyle, A. C., Sir. *The Boscombe Valley Mystery.* (1928) John Murray, London.

Drew, W. L., J. Mills, J. Levy, J. Dylewski, C. Casavant, A. J. Ammann, H. Brodie, and T. Merigan. Cytomegalovirus infection and abnormal T-lymphocyte subset ratios in homosexual men. *Ann. Intern. Med.* 103 (1985): 61–63.

Dubin, G. and M. N. Braffman. Zidovudine-induced hepatotoxicity. *Ann. Intern. Med.* 110 (1989): 85–86.

Duckett, M. and A. J. Orkin. AIDS-related migration and travel policies and restrictions: a global survey. *AIDS* 3 (Suppl. 1) (1989): S231–S252.

Duesberg, P. H. The RNAs of influenza virus. *Proc. Natn. Acad. Sci. U.S.A.* 59 (1968): 930–937.

———. Retroviruses as carcinogens and pathogens: expectations and reality. *Cancer Res.* 47 (1987): 1199–1220.

———. Defective viruses and AIDS. *Nature (Lond.)* 340 (1989a): 515.

———. Does HIV cause AIDS? *J. AIDS* 2 (1989b): 514–515.

———. Human immunodeficiency virus and acquired immunodeficiency syndrome: correlation but not causation. *Proc. Natn. Acad. Sci. U.S.A.* 86 (1989c): 755–764.

———. AIDS: noninfectious deficiencies acquired by drug consumption and other risk factors. *Res. Immun.* 141 (1990a): 5–11.

———. Responding to the AIDS debate. *NaturwissenschaSten* 77 (1990b): 97–102.

———. Quantitation of Human Immunodeficiency Virus in the blood. *New Engl. J. Med.* 322 (1990c): 1466.

———. AIDS epidemiology: inconsistencies with Human Immunodeficiency Virus and with infectious disease. *Proc. Natn. Acad Sci. U.S.A.* 85 (1991a): 1575–1579.

———. Can alternative hypotheses survive in this era of megaprojects? *The Scientist*, July 8, 1991(b).

———. Defense says only AIDS not infectious (letter). *San Francisco Examiner*, August 30, 1991(c).

———. AIDS: the alternative view (letter). *Lancet* 339 (1992a): 1547.

———. A giant hole in the HIV–AIDS hypothesis. *The Sunday Times*, May 31, 1992(b).

———. HIV as target for zidovudine. *Lancet* 339 (1992c): 551.

———. HIV, AIDS, and zidovudine. *Lancet* 339 (1992d): 805–806.
———. Questions about AIDS (letter). *Nature (London)* 358 (1992e): 10.
———. The role of drugs in the origin of AIDS. *Biomed. Pharmacother.* 46 (1992f): 3–15.
Duesberg, P. H. and B. J. Ellison. Is the AIDS virus a science fiction? *Policy Rev.* 53 (1990): 40–51.
Duesberg, P. H. and J. R. Schwartz. Latent viruses and mutated oncogenes: no evidence for pathogenicity. *Prog. Nucleic Acid Res. Molec. Biol.* 43 (1992): 135–204.
Duesberg, P. H. and P. K. Vogt. Differences between the ribonucleic acids of transforming and non-transforming avian tumor viruses. *Proc. Natn. Acad. Sci. U.S.A.* 67 (1970): 1673–1680.
Editorial: AIDS in Africa. *Lancet* ii (1987): 192–194.
———. Zidovudine for symptomless HIV infection. *Lancet* 335 (1990): 821–822.
———. The impact of the National Cancer Act. *J. Natn. Cancer Inst.* 83 (1991): S1–S16.
Eggers, H. J. and J. J. Weyer. Linkage and independence of AIDS Kaposi disease: the interaction of Human Immunodeficiency Virus and some coagents. *Infection* 19 (1991): 115–122.
Eigen, M. The AIDS debate. *Naturwissenschaften* 76 (1989): 341–350.
Elwell, L. P., R. Ferone, G. A. Freeman, J. A. Fyfe, J. A. Hill, P. H. Ray, C. A. Richards, S. C. Singer, C. B. Knick, J. L. Rideout, and T. P. Zimmerman. Antibacterial activity and mechanism of action of 3'-azido-3'-deoxythymidine (BW A509U). *Antimicrob. Agents Chemother.* 31 (1987): 274–280.
Ensoli, B., G. Barillari, S. Z. Salahuddin, R. C. Gallo, and F. Wong-Staal. Tat protein of HIV-1 stimulates growth of cells derived from Kaposi's sarcoma lesions of AIDS patients. *Nature (London)* 345 (1990): 84–86.
Espinoza, P., I. Bouchard, C. Buffet, V. Thiers, J. Pillot, and J. P. Etienne. High prevalence of infection by hepatitis B virus and HIV in incarcerated French drug addicts. *Gastroenterol. Clin. Biol.* 11 (1987): 288–292.
Ettinger, N. A. and R. J. Albin. A review of the respiratory effects of smoking cocaine. *Am. J. Med.* 87 (1989): 664–668.
European Collaborative Study. Children born to women with HIV-1 infection: natural history and risk of transmission. *Lancet* 337 (1991): 253–260).
Evans, A. S. Author's reply. *J. AIDS* 2 (1989a): 515–517.
———. Does HIV cause AIDS? An historical perspective. *J. AIDS* 2 (1989b): 107–113.

———. *Viral Infections of Humans, Epidemiology and Control.* (1989c) Plenum Publishing Corporation, New York.

———. AIDS: the alternative view (letter). *Lancet* 339 (1992): 1547.

Evatt, B. L., R. B. Ramsey, D. N. Lawrence, L. D. Zyla, and J. W. Curran. The acquired immunodeficiency syndrome in patients with hemophilia. *Ann. Intern. Med.* 100 (1984): 499–505.

Evatt, B. L., E. D. Gomperts, J. S. McDougal, and R. B. Ramsey. Coincidental appearance of LAV/HTLV-III antibodies in hemophiliacs and the onset of the AIDS epidemic. *New Engl. J. Med.* 312 (1985): 483–486.

Eyster, M. E., D. A. Whitehurst, P. M. Catalano, C. W. McMillan, S. H. Goodnight, C. K. Kasper, J. C. Gill, L. M. Aledort, M. W. Hilgartner, P. H. Levine, J. R. Edson, W. E. Hathaway, J. M. Lusher, F. M. Gill, W. K. Poole, and S. S. Shapiro. Long-term follow-up of hemophiliacs with lymphocytopenia or thrombocytopenia. *Blood* 66 (1985): 1317–1320.

Farber, C. Fatal distraction. *Spin* 3 (1992): 36–45, 84, 90–91.

Fauci, A. Writing for my sister Denise. *The AAAS Observer, Supplement to Science*, September 1 (1989): 4.

Fauci, A. S. The role of the endogenous cytokines in the regulation of HIV expression. *HIV Adv. Res. Ther.* 1 (1991): 3–7.

Fenner, F., B. R. McAuslan, C. A. Mims, J. Sambrook, and D. O. White. *The Biology of Animal Viruses*. Academic Press, Inc., New York (1974).

Fineberg, H. V. The social dimensions of AIDS. *Sci. Am.* 259 (1988): 128–134.

Finnegan, L. P., J. M. Mellot, L. R. Williams, and R. J. Wapner. Perinatal exposure to cocaine: human studies. In: *Cocaine: Pharmacology, Physiology and Clinical Strategies* (1992), pp. 391–409, J. M. Lakoski, M. P. Galloway, and F. J. White (eds) CRC Press, Boca Raton, FL.

Fischl, M. A., D. D. Richman, M. H. Grieco, M. S. Gottlieb, P. A. Volberding, O. L. Laskin, J. M. Leedon, J. E. Groopman, D. Mildvan, R. T. Schooley, G. G. Jackson, D. T. Durack, D. King, and the AZT Collaborative Working Group. The efficacy of azidothymidine (AZT) in the treatment of patients with AIDS and AIDS-related complex. *New Engl. J. Med.* 317 (1987): 185–191.

Fischl, M. A., D. D. Richman, D. M. Causey, M. H. Greco, Y. Bryson, D. Mildvan, O. L. Laskin, J. E. Groopman, P. A. Volberding, R. T. Schooley, G. G. Jackson, D. T. Durack, J. C. Andrews, S. Nuslnoff-Lehrman, D. W. Barry, and the AZT Collaborative Working Group. Prolonged zidovudine therapy in patients with AIDS and advanced AIDS-related

complex. *J. Am. Med. Ass.* 262 (1989): 2405–2410.
Flanagan, T. J. and K. Maguire. Sourcebook of Criminal Justice Statistics (1989)—Bureau of Justice Statistics NCJ-124224. U.S. Department of Justice, U.S. Government Printing Office, Washington, D.C. (1989).
Folkart, B. A. Paul Gann dies—Tax-crusading Prop. 13 author. *Los Angeles Times,* September 12, 1989.
Francis, D. P. The search for the cause. In: *The AIDS Epidemic,* pp. 137–148, Cahill, K. M. (ed.) St Martin's Press, New York (1983).
Francis, D. P., J. W. Curran, and M. Essex. Epidemic acquired immune deficiency syndrome: epidemiologic evidence for a transmissible agent. *J. Natn. Cancer Inst.* 71 (1983): 1–4.
Freeman, B. A. *Burrows Textbook of Microbiology.* W. B. Saunders Co., Philadelphia (1979).
Freestone, D. S. Zidovudine. *Lancet* 339 (1992): 626.
Fricker, H. S. and S. Segal. Narcotic addiction, pregnancy, and the newborn. *Am. J. Dis. Child.* 132 (1978): 360–366.
Friedland, G. H., B. R. Satzman, M. F. Rogers, P. A. Kahl, M. L. Lesser, M. M. Mayers, and R. S. Klein. Lack of transmission of HTLV-III/LAV infection to household contacts of patients with AIDS or AIDS-related complex with oral candidiasis. *New Engl. J. Med.* 314 (1986): 344–349.
Friedman–Kien, A. E., B. R. Saltzman, Y. Cao, M. S. Nestor, M. Mirabile, J. J. Li, and T. A. Peterman. Kaposi's sarcoma in HIV-negative homosexual men. *Lancet* 335 (1990): 168–169.
Froesner, G. Congress report: VII Internationale AIDS-Konferenz in Florenz, 1991. *AIDS-Forschung* 9 (1991): 477–486.
Fry, T. C. *The Great AIDS Hoax.* Life Science Inst., Austin, Tex. (1989).
Fultz, P. N., R. B. Stricker, H. M. McClure, D. C. Anderson, W. M. Switzer, and C. Horaist. Humoral response to SIV/SMM infection in macaque and mangabey monkeys. *J. AIDS* 3 (1990): 319–329.
Furman, P. A., J. A. Fyfe, M. St. Clair, K. Weinhold, J. L. Rideout, G. A. Freeman, S. Nuslnoff Lehrman, D. P. Bolognesi, S. Broder, H. Mitsuya, and D. W. Barry. Phosphorylation of 3′-azido-3′deoxythymidine and selective interaction of the 5′-triphosphate with Human Immunodeficiency Virus reverse transcriptase. *Proc. Natn. Acad. Sci. U.S.A.* 83 (1986): 8333–8337.
Gajdusek, D. C. Unconventional viruses and the origin and disappearance of kuru. *Science* 197 (1977): 943–960.
Gallo, R. C. The AIDS virus. *Sci. Am.* 256 (1987): 45–46.
———. Mechanism of disease induction by HIV. *J. AIDS* 3 (1990): 380–389.
———. *Virus-Hunting—AIDS, Cancer, and the Human Retrovirus: A Story of Scientific Discovery.* Basic Books, New York (1991).

———. Self-deluding (letter). *The Sunday Times*, May 24, 1992.
Gallo, R. C. and L. Montagnier, The chronology of AIDS research. *Nature (London)* 326 (1987): 435–436.
———. AIDS in 1988. *Sci. Am.* 259 (1988): 41–48.
Gallo, R. C., P. S. Sarin, E. P. Gelmann, M. Robert-Guroff, and E. Richardson. Isolation of human T-cell leukemia virus in acquired immune deficiency syndrome (AIDS). *Science* 220 (1983): 865–867.
Gallo, R. C., S. Z. Salahuddin, M. Papovic, G. M. Shearer, M. Kaplan, B. F. Haynes, T. J. Palker, R. Redfield, J. Oleske, B. Safai, G. White, P. Foster, and P. D. Markham. Frequent detection and isolation of cytopathic retrovirus (HTLV-III) from patients with AIDS and at risk for AIDS. *Science* 224 (1984): 500–503.
Garry, R. F. and G. Koch. Tat contains a sequence related to snake neurotoxins. *AIDS* (1992).
Garry, R. F., M. H. Witte, A. A. Gottlieb, M. Elvin-Lewis, M. S. Gottlieb, C. L. Witte, S. S. Alexander, W. R. Cole, and W. L. Drake. Documentation of an AIDS virus infection in the United States in 1968. *J. Am. Med. Ass.* 260 (1988): 2085–2087.
Garry, R. F., J. J. Kort, F. Koch-Nolte, and G. Koch. Similarities of viral proteins to toxins that interact with monovalent cation channels. *AIDS* 5 (1991): 1381–1384.
Geller, S. A. and B. Stimmel. Diagnostic confusion from lymphatic lesions in heroin addicts. *Ann. Intern. Med.* 78 (1973): 703–705.
Gilks, C. What use is a clinical case definition for AIDS in Africa? *Br. Med. J.* 303 (1991): 1189–1190.
Gill, J. C., M. D. Menitove, P. R. Anderson, J. T. Casper, S. G. Devare, C. Wood, S. Adair, J. Casey, C. Scheffel, and M. D. Montgomery. HTLV-III serology in hemophilia: relationship with immunologic abnormalities. *J. Pediat.* 108 (1986): 511–516.
Gill, P. S., M. Rarick, R. K. Byrnes, D. Causey, C. Loureiro, and A. M. Levine. Azidothymidine associated with bone marrow failure in the acquired immunodeficiency syndrome (AIDS). *Ann. Intern. Med.* 107 (1987): 502–505.
Ginzburg, H. M. Acquired immune deficiency syndrome (AIDS) and drug abuse. In: *AIDS and I. V. Drug Abusers Current Perspectives*, pp. 61–73, Galea, R. P., B. F. Lewis, and L. Baker (eds.) National Health Publishing, Owings Mills, MD (1988).
Goedert, J. J., C. Y. Neuland, W. C. Wallen, M. H. Greene, D. L. Mann, C. Murray, D. M. Strong, J. F. Fraumeni, Jr., and W. A. Blattner. Amyl nitrite may alter T lymphocytes in homosexual men. *Lancet* i (1982): 412–416.
Goedert, J. J., C. M. Kessler, L. M. Aledort, R. J. Biggar, W. A. Andes, G. C. White, II, J. E. Drummond, K. Vaidya, D. L. Mann, M. E. Eyster,

M. V. Ragni, M. M. Lederman, A. R. Cohen, G. L. Bray, P. S. Rosenberg, R. M. Friedman, M. W. Hilgartner, W. A. Blattner, B. Kroner, and M. H. Gail. A prospective study of Human Immunodeficiency Virus type I infection and the development of AIDS in subjects with hemophilia. *New Engl. J. Med.* 321 (1989): 1141–1148.

Goldsmith, M. S. Science ponders whether HIV acts alone or has another microbe's aid. *J. Am. Med. Ass.* 264 (1990): 265–266.

Goodgame, R. W. AIDS in Uganda—clinical and social features. *New Engl. J. Med.* 323 (1990): 383–389.

Gorard, D. A. and R. J. Guilodd. Necrotising myopathy and zidovudine. *Lancet* i (1988): 1050.

Gottlieb, M. S., H. M. Schanker, P. T. Fan, A. Saxon, J. D. Weisman, and J. Pozalski. *Pneumocystis* pneumonia—Los Angeles. *Morb. Mort. Weekly Rep.* 30 (1981): 250–252.

Goudsmit, J. Alternative view on AIDS. *Lancet* 339 (1992): 1289–1290.

Graham, N. M. H., S. L. Zeger, L. P. Park, J. P. Phair, R. Detels, S. H. Vermund, M. Ho, A. J. Saah, and the Multicenter Aids Cohort Study. Effect of zidovudine and *Pneumocystis carinii* pneumonia prophylaxis on progression of HIV-1 infection to AIDS. *Lancet* 338: 265–269.

Greenberg, D. S. (1986) What ever happened to the War on Cancer? *Discover*, March 1991, 47–66.

Grimshaw, J. Being HIV antibody-positive. *Br. Med. J.* 295 (1987): 256–257.

Guinan, M. E. and A. Hardy. Epidemiology of AIDS in women in the United States, 1981 through 1986. *J. Am. Med. Ass.* 257 (1987): 2039–2042.

Guyton, A. C. *Textbook of Medical Physiology.* W. B. Saunders, New York (1987).

Haas, M. The need for a search for a proximal principle of human AIDS. *Cancer Res.* 49 (1989): 2184–2187.

Hahn, B. H., G. M. Shaw, M. E. Taylor, R. R. Redfield, P. D. Markham, S. Z. Salahuddin, F. Wong-Staal, R. C. Gallo, E. S. Parks, and W. P. Parks. Genetic variation in HTLV-III/LAV over time in patients with AIDS or at risk for AIDS. *Science* 232 (1986): 1548–1553.

Hamilton, D. P. What next in the Gallo case? *Science* 254 (1991): 941–945.

Hamilton, J. D., P. M. Hartigan, M. S. Simberkoff, P. L. Day, G. R. Diamond, G. M. Dickinson, G. L. Drusano, M. J. Egorin, W. L. George, F. M. Gordin, and the Veterans Affairs Cooperative Study Group on AIDS Treatment. A controlled trial of early versus late treatment with zidovudine in symptomatic Human Immunodeficiency Virus infection. *New Engl. J. Med.* 326 (1992): 437–443.

Hardy, A. M., J. R. Allen, W. M. Morgan, and J. W. Curran. The incidence rate of acquired immunodeficiency syndrome in selected populations. *J. Am. Med. Ass.* 253 (1985): 215–220.

Harris, P. D. and R. Garret. Susceptibility of addicts to infection and neoplasia. *New Engl. J. Med.* 287 (1972): 310.

Haseltine, W. A. and F. Wong-Staal. The molecular biology of the AIDS virus. *Sci. Am.* 259 (1988): 52–62.

Haverkos, H. W. Epidemiologic studies—Kaposi's sarcoma vs opportunistic infections among homosexual men with AIDS. In: *Health Hazards of Nitrite Inhalants,* pp. 96–105, Haverkos, H. W. and J. A. Dougherty (eds) NIDA Research Monograph 83, National Institute on Drug Abuse, Washington, D.C. (1988a).

———. Kaposi's sarcoma and nitrite inhalants. In: *Psychological, Neuropsychiatric and Substance Abuse Aspects Of AIDS,* pp. 165–172, Bridge, T. P., Mirsky, A. F. and Goodwin, F. K. (eds) Raven Press, New York (1988b).

———. Nitrite inhalant abuse and AIDS-related Kaposi's sarcoma. *J. AIDS* 3 (Suppl. 1) (1990): S47–S50.

Haverkos, H. W. and J. A. Dougherty (eds) *Health Hazards of Nitrite Inhalants.* NIDA Research Monograph 83, National Institute on Drug Abuse, Washington, D.C. (1988).

Haverkos, H. W., P. F. Pinsky, D. P. Drotman, and D. J. Bregman. Disease manifestation among homosexual men with acquired immunodeficiency syndrome: a possible role of nitrites in Kaposi's sarcoma. *J. Sexually Transmitted Dis.* 12 (1985): 203–208.

Hearst, N. and S. Hulley. Preventing the heterosexual spread of AIDS: are we giving our patients the best advice? *J. Am. Med. Ass.* 259 (1988): 2428–2432.

Helbert, M., T. Fletcher, B. Peddle, J. R. W. Harris, and A. J. Pinching. Zidovudine-associated myopathy. *Lancet* ii (1988): 689–690.

Hersh, E. M., J. M. Reuben, H. Bogerd, M. Rosenblum, M. Bielski, P. W. A. Mansell, A. Rios, G. R. Newell, and G. Sonnenfeld. Effect of the recreational agent isobutyl nitrite on human peripheral blood leukocytes and on *in vitro* interferon production. *Cancer Res.* 43 (1983): 1365–1371.

Heyward, W. L. and J. W. Curran. The epidemiology of AIDS in the U.S. *Sci. Am.* 259 (1988): 72–81.

Hilts, P. J. Monkey's HIV vulnerability portends AIDS breakthrough. *New York Times,* June 12, 1992.

Hishida, O., E. Ido, T. Igarashi, M. Hayami, M. Miyazaki, N. K. Ayisi, and M. Osei-Kwasi. Clinically diagnosed AIDS cases without evident association with HIV type 1 and 2 infection in Ghana. *Lancet* 340 (1992): 971–972.

Hitchcock, M. J. M. Review: antiviral portrait series, Number 1; 2',3'-didehydro-2',3'dideoxythymidine (D4T), and anti-HIV agent. *Antiviral Chem. Chemother.* 2 (1991): 125–132.
Ho, D. D., T. Moudgil, and M. Alam. Quantification of Human Immunodeficiency Virus type 1 in the blood of infected persons. *New Engl. J. Med.* 321 (1989a): 1621–1625.
Ho, D. D., T., Moudgil, H. S. Robin, M. Alam, B. J. Wallace, and Y. Mizrachi. Human immunodeficiency virus type 1 in a seronegative patient with visceral Kaposi's sarcoma and hypogammaglobulinemia. *Am. J. Med.* 86 (1989b): 349–351.
Ho, H.-T. and M. J. M. Hitchcock. Cellular pharmacology of 2_,3_-dideoxy-2,3_-didehydrothymidine, a nucleotide analog active against Human Immunodeficiency Virus. *Antimicrob. Agents Chemother.* 33 (1989): 844–849.
Hodgkinson, N. Experts mount startling challenge to AIDS orthodoxy. *Sunday Times (Focus)*, April 26, 1992.
Hoffman, M. S. (ed.) *World Almanac and Book of Facts.* Scripps Howard Co., New York (1992).
Hoffmann, G. W. A response to P. H. Duesberg with reference to an idiotypic network model of AIDS immunopathogenesis. *Res. Immun.* 141 (1990): 701–709.
Holmberg, S. D., L. J. Conley, S. P. Buchbinder, F. N. Judson, M. H. Katz, K. A. Penley, T. J. Bush, R. C. Hershow, and A. R. Lifson. Use of therapeutic and prophylactic drugs for AIDS by homosexual and bisexual men in three U.S. cities. *Lancet* (1992).
Holub, W. R. AIDS: a new disease? *Am. Clin. Prod. Rev.* 7 (1988): 28–37.
Hoxie, J. A., B. S. Haggarty, J. L. Rakowski, N. Pillsbury, and J. A. Levy. Persistent noncytopathic infection of normal human T lymphocytes with AIDS-associated retrovirus. *Science* 229 (1985): 1400–1402.
Huang, Y. Q., J. J. Li, M. G. Rush, B. J. Poiesz, A. Nicolaides, M. Jacobson, W. G. Zhang, E. Coutavas, M. A. Abbott, and A. E. Friedman–Kien. HPV-16–related DNA sequences in Kaposi's sarcoma. *Lancet* 339 (1992): 515–518.
Institute of Medicine *Confronting AIDS*. National Academy Press, Washington, D.C. (1986).
Institute of Medicine. *Confronting AIDS—Update 1988*. National Academy Press, Washington, D.C. (1988).
Jaffe, H. W., D. J. Bregman, and R. M. Selik. Acquired immune deficiency syndrome in the United States: the first 1,000 cases. *J. Infect. Dis.* 148 (1983a): 339–345.
Jaffe, H. W., K. Choi, P. A. Thomas, H. W. Haverkos, D. M. Auerbach,

M. E. Guinan, M. F. Rogers, T. J. Spira, W. W. Darrow, M. A. Kramer, S. M. Friedman, J. M. Monroe, A. E. Friedman-Kien, L. J. Laubenstein, M. Marmor, B. Safai, S. K. Dritz, S. J. Crispi, S. L. Fannin, J. P. Orkwis, A. Kelter, W. R. Rushing, S. B. Thacker, and J. W. Curran. National case-control study of Kaposi's sarcoma and *Pneumocystis carinii* pneumonia in homosexual men: Part I, epidemiologic results. *Ann. Intern. Med.* 99 (1983b): 145–151.

Jason, J., R. C. Holman, B. L. Evatt, and the Hemophilia–AIDS Collaborative Study Group. Relationship of partially purified factor concentrates to immune tests and AIDS. *Am. J. Hematol.* 34 (1990): 262–269.

Jin, Z., R. P. Cleveland, and D. B. Kaufman. Immunodeficiency in patients with hemophilia: an underlying deficiency and lack of correlation with factor replacement therapy or exposure to Human Immunodeficiency Virus. *Allergy Clin. Immun.* 83 (1989): 165–170.

Johnson, R. E., D. N. Lawrence, B. L. Evatt, D. J. Bregman, L. D. Zyla, J. W. Curran, L. M. Aledort, M. E. Eyster, A. P. Brownstein, and C. J. Carman. Acquired immunodeficiency syndrome among patients attending hemophilia treatment centers and mortality experience of hemophiliacs in the United States. *Am. J. Epidemiol.* 121 (1985): 797–810.

Jorgensen, K. A. and S.-O. Lawesson. Amylnitrite and Kaposi's sarcoma in homosexual men. *New Engl. J. Med.* 307 (1982): 893–894.

Judson, F. N., K. A. Penley, M. E. Robinson, and J. K. Smith. Comparative prevalence rates of sexually transmitted diseases in heterosexual and homosexual men. *Am. J. Epidemiol.* 112 (1980): 836–843.

Kaslow, R. A., W. C. Blackwelder, D. G. Ostrow, D. Yerg, J. Palenneck, A.H. Goulson, and R. O. Valdiserri. No evidence for a role of alcohol or other psychoactive drugs in accelerating immunodeficiency in HIV-1 positive individuals. *J. Am. Med Ass.* 261 (1989): 3424–3429.

Kestler, H., T. Kodoma, D. Ringler, M. Marthas, N. Pederson, A. Lackner, D. Regier, P. Sehgal, M. Daniel, N. King, and R. Desrosiers. Induction of AIDS in rhesus monkeys by molecularly cloned simian immunodeficiency virus. *Science* 248 (1990): 1109–1112.

Kestler, H. W., D. J. Ringler, M. D. Mori, D. L. Panicalli, P. K. Sehgal, M. D. Daniel, and R. C. Desrosiers. Importance of the nef gene for maintenance of high virus loads and for development of AIDS. *Cell* 65 (1991): 651–662.

Klein, G. The role of science. *J. AIDS* 1 (1988): 611–615.

Koch, T. Uninfected children of HIV-infected mothers may still suffer nervous problems. *CDC AIDS Weekly*, July 30, 1990: 9.

Koch, T. K., R. Jeremy, E. Lewis, P. Weintrub, and M. Cowan. Developmental abnormalities in uninfected infants born to Human Immunodeficiency Virus-infected mothers. *Ann. Neurol.* 28 (1990): 456–457.

Koerper, M. A. AIDS and hemophilia. In: *AIDS: Pathogenesis and Treatment*, pp. 79–95, Levy, J. A. (ed.) Marcel Dekker, Inc., New York (1989).
Kolata, G. Imminent marketing of AZT raises problems; marrow suppression hampers AZT use in AIDS victims. *Science* 235 (1987): 1462–1463.
———. After 5 years of use, doubt still clouds leading AIDS drug. *New York Times*, June 2, 1992.
Kono, R. Introductory review of subacute myelo-optico-neuropathy (SMON) and its studies done by the SMON research commission. *Jap. J. Med. Sci. Biol.* 28 (Suppl.) (1975): 1–21.
Konotey-Ahulu, F. I. D. AIDS in Africa: misinformation and disinformation. *Lancet* ii: (1987) 206–207.
———. *What Is AIDS?* Tetteh-A'Domenco Co., Watford, England (1989).
Kozel, N. J. and E. H. Adams. Epidemiology of drug abuse: an overview. *Science* 234 (1986): 970–974.
Kreek, M. J. Methadone maintenance treatment for harm reduction approach to heroin addiction. In: *Drug Addiction and AIDS*, pp. 153–178, Loimer, N., Schmid, R. and Springer, A. (eds) Springer-Verlag, New York (1991).
Kreiss, J. K., C. K. Kasper, J. L. Fahey, M. Weaver, B. R. Visscher, J. A. Stewart, and D. N. Lawrence. Nontransmission of T-cell subset abnormalities from hemophiliacs to their spouses. *J. Am. Med. Ass.* 251 (1984): 1450–1454.
Kreiss, J. K., L. W. Kitchen, H. E. Prince, C. K. Kasper, A. L. Goldstein, P. H. Naylor, O. Preble, J. A. Stewart, and M. Essex. Human T cell leukemia virus type III antibody, lymphadenopathy, and acquired immune deficiency syndrome in hemophiliac subjects. *Am. J. Med* 80 (1986): 345–350.
Lambert, B. Kimberly Bergalis is dead of AIDS at 23 (contracted from a health care worker) (Obituary). *New York Times*, December 9, 1991.
Landesmann, S., H. Minkoff, S. Holman, S. McCalla, and O. Sijin. Serosurvey of Human Immunodeficiency Virus infection in parturients. *J. Am. Med. Ass.* 258 (1987): 2701–2703.
Lang, D. J., A. A. S. Kovacs, J. A. Zaia, G. Doelkin, J. C. Niland, L. Aledort, S. P. Azen, M. A. Fletcher, J. Gauderman, G. J. Gjerst, J. Lusher, E. A. Operskalski, J. W. Parker, C. Pegelow, G. N. Vyas, J. W. Mosley, and The Transfusion Safety Group. Seroepidemiologic studies of cytomegalovirus and Epstein-Barr virus infections in relation to Human Immunodeficiency Virus type I infection in selected recipient populations. *J. AIDS* 2 (1989): 540–549.

Lang, W., H. Perkins, R. E. Anderson, R. Royce, N. Jewell, and W. Winkelstein, Jr. Patterns of T lymphocyte changes with Human Immunodeficiency Virus infection: from seroconversion to the development of AIDS. *J. AIDS* 2 (1989): 63–69.

Lang, W., D. Osmond, M. Samuel, A. Moss, L. Schrager, and W. Winkelstein, Jr. Population-based estimates of zidovudine and aerosol pentamidine use in San Francisco: 1987–1989. *J. AIDS* 4 (1991): 713–716.

Lange, W. R., E. M. Dax, C. A. Haertzen, F. R. Snyder, and J. H. Jaffe. Nitrite inhalants: contemporary patterns of abuse. In: *Health Hazards of Nitrite Inhalants*, pp. 86–94, Haverkos, H. W. and Dougherty, J. A. (eds) National Institute on Drug Abuse, Washington, D.C. (1988).

Langhof, E., J. McElrath, H. J. Bos, J. Pruett, A. Granelli Piperno, Z. A. Cohn, and R. M. Steinman. Most CD4+ T-cells from Human Immunodeficiency Virus-1 infected patients can undergo prolonged clonal expansion. *J. Clin. Invest.* 84 (1989): 1637–1643.

Laurence, J., F. P. Siegal, E. Schattner, I. H. Gelman, and S. Morse. Acquired immunodeficiency without evidence of infection with human immunodeficiency virus types 1 and 2. *Lancet* 340 (1992): 273–274.

Laurent-Crawford, A. G., B. Krust, S. Muller, Y. Riviere, M.-A. Rey-Cuille, J.-M. Bechet, L. Montagnier, and A. G. Hovanessian. The cytopathic effect of HIV is associated with apoptosis. *Virology* 185 (1991): 829–839.

Lauritsen, J. *Poison by Prescription—The AZT Story*. Asklepios Press, New York (1990).

———. The "AIDS" war: censorship and propaganda dominate media coverage of the epidemic. *New York Native*, August 12, 1991.

———. FDA documents show fraud in AZT trials. *New York Native*, March 30, 1992.

Lauritsen, J. and H. Wilson. *Death Rush, Poppers and AIDS*. Pagan Press, New York (1986).

Lawrence, D. N., J. M. Jason, R. C. Holman, and J. J. Murphy. HIV transmission from hemophilic men to their heterosexual partners. In: *Heterosexual Transmission of AIDS*, pp. 35–53, Alexander, N. J., Gabelnick, H. L., and Spieler, J. M. (eds) Wiley–Liss, New York (1990).

Layon, J., A. Idris, M. Warzynski, R. Sherer, D. Brauner, O. Patch, D. McCulley, and P. Orris. Altered T-lymphocyte subsets in hospitalized intravenous drug abusers. *Arch. Intern. Med.* 144 (1984): 1376–1380.

Learmont, J., B. Tindall, L. Evans, A. Cunningham, P. Cunningham, J. Wells, R. Penny, J. Kaldor, and D. A. Cooper. Long-term symptomless HIV-1 infection in recipients of blood products from a single donor. *Lancet* 340 (1992): 863–867.

Lemaitre, M., D. Guetard, Y. Henin, L. Montagnier, and A. Zerial. Protective activity of tetracycline analogs against the cytopathic effect of the Human Immunodeficiency Viruses in CEM cells. *Res. Virol.* 141 (1990): 5–16.
Lemp, G. F., S. F. Payne, G. W. Rutherford, N. A. Hessol, W. Winkelstein, Jr., J. A. Wiley, A. R. Moss, R. E. Chaisson, R. T. Chen, D. W. Feigel, P. A. Thomas, and D. Werdegar. Projections of AIDS morbidity and mortality in San Francisco. *J. Am. Med. Ass.* 263 (1990): 1497–1501.
Leonhard, H.-W. Alles nur ein Irrtum? *neue praxis: Zeitschrift fur Sozialarbeit, Sozialpadagogik und Sozialpolitik* 22 (1992): 14–29.
Lepage, P., F. Dabis, D.-G. Hitimana, P. Msellati, C. Van Goethem, A.-M. Stevens, F. Nsengumuremyi, A. Bazubargira, A. Serufilira, A. De Clercq, and P. Van De Perre. Perinatal transmission of HIV-1: lack of impact of maternal HIV infection on characteristics of livebirths and on neonatal mortality in Kigali, Rwanda. *AIDS* 5 (1991): 295–300.
Lerner, W. D. Cocaine abuse and acquired immunodeficiency syndrome: tale of two epidemics. *Am. J. Med.* 87 (1989): 661–663.
Lesbian and Gay Substance Abuse Planning Group. *Gay Men, Lesbians and their Alcohol and other Drug Use: A Review of the Literature.* San Francisco Department of Public Health, San Francisco, CA (1991a).
Lesbian and Gay Substance Abuse Planning Group. *San Francisco Lesbian, Gay and Bisexual Substance Abuse Needs Assessment.* San Francisco Department of Public Health, San Francisco, CA (1991b).
Levy, J. Mysteries of HIV: challenges for therapy and prevention. *Nature (London)* 333 (1988): 519–522.
Lifschitz, M. H., G. S. Wilson, E. O. Smith, and M. M. Desmond. Fetal and postnatal growth of children born to narcotic-dependent women. *J. Pediat.* 102: (1983) 686–691.
Liversidge, A. AIDS: words from the front. *Spin* 12 (1989): 54–56, 60–61, 81.
Lo, S.-C., S. Tsai, J. R. Benish, J. W.-K. Shih, D. J. Wear, and D. M. Wong. Enhancement of HIV-1 cytocidal effects in CD4+ lymphocytes by the AIDS-associated mycoplasma. *Science* 251 (1991): 1074–1076.
Louria, D. B. Infectious complications of nonalcoholic drug abuse. *A. Rev. Med.* 25 (1974): 219–231.
Lowdell, C. P. and M. G. Glaser. Long term survival of male homosexual patients with Kaposi's sarcoma. *J. R. Soc. Med.* 82 (1989): 226–227.
Lu, W. and J.-M. Andrieu. Similar replication capacities of primary Human Immunodeficiency Virus type 1 isolates derived from a wide range of clinical sources. *J. Virol.* 66 (1992): 334–340.
Luca-Moretti, M. Specific behavioral factors among intravenous drug

users have been shown to influence HIV seroconversion. *J. Intern. Med. Health Ass.* 1: (1992) 1–9.
Ludlam, C. A. AIDS: the alternative view (letter). *Lancet* 339 (1992): 1547–1548.
Ludlam, C. A., J. Tucker, C. M. Steel, R. S. Tedder, R. Cheingsong-Popov, R. Weiss, D. B. L. McClelland, I. Phillip, and R. J. Prescott. Human T-lymphotropic virus type III (HTLV-III) infection in seronegative hemophiliacs after transfusion of factor VIII. *Lancet* ii (1985): 233–236.
Lui, K.-J., W. W. Darrow, and G. W. Rutherford, III. A model-based estimate of the mean incubation period for AIDS in homosexual men. *Science* 240 (1988): 1333–1335.
Lusso, P., A. De Maria, M. Mainati, F. Lori, S. E. Derocco, M. Baseler, and R. C. Gallo. Induction of CD4 and susceptibility to HIV-1 infection in human CD8 + T lymphocytes by human herpes virus 6. *Nature (London)* 349 (1991): 533–535.
Maddox, J. AIDS research turned upside down. *Nature (London)* 353 (1991a): 297.
———. Basketball, AIDS and education. *Nature (London)* 354 (1991b): 103.
———. Rage and confusion hide role of HIV. *Nature (London)* 357 (1992a): 188.
———. (1992b) Humbling of world's AIDS researchers. *Nature (London)* 358: 367.
Madhok, R., A. Gracie, G. D. O. Lowe, A. Burnett, K. Froebel, E. Follet, and C. D. Forbes. Impaired cell mediated immunity in haemophilia in the absence of infection with Human Immunodeficiency Virus. *Br. Med. J.* 293 (1986): 978–980.
Mahir, W. S., R. E. Millard, J. C. Booth, and P. T. Flute. Functional studies of cell-mediated immunity in haemophilia and other bleeding disorders. *Br. J. Haemat.* 69 (1988): 367–370.
Maikel, R. P. The fate and toxicity of butyl nitrites. In: *Health Hazards of Nitrite Inhalants*, pp. 15–27, Haverkos, H. W. and J. A. Dougherty (eds) NIDA Research Monograph 83, National Institute on Drug Abuse, Washington, D.C. (1988).
Mann, J., J. Chin, P. Piot, and T. Quinn. The international epidemiology of AIDS. *Sci. Am.* 259 (1988): 82–89.
Mann, J. and Global Aids Policy Coalition. *AIDS in the World.* Harvard International AIDS Center (1992).
Mansuri, M. M., M. J. M. Hitchcock, R. A. Buroker, C. L. Bregman, I. Ghazzouli, J. V. Desiderio, J. E. Starrett, R. Z. Sterzycki, and J. C. Martin. Comparison of *in vitro* biological properties and mouse toxi-

cities of three thymidine analogs active against Human Immunodeficiency Virus. *Antimicrob. Agents Chemother.* 34 (1990): 637–641.
Marion, S. A., M. T. Schechter, M. S. Weaver, W. A. McLeod, W. J. Boyko, B. Willoughby, B. Douglas, K. J. P. Craib, and M. O'Shaughnessy. Evidence that prior immune dysfunction predisposes to Human Immunodeficiency Virus infection in homosexual men. *J. AIDS* 2 (1989): 178–186.
Marmor, M., A. E. Friedman-Kien, L. Laubenstein, R. D. Byrum, D. C. William, S. D'Onofrio, and N. Dublin. Risk factors for Kaposi's sarcoma in homosexual men. *Lancet* i (1982): 1083–1087.
Marquart, K.-H., R. Engst, and G. Oehlschlaegel. An 8-year history of Kaposi's sarcoma in an HIV-negative bisexual man. *AIDS* 5 (1991): 346–348.
Martin, M. A., T. Byran, S. Rasheed, and A. S. Khan. Identification and cloning of endogenous retroviral sequences present in human DNA. *Proc. Natn. Acad. Sci. U.S.A.* 78 (1981): 4892–4896.
Marx, J. Circumcision may protect against the AIDS virus. *Science* 245 (1989): 470–471.
Mathe, G. Is the AIDS virus responsible for the disease? *Biomed. Pharmacother.* 46 (1992): 1–2.
Matheson, D. S., B. J. Green, M. J. Fritzler, M.-C. Poon, T. J. Bowen, and D. I. Hoar. Humoral immune response in patients with hemophilia. *Clin. Immun. Immunopath.* 4(1987): 41–50.
Mathur-Wagh, U., R. W. Enlow, I. Spigland, R. J. Winchester, H. S. Sacks, E. Rorat, S. R. Yancovitz, M. J. Klein, D. C. William, and D. Mildwan. Longitudinal study of persistent generalized lymphadenopathy in homosexual men: relation to acquired immunodeficiency syndrome. *Lancet* i (1984): 1033–1038.
Mathur-Wagh, U., D. Mildvan, and R. T. Senie. Follow-up of 4 1/2 years on homosexual men with generalized lymphadenopathy. *New Engl. J. Med.* 313 (1985): 1542–1543.
McDonough, R. J., J. J. Madden, A. Falek, D. A. Shafer, M. Pline, D. Gordon, P. Bokof, J. C. Kuehnle, and J. Mandelson. Alteration of T and null lymphocyte frequencies in the peripheral blood of human opiate addicts: *in vivo* evidence of opiate receptor sites on T lymphocytes. *J. Immun.* 125 (1980): 2539–2543.
McDunn, S. H., J. N. Winter, D. Variakojis, A. W. Rademaker, J. H., Von Roenn, M. S. Tallman, L. I. Gordon, and K. D. Bauer. Human immunodeficiency virus-related lymphomas: a possible association between tumor proliferation, lack of ploidy anomalies and immune deficiency. *J. Clin. Oncol.* 9 (1991): 1334–1340.
McGrady, G. A., J. M. Jason, and B. L. Evatt. The course of the epidemic

of acquired immunodeficiency syndrome in the United States hemophilia population. *Am. J. Epidemiol.* 126 (1987): 25–30.

McKee, M. Defense puts AIDS virus on trial. *The Recorder*, January 17, 1992.

McKegney, F. P., M. A. O'Dowd, C. Feiner, P. Selwyn, E. Drucker, and G. H. Friedland. A prospective comparison of neuropsychologic function in HIV-seropositive and seronegative methadonemaintained patients. *AIDS* 4 (1990): 565–569.

McKeown, T. *The Role of Medicine: Dream, Mirage, or Nemesis?* Princeton University Press, Princeton, NJ (1979).

Medical Economics Data. *Physician's Desk Reference 1992.* Medical Economics Co., Montvale, NJ (1992).

Menitove, J. E., R. H. Aster, J. T. Casper, S. J. Lauer, J. L. Gottschall, J. E. Williams, J. C. Gill, D. V. Wheeler, V. Piaskowski, P. Kirchner, and R. R. Montgomery. T-lymphocyte subpopulations in patients with classic hemophilia treated with cryoprecipitate and lyophilized concentrates. *New Engl. J. Med.* 308 (1983): 83–86.

Merriam-Webster (eds). *Webster's Third International Dictionary.* G. and C. Merriam Co., Springfield, MA (1965).

Messiah, A., J. Y. Mary, J. B. Brunet, W. Rozenbaum, M. Gentilini, and A. J. Valleron. Risk factors for AIDS among homosexual men in France. *Eur. J. Epidemiol.* 4 (1988): 68–74.

Mientjes, G. H., E. Miedema, E. J. Van Ameijden, A. A. Van Den Hoek, P. T. A. Schellekens, M. T. Roos, and R. A. Coutinho. Frequent injecting impairs lymphocyte reactivity in HIV-positive and HIV-negative drug users. *AIDS* 5 (1991): 35–41.

Mientjes, G. H., E. J. Van Ameijden, A. J. A. R. Van Den Hoek, and R. A. Coutinho. Increasing morbidity without rise in non-AIDS mortality among HIV-infected intravenous drug users in Amsterdam. *AIDS* 6 (1992): 207–212.

Mills, J. and H. Masur. AIDS-related infections. *Sci. Am.* 263 (1990): 5–57.

Mims, C. and D. O. White. *Viral Pathogenesis and Immunology.* Blackwell Scientific Publications, Oxford (1984).

Mirvish, S. S. and H. W. Haverkos. Butylnitrite in the induction of Kaposi's sarcoma in AIDS. *New Engl. J. Med.* 317 (1987): 1603.

Mirvish, S. S., M. D. Ramm, and D. M. Babcock. Indications from animal and chemical experiments of a carcinogenic role for isobutyl nitrite. In: *Health Hazards of Nitrite Inhalants*, pp. 39–49, Haverkos, H. W. and J. A. Dougherty (eds) NIDA Research Monograph 83, National Institute on Drug Abuse, Washington, D.C. (1988).

Moberg, C. L. and Z. A. Cohn. Rene Jules Dubos. *Sci. Am.* 264 (1991): 66–74.

Mok, J. Q., A. De Rossi, A. E. Ades, C. Giaquinto, I. Grosch-Woerner, and C. S. Peckham. Infants born to mothers seropositive for Human Immunodeficiency Virus. *Lancet* i (1987): 1164–1168.

Montagnier, L., J. C. Chermann, F. Barre-Sinoussi, S. Chamaret, J. Gruest, M. T. Nugeyre, F. Rey, C. Dauguet, C. Axler-Blin, F. Vezinet–Brun, C. Rouzloux, G.-A. Saimot, W. Rozenbaum, J. C. Gluckman, D. Klatzmann, E. Vilmer, C. Griscelli, C. Foyer-Gazengel, and J. B. Brunet. A new human T-lymphotropic retrovirus: characterization and possible role in lymphadenopathy and acquired immune deficiency syndromes. In: *Human T-Cell Leukemia/Lymphoma Virus; The Family of Human T-Lymphotropic Retroviruses: Their Role in Malignancies and Association with AIDS*, pp. 363–379, Gallo, R. C., Essex, M. E. and Gross, L. (eds) Cold Spring Harbor Laboratory, Cold Spring Harbor, NY (1984).

Moore, J. D. M., E. J. Cone, and S. S. Alexander. HTLV-III seropositivity in 1971–1972 parental drug abusers—a case of false positives or evidence of viral exposure? *New Engl. J. Med.* 314 (1986): 1387–1388.

Morgan, M., J. W. Curran, and R. L. Berkelman. The future course of AIDS in the United States. *J. Am. Med. Ass.* 263 (1990): 1539–1540.

Moss, A. R. AIDS and intravenous drug use: the real heterosexual epidemic. *Br. Med. J.* 294 (1987): 389–390.

Moss, A. R., D. Osmond, and P. Bacchetti. The cause of AIDS. *Science* 242 (1988): 997.

Moss, A. R., D. Osmond, P. Bacchetti, J. Chermann, F. Barre-Sinoussi, and J. Carlson. Risk factors for AIDS and HIV seropositivity in homosexual men. *Am. J. Epidemiol.* 125 (1987): 1035–1047.

Muñoz, A., D. Vlahov, L. Solomon, J. B. Margolick, J. C. Bareta, S. Cohn, J. Astemborski, and K. E. Nelson. Prognostic indicators for development of AIDS among intravenous drug users. *J. AIDS* 5 (1992): 694–700.

Murray, H. W., D. A. Scavuzzo, C. D. Kelly, B. Y. Rubin, and R. B. Roberts. T4 + cell production of interferon gamma and the clinical spectrum of patients at risk for and with acquired immunodeficiency syndrome. *Arch. Intern. Med.* 148 (1988): 1613–1616.

Nakamura, N., H. Sugino, K. Takahara, C. Jin, S. Fukushige, and K. Matsubara. Endogenous retroviral LTR DNA sequences as markers for individual human chromosomes. *Cytogenet. Cell Genet.* 57 (1991): 18–22.

National Center for Health Statistics. *Monthly Vital Statistics Report.* Department of Health and Human Services—Public Health Service—Publication No. (PMS) 89-1120, Hyattsville, MD (1989).

National Center for Health Statistics. *Health, United States, 1991.* Public Health Service, Hyattsville, MD (1992).

National Commission on AIDS. *The Twin Epidemics of Substance Use and HIV.* National Commission on AIDS, Washington, D.C. (1991).

———. *Trends in Drug Abuse Related Hospital Emergency Room Episodes and Medical Examiner Cases for Selected Drugs: DAWN 1976-1985.* National Institute on Drug Abuse, Bethesda, MD (1987).

———. *Annual Emergency Room Data 1990.* U.S. Department of Health and Human Services (1990a).

———. *Annual Medical Examiner Data 1990.* U.S. Department of Health and Human Services (1990b).

Nelson, J., J. Rodack, R. Fitz, and A. B. Smith. Magic reeling as worst nightmare comes true—he's getting sicker. *National Enquirer,* December 10, 1991.

Newell, G. R., P. W. A. Mansell, M. B. Wilson, H. K. Lynch, M. R. Spitz, and E. M. Hersh. Risk factor analysis among men referred for possible acquired immune deficiency syndrome. *Prevent. Med.* 14 (1985a): 81–91.

Newell, G. R., P. W. A. Mansell, M. R. Spitz, J. M. Reuben, and E. M. Hersh. Volatile nitrites: use and adverse effects related to the current epidemic of the acquired immune deficiency syndrome. *Am. J. Med.* 78 (1985b): 811–816.

Newell, G. R., M. R. Spitz, and M. B. Wilson. Nitrite inhalants: historical perspective. In: *Health Hazards of Nitrite Inhalants,* pp. 1–14, Haverkos, H. W. and J. A. Dougherty (eds) NIDA Research Monograph 83, National Institute on Drug Abuse, Washington, D.C. (1988).

News Report. HHS Secretary Sullivan: no cause for alarm over HIV-negative immune deficiency. *AIDS Weekly,* August 3, 1992: 2–5.

Nicholson, J. Rebel scientists: HIV doesn't cause AIDS. *New York Post,* July 23, 1992.

Novick, D. M., D. J. C. Brown, A. S. F. Lok, J. C. Lloyd, and H. C. Thomas. Influence of sexual preference and chronic hepatitis B virus infection on T lymphocyte subsets, natural killer activity, and suppressor cell activity. *J. Hepatol.* 3 (1986): 363–370.

Nussbaum, B. *Good Intentions: How Big Business, Politics, and Medicine Are Corrupting the Fight Against AIDS.* Atlantic Monthly Press, New York (1990).

O'Brien, T. R., J. R. George, and S. D. Holmberg. Human immunodeficiency virus type 2 infection in the United States. *J. Am. Med. Ass.* 267 (1992): 2775–2779.

Office of National Drug Control Policy. *The National Narcotics*

Intelligence Consumers Committee Reports. Executive Office of the President, Washington, D.C. (1988).
Oppenheimer, G. M. Causes, cases, and cohorts: the role of epidemiology in the historical construction of AIDS. In: *AIDS: The Making of a Chronic Disease,* pp. 49–83, Fee, E. and D. M. Fox (eds) University of California Press, Berkeley (1992).
Osterloh, J. and Olson, K. Toxicities of alkylnitrites. *Ann. Intern. Med* 104 (1986): 727.
Ostrow, D. G., M. J. Van Raden, R. Fox, L. A. Kingsley, J. Dudley, R. A. Kaslow, and the Multicenter AIDS Cohort Study (MACS). Recreational drug use and sexual behavior change in a cohort of homosexual men. *AIDS* 4 (1990): 759–765.
Ou, C. Y., C. A. Ciesielski, G. Myers, C. I. Bandea, C.-C. Luo, B. T. M. Korber, J. I. Mullins, G. Schochetman, R. L. Berkelman, A. N. Economou, J. J. Witte, L. J. Furman, G. A. Satten, K. A. Macinnes, J. W. Curran, H. W. Jaffe, and the Laboratory Investigation Group and Epidemiologic Investigation Group. Molecular epidemiology of HIV transmission in a dental practice. *Science* 256 (1992): 1165–1171.
Palca, J. (1990) A reliable animal model for AIDS. *Science* 248: 1078.
———. Hints emerge from the Gallo probe. *Science* 253 (1991a): 728–731.
———. The sobering geography of AIDS. *Science* 252 (1991b): 372–373.
———. The case of the Florida dentist. *Science* 255 (1992a): 392–394.
———. CDC closes the case of the Florida dentist. *Science* 256 (1992b): 1130–1131.
Pallangyo, K. J., I. M. Mbaga, F. Mugusi, E. Mbena, F. S. Mhalu, U. Bredberg, and G. Biberfeld. Clinical case definition of AIDS in African adults. *Lancet* ii (1987): 972.
Papadopulos-Eleopulos E. Reappraisal of AIDS—Is the oxidation induced by the risk factors the primary cause? *Med. Hypotheses* 25 (1988): 151–162.
Peterman, T. A., R. L. Stoneburner, J. R. Allen, H. W. Jaffe, and J. W. Curran. Risk of Human Immunodeficiency Virus transmission from heterosexual adults with transfusion-associated infections. *J. Am. Med. Ass.* 259 (1988): 55–58.
Peters, B. S., E. J. Beck, D. G. Coleman, M. J. H. Wadsworth, O. McGuinness, J. R. W. Harris, and A. J. Pinching. Changing disease patterns in patients with AIDS in a referral centre in the United Kingdom: the changing face of AIDS. *Br. Med J.* 302 (1991): 203–207.
Pifer, L. L. *Pneumocystis carinii*: a misunderstood opportunist. *Eur. J. Clin. Microbiol.* 3 (1984): 169–173.

Pillai, R., B. S. Nair, and R. R. Watson. AIDS, drugs of abuse and the immune system: a complex immunotoxicological network. *Arch. Toxicol.* 65 (1991): 609–617.
Piot, P., F. A. Plummer, F. S. Mhalu, J.-L. Lamboray, J. Chin, and J. M. Mann. AIDS: an international perspective. *Science* 23 (1988): 573–579.
Pitchenik, A. E., J. Burr, M. Suarez, D. Fertel, G. Gonzalez, and C. Moas. Human T-cell lymphotropic virus-III (HTLV-III) seropositivity and related disease among 71 consecutive patients in whom tuberculosis was diagnosed: a retrospective study. *Am. Rev. Resp. Dis.* 135 (1987): 875–879.
Pitchenik A. E., P. Burr, M. Laufer, G. Miller, R. Cacciatore, W. J. Bigler, J. J. Witte, and T. Cleary. Outbreaks of drug-resistant tuberculosis at AIDS centre. *Lancet* 336 (1990): 440–441.
Pluda, J. M., R. Yarchoan, E. S. Jaffe, I. M. Feuerstein, D. Solomon, S. Steinberg, K. M. Will, A. Raubitschek, D. Katz, and S. Broder. Development of non-Hodgkin lymphoma in a cohort of patients with severe Human Immunodeficiency Virus (HIV) infection on long-term antiretroviral therapy. *Ann. Intern. Med.* 113 (1990): 276–282.
Pollack, S., D. Atias, G. Yoffe, R. Katz, Y. Shechter, and I. Tatarsky. Impaired immune function in hemophilia patients treated exclusively with cryoprecipitate: relation to duration of treatment. *Am. J. Hematol.* 20 (1985): 1–6.
Prince, H. The significance of T lymphocytes in transfusion medicine. *Transfusion Med. Rev.* 16 (1992): 32–3.
Project Inform. Is HIV the cause of AIDS? *Project Discussion Paper #5*, San Francisco, May 27, 1992.
Püschel, K. and F. Mohsenian. HIV-1 prevalence among drug deaths in Germany. In: *Drug Addiction and AIDS*, pp. 89–96, Loimer, N., Schmid, R. and Springer, A. (eds) Springer-Verlag, New York (1991).
Quinn, T. C., J. M. Mann, J. W. Curran, and P. Piot. AIDS in Africa: an epidemiological paradigm. *Science* 234 (1986): 955–963.
Quinn, T. C., P. Piot, J. B. McCormick, F. M. Feinsod, H. Taelman, B. Kapita, W. Stevens, and A. S. Fauci. Serologic and immunologic studies in patients with AIDS in North America and Africa: the potential role of infectious agents as cofactors in Human Immunodeficiency Virus infection. *J. Am. Med. Ass.* 257 (1987): 2617–2621.
Rappoport, J. *AIDS INC.* Human Energy Press, San Bruno, CA (1988).
Ratner, R. A. Editor's Notes; Duesberg: An enemy of the people? *MSDC Physician*, June/July 6 (1992): 4–5.
Raymond, C. A. Combating a deadly combination: intravenous drug abuse, acquired immunodeficiency syndrome. *New Engl. J. Med* 259 (1988): 329–332.

Rezza, G., A. Lazzarin, G. Angarano, R. Zerboni, A. Sinicco, B. Salassa, R. Pristera, M. Barbanera, L. Ortona, and F. Aiuti. Risk of AIDS in HIV seroconverters: a comparison between intravenous drug users and homosexual males. *Eur. J. Epidemiol.* 6 (1990): 99–101.
Richman, D. D., M. A. Fischl, M. H. Greco, M. S. Gottlieb, P. A. Volberding, O. L. Laskin, J. M. Leedom, J. E. Groopman, D. Mildvan, M. S. Hirsch, G. G. Jackson, D. T. Durack, Nusinoff-Lehrmans, and the AZT Collaborative Working Group. The toxicity of azidothymidine (AZT) in the treatment of patients with AIDS and AIDS-related complex. *New Engl. J. Med.* 317 (1987): 192–197.
Rogers, M. F., C.-Y. Ou, M. Rayfield, P. A. Thomas, E. E. Schoenbaum, E. Abrams, K. Krasinski, P. A. Selwyn, J. Moore, A. Kaul, K. T. Grimm, M. Bamji, G. Schochetman, and the New York City Collaborative Study of Maternal HIV Transmission and Montefiori Medical Center HIV Perinatal Transmission Study Group. Use of the polymerase chain reaction for early detection of the proviral sequences of Human Immunodeficiency Virus in infants born to seropositive mothers. *New Engl. J. Med* 320 (1989): 1649–1654.
Root–Bernstein, R. S. Do we know the cause(s) of AIDS? *Perspect. Biol. Med.* 33 (1990a): 480–500.
———. Multiple-antigen–mediated autoimmunity (MAMA) in AIDS: a possible model for postinfectious autoimmune complications. *Res. Immun.* 141 (1990b): 321–339.
———. Non-HIV immunosuppressive factors in AIDS: a multifactorial, synergistic theory of AIDS etiology. *Res. Immun.* 141 (1990c): 815–838.
Rosenberg, M. J. and J. M. Weiner. Prostitutes and AIDS: a health department priority? *Am. J. Pub. Health* 78 (1988): 418–423.
Rowe, W. P. Genetic factors in the natural history of murine leukemia virus infection: G. H. A. Clowes Memorial Lecture. *Cancer Res.* 33 (1973): 3061–3068.
Rubin, H. and Temin, H. A radiological study of cell–virus interaction in the Rous sarcoma. *Virology* 7 (1958): 75–91.
Rubinstein, E. II: The untold story of HUT78. *Science* 248 (1990): 1499–1507.
Safai, B., H. Peralta, K. Menzies, H. Tizon, P. Roy, N. Flomberg, and S. Wolinsky. Kaposi's sarcoma among HIV-negative high risk population. *VII International Conference on AIDS*, Florence, Italy (1991).
Salahuddin, S. K., S. Naecamura, P. Biberfeld, M. H. Kaplan, P. D. Markham, L. Larsson, and R. C. Gallo. Angiogenic properties of Kaposi's sarcoma-derived cells after long-term culture *in vitro*. *Science* 242 (1988): 430–433.

Sande, M. A. Transmissions of AIDS. The case against casual contagion. *New Engl. J. Med.* 314 (1986): 380–382.

Sapira, J. D. The narcotic addict as a medical patient. *Am. J. Med.* 45 (1968): 555–588.

Sarngadharan, M. G., M. Popovic, L. Bruch, J. Schupach, and R. Gallo. Antibodies reactive with Human T-lymphotropic retroviruses (HTLV-III) in the serum of patients with AIDS. *Science* 22 (1984): 506–508.

Savitz, E. J. No magic cure, the war on AIDS produces few gains except on Wall Street. *Barron's* December 16, 1991: 10–29.

Savona, S., M. A. Nardi, E. T. Lenette, and S. Karpatkin. Thrombocytopenic purpura in narcotics addicts. *Ann. Intern. Med.* 102 (1985): 737–741.

Scheper-Hughes, N. and R. Herrick. Ethical tangles: Cuba's highly controversial AIDS and HIV program raises thorny ethical issues. *New Internationalist (Oxford, U.K)*, May (1992): 35.

Schnittman, S. M., M. C. Psallidopoulos, H. C. Lane, L. Thompson, M. Baseler, F. Massari, C. H. Fox, N. P. Salzman, and A. Fauci. The reservoir for HIV-1 in human peripheral blood is a T cell that maintains expression of CD4. *Science* 245 (1989): 305–308.

Schoch, R. Dad, I'm HIV positive. *Newsweek*, August 17 (1992): 9.

Schüpach, J. First isolation of HTLV-III. *Nature (London)* 321 (1986): 119–120.

Schuster, C. R. Foreword. In: *Cocaine: Pharmacology, Effects and Treatment of Abuse*, pp. 7–8, Grabowski, J. (ed.) NIDA Research Monograph 50, National Institute on Drug Abuse, Washington, D.C. (1984).

Schwartz, R. H. Deliberate inhalation of isobutyl nitrite during adolescence: a descriptive study. In: *Health Hazards of Nitrite Inhalants*, pp. 81–85, Haverkos, H. W. and Dougherty, J. A. (eds) NIDA Research Monograph 83, National Institute on Drug Abuse, Washington, D.C. (1988).

Schwitzer, G. The magical medical media tour. *J. Am. Med. Ass.* 267 (1992): 1969–1971.

Scolaro, M., R. Durham, and G. Pieczenik. Potential molecular competitor for HIV. *Lancet* 337 (1991): 731–732.

Seage, G. R., K. H. Mayer, C. R. Hornsburgh, S. D. Holmburg, M. W. Moon, and G. A. Lamb. The relation between nitrite inhalants, unprotected receptive anal intercourse, and the risk of Human Immunodeficiency Virus infection. *J. Am. Epidemiol.* 135 (1992): 1–11.

Seligmann, M., L. Chess, J. L. Fahey, A. S. Fauci, P. J. Lachmann, J. L. Age-Stehr, J. Ngu, A. J. Pinching, F. S. Rosen, T. J. Spira, and J. Wybran. AIDS—an immunologic reevaluation *New Engl. J. Med.* 311 (1984): 1286–1292.

Selik, R. M., E. T. Starcher, and J. W. Curran. Opportunistic diseases reported in AIDS patients: frequencies, associations, and trends. *AIDS* 1 (1987): 175–182.

Selik, R. M., J. W. Buehler, J. M. Karon, M. E. Chamberland, and R. L. Berkelman. Impact of the 1987 revision of the case definition of acquired immune deficiency syndrome in the United States. *J. AIDS* 3 (1990): 73–82.

Selwyn, P. A., A. R. Feingold, D. Hartel, E. E. Schoenbaum, M. H. Adderman, R. S. Klein, and S. H. Freidland. Increased risk of bacterial pneumonia in HIV-infected intravenous drug users without AIDS. *AIDS* 2 (1988): 267–272.

Selwyn, P. A., D. Hartel, W. Wasserman, and E. Drucker. Impact of the AIDS epidemic on morbidity and mortality among intravenous drug users in a New York City methadone maintenance program. *Am. J. Publ. Health* 79 (1989): 1358–1362.

Semple, M., C. Loveday, I. Weller, and R. Tedder. Direct measurement of viraemia in patients infected with HIV-1 and its relationship to disease progression and zidovudine therapy. *J. Med. Virol.* 35 (1991): 38–45.

Shannon, E., C. Booth, D. Fowler, and M. McBride. A losing battle. *Time*, December 3, 24 (1990): 44–48.

Sharp, R. A., S. M. Morley, J. S. Beck, and G. E. D. Urquhart. Unresponsive to skin testing with bacterial antigens in patients with hemophilia A not apparently infected with Human Immunodeficiency Virus (HIV) *J. Clin. Path.* 40 (1987): 849–852.

Shaw, G. M., B. H. Hahn, S. K. Arya, J. E. Groopman, R. C. Gallo, and F. Wong-Staal. Molecular characterization of human T-cell leukemia (lymphotropic) virus type III in the acquired immune deficiency syndrome. *Science* 226 (1984): 1165–1171.

Shenton, J. HIV, AIDS, and zidovudine. *Lancet* 339 (1992): 806.

Shigematsu, I., H. Yanagawa, S.-I. Yamamoto, and K. Nakae. Epidemiological approach to SMON (Subacute Myelo-optico-neuropathy). *Jap. J. Med. Sci. Biol.* 28 (Suppl.) (1975): 23–33.

Shilts, R. *And the Band Played On.* St. Martin's Press, New York (1985).

———. Gay troops in the Gulf War can't come out. *San Francisco Chronicle*, February 2, 1991.

Shorter, E. *The Health Century.* Doubleday, New York (1987).

Simmonds, P., P. Balfe, J. F. Peutherer, C. A. Ludlam, I. O. Bishop, and A. J. Leigh-Brown. Human immunodeficiency virus-infected individuals contain provirus in small numbers of peripheral mononuclear cells and at low copy numbers. *J. Virol.* 64 (1990): 864–872.

Smith, D. G. Thailand: AIDS crisis looms. *Lancet* 335 (1990): 781–782.

Smith, G. D. and A. N. Phillips. Confounding in epidemiological studies:

why "independent" effects may not be all they seem. *Br. Med. J.* 305 (1992): 757–759.

Smothers, K. Pharmacology and toxicology of AIDS therapies. *The AIDS Reader* 1 (1991): 29–35.

Sonnabend, I. A., S. S. Witkin, and D. T. Purtillo. A multifactorial model for the development of AIDS in homosexual men. *Ann. NY Acad. Sci.* 437 (1983): 177–183.

Spira, T. J. and B. M. Jones. Is there another agent that causes low CD4 counts and AIDS? *J. Cell. Biochem.* (Suppl.) 16E (1992): 56.

Sprent, J. *Migration and Lifespan of Lymphocytes, in B and T-cells in Immune Recognition.* John Wiley and Sons, New York (1977).

Spornraft, P., M. Froschl, J. Ring, M. Meurer, F. D. Goebel, H. W. Ziegler–Heitbrock, G. Riethmuller, and O. Braun–Falco. T4/T8 ratio and absolute T4 cell numbers in different clinical stages of Kaposi's sarcoma in AIDS. *Br. J. Derm.* 119 (1988): 1–9.

Stark, K., R. Muller, I. Gugenmoos-Holzmann, S. Deiniger, E. Meyer, and U. Bienzle. HIV infection in intravenous drug abusers in Berlin: risk factors and time trends. *Klin. Wochenschr.* 68 (1990): 415–420.

Stehr–Green, J. K., R. C. Holman, J. M. Jason, and B. L. Evatt. Hemophilia-associated AIDS in the United States, 1981 to September 1987. *Am. J. Pub. Health* 78 (1988): 439–442.

Stehr–Green, J. K., J. M. Jason, B. L. Evatt, and the Hemophilia-Associated AIDS Study Group. Geographic variability of hemophilia-associated AIDS in the United States: effect of population characteristics. *Am. J. Hematol.* 32 (1989): 178–183.

Steinbrook, R. Scientists infect monkeys with human AIDS virus. *Los Angeles Times*, June 12, 1992.

Stewart, G. T. Limitations of the germ theory. *Lancet* i (1968): 1077–1081.

———. Uncertainties about AIDS and HIV. *Lancet* i (1989): 1325.

St. Louis, M. E., G. A. Conway, C. R. Hayman, C. Miller, L. R. Peterson, and T. J. Dondero. Human immunodeficiency virus infection in disadvantaged adolescents. *J. Am. Med. Ass.* 266 (1991): 2387–2391.

Stoneburner, R. L., D. C. Des Jarlais, D. Benezra, L. Gorelkin, J. L. Sotheran, S. R. Friedman, S. Schultz, M. Marmor, D. Mildvan, and R. Maslansky. A larger spectrum of severe HIV-1–related disease in intravenous drug users in New York City. *Science* 242 (1988): 916–919.

Stutman, O. Immunodepression and malignancy. *Adv. Cancer Res.* 22 (1975): 261–422.

Sullivan, J. L., F. E. Brewster, D. B. Brettler, A. D. Forsberg, S. H. Cheeseman, K. S. Byron, S. M. Baker, D. L. Willitts, R. A. Lew, and P. H. Levine. Hemophiliac immunodeficiency: influence of exposure to

factor VIII concentrate, LAV/HTLV-III, and herpes viruses. *J. Pediat.* 108 (1986): 504–510.
Swanson, C. E., D. A. Cooper, and the Australian Zidovudine Study Group. Factors influencing outcome of treatment with zidovudine of patients with AIDS in Australia. *AIDS* 4 (1990): 749–757.
Taelman, H., A. Kagame, J. Batungwanayo, J. Bogaerts, J. Clerinx, S. Allen, and P. Van De Perre. Tuberculosis and HIV infection (letter). *Br. Med. J.* 302 (1991): 1206.
Tedder, R. S., M. G. Semple, M. Tenant-Flowers, and C. Loveday. HIV, AIDS, and zidovudine. *Lancet* 339 (1992): 805.
Temin, H. M. Proof in the pudding. *Policy Review* 54 (1990): 71–72.
Temin, H. and S. Mitzutani. RNA-dependent DNA polymerase in virions of Rous sarcoma virus. *Nature (London)* 226 (1970): 1211–1213.
Terry, C. E. and M. Pellens. *The Opium Problem.* Bureau of Social Hygiene of New York (1928).
The Software Toolworks World Atlas.™ *Population Growth Rate.* Chatsworth, CA (1992).
Thompson, D. A long battle with AIDS. *Time*, July 2 (1990): 42–43.
Till, M. and K. B. MacDonnell. Myopathy with Human Immunodeficiency Virus type 1 (HIV-1) infection: HIV-1 or zidovudine? *Ann. Intern. Med.* 113 (1990): 492–494.
Toufexus, A. Innocent victims. *Time*, May 13, 19 (1991): 56–60.
Treichler, P. A. The country and the city: dreams of Third World AIDS. In: *AIDS: The Making of a Chronic Disease*, pp. 386–412, E. Fee, and D. M. Fox (eds) University of California Press, Berkeley (1992).
Tsoukas, C., F. Gervais, J. Shuster, P. Gold, M. O'Shaughnessy, and M. Rosert-Guroff. Association of HTLV-III antibodies and cellular immune status of hemophiliacs. *New Engl. J. Med.* 311 (1984): 1514–1515.
Turner, C. F., H. G. Miller, and L. E. Moses. *AIDS, Sexual Behavior and Intravenous Drug Use.* National Academy Press, Washington, D.C. (1989).
U.S. Department of Health and Human Services. *National HIV Serosurveillance Summary.* Centers for Disease Control, Atlanta (1990).
Valentine, C. B., R. Weston, V. Kitchen, J. Main, K. C. Moncrieff, and V. R. Aber. Anonymous questionnaire to assess consumption of prescribed and alternative medication and patterns of recreational drug use in a HIV population. *AIDS Weekly*, August 10 (1992): 18.
Vandenbroucke, J. P. and V. P. A. M. Pardoel. An autopsy of epidemiologic methods: the case of "poppers" in the early epidemic of the acquired immunodeficiency syndrome (AIDS). *J. Epidemiol.* 129 (1989): 455–457.

van Griensven, G. J. P., R. A. P. Telman, J. Goudsmit, J. Van Der Noorda, F. De Wolf, E. M. M. De Vroome, and R. A. Coutinho. Risk factors and prevalence of HIV antibodies in homosexual men in The Netherlands. *Am. J. Epidemiol.* 125 (1987): 1048–1057.
van Griensven, G. J. P., E. M. M. Vroome, H. De Wolf, J. Goudsmit, M. Roos, and R. A. Coutinho. Risk factors for progression of Human Immunodeficiency Virus (HIV) infection among seroconverted and seropositive homosexual men. *Am. J. Epidemiol.* 132 (1990): 203–210.
van Leeuwen, R., P. J. Van Den Hurk, G. J. Jobis, P. A. Van Der Wouw, P. Reiss, J. K. M. Eeftinck Schattenkerk, S. A. Danner, and J. M. A. Lange. Failure to maintain high-dose treatment regimens during long-term use of zidovudine in patients with symptomatic Human Immunodeficiency Virus type 1 infection. *Genitourin Med.* 66 (1990): 418–22.
Van Voorhis, B. J., A. Martinez, K. Mayer, and D. J. Anderson. Detection of Human Immunodeficiency Virus type 1 in semen from seropositive men using culture and polymerase chain reaction deoxyribonucleic acid amplification techniques. *Fertil. Steril.* 55 (1991): 588–594.
Varmus, H. Regulation of HIV and HTLV gene expression. *Genes Deu.* 2 (1988): 1055–1062.
Vermund, S. Changing estimates of HIV-1 seroprevalence in the United States. *J. NIH Res.* 3: 77–81 (1991).
Villinger, F., J. D. Powell, T. Jehuda-Cohen, N. Neckelmann, M. Vuchetich, B. De, T. M. Folks, H. M. McClure, and A. A. Ansari. Detection of occult simian immunodeficiency virus SIV smm infection in asymptomatic seronegative nonhuman primates and evidence for variation in SIV *gag* sequence between *in vivo-* and *in vitro-*propagated virus. *J. Virol.* 65 (1991): 1855–1862.
Voevodin, A. HIV screening in Russia (letter). *Lancet* 339 (1992): 1548.
Volberding, P. A., S. W. Lagakos, M. A. Koch, C. Pettinelli, M. W. Myers, D. K. Booth, H. H. Balfour, Jr., R. C. Reichman, J. A. Bartlett, M. S. Hirsch, R. L. Murphy, W. D. Hardy, R. Soeiro, M. A. Fischl, J. G. Bartlett, T. C. Merigan, N. E. Hyslop, D. D. Richman, F. T. Valentine, L. Corey, and the AIDS Clinical Trial Group. Zidovudine in asymptomatic Human Immunodeficiency Virus infection: a controlled trial in persons with fewer than 500 CD4-positive cells per cubic millimeter. *New Engl. J. Med.* 322 (1990): 941–949.
Waldholz, M. Stymied Science: New Discoveries Dim Drug Makers' Hopes for Quick AIDS Cure. *Wall Street Journal,* May 26, 1992.
Walker, R. E., R. I. Parker, J. A. Kovacs, H. Masur, H. C. Lane, S. Carleton, L. E. Kirk, H. R. Gralnick, and A. S. Fauci. Anaemia and

erythropoiesis in patients with the acquired immunodeficiency syndrome (AIDS) and Kaposi sarcoma treated with zidovudine. *Ann. Intern. Med* 108 (1988): 372–376.
Walton, J., P. B. Beeson, and R. B. Scott (eds). *The Oxford Companion to Medicine*. Oxford University Press, Oxford, New York (1986).
Wang, L.-H., D. Galehouse, P. Mellon, P. Duesberg, W. Mason, and P. K. Vogt. Mapping oligonucleotides of Rous sarcoma virus RNA that segregate with polymerase and group-specific antigen markers in recombinants. *Proc. Natn. Acad. Sci. U.S.A.* 73 (1976): 3952–3956.
Ward, J. W., T. J. Bush, H. A. Perkins, L. E. Lieb, J. R. Allen, D. Goldfinger, S. M. Samson, S. H. Pepkowitz, L. P. Fernando, P. V. Holland, and the Study Group from the AIDS Program. The natural history of transfusion-associated infection with Human Immunodeficiency Virus. *New Engl. J. Med.* 321 (1989): 947–952.
Weber, J. N., L. A. Rogers, K. Scott, E. Berrie, J. R. W. Harris, J. Wadsworth, O. Moshtael, T. McManus, D. J. Jeffries, and A. J. Pinching. Three-year prospective study of HTLV-III/LAV infection in homosexual men. *Lancet* i (1986): 1179–1182.
Weber, R., W. Ledergerber, M. Opravil, W. Siegenthaler, and R. Luthy. Progression of HIV infection in misusers of injected drugs who stop injecting or follow a programme of maintenance treatment with methadone. *Br. Med J.* 301 (1990): 1361–1365.
Weiss, R. A. A virus in search of a disease. *Nature (London)* 333 (1988): 497–498.
———. AIDS: defective viruses to blame? *Nature (London)* 338 (1989): 458.
Weiss, R. Provenance of HIV strains. *Nature (London)* 349 (1991): 374.
Weiss R. and H. Jaffe. Duesberg, HIV and AIDS. *Nature (London)* 345 (1990): 659–660.
Weiss R., N. Teich, H. Varmus, and J. Coffin. *Molecular Biology of RNA Tumor Viruses*. Cold Spring Harbor Press, Cold Spring Harbor, NY (1985).
Weiss, S. H. Links between cocaine and retroviral infection. *J. Am. Med. Ass.* 261 (1989): 607–609.
Weiss, S. H., J. J. Goedert, S. Gartner, M. Popovic, D. Waters, P. Maricam, F. Di Marzoveronesi, M. H. Gail, R. C. Gallo, and W. A. Blattner. Risk of Human Immunodeficiency Virus (HIV-1) infection among laboratory workers. *Science* 239 (1988): 68–71.
Weller, R. Zur Erzeugung von Pneumocystosen im Tierversuch. *Zeitschrift fur Kinderheilkunde* 76 (1955): 366–378.
Weniger, B. G., K. Limpakarnjanarat, K. Ungchusak, S. Thanprasertsuk, K. Choopanya, S. Vanichseni, T. Uneklabh, P. Thongcharoen, and C.

Wasi. The epidemiology of HIV infection and AIDS in Thailand. *AIDS* 5 (Suppl. 2) (1991): S71–S85.

Weyer, I. and H. I. Eggers. On the structure of the epidemic spread of AIDS: the influence of an infectious coagent. *Zentralbl. Bakteriol.* 273 (1990): 52–67.

Widy-Wirski, R., S. Berkley, R. Downing, S. Okware, U. Recine, R., Mugerwas, A. Lwegaba, and S. Sempala. Evaluation of the WHO clinical case definition for AIDS in Uganda. *J. Am. Med. Ass.* 260 (1988): 3286–3289.

Williams, A. E., C. T. Fang, and G. Sandler. HTLV-I/II and blood transfusion in the United States. In: *Human Retrovirology: HTLV*, pp. 349–362, Blattner, W. A. (ed.) Raven Press, New York (1990).

Williford Pifer, L. L., D. R. Woods, C. C. Edwards, R. E. Joyner, F. I. Anderson, and K. Arheart. Pneumocystis carinii serologic study in pediatric acquired immunodeficiency syndrome. *Am. J. Dis. Child* 142 (1988): 36–39.

Winkelstein, W. Jr., D. H. Lyman, N. Padian, R. Grant, M. Samuel, I. A. Wiley, R. E. Anderson, W. Lang, J. Riggs, and I. A. Levy. Sexual practices and risk of infection by the Human Immunodeficiency Virus: the San Francisco men's health study. *J. Am. Med. Ass.* 257 (1987): 321–325.

Witte, J. J. and K. R. Wilcox. Update: transmission of HIV infection during invasive dental procedures—Florida. *Morb. Mort. Weekly Rep.* 40 (1991): 377–381.

Wood, R. W. The acute toxicity of butyl nitrites. In: *Health Hazards of Nitrite Inhalants*, pp. 28–38, Haverkos, H. W. and Dougherty, I. A. (eds.) NIDA Research Monograph 83, National Institute on Drug Abuse, Washington, D.C. (1988).

World Health Organization. *WHO–Report No. 32: AIDS Surveillance in Europe (Situation by 31st December 1991)*. World Health Organization, Geneva (1992a).

World Health Organization. *Acquired Immunodeficiency Syndrome (AIDS)—Data as of 1 January 1992*. World Health Organization, Geneva (1992b).

Yarchoan, R. and S. Broder. Antiretroviral therapy for AIDS. *New Engl. J. Med.* 317 (1987a): 630.

———. Development of antiretroviral therapy for the acquired immunodeficiency syndrome and related disorders. *New Engl. J. Med.* 316 (1987b): 557–564.

Yarchoan, R., J. M. Pluda, C.-F. Perno, H. Mitsuya, and S. Broder. Antiretroviral therapy of Human Immunodeficiency Virus infection: current strategies and challenges for the future. *Blood* 78 (1991): 859–884.

APPENDIX C

The HIV Gap in National AIDS Statistics*

Peter H. Duesberg
Dept. of Molecular and Cell Biology
University of California at Berkeley

The HIV-AIDS hypothesis rests on the assertion that all AIDS cases are associated with HIV (*Confronting AIDS—Update* 1988, Natl. Acad. Sci. Press, Wash., D.C.; Blattner, W., et al., 1988, *Science* 241, 514–515; Weiss, R. and Jaffe, H., 1990, *Nature* 345, 659–660; Duesberg, P. H., 1992, *Pharmacol. Ther.* 55, 201–277). Therefore, the Centers for Disease Control (CDC) groups American AIDS cases in its *HIV/AIDS Surveillance* into "exposure (to HIV) categories." However, there are no national AIDS statistics that document the natural coincidence between AIDS diseases and HIV. Contrary to its title, the *HIV/AIDS Surveillance* of the CDC does not report HIV tests. Correlations between HIV and AIDS can only be determined from individual studies and from those CDC AIDS case report forms that include HIV tests.

But most "HIV tests" measure antibodies against HIV rather than measuring the virus itself. And antibodies are not unambiguous evidence for the presence of a virus, nor are they rational predictors for viral disease. Instead, antibodies neutralize HIV and restrict the virus to latency. This is the reason that leading AIDS researchers have had notorious difficulties in isolating HIV, even in people dying from AIDS (Weiss, R., 1991, *Nature* 349, 374; Cohen, J., 1993, *Science* 259, 168–170).

Moreover, antibody tests generate false-positive results if an epitope is shared between different organisms. According to a recent review entitled

---

* Article originally appeared in *Bio/Technology*, 11 (1993). Reprinted by permission of the publisher.

"HIV testing: State of the Art," "depending on the population tested, 20 to 70% of... two successive positive ELISAs [enzyme-linked immunosorbent assay] are confirmed by Western blot [an alternative antibody assay]." (Sloand, E. M., et al., 1991, *JAMA* 266, 2861-2866).

In a population with a low probability of infection, the false-positive rate is high. According to the widely cited study of applicants to the U.S. Army by Burke et al., 83 percent of all initially positive ELISAs (10,000 of 12,000) were false-positives (*New Eng. J. Med.* 319, 961-964, 1988).

In a population with a high incidence of infection, however, the false-positive rate is expected to be low. Therefore the CDC assumes that "the tiny proportion of possibly false-positive screening tests in persons with AIDS-indicative diseases is of little consequence " (*Confronting AIDS—Update 1988*). But this is not observed.

For example, one study documented 131 repeatedly ELISA-positive homosexual men with negative Western blots in a cohort of 4,994 homosexuals of which 37 percent were HIV-positive (Phair, J., et al., 1992, *J. AIDS* 5,988–5,992). Another study "found HIV-1 infection in only 4 (12.5%) of 32 high-risk cases" with repeatedly positive ELISAs (Celum, C. L., et al., 1991, *J. Infect. Dis.* 164, 656–664). HIV infection was negative by Western blot, provirus amplification with the polymerase chain reaction (PCR), and virus isolation tests. Another study identified 33 ELISA-positive and even Western blot-positive subjects who were HIV-negative based on the PCR test for HIV DNA (Schechter, M., et al., 1991, *AIDS* 5, 373–379). These subjects were from a group of 316 homosexuals of which 158 (50 percent) were PCR-positive.

The relatively high incidence of false-positive HIV antibody tests in these HIV risk groups probably reflects the presence of antibodies to other viruses and microbes that may cross-react with HIV. For example, seven out of ten blood donors treated with an influenza virus vaccine in 1991 became HIV ELISA-positive. Each of these proved to be false-positives upon confirmation with a Western blot (MacKenzie, W. R., et al., 1992, *JAMA* 268, 1015–1017). Since the CDC "...accepts a reactive screening test for HIV antibody without confirmation by a supplemental test..." (*Confronting AIDS—Update 1988*) and does not request a repeatedly positive antibody test in its "AIDS adult confidential case report" forms, it includes false-positives in its *HIV/AIDS Surveillance*.

In fact, the CDC even includes AIDS cases in its *HIV/AIDS Surveillance* "without laboratory evidence regarding HIV infection" (*Confronting Aids—Update 1988*). Upon request, the CDC's director of the HIV/AIDS division, Harold Jaffe, stated that the HIV status of 43,606 out of the 253,448 American AIDS cases recorded by the end of 1992 was "not tested" (personal communication, 1993). However, this figure seems to be an understatement. Obviously, all 10,360 American AIDS cases diagnosed before the HIV antibody test, i.e., before 1985, were not tested [*HIV/AIDS Surveillance*, February 1993]. In addition, the CDC published that "[a]pproximately one third of AIDS patients in the United States have been from New York and San Francisco, where, since 1985, <7% have been reported with HIV-antibody test results, compared with >60% in other areas" (*Confronting AIDS—Update 1988*). Thus, between 1985 and 1987, 58 percent (93% x 1/3 + 40% x 2/3) of the 56,807 AIDS cases recorded in that period, or 32,948, have not been tested. For 1988, the CDC reported that 27 percent, or 9,039 of the 33,480 AIDS cases recorded for that year, were not tested for HIV (Selik, R. M., et al., 1990, *J. AIDS* 3, 73–82). According to the CDC's Technical Information Activity, 3,682 AIDS cases without an HIV test were recorded in 1989, 2,888 in 1990, 1,960 in 1991, and 1,395 in 1992 (personal communication, 1993). Thus, at least 62,272, or 18,666 more than Jaffe reports, were not tested.

Determination of the HIV-AIDS correlation is further obscured because HIV-free AIDS cases are not recorded in the CDC's *HIV/AIDS Surveillance*. By 1993, at least 4,621 HIV-free AIDS cases had been documented in the U.S., Europe, and Africa with the clinical AIDS definition (Table 1). Even Jaffe, again upon request, reported eighty-nine HIV-free AIDS cases (personal communication, 1993). The cases recorded in Table 1 suffered from one or more of the over twenty-five heterogeneous AIDS-defining diseases and from AIDS-defining immunodeficiencies without diseases. Some of these proved to be HIV-free even by PCR amplification of viral RNA and DNA.

Table 1 includes some American and European immunodeficiencies that may not exactly fit the current definition of AIDS-defining immunodeficiency without disease, which is <200 T-cells per microliter (CDC, 1992, *MMWR* 41, RR17, 1–19), as for example, HIV-free male homosexuals on various recreational drugs with "<600 cells per cubic millimeter" (Table 1, ref. 14) or HIV-negative hemophiliacs with T4/T8-cell ratios of about 1 or <1 (Table 1, refs. 46–61). But even if not all of these cases fit the current definition of AIDS-defining immunodeficiency exactly, they do so prospectively. This is because their T-cells typically continue to decline either because of risk behavior, such as the consumption of recreational drugs, or because of clinical AIDS risks, such as chronic transfusion of foreign proteins as prophylaxis against hemophilia (Duesberg, P. H. 1992, op. cit.).

Since a clinical definition is used in Africa, statistics from this continent are not biased against HIV-free AIDS. For example, 2,215 out of 4,383 (50.5%) African AIDS patients from Abidjan, Ivory Coast;

Table 1. HIV-free AIDS-defining diseases and immunodeficiencies

Risk Group	U.S./Canada	Europe	Africa	References
Homosexuals	722	37		1-22/ 23-26, 74
Intravenous (IV) drug users	251	335		18-20, 27-35, 75/36-39, 74
Infants of IV drug users	55	11		40-43/44, 45
Hemophiliacs	256	78		46-56/57-61
None/unreported	307	14	2,555	16-21, 62-67/ 21, 68/26, 69-73
Totals	1,591	475	2,555	
Sum total	4,621			

Lusaka, Zambia; and Kinshasa, Zaire, were HIV-antibody negative (Table 1, refs. 70, 71). Another study using antibody tests and supplementary PCR tests for HIV reports 135 (59%) HIV-free AIDS patients from Ghana out of 227 suffering from weight loss, diarrhea, chronic fever, tuberculosis, and neurological diseases (Table 1, ref. 72). Only 37 (30%) of a group of 122 African tuberculosis patients were HIV-positive, according to a study published in 1993 (Table 1, ref. 73). An earlier study documents 116 HIV-negatives among 424 African patients, and Montagnier et al. diagnosed HIV in four out of eight (Table 1, refs. 26, 69). It follows that about 50 percent of the African AIDS cases, or 65,000 of the 129,000 diagnosed by 1992 (Duesberg, P. H., 1992, op. cit.), may be HIV-free and thus not caused by HIV.

Instead of considering the potential usefulness of HIV-free AIDS cases in the search for the cause of AIDS, the CDC and the NIH's director for AIDS research hid in 1992 the then-rapidly growing numbers of HIV-free AIDS cases (Duesberg, P. H., 1992, op. cit.) under a new name, "idiopathic CD4 lymphocytopenia," or ICL. Indeed, the new name has sent HIV-free AIDS cases into obscurity. But efforts to set apart HIV-free from HIV-positive AIDS cases by the new term are not based on clinical or scientific arguments. According to an editorial by Anthony Fauci, HIV-free AIDS or ICL cases are unlike the HIV-positive cases because (1) "[g]iven the heterogeneity of the [ICL] syndrome, it is highly likely that there is no common cause," and because (2) "[a]pproximately one-third of the patients are women, as compared with 11 percent among those with HIV... [in America]" (Fauci, A., 1993, New Eng. J. Med. 328, 429–431).

Yet proponents of the HIV hypothesis, including Fauci, insist that HIV is the common cause of the more than twenty-five heterogeneous AIDS diseases and that HIV causes African AIDS, although about 50 percent of the African patients are women (Duesberg, P. H., 1992, op. cit.).

In view of the above, I submit that the natural coincidence between HIV and AIDS in America and Europe remains unknown, and is certainly less than perfect. Thus, arguments for the etiological role of HIV in AIDS, which assume a perfect correlation, are fundamentally flawed.

REFERENCES

1. Drew, W. L., et al. *Ann. Intern. Med.* 103, 61–63 (1985).
2. Weber, J. N., et al. *Lancet* i, 1179–1182 (1986).
3. Novick, D. M., et al. *J. Hepatol.* 3, 363–370 (1986).
4. Collier, A. C., et al. *Ant. J. Med.* 82, 593–601 (1987).
5. Bartholomew, C., et al. *J. Am. Med. Assoc.* 257, 2604–2608 (1987).
6. Buimovici-Klein, E., et al. *J. Med. Virology* 25, 371–385 (1988).
7. Afrasiabi, R., et al. *Am. J. Med.* 81, 969–973 (1986).
8. Bowden, F. J., et al. *Lancet* 337, 1313–1314 (1991).
9. Safai, B., et al. "Kaposi's Sarcoma among HIV-negative High Risk Population." VII International Conference on AIDS (Florence, Italy) (1991).
10. Castro, A., et al. *Lancet* 339, 868 (1992).
11. Huang, Y. Q., et al. *Lancet* 339, 515–518 (1992).
12. Friedman-Kien, A. E., et al. *Lancet* 335, 168–169 (1990).
13. Marion, S. A., et al. *J. Acquir. Immune Defic. Syndr.* 2, 178–186 (1989).
14. Kaslow, R. A., et al. *JAMA* 261, 3424–3429 (1989).
15. Macon, W. R., et al. *AIDS Weekly* April, 12, 15 (1993).
16. Laurence, J., et al. *Lancet* 340, 273–274 (1992).
17. Smith, D. K., et al. *New Eng. J. Med.* 328, 373–379 (1993).
18. Ho, D. D., et al. *New Eng. J. Med.* 328, 380–385 (1993).
19. Theuer, C. P., et al. *J. Infect. Dis.* 5, 399–405 (1990).
20. Shafer, R. W. et al. *AIDS* 5, 399–405 (1991).
21. Centers for Disease Control. *Morbid. Mort. Weekly Report.* 41, 541–545 (1992).
22. Lang, W., et al. *J. Acquir. Immune Defic. Syndr.* 2, 63–69 (1989).
23. Marquart, K.-H., et al. *AIDS* 5, 346–348 (1991).
24. Archer, C. B., et al. *Clin. Exper. Dermatol.* 14, 233–236 (1989).
25. Lowdell, C. P., et al. *J. Royal Soc. Med.* 82, 226–227 (1989).
26. Brun-Vezinet, F., et al. *The Lancet* i, 1253–1256 (1984).

27. Stoneburner, R. L., et al. *Science* 242, 916–919 (1988).
28. Selwyn, P. A., et al. *AIDS* 2, 267–272 (1988).
29. Braun, M. M., et al. *JAMA* 261, 393–397 (1989).
30. Donahoe, R. M., et al. *Ann. NY Acad. Sci.* 496, 711–721 (1987).
31. Savona, S., et al. *Ann. Intern. Med.* 102, 737–741 (1985).
32. Des Jarlais, D. C., et al. in *AIDS and IV Drug Abusers: Current Perspectives*, R. P. Galea, B. F. Lewis, L. Baker, Eds. (National Health Publishing, Owings Mills, MD), pp. 97–109 (1988).
33. Munoz, A., et al. *J. Acquir. Immune Defic. Syndr.* 5, 694–700 (1992).
34. Brudney, K., et al. *Am. Rev. Respir. Dis.* 144, 744–749 (1991).
35. Weiss, S. H., et al. *The Lancet* 340, 608–609 (1992).
36. Espinoza, P., et al. *Gastroenterologie Clinique et Biologique* 11, 288–292 (1987).
37. Mienjtes, G. H., et al. *AIDS* 5, 35–41 (1991).
38. Mientjes, G. H., et al. *AIDS* 6, 207–212 (1992).
39. Mientjes, G. H. C., et al. *Br. Med. J.* 306, 371–373 (1993).
40. Koch, T. K., et al. *Ann. Neurol.* 28, 456–457 (1990).
41. Duesberg, P. H. *Pharmacol. Ther.* 55, 201–277 (1992).
42. Rogers, M. F., et al. *New Eng. J. Med.* 320, 1649–1654 (1989).
43. Culver, K. W., et al. *J. Pediatr.* 111, 230–235 (1987).
44. Bianche, S., et al. *New Eng. J. Med.* 320, 1643–1648 (1989).
45. Laurence, J., et al. *New Eng. J. Med.* 311, 1269–1273 (1984).
46. Tsoukas, C., et al. *New Eng. J. Med.* 311, 1514–1515 (1984).
47. Sullivan, J. L., et al. *J. Pediatr.* 108, 504–510 (1986).
48. Kreiss, J. K., et al. *Am. J. Med.* 80, 345–350 (1986).
49. Gill, J. C., et al. *J. Pediatr.* 108, 511–516 (1986).
50. Sharp, R. A., et al. *J. Clin. Pathol.* 40, 849–852 (1987).
51. Matheson, D. S., et al. *Clin. Immunol. Immunopathol.* 4, 41–50 (1987).
52. Aledort, L. M. *Seminars in Hematology* 25, 14–19 (1988).
53. Jin, Z., et al. *Allergy Clin. Immunol.* 83, 165–170 (1989).
54. Lang, D. J., et al. *J. AIDS* 2, 540–549 (1989).
55. Becherer, P. R., et al. *Am. J. Hematol.* 34, 204–209 (1990).
56. Jason, J., et al. *Am. J. Hematol.* 34, 262–269 (1990).
57. Ludlam, C. A., et al. *Lancet* ii, 233–236 (1985).
58. AIDS–Hemophilia French Study Group, *Blood* 66, 896–901 (1985).
59. Madhok, R., et al. *Br. Med. J.* 293, 978–980 (1986).
60. Mahir, W. S., et al., *Br. J. Haem.* 69, 367–370 (1988).
61. Antonaci, S., et al. *Diagn. Clin. Immunol.* 5, 318–325 (1988).
62. Pitchenik. A. E., et al. *Am. Rev. Respir. Dis.* 135, 875–879 (1987).

63. Pitchenik, A. E., et al. *Lancet* 336, 440–441 (1990).
64. Dube, M. P., et al. *Am. J. Med.* 93, 520–524 (1992).
65. Gupta, S., et al. *Proc. Natl. Acad. Sci. USA* 89, 7831–7835 (1992).
66. Spira, T. J., et al. *New Eng. J. Med.* 328, 386–392 (1993).
67. Duncan, R. A., et al. *N. Eng. J. Med.* 328, 393–398 (1993).
68. Kaczmarski, R. S., et al. *Lancet* 340, 608 (1992).
69. Widy-Wirski R., et al. *JAMA* 260, 3286–3289 (1988).
70. De Cock, K. M., et al. *Br. Med. J.* 303, 1185–1188 (1991).
71. Taelman H., et al. *Br. Med. J.* 302, 1206 (1991).
72. Hishida, O., et al. *Lancet* 340, 971–972 (1992).
73. Brindle, R. J., et al. *Am Rev. Respir. Dis.* 147, 958–961 (1993).
74. Mientjes, G. H. C., et al. *Br. J. Haem.* 82, 615–619 (1992).
75. Koury, M. J. *Am. J. Hem.* 35, 134–135 (1990).

APPENDIX D

"The Duesberg Phenomenon": Duesberg and Other Voices*

Peter H. Duesberg
Dept. of Molecular and Cell Biology
University of California at Berkeley

In the Special News Report of 9 December (p. 1642) by Jon Cohen, *Science* struggles with what is called "The Duesberg phenomenon"—"a Berkeley virologist and his supporters continue to argue that HIV [human immunodeficiency virus] is not the cause of AIDS [acquired immunodeficiency syndrome]." Cohen tries to explain why "mainstream AIDS researchers" believe that HIV causes AIDS and why "HIV now fulfills the classic postulates... by Robert Koch." One week later (16 Dec., p. 1803), Cohen himself appears to become part of the phenomenon, when he writes: "Is a new virus the cause of KS [Kaposi's sarcoma]?" One should realize the heresy of this question. *KS has been and still is the signal disease of the AIDS syndrome.* The Centers for Disease Control include it in its list of 29 diseases defining AIDS in the presence of HIV (1). No other AIDS-defining disease has increased more than KS over its long-established background. It was so rare before AIDS that many doctors told me that they had never seen it before in young men. This is the reason why KS has become a hallmark for AIDS. And now, according to Cohen, "solid headway will have been made... " if HIV is found *not* to be the cause of KS.

Since "mainstream AIDS researchers" now consider one non-HIV cause for AIDS, why not consider others? Accordingly, I submit two experimental tests to find such causes.

1) Cohen wonders (16 Dec., p. 1803) about the "mystery" that "KS is almost exclusively confined to male homosexuals," but he reports (9

---

* Article originally appeared in *Science*, 267 (1995): 313–314. Reprinted by permission of the publisher.

Dec., p. 1648) that "use of nitrite inhalants, known as 'poppers'... has been high among some subgroups in the homosexual population" and that "nitrite inhalants [are] popular among gay men" (16 Dec., p. 1803). Cohen also interviewed the authors of a study that had shown in 1993 that every one of 215 homosexual AIDS patients from San Francisco had used poppers in addition to other recreational drugs and AZT (2).

Since nitrites are some of the best known mutagens and carcinogens (3) and AIDS KS typically occurs on the skin and in the lungs, the primary site of nitrite inhalant exposure, I propose to solve the "mystery": Expose 100 mice, or cats, or monkeys to nitrite inhalants at doses comparable with human recreational use and for time periods approximating the so-called 10-year latent period between infection by HIV to the onset of AIDS—possibly a euphemism for the time of drug use necessary for AIDS to develop. (It takes 10 to 20 years of smoking for emphysema or lung cancer to develop.) I would predict this result: immunodeficiency, pneumonia, and pulmonary KS in animals.

2) According to Cohen, mainstream AIDS researchers argue that it is "impossible" to eliminate confounding factors from HIV in typical AIDS risk groups, as for example in hemophiliacs "because [they] do not keep track of each factor VIII treatment" (9 Dec., p. 1645). Therefore, we are asked to accept confounded epidemiological studies of HIV-positives—who are either male homosexuals using immunotoxic nitrites (2), or are intravenous drug users, or are hemophiliacs subject to immunosuppressive transfusions, or are being treated with AZT, or are subject to exotic lifestyles—as evidence that HIV causes AIDS.

In view of this, I propose a very possible epidemiological test of whether HIV or non-HIV factors cause AIDS: Compare the incidence of AIDS-defining diseases in 3650 homo- or heterosexual American men, who are not on transfusions and recreational drugs or AZT, but are HIV-positive, to the incidence in 3650 HIV-negative counterparts. These healthy subjects could be found by the U.S. Army, which tests more than 2.5 million per year, or among those contributing to the blood banks, which test more than 12 million a year. If the 3650-day latent period is correct, every 2 days one of the people that are HIV-positive would develop AIDS. I would predict this result: The percentage incidence in the HIV-positive group will be the same as in the HIV-negative group.

If the mainstream AIDS researchers are not already doing these experiments, I would be delighted to do them provided I can get funded.

Peter H. Duesberg
Department of Molecular and Cell Biology,
University of California, Berkeley, CA 94720, USA

REFERENCES

1. Centers for Disease Control and Prevention, *Morb Mort Weekly Rep* 41 (No. RR17) (1992): 1–19.
2. M. S. Ascher, H. W. Sheppard, W. Winkelstein Jr., E. Vittinghoff, *Nature (London)* 362 (1993): 103–104.
3. H. W. Haverkos, J. A. Dougherty, Eds., *Health Hazards of Nitrite Inhalants*, NIDA Research Monograph 83 (U.S. Dept. Health & Human Services, Washington, D.C., 1988).

Notes

CHAPTER 1

1. P. H. Duesberg, "AIDS Acquired by Drug Consumption and Other Noncontagious Risk Factors," *Pharmacology and Therapeutics*, 55 (1992): 201–277.
2. D. M. Auerbach, W. W. Darrow, H. W. Jaffe, and J. W. Curran, "Cluster of Cases of the Acquired Immune Deficiency Syndrome Patients Linked by Sexual Contact," *American Journal of Medicine*, 76 (1984): 487–492.
3. National Institute of Allergy and Infectious Diseases, *NIAID Backgrounder: How HIV Causes AIDS* (National Institutes of Health, 1994).
4. Centers for Disease Control and Prevention, "U.S. HIV and AIDS Reported Through December 1993; Year-End Edition," *HIV/AIDS Surveillance Report*, 5 (1993): 1–33.
5. Duesberg, "AIDS Acquired by Drug Consumption," 201–277; P. H. Duesberg, "Human Immunodeficiency Virus and Acquired Immunodeficiency Syndrome: Correlation but Not Causation," *Proceedings of the National Academy of Sciences*, 86 (1989): 755–764.
6. J. Maddox, *Nature* (London), 333 (1988): 11.
7. L. K. Altman, "Researchers Believe AIDS Virus Is Found," *New York Times*, 24 April 1984, C1, C3.
8. J. Cairns, *Cancer: Science and Society* (San Francisco: W. H. Freeman and Company, 1978).
9. Altman, "Researchers Believe."
10. R. Kono, "The SMON Virus Theory," *Lancet*, ii (1975): 370–371; I. Shigematsu, H. Yanagawa, S. I. Yamamoto, and K. Nake, "Epidemiological Approach to SMON (Subacute Myelo-Optico-Neuropathy)," *Japanese Journal of Medicine, Science, and Biology*, 28 Supplement (1975): 23–33.
11. Kono, "SMON Virus Theory," 370–371.
12. E. Totsuka, personal communication, 1 May 1992.
13. T. E. Soda, *Drug-Induced Sufferings: Medical, Pharmaceutical, and Legal Aspects* (Amsterdam: Excerpta Medica, 1980).

14. Totsuka, personal communication, 1 May 1992.
15. Ibid.
16. Soda, *Drug-Induced Sufferings*.
17. Ibid.
18. Totsuka, personal communication, 1 May 1992.
19. Kono, "SMON Virus Theory," 370–371.
20. Totsuka, personal communication, 1 May 1992.
21. Soda, *Drug-Induced Sufferings*.

CHAPTER 2

1. T. D. Brock, *Robert Koch: A Life in Medicine and Bacteriology* (Madison, Wis.: Science Tech Publishers, 1988), 75.
2. Ibid., 121.
3. T. McKeown, *The Role of Medicine: Dream, Mirage, or Nemesis?* (Princeton, N.J.: Princeton University Press, 1979).
4. K. J. Carpenter, *The History of Scurvy and Vitamin C* (Cambridge, England: Cambridge University Press, 1986), 11.
5. A. Von Haller, *The Vitamin Hunters* (Philadelphia and New York: Chilton Co., 1962).
6. H. Bailey, *The Vitamin Pioneers* (Emmaus, Pa.: Rodale Books, 1968).
7. Carpenter, *History of Scurvy*.
8. Ibid.
9. Carpenter, *History of Scurvy*.
10. C. P. Stewart and D. Guthrie, eds., *Lind's Treatise on Scurvy* (Edinburgh: Edinburgh University Press, 1953), 408–409.
11. R. R. Williams, *Toward the Conquest of Beriberi* (Cambridge, Mass.: Harvard University Press, 1961), 18.
12. Ibid.
13. Bailey, *Vitamin Pioneers*.
14. E. W. Etheridge, *The Butterfly Caste: A Social History of Pellagra in the South* (Westport, Conn.: Greenwood Publishing Co., 1972), 11.
15. Ibid.
16. Ibid.
17. W. C. Winn, "Legionnaires' Disease: Historical Perspective." *Clinical Microbiology Reviews*, 1 (1988): 60–81.
18. Ibid.
19. B. J. Culliton, "Legion Fever: Postmortem on an Investigation that Failed," *Science*, 194 (1976): 1025–1027.
20. House Subcommittee on Consumer Protection and Finance, *Legionnaires' Disease*, 23–24 Nov. 1976, 5.
21. Ibid.

CHAPTER 3

1. R. E. Lapp. *The New Priesthood: The Scientific Elite and the Uses of Power* (New York: Harper & Row, 1965), 39–40.
2. D. C. Greenwood, *Solving the Scientist Shortage* (Washington, D.C.: Public Affairs Press, 1958), 23; National Science Board, *Science and Engineering Indicators—1989*, 1 Dec. 1989; "The Month in AIDS," *Rethinking AIDS* 1, no. 7 (1 July 1993).
3. Lapp, *New Priesthood*, 39–40.
4. Greenwood, *Solving the Scientist Shortage*.
5. Lapp, *New Priesthood*, 39–40.
6. Ibid.
7. National Research Council, *Postdoctoral Appointments and Disappointments* (National Research Council, 1981), 79.
8. National Science Foundation, *Doctoral Scientists and Engineers: A Decade of Change* (National Science Foundation, March 1988).
9. M. D. Reagan, *Science and the Federal Patron* (New York: Oxford University Press, 1969), 11.
10. Association of American Universities, *The Ph.D. Shortage: The Federal Role* (Association of American Universities, 11 Jan. 1990), IV.
11. "Careers '95: The Future of the Ph.D.," *Science*, 270: 121–136.
12. Reagan, *Science and the Federal Patron*.
13. National Science Board, *Science and Engineering Indicators*.
14. Lapp, *New Priesthood*, 39–40.
15. M. Shodel, "NIH Overview," *Journal of NIH Research*, 2 (1990): 22.
16. J. A. Shannon, "The National Institutes of Health: Some Critical Years, 1955–1957," *Science* 237 (1987): 865–868.
17. Ibid.
18. B. Moseley, "Interview: D. Carleton Gajdusek," *Omni*, March 1986, 62–68, 104–106.
19. D. C. Gajdusek, C. J. Gibbs, and M. Alpers, eds., *Slow, Latent, and Temperate Virus Infections* (Washington, D.C.: Government Printing Office, 1965), 5.
20. P. H. Duesberg and J. R. Schwartz, "Latent Viruses and Mutated Oncogenes: No Evidence for Pathogenicity." *Progress in Nucleic Acid Research and Molecular Biology*, 43 (1992): 135–204.
21. G. Kolata. "Anthropologists Suggest Cannibalism Is a Myth," *Science*, 232 (1986): 1497–1500.
22. Ibid.
23. Ibid.
24. D. R. Schryer, "Existence of Cannibalism," *Science*, 233 (1986): 926.
25. Duesberg and Schwartz, "Latent Viruses," 135–204.
26. G. Williams, *Virus Hunters* (New York: Alfred A. Knopf, 1959), 346.
27. Duesberg and Schwartz, "Latent Viruses," 135–204.
28. Ibid.
29. Ibid.

30. Barnum. "U.S. Biotech Is Thriving in Japan," *San Francisco Chronicle*, 12 May 1992, 1, 4.
31. Ibid.

CHAPTER 4

1. Koelner Stadtanzeiger, 1993.
2. H.D. Stetten and W. T. Carrigan, eds., *NIH: An Account of Research in Its Laboratories and Clinics* (New York: Academic Free Press [Harcourt Brace Jovanovich], 1984), 350.
3. All such viruses work only in immune-deficient animals that do not suppress the virus, including newborn mice or chickens. Some of these viruses contain special "cancer genes," which create tumors within days of inoculation, but most tumor viruses can only barely promote cancer in already-susceptible animals. The leukemia viruses, for example, only cause cancer in selected strains of laboratory animals that had become progressively more sickly through many generations of careful inbreeding. Unlike healthier animals in the wild, such lab animals more easily develop cancer or other disease in response to mild insults. Often a successful experiment required putting a lab animal through extraordinary conditions, such as forcing female mice through many extra rounds of pregnancy. The very fact that such "tumor viruses" can only affect these weak animals under odd conditions indicates that they would never cause cancer in humans with normal immune systems.
4. P. H. Duesberg and J. R. Schwartz, "Latent Viruses and Mutated Oncogenes: No Evidence for Pathogenicity," *Progress in Nucleic Acid Research and Molecular Biology*, 43 (1992): 135–204.
5. Stetten and Carrigan, *NIH: An Account of Research*; K. E. Studer and D. E. Chubin, *The Cancer Mission: Social Contexts of Biomedical Research* (Beverly Hills: Sage Publications, 1980).
6. G. Williams, *Virus Hunters* (New York: Alfred A. Knopf, 1959).
7. W. M. Stanley, "The Virus Etiology of Cancer," *Proceedings of the Third National Cancer Conference of the American Cancer Society* (1957), 42–51.
8. A. Lwoff, introduction to *RNA Viruses and Host Genome in Oncogenesis*, eds. P. Emmelot and P. Bentvelzen (New York: American Elsevier Publishing, 1972), 1–11.
9. J. Tooze, ed., *The Molecular Biology of Tumor Viruses* (Cold Spring Harbor, N.Y.: Cold Spring Harbor Laboratory, 1973), 56.
10. Duesberg and Schwartz, "Latent Viruses," 135–204.
11. D. S. Greenberg. "What Ever Happened to the War on Cancer?" *Discover*, March 1986, 55.
12. P. Rous, "The Challenge to Man of the Neoplastic Cell," *Science*, 157 (1967): 26.

13. Studer and Chubin, *Cancer Mission*.
14. J. Scott, "Dangerous Liaisons," *Los Angeles Times Magazine*, 11 March 1990, 12, 16.
15. Ibid., 12.
16. Ibid., 12–13.
17. Ibid., 12.
18. Ibid., 14.
19. Ibid., 16.
20. D.A. Galloway and J.K.McDougal, "The Oncogenic Potential of Herpes Simplex Viruses: Evidence for a 'Hit-and-Run' Mechanism," *Nature*, 302 (1983): 21–24, cited in Duesberg and Schwartz, "Latent Viruses," 135–204.
21. H. zur Hausen, "Papilloma Viruses in Anogenital Cancer as a Model to Understand the Role of Viruses in Human Cancers," *Cancer Research*, 49 (1989): 4677.
22. Rous, "Challenge to Man," 24–28.
23. Duesberg and Schwartz, "Latent Viruses," 135–204.
24. A. S. Evans, ed., *Viral Infections of Humans: Epidemiology and Control* (New York: Plenum Publishing Corporation, 1989).
25. C. E. Rogler, "Cellular and Molecular Mechanisms of Hepatocarcinogenesis Associated with Hepatovirus Infection," *Current Topics Microbiology and Immunology*, 168 (1991): 103–140.
26. J. Cairns, *Cancer: Science and Society* (San Francisco: W. H. Freeman and Company, 1978).
27. R. Yuan and M. Hsu, "Vaccines in Asia," *Genetic Engineering News*, 1 Oct. 1992, 14.
28. Studer and Chubin, *Cancer Mission*.
29. R. A. Weiss, N. Teich, and J. Coffin, eds., *Molecular Biology of RNA Tumor Viruses* (Cold Spring Harbor, N.Y.: Cold Spring Harbor Press, 1985).
30. "Central Dogma Reversed," *Nature* 226 (1970): 1198–1199.
31. Weiss, Teich, and Coffin, *Molecular Biology*.
32. N. Wade, "Scientists and the Press: Cancer Scare Story That Wasn't," *Science*, 174 (1971): 679–680.
33. R. C. Gallo, *Virus Hunting—AIDS, Cancer, and the Human Retrovirus: A Story of Scientific Discovery* (New York: Basic Books, 1991).
34. W. A. Blattner, ed., *Human Retrovirology: HTLV* (New York: Raven Press, 1990).
35. R. C. Gallo, "The First Human Retrovirus," *Scientific American*, 256 (1987): 45–46.
36. Duesberg and Schwartz, "Latent Viruses," 135–204.
37. Blattner, *Human Retrovirology*.
38. S. S. Epstein, E. Bingham, D. Rail, and I. D. Bross, "Losing the 'War Against Cancer': A Need for Public Policy Reforms," *International Journal of Health Services and Molecular Biology*, 22 (1992): 455–469.

660 ■ Notes

39. P. Gunby, "Battles Against Many Malignancies Lie Ahead as Federal 'War on Cancer' Enters Third Decade," *Journal of the American Medical Association*, 267 (1992): 1891.
40. Greenberg, "What Ever Happened," 47–66.

CHAPTER 5

1. E. Shorter, *The Health Century*, (New York: Doubleday & Co., 1987), 100.
2. A. D. Langmuir, "Biological Warfare Defense," *American Journal of Public Health* 42 (1952): 235–238.
3. T. H. Denetclaw and W. F. J. Denetclaw, "Is 'Southwest U.S. Mystery Disease' Caused by Hantavirus?" *Lancet*, 343 (1994b): 53–54.
4. S. Russel, "On the Trail of Hantavirus," *San Francisco Chronicle*, 4 July 1995, A1, A12.
5. M. D. Lemonick, "Return to the Hot Zone." *Time International*, 145, 22 May 1995, 56–57.
6. "Signs that Ebola Virus Is Fading Away," *San Francisco Chronicle*, 24 May 1995, A6.
7. K. Day, "After AIDS, Superbugs Give Medicine the Jitters," *International Herald Tribune*, 28 June 1995, 2.
8. Ibid.
9. G. Thomas and M. Morgan-Witts, *Anatomy of an Epidemic* (Garden City, N.Y.: Doubleday & Co., 1982), 105.
10. C. H. Wecht, "The Swine Flu Immunization Program: Scientific Venture or Political Folly?" *Legal Medicine Annual* (1978), 231.
11. Ibid., 227–244; P. Cotton, "CDC Nears Close of First Half-Century," *Journal of the American Medical Association*, 263 (1990): 2579–2580.
12. Thomas and Morgan-Witts, *Anatomy of an Epidemic*, 5–6.
13. Ibid., 3.
14. B. Culliton, "Legion Fever: Postmortem on an Investigation that Failed," *Science*, 194 (1976): 1025.
15. House Subcommittee on Consumer Protection and Finance, *Legionnaires' Disease*, 23–24 Nov. 1976.
16. Wecht, "Swine Flu Immunization," 227–244.
17. "Red Cross Knew of AIDS Blood Threat," *San Francisco Chronicle*, 16 May 1994.
18. R. Shilts, *And the Band Played On* (New York: St. Martin's Press, 1987), 63.
19. Ibid., 67.
20. M.S. Gottlieb, H.M. Schanker, P.T. Fan, A. Saxon, O.D. Weisman, and J. Pozalski, "Pneumocystis Pneumonia—Los Angeles," *Morbidity and Mortality Weekly Report* 30 (5 June 1981): 250–252.
21. Shilts, *And the Band Played On*.

22. E. W. Etheridge, *Sentinel for Health: A History of the Centers for Disease Control* (Berkeley, Calif.: University of California Press, 1992), 326.
23. D. M. Auerbach, W. W. Darrow, H. W. Jaffe, and J. W. Curran, "Cluster of Cases of the Acquired Immunity Deficiency Syndrome Patients Linked by Sexual Contact," *American Journal of Medicine*, 76 (1984): 487–492.
24. Ibid.
25. Shilts, *And the Band Played On*.
26. Ibid.
27. D. P. Drotman, T. A. Peterman, and A. E. Friedman-Kein, "Kaposi's Sarcoma. How Can Epidemiology Help Find the Cause?" *Dermatoepidemiology*, 13 (1995): 575–582.
28. Auerbach, Darrow, Jaffe, and Curran, "Cluster of Cases," 487–492.
29. Shilts, *And the Band Played On*.
30. Ibid.
31. Ibid.
32. Etheridge, *Sentinel for Health*.
33. Ibid.
34. Shilts, *And the Band Played On*.
35. Ibid.
36. Ibid.
37. H. W. Haverkos and J. A. Dougherty, eds., *Health Hazards of Nitrite Inhalants*, NIDA Research Monograph 83 (Washington, D.C.: U.S. Department of Health Services, 1988).
38. Shilts, *And the Band Played On*.
39. Ibid.
40. Etheridge, *Sentinel for Health*.
41. Shilts, *And the Band Played On*.
42. Ibid.
43. Ibid.
44. J. Crewdson, "The Great AIDS Quest," *Chicago Tribune*, 19 Nov. 1989.
45. L. K. Altman, "Researchers Believe AIDS Virus Is Found." *New York Times*, 24 April 1984, C1, C3.
46. Crewdson, "The Great AIDS Quest."
47. A. Karpas, letter to Serge Lang, 3 Feb. 1993.
48. K. Malik, J. Even, and A. Karpas, "Molecular Cloning and Complete Nucleotide Sequence of an Adult T Cell Leukemia Virus/Human T Cell Leukemia Virus Type I (ATLV/HTLV-I) Isolate of Caribbean Origin: Relationship to Other Members of the ATLV/HTLV-I Subgroup," *J.Gen.Viral.*, 69 (1988): 1695–1710, as cited in Karpas letter to Lang.
49. Ibid.
50. R. C. Gallo, "...And His Response," *Nature* 351 (1991): 358; M.S. Reitz, H.Z. Streicher, and R.C. Gallo, "Gallo's Virus Sequence," *Nature* 351 (1991) 358; Rep. J. Dingell, chair, *Report by the*

Subcommittee on Oversight and Investigations of the House of Representatives (1994).
51. Gallo, "...And His Response"; Reitz "Gallo's Virus Sequence"; Dingell, *Report by the Subcommittee*.
52. R. C. Gallo, *Virus Hunting—AIDS, Cancer, and the Human Retrovirus: A Story of Scientific Discovery* (New York: Basic Books, 1991), 210.
53. J. Crewdson, "Burden of Proof," *Chicago Tribune*, 6 Dec. 1992, section 4.
54. E. Rubinstein, "The Untold Story of HUT78," *Science* 248 (1990): 1499–1507; J. Dingell, "Shattuck Lecture—Misconduct in Medical Research," *New England Journal of Medicine*, 328 (1993): 1613.
55. J. Crewdson, "Virus From AIDS Lab Sold, Probe Suggests," *Chicago Tribune*, 29 April 1990.
56. J. Crewdson, "Ex-Gallo Aide Guilty of Pocketing $25,000," *Chicago Tribune*, 8 July 1992.
57. Crewdson, "Burden of Proof."
58. B. J. Culliton, "Gallo Reports Mystery Break-in," *Science*, 250 (1990): 502.
59. B. Werth, "By AIDS Obsessed," *Gentlemen's Quarterly*, August 1991, 207–208.
60. R. Weiss, "Provenance of HIV Strains," *Nature* (London), 349 (1991): 374.
61. Werth, "By AIDS Obsessed," 141–151, 205–208.
62. Mulder, "A Case of Mistaken Non-Identity," *Nature*, 331 (1988): 562–563, as cited in Karpas letter to Lang.
63. Karpas, letter to Lang.
64. D. S. Greenberg, "Saint or Scoundrel? The Gallo Controversy Goes On," *Science and Government Report*, 1 Feb. 1994, 6.
65. Ibid.
66. J. Crewdson, "In Gallo Case, Truth Termed a Casualty," *Chicago Tribune*, 1 Jan. 1995.
67. "Defending the Indefensible Dr. Gallo," *Chicago Tribune*, 6 Jan. 1995.
68. Serge Lang, personal communication.

CHAPTER 6

1. "Projections Show Rising Worldwide AIDS Toll," *American Medical News*, 8 May 1987, 40, cited in L. J. McNamee and B. F. McNamee, *AIDS: The Nation's First Politically Protected Disease* (La Habra, Calif.: National Medical Legal Publishing House, 1988), 6–7.
2. J. Groopman et al., "HTLV-III in Saliva of People with AIDS-Related Complex and Healthy Homosexual Men at Risk for AIDS," *Science* (24 Oct. 1984): 447–449, cited in McNamee and McNamee, *AIDS:*

The Nation's First Politically Protected Disease, 75–76; Centers for Disease Control announcement, 11 Jan. 1985, as quoted in G. Antonio, *The AIDS Cover-Up: The Real and Alarming Facts About AIDS* (San Francisco, Calif.: Ignatius Press, 1986), 108.
3. Antonio, *AIDS Cover-Up,* 109–115; see also McNamee and McNamee, *AIDS: The Nation's First Politically Protected Disease.*
4. Antonio, *AIDS Cover-Up,* 105–107; see also McNamee and McNamee, *AIDS: The Nation's First Politically Protected Disease,* 83–87.
5. L. C. Humphrey, ed., *America Living with AIDS* (National Commission on AIDS, 1991), 13.
6. Ibid., 2.
7. Ibid., 7.
8. Staff and Brubaker, *The AIDS Epidemic* (New York: Warner Books, 1985), 162–163, quoted in Antonio, *AIDS Cover-Up.*
9. Antonio, *AIDS Cover-Up,* 189.
10. P. H. Duesberg, "AIDS Acquired by Drug Consumption and Other Noncontagious Risk Factors," *Pharmacology and Therapeutics,* 55 (1992): 201–277.
11. J. D. Dingell, *AIDS Prevention Act of 1990,* House Committee on Energy and Commerce (31 May 1990), 189.
12. M. Piatak et al., "High Levels of HIV-1 in Plasma During All Stages of Infection Determined by Competitive PCR," *Science,* 259 (1993): 1749–1754.
13. J. Maddox, "Where the AIDS Virus Hides Away," *Nature* (London), 362 (1993b): 287; G. Pantello, C. Graziosi, J. F. Demarest, L. Butini, M. Montroni, C. H. Fox, J. M. Orenstein, D. P. Kotler, and A. S. Fauci, "HIV Infection Is Active and Progressive in Lymphoid Tissue During the Clinically Latent Stage of Disease," *Nature* (London), 362 (1993): 355–358; J. Maddox, "Duesberg and the New View of HIV," *Nature* (London), 373 (1995a): 189; S. Wain-Hobson, "Virological Mayhem," *Nature* (London), 373 (1995): 102.
14. Duesberg, "AIDS Acquired by Drug Consumption."
15. World Health Organization, *The Current Global Situation of the HIV-AIDS Pandemic* (1995).
16. E. Papadopulos-Eleopulos, V. F. Turner, and J. M. Papadimitriou, "Is a Positive Western Blot Proof of HIV Infection?" *Bio/Technology,* 11 (1993): 696–707; P. H. Duesberg, "The HIV Gap in National AIDS Statistics," *Bio/Technology,* 11 (1993).
17. Duesberg, "HIV Gap."
18. P. H. Duesberg, "AIDS Acquired by Drug Consumption," 201–277; J. A. Jacquez, J. S. Koopman, C. P. Simon, and I. M. Longini Jr., "Role of the Primary Infection in Epidemics of HIV Infection in Gay Cohorts," *Journal of Acquired Immune Deficiency Syndromes,* 7 (1994): 1169–1184.
19. Duesberg, "AIDS Acquired by Drug Consumption."

20. Ibid.
21. W. A. Blattner, R. C. Gallo, and H. M. Temin, "HIV Causes AIDS," *Science*, 241 (1988): 516–517.
22. Duesberg, "AIDS Acquired by Drug Consumption."
23. "When a House Officer Gets AIDS," *New England Journal of Medicine* (1989).
24. Duesberg, "AIDS Acquired by Drug Consumption."
25. P. H. Duesberg, "Foreign-Protein-Mediated Immunodeficiency in Hemophiliacs With and Without HIV," *Genetica* 95 (1995b): 51–70.
26. World Health Organization, *Current Global Situation*.
27. Duesberg, "AIDS Acquired by Drug Consumption."
28. Ibid.
29. Ibid.
30. Duesberg, "HIV Gap"; H. Bacellar, A. Muñoz, E. N. Miller, E. A. Cohen, D. Besley, O. A. Selnes, J. T. Becker, and J. C. McArthur, "Temporal Trends in the Incidence of HIV-1–Related Neurological Diseases: Multicenter AIDS Cohort Study: 1985-1992," *Neurology*, 44 (1994): 1892–1900.
31. J. Bendit and B. Jasny, "AIDS: The Unanswered Questions," *Science*, 260 (1993): 1219, 1253–1293; J. Cohen, "'The Duesberg Phenomenon': Duesberg and Other Voices," *Science*, 266 (1994a): 1642–1649.
32. J. Cohen, "Keystone's Blunt Message: 'It's the Virus Stupid,'" *Science*, 260 (1993): 292–293; Maddox, "Where the AIDS Virus Hides Away," 287; Cohen, "'Duesberg Phenomenon'"; J. Cohen, "Researchers Air Alternative Views on How HIV Kills Cells," *Science*, 269 (1995): 1044–1045; J. Maddox, "More Conviction on HIV and AIDS," *Nature* (London), 377 (1995b): 1; Maddox, "Duesberg and the New View of HIV."
33. Maddox, "Where the AIDS Virus Hides Away."
34. P. H. Duesberg, "Infectious AIDS—Stretching the Germ Theory Beyond Its Limits," *Int. Arch. Allergy Immuno.*, 103 (1994): 131–142.
35. Duesberg, "AIDS Acquired by Drug Consumption"; K. Mullis, "A Hypothetical Disease of the Immune System that May Bear Some Relation to the Acquired Immune Deficiency Syndrome," *Genetica*, 95 (1995): 195–197.
36. Duesberg, "AIDS Acquired by Drug Consumption."
37. P. H. Duesberg, "How Much Longer Can We Afford the AIDS Virus Monopoly?" in *AIDS: Virus or Drug-Induced?*, eds. Kluwer and Dordrecht (The Netherlands: Genetica, in press).
38. A. S. Fauci, "CD+ T-Lymphocytopenia Without AIDS Infection—No Lights, No Camera, Just Facts," *New England Journal of Medicine*, 328 (1993): 429–431.
39. J. Cohen, "Is a New Virus the Cause of KS?" *Science*, 266 (1994b): 1803–1804.
40. D. J. Bergman and A. D. Langmuir, "Farr's Law Applied to AIDS Pro-

jections," *Journal of the American Medical Association* 263 (1990): 50–57.
41. Ibid.
42. Cohen, "'Duesberg Phenomenon,'" 1642–1649.
43. Ibid.
44. R. C. Gallo, *Virus Hunting—AIDS, Cancer, and the Human Retrovirus: A Story of Scientific Discovery* (New York: Basic Books, 1991), 277.
45. R. A. Weiss and H. W. Jaffe, "Duesberg, HIV, and AIDS," *Nature*, 345 (1990): 659–660.
46. R. C. Gallo and L. Montagnier, "AIDS in 1988," *Scientific American*, 259 (1988): 40–48.
47. G. M. Oppenheimer, "Causes, Cases, and Cohorts: The Role of Epidemiology in the Historical Construction of AIDS," in *AIDS: The Making of a Chronic Disease*, eds. E. Fee and D. M. Fox (Berkeley: University of California Press, 1992), 49–83.
48. Duesberg, "AIDS Acquired by Drug Consumption."
49. Blattner, Gallo, and Temin, "HIV Causes AIDS," 516–517.
50. Weiss and Jaffe, "Duesberg, HIV, and AIDS."
51. Blattner, Gallo, and Temin, "HIV Causes AIDS."
52. Weiss and Jaffe, "Duesberg, HIV, and AIDS."
53. Blattner, Gallo, and Temin, "HIV Causes AIDS."
54. Weiss and Jaffe, "Duesberg, HIV, and AIDS."
55. Gallo, *Virus Hunting—AIDS, Cancer, and the Human Retrovirus.*
56. Blattner, Gallo, and Temin, "HIV Causes AIDS."
57. For the various examples of HIV-free AIDS, see Duesberg, "AIDS Acquired by Drug Consumption"; see also Duesberg, "HIV Gap."
58. World Health Organization, *Current Global Situation.*
59. P. H. Duesberg, "Can Epidemiology Determine Whether Drugs or HIV Cause AIDS?" *AIDS-Forschung*, 12 (1993): 627–635. See also Duesberg, "AIDS Acquired by Drug Consumption," and Duesberg, "HIV Gap."
60. S. Stolberg, "AIDS Tally to Increase Due to New Definition," *Los Angeles Times*, 31 Dec. 1992, A3, A29.
61. Blattner, Gallo, and Temin, "HIV Causes AIDS."
62. M. Fumento, *The Myth of Heterosexual AIDS* (New York: Basic Books, 1989).
63. Bergman and Langmuir, "Farr's Law," 1522–1525.
64. Centers for Disease Control and Prevention, "U.S. HIV and AIDS Cases Reported Through December 1994," *HIV/AIDS Surveillance Report*, 6 (1994): 1–39.
65. L. K. Altman, "Obstacle-Strewn Road to Rethinking the Numbers on AIDS," *New York Times*, 1994, C3; C. Farber, "AIDS: Words From the Front," *Spin*, 11 July 1995, 69.
66. D. M. Auerbach, W. W. Darrow, H. W. Jaffe, and J. W. Curran, "Cluster of Cases of the Acquired Immune Deficiency Syndrome Patients

Linked by Sexual Contact," *American Journal of Medicine,* 76 (1984): 487–492.
67. Duesberg, "Foreign-Protein-Mediated Immunodeficiency," 51–70.
68. For various examples of HIV-free AIDS, see Duesberg, "AIDS Acquired by Drug Consumption," 223.
69. Duesberg, "HIV Gap."
70. P. H. Duesberg, "'The Duesberg-Phenomenon': Duesberg and Other Voices" (letter) *Science,* 267 (1995a): 313.
71. M. S. Ascher, H. W. Sheppard, W. Winkelstein Jr., and E. Vittinghoff, "Does Drug Use Cause AIDS?" *Nature* (London), 362 (1993): 103–104.

CHAPTER 7

1. J. A. Sonnabend, "The Etiology of AIDS," *AIDS Research,* 1 (1983): 1–12.
2. J. A. Sonnabend, "Caution on AIDS Viruses," *Nature,* 310 (1984): 103.
3. J. A. Sonnabend, letter to the editor, *Wall Street Journal,* 18 Nov. 1985.
4. J. A. Sonnabend, letter to Peter Duesberg, 24 May 1988.
5. D. Hopkins, "Dr. Joseph Sonnabend," *Interview Magazine,* December 1992, 124–126, 142–143.
6. Ibid.
7. G. Stewart, R. Root-Bernstein, Luca-Moretti, M. Brands, M. Callen, J. A. Sonnabend, E. Papadopulos-Eleopulos, M. Bastide, and J. Leiphart, "AIDS—A Different View," press release, International Symposium, Amsterdam, 6 May 1992.
8. J. A. Sonnabend, letter to the editor, *New York Native,* 29 June 1992, 4; J. Lauritsen, letter to the editor, *New York Native,* 29 June 1992, 4.
9. J. Lauritsen, "AZT: Iatrogenic Genocide," *New York Native,* 28 March 1988, 13–17.
10. M. Callen, "A Dangerous Talk With Dr. Sonnabend," *QW,* 27 Sept. 1992. 42–46, 71–72.
11. J. A. Sonnabend, letter, *Science,* 267 (1995): 159.
12. N. Lehrman, endorsement for Lauritsen, *Poison by Prescription: The AZT Story,* by J. Lauritsen (New York: Asklepios, 1990), back cover.
13. J. Lauritsen, personal communication, 28 May 1993.
14. J. Lauritsen, personal communication, 21 April 1993.
15. J. Lauritsen, notes on conversations with CDC/NCI personnel, 11–12 June 1987.
16. A. S. Relman, letter to John Lauritsen, 11 July 1988.
17. H. Coulter, *AIDS and Syphilis: The Hidden Link* (Berkeley, Calif: North Atlantic Books, 1987).
18. K. Wright, "Mycoplasmas in the AIDS Spotlight," *Science,* 248 (1990): 682–683.

19. S.-C. Lo and D. J. Wear, letter to the editor, *Policy Review*, (Fall 1990): 76–77.
20. R. Rapoport, "Dissident Scientist's AIDS Theory Angers Colleagues," *Oakland Tribune*, 31 Jan. 1988, B1.
21. N. Regush, "AIDS Risk Limited, Studies Suggest," *Montreal Gazette*, 15 Aug. 1987, B1, B4.
22. Ibid.
23. R. Rapoport, "AIDS: The Unanswered Questions," *Oakland Tribune*, 21 May 1989, A1, A2.
24. S. S. Hall, "Gadfly in the Ointment," *Hippocrates* (Sept./Oct. 1988): 80.
25. A. Liversidge, "Heresy! Modern Galileos," *Omni*, June 1993, 48.
26. "The AIDS Catch," Dispatches, Channel Four Television, (London: Meditel Productions), 13 June 1990.
27. Liversidge, "Heresy!" 50.
28. M. Eigen, "The AIDS Debate," *Naturwissenschaften*, 76 (1989): 341–350.
29. P. H. Duesberg, "Responding to 'The AIDS Debate,'" *Naturwissenschaften*, 77 (1990): 97–102.
30. M. Balter, "Montagnier Pursues the Mycoplasm–AIDS Link," *Science*, 251 (1991): 271.
31. P. Cotton, "Cofactor Question Divides Codiscoverers of HIV," *Medical News and Perspectives*, 264 (1990): 3111–3112.
32. R. C. Gallo, *Virus Hunting—AIDS, Cancer, and the Human Retrovirus: A Story of Human Discovery* (New York: Basic Books, 1991), 286, 297.
33. Balter, "Montagnier Pursues."
34. M. Lemaitre, D. Guetard, Y. Henin, L. Montagnier, and A. Zerial, "Protective Activity of Tetracycline Analogs Against the Cytopathic Effect of the Human Immunodeficiency Viruses in CEM Cells," *Research in Virology*, 141 (1990): 5–16.
35. D. E. Koshland, letter to Charles A. Thomas Jr., 11 June 1991.
36. E. Bauman, T. Bethell, H. Bialy, C. Farber, C. Geshekter, P. Johnson, R. Maver, R. Schoch, G. Stewart, R. Strohman, C. Thomas Jr. (for the Group for the Scientific Reappraisal of the HIV/AIDS Hypothesis), "AIDS Proposal," *Science*, 267 (1995a): 945–946.
37. E. Bauman, T. Bethell, H. Bialy, C. Farber, C. Geshekter, P. Johnson, R. Maver, R. Schoch, G. Stewart, R. Strohman, C. Thomas Jr. (for the Group for the Scientific Reappraisal of the HIV/AIDS Hypothesis), "AIDS Proposal," *AIDS–Forschung*, 10 (1995b): 280.
38. R. S. Root-Bernstein, "Do We Know the Cause(s) of AIDS?" *Perspectives in Biology and Medicine*, 33 (1990): 480–500.
39. R. S. Root-Bernstein, *Rethinking AIDS: The Tragic Cost of Premature Consensus* (New York: The Free Press, 1993).
40. "The AIDS Catch."
41. Root-Bernstein, *Rethinking AIDS*.
42. P. H. Duesberg, "AIDS Acquired by Drug Consumption and Other

668 ▪ Notes

Noncontagious Risk Factors," *Pharmacology and Therapeutics*, 55 (1992): 201-227.
43. R. S. Root-Bernstein, "'The Duesberg Phenomenon': What Does It Mean?" *Science*, 267 (1995): 159.
44. "The AIDS Catch."
45. J. Miller, "AIDS Heresy," *Discover*, June 1988, 66.
46. N. Hodgkinson, "Experts Mount Startling Challenge to AIDS Orthodoxy," *Sunday Times of London*, 26 April 1992, 1, 12–13.
47. P. H. Duesberg, "The HIV Gap in National AIDS Statistics," *Bio/Technology*, 11 (1993).
48. D. McCormick, "Right of Reply" *Bio/Technology*, 11 (1993): 955.
49. B. E. Griffin, "Burden of Proof," *Nature*, 338 (1989): 670.
50. Hodgkinson, "Experts Mount Startling Challenge."
51. See Mullis's foreword to this book.
52. Carroll, "The Weird Way to Win a Nobel Prize," *San Francisco Chronicle*, 21 Oct. 1993, E9.
53. E. Papadopulos-Eleopolos, V. F. Turner, and J. M. Papadimitriou, "Is a Positive Western Blot Proof of HIV Infection?" *Bio/Technology*, 11 (1993): 696–707.
54. Ibid.
55. E. Papadopulos-Eleopolos, V. F. Turner, J. M. Papadimitriou, and D. Causer, "Factor VIII, HIV, and AIDS in Haemophiliacs: An Analysis of Their Relationship," *Genetica*, 95 (1995a): 5–24; E. Papadopulos-Eleopolos, V. F. Turner, J. M. Papadimitriou, D. Causer, B. Hedland-Thomas, and B. A. P. Page, "A Critical Analysis of the HIV = T4-Cell–AIDS Hypothesis," *Genetica*, 95 (1995b): 5–24.

CHAPTER 8

1. M. S. Gottlieb, H. M. Shanker, P. T. Fan, A. Saxon, J. D. Weisman, and J. Pozalski, "Pneumocystis Pneumonia—Los Angeles," *Morbidity and Mortality Weekly Report*, 30 (1981): 250–252.
2. F. I. D. Konotey-Ahulu, *What Is AIDS?* (Watford, England: Tetteh-A'Domeno Co., 1989), 109.
3. P. H. Duesberg, "The HIV Gap in National AIDS Statistics," *Bio/Technology*, 11 (1993): 955–956.
4. Centers for Disease Control and Prevention, "U.S. HIV and AIDS Cases Reported Through June 1994; Mid-Year Edition," *HIV/AIDS Surveillance Report*, 6 (1994): 1–27.
5. Ibid.
6. Ibid.
7. M. H. Merson, "Slowing the Spread of HIV: Agenda for the 1990's," *Science*, 260 (1993): 1266–1268.
8. World Health Organization, *The Current Global Situation of the HIV/AIDS Pandemic* (1995).

9. P. H. Duesberg, "AIDS Acquired by Drug Consumption and Other Noncontagious Risk Factors," *Pharmacology and Therapeutics*, 55 (1992): 201–277.
10. Ibid.
11. P. H. Duesberg, "AIDS: Virus or Drug Induced?" *American J. Continuing Education Nursing*, 7 (1995): 31–44; P. H. Duesberg, "How Much Longer Can We Afford the AIDS Virus Monopoly?" in *AIDS: Virus or Drug Induced?* eds. Kluwer and Dordrecht (The Netherlands: Genetica, in press).
12. Duesberg, "AIDS Acquired by Drug Consumption."
13. S. V. Meddis, "Heroin Use Said to Near Crisis Level," *USA Today*, 25 May 1994, 1, 3A; J. Gettman, "Heroin Returning to Center Stage," *High Times*, December 1994, 23.
14. L. P. Finnegan, J. M. Mellot, L. R. Williams, and R. J. Wapner, "Perinatal Exposure to Cocaine: Human Studies," in *Cocaine: Pharmacology, Physiology, and Clinical Studies*, eds. J. M. Lakowski, M. P. Galloway, and F. J. White (Boca Raton, Fla.: CRC Press, 1992), 391–409.
15. B. R. Edlin, K. L. Irwin, S. Farouque, C. B. McCoy, C. Word, Y. Serrano, J. A. Inciardi, B. P. Bowser, R. F. Schilling, S. D. Holmberg, and Multicenter Crack Cocaine and HIV Infection Study Team, "Intersecting Epidemics—Crack Cocaine Use and HIV Infection Among Inner-City Young Adults," *New England Journal of Medicine*, 331 (1994): 1422–1427.
16. Ibid.
17. G. R. Newell, M. R. Spitz, and M. B. Wilson, "Nitrite Inhalants: Historical Perspective," in H. W. Haverkos and J. A. Dougherty, *Health Hazards of Nitrite Inhalants*, NIDA Research Monograph 83 (Rockville, Md.: National Institute on Drug Abuse, 1988), 5; J. Lauritsen, *The AIDS War: Propaganda, Profiteering, and Genocide from the Medical-Industrial Complex* (New York: Asklepios, 1993), 104.
18. J. Lauritsen and H. Wilson, *Death Rush: Poppers and AIDS* (New York: Pagan Press, 1986), 6, 15.
19. Ibid., 48, 53.
20. Duesberg, "AIDS Acquired by Drug Consumption."
21. Ibid.
22. Ibid.
23. Ibid.
24. R. S. Root-Bernstein, "Do We Know the Cause(s) of AIDS?" *Perspectives in Biology and Medicine*, 33 (1990): 484; Lauritsen, *AIDS War*, 197–199; H. Coulter, *AIDS and Syphilis: The Hidden Link* (Berkeley, Calif.: North Atlantic Books, 1987), 47.
25. F. Buianouckas, letter to Peter Duesberg, 29 April 1993.
26. C. A. Hill, letter to Peter Duesberg, 4 June 1992.
27. W. B. Coyle, "In Memory of... The Way it Was" (an open letter, unpublished), 14 Jan. 1992.

28. J. Adams, *AIDS: The HIV Myth* (New York: St. Martin's Press, 1989), 129.
29. L. M. Krieger and C. A. Caceres, "The Unnoticed Link in AIDS Cases," *Wall Street Journal*, 24 Oct. 1985.
30. J. J. McKenna, R. Miles, D. Lemen, S. H. Danford, and R. Renirie, "Unmasking AIDS: Chemical Immunosuppression and Seronegative Syphilis," *Medical Hypotheses*, 21 (1986): 421–430.
31. Lauritsen and Wilson, *Death Rush*, 11.
32. Duesberg, "AIDS Acquired by Drug Consumption."
33. M. S. Ascher, H. W. Sheppard, W. Winkelstein Jr., and E. Vittinghoff, "Does Drug Use Cause AIDS?" *Nature* (London), 362 (1993): 103–104; P. H. Duesberg, "Aetiology of AIDS," *Lancet*, 341 (1993): 1544; P. H. Duesberg, "HIV and the Aetiology of AIDS," *Lancet*, 341 (1993): 957–958; P. H. Duesberg, "Can Epidemiology Determine Whether Drugs or HIV Cause AIDS?" *AIDS-Forschung*, 12 (1993): 627–635.
34. Duesberg, "Can Epidemiology Determine," 627–635; M. T. Schechter, K. J. P. Craib, K. A. Gelmon, J. S. G. Montaner, T. N. Le, and M. V. O'Shaughnessy, "HIV-1 and the Aetiology of AIDS," *Lancet*, 341 (1993): 658–659.
35. Duesberg, "Can Epidemiology Determine," 627–635; Schechter, Craib, Gelmon, Montaner, Le, and O'Shaughnessy, "HIV-1 and the Aetiology of AIDS," 658–659.
36. R. Shilts, *And the Band Played On* (New York: St. Martin's Press, 1987), 104.
37. Duesberg, "AIDS Acquired by Drug Consumption," 237–238, 241, 247.
38. T. Koch, R. Jeremy, E. Lewis, P. Weintrub, C. Rumsey, and M. Cowan, UC-San Francisco, censored manuscript cited in Duesberg, "AIDS Acquired by Drug Consumption," 201–277.
39. E. H. Aylward, A. M. Butz, N. Hutton, M. L. Joyner, and J. W. Vogelhut, "Cognitive and Motor Development in Infants at Risk for Human Immunodeficiency Virus," *Am. J. Dis. Child*, 146 (1992): 218–222.
40. M. F. Rogers, C.-Y. Ou, M. Rayfield, P. A. Thomas, E. E. Schoenbaum, E. Abrams, K. Krasinski, P. A. Selwyn, J. Moore, A. Kaul, K. T. Grimm, M. Bamji, G. Schochetman, and the New York City Collaborative Study of Maternal HIV Transmission and Montefiori's Medical Center HIV Perinatal Transmission Study Group, "Use of the Polymerase Chain Reaction for Early Detection of the Proviral Sequences of Human Immunodeficiency Virus in Infants Born to Seropositive Mothers," *New England Journal of Medicine*, 320 (1989): 1649–1654.
41. K. W. Culver, A. J. Ammann, J. C. Patridge, D. F. Wong, D. W. Wara, and M. J. Cowan, "Lymphocyte Abnormalities in Infants Born to Drug-Abusing Mother," *Journal of Pediatricians*, 111 (1987): 230–235.

42. Duesberg, "AIDS Acquired by Drug Consumption," 201–277; P. H. Duesberg, "How Much Longer Can We Afford the AIDS Virus Monopoly?" in *AIDS: Virus or Drug-Induced?*, eds. Kluwer and Dordrecht (The Netherlands: Genetica, in press).
43. Duesberg, "HIV Gap," 955–956.
44. P. H. Duesberg, "'The Duesberg Phenomenon': Duesberg and Other Voices" (letter), *Science*, 267 (1995a): 313.
45. Duesberg, "AIDS Acquired by Drug Consumption," 240.
46. Lauritsen and Wilson, *Death Rush*, 5, 33.
47. L. T. Sigell, F. T. Kapp et al., "Popping and Snorting Volatile Nitrites: A Current Fad for Getting High," *American Journal of Psychiatry*, 135 (1978): 1216–1218.
48. Lauritsen and Wilson, *Death Rush*, 5.
49. G. R. Newell, P. W. A. Mansell, M. R. Spitz, J. M. Reuben, and E. M. Hersh, "Volatile Nitrites: Use and Adverse Effects Related to the Current Epidemic of the Acquired Immune Deficiency Syndrome," *American Journal of Medicine*, 78 (1985): 811–816; Lauritsen, *AIDS War*, 104.
50. S. S. Mirvish, J. Williamson, D. Badcook, and C. Sheng-Chong, "Mutagenicity of Iso-butyl Nitrite Vapor in the Ames Test and Some Relevant Chemical Properties, Including the Reaction of Iso-butyl Nitrite with Phosphate," *Environ. Mol. Mutagen.*, 21 (1993): 247–252.
51. J. J. Goedert, C. Y. Neuland, W. C. Wallen, M. H. Greene, D. L. Mann, C. Murray, D. M. Strong, J. F. Fraumeni Jr., and W. A. Blattner, "Amyl Nitrite May Alter T Lymphocytes in Homosexual Men," *Lancet* (1982): 412–414; Haverkos and Dougherty, "Health Hazards of Nitrite Inhalants."
52. Newell, Mansell, Spitz, Reuben, and Hersh, "Volatile Nitrites," 811–816.
53. R. J. S. Lewis, *Food Additives Handbook* (New York: Van Nostrand Reinhold, 1989).
54. G. D. Cox, "County Health Panel Urges 'Poppers' Ban, Cites AIDS Link," *Los Angeles Daily Journal*, 24 March 1986, Section II, p. 1.
55. H. W. Haverkos, "Nitrite Inhalant Abuse and AIDS-Related Kaposi's Sarcoma," *Journal of Acquired Immune Deficiencies*, 3 (1990): Supplement 1, S47–S50.
56. Duesberg, "AIDS Acquired by Drug Consumption," 201–277.
57. T. Bethell, "Do 'Poppers' Hold the Secret to One of the Great Mysteries of AIDS?" *Spin*, 10 (1994), 87–89, 116.
58. Project Inform, "Is HIV the Cause of AIDS?" *Project Discussion Paper #5* (San Francisco, 27 May 1992), 1–6; Ascher, Sheppard, Winkelstein, and Vittinghoff, "Does Drug Use Cause AIDS?" 103–104; Schechter, Craib, Gelmon, Montaner, Le, and O'Shaughnessy, "HIV-1 and the Aetiology of AIDS," 658–659; J. Cohen, "'The Duesberg Phenomenon': Duesberg and Other Voices," *Science* 266 (1994): 1642–1649.

59. J. Cohen, "Is a New Virus the Cause of KS?" *Science*, 266 (1994): 1803–1804.
60. Ascher, Sheppard, Winkelstein, and Vittinghoff, "Does Drug Use Cause AIDS?" 103–104; Duesberg, "Can Epidemiology Determine," 627–635; S. Mansfield and G. Owen, "The Use of Ecstasy and Other Recreational Drugs in Patients Attending an HIV Clinic in London and Its Association with Sexual Behavior" (IX International Conference on AIDS, Berlin); Schechter, Craib, Gelmon, Montaner, Le, and O'Shaughnessy, "HIV-1 and the Aetiology of AIDS," 658–659; M. T. Schechter, K. J. P. Craib, J. S. G. Montaner, T. N. Le, M. V. O'Shaughnessy, and K. A. Gelmon, "Aetiology of AIDS," *Lancet*, 341 (1993): 1222–1223; Bethell, "Do 'Poppers' Hold the Secret," 87–89, 116; P. Gorman, "Peter Duesberg: Visionary or Public Menace?" *High Times*, December 1994, 58–61, 66; N. Hodgkinson, "New Evidence Links Gay Sex Drugs to AIDS," *Sunday Times of London*, 10 April 1994, 1–2; J. Lauritsen, "NIH Reconsiders Nitrites' Link to AIDS," *Bio/Technology*, 12 (1994): 762–763; D. Sadownick, "Kneeling at the Crystal Cathedral," *Genre*, December/January 1994, 40–45, 86–90; A. Vollbrechtshausen, "Drogen: Poppers, Speed and XTC: Am Wochenende bin ich nicht auf dieser Welt," *Magnus* (Germany), February 1994, 48–53; H. W. Haverkos and D. P. Drotman, "NIDA Technical Review: Nitrite Inhalants," (Washington, D.C.: National Institute on Drug Abuse and Atlanta, Ga.: Centers for Disease Control, 1995, unpublished).
61. Duesberg, "AIDS Acquired by Drug Consumption," 248–249.
62. P. H. Duesberg, "AIDS Epidemiology: Inconsistencies with Human Immunodeficiency Virus and with Infectious Disease, *Proc. Natl. Acad. Sci. U.S.A*, 88 (1991): 1575–1579 (see especially references 24–26).
63. Newell, Mansell, Spitz, Reuben, and Hersh, "Volatile Nitrites, 811–816; A. R. Lifson, W. W. Darrow, N. A. Hessol, P. M. O'Malley, J. L. Barnhart, H. W. Jaffe, and G. W. Rutherford, "Kaposi's Sarcoma in a Cohort of Homosexual and Bisexual Men: Epidemiology and Analysis for Cofactors," *American Journal of Epidemiology*, 131 (1990): 221–231; Haverkos and Dougherty, "Health Hazards of Nitrite Inhalants"; V. Beral, T. A. Peterman, R. L. Berkelman, and H. W. Jaffe. "Kaposi's Sarcoma Among Persons with AIDS: A Sexually Transmitted Infection?" *Lancet*, 335 (1990): 123–128.
64. L. M. Krieger, "Kaposi's Sarcoma, AIDS Link Questioned," *San Francisco Examiner*, 5 June 1992, A-1, A-17.
65. Duesberg, "AIDS Acquired by Drug Consumption," 201–277; Duesberg, "HIV Gap," 955–956; A. E. Friedman-Kien, B. R. Saltzman, Y. Cao, M. S. Nestor, M. Mirabile, J. J. Li, and T. A. Peterman, "Kaposi's Sarcoma in HIV-negative Homosexual Men," *Lancet*, 355 (1990): 168–169.
66. Duesberg, "HIV Gap," 955–956.

67. Friedman-Kien, Saltzman, Cao, Nestor, Mirabile, Li, and Peterman, "Kaposi's Sarcoma," 168–169.
68. Krieger, "Kaposi's Sarcoma, AIDS Link Questioned," A–1, A–17.
69. A. S. Fauci, "CD4+ T-Lymphocytopenia Without HIV Infection—No Lights, No Camera, Just Facts," *New England Journal of Medicine*, 328 (1993): 429–431.
70. L. K. Altman, "Going off the Beaten Path to Track Down Clues About AIDS," *New York Times*, 20 Dec. 1994, 3; Cohen, "Is a New Virus the Cause of KS?" 1803–1804.
71. L. K. Altman, "AIDS Cancer Said to Have Viral Source," *New York Times*, 1 Feb. 1995, 22A.
72. Haverkos and Dougherty, *Health Hazards of Nitrite Inhalants*; Duesberg, "AIDS Acquired by Drug Consumption," 201–277.
73. R. B. Nieman, J. Fleming, R. J. Coker, J. R. Harris, and D. M. Mitchell, "The Effects of Cigarette Smoking on the Development of AIDS in HIV-1–Seropositive Individuals," *AIDS*, 7 (1993): 705–710.
74. Haverkos and Dougherty, *Health Hazards of Nitrite Inhalants*.
75. C. Achard, H. Bernard, and C. Cagneux, "Action de la Morphine sur les Proprietes Leucocytaires; Leuco-diagnostic du Morphinisme," *Bulletin et Memoires de la Societe Medicale des Hopitaux de Paris*, 28, 3rd Series (1909): 958–966.
76. C. E. Terry and M. Pellens, *The Opium Problem* (Bureau of Social Hygiene of New York, 1928); J. H. Briggs, C. G. McKerron, R. L. Souhami, D. J. E. Taylor, and H. Andrews, "Severe Systemic Infections Complicating 'Mainline' Heroin Addiction," *Lancet*, ii (1967): 1227–1231; W. E. Dismuskes, A. W. Karchmer, R. F. Johnson, and W. J. Dougherty, "Viral Hepatitis Associates with Illicit Parenteral Use of Drugs," *Journal of the American Medical Association*, 206 (1968): 1048–1052; P. D. Harris, and R. Garret, "Susceptibility of Addicts to Infection and Neoplasia," *New England Journal of Medicine*, 287 (1972): 310; S. A. Geller and B. Stimmel, "Diagnostic Confusion from Lymphatic Lesion in Heroin Addicts," *Ann. Intern. Med.*, 78 (1973): 703–705; S. M. Brown, B. Stimmel, R. N. Taub, S. Kochwa, and R. E. Rosenfield, "Immunologic Dysfunction in Heroin Addicts," *Arch. Intern. Med.*, 134 (1974): 1001–1006; D. B. Louria, "Infectious Complication of Nonalcoholic Drug Abuse," *Annu. Rev. Med.*, 25 (1974): 219–231; R. J. McDonough, J. J. Madden, A. Falek, D. A. Shafer, M. Pline, D. Gordon, P. Bokof, J. C. Kueknle, and J. Mandelson, "Alternation of T and Null Lymphocyte Frequencies in the Peripheral Blood of Human Opiate Addicts: In Vivo Evidence of Opiate Receptor Sites on T Lymphocytes," *J. Immunol.*, 125 (1980): 2539–2543; J. Layon, A. Idris, M. Warzynski, R. Sherer, D. Brauner, O. Patch, D. McCulley, and P. Orris, "Altered T-Lymphocyte Subsets in Hospitalized Intravenous Drug Abusers," *Arch. Intern. Med.*, 144 (1984): 1376–1380; M. J. Kreek, "Methadone Maintenance Treatment for Harm Reduction Approach to Heroin Addiction," in *Drug Addiction and AIDS*,

eds. N. Loimer, R. Schmidt, and A. Springer (New York: Springer-Verlag, 1991), 153–178; G. H. C. Mientjes, E. J. C. van Ameijden, H. M. Weigel, J. A. R. van den Hoek, and R. A. Coutinho, "Clinical Symptoms Associated with Seroconversion for HIV-1 Among Misusers of Intravenous Drugs: Comparison with Homosexual Seroconverters and Infected and Non-Infected Intravenous Drug Misusers," *Br. Med J.*, 306 (1993): 371–373.
77. Duesberg, "AIDS Acquired by Drug Consumption," 201–277.
78. Cohen, "'Duesberg Phenomenon,'" 1642–1649.
79. Edlin, Irwin, Farouque, McCoy, Word, Serrano, Inciardi, Bowser, Schilling, Holmberg, and Multicenter Crack Cocaine and HIV infection Study Team, "Intersecting Epidemics," 1422–1427.
80. T. C. Cox, M. R. Jacobs, A. E. Leblanc, and J. A. Marshman, *Drugs and Drug Abuse* (Toronto, Canada: Addiction Research Foundation, 1983); Layon, Idris, Warzynski, Sherer, Brauner, Patch, McCulley and Orris, "Altered T-Lymphocyte Subsets in Hospitalized Intravenous Drug Abusers," 1376–1380; N. J. Kozel and E. H. Adams, "Epidemiology of Drug Abuse: An Overview," *Science*, 234 (1986): 970–974; W. D. Lerner, "Cocaine Abuse and Acquired Immunodeficiency Syndrome: A Tale of Two Epidemics," *American Journal of Medicine*, 87 (1989): 661–663; P. A. Selwyn, D. Hartel, W. Wasserman, and E. Drucker, "Impact of the AIDS Epidemic on Morbidity and Mortality Among Intravenous Drug Users in a New York City Methadone Maintenance Program," *American Journal of Public Health*, 79 (1989): 1358–1362; C. F. Turner, H. G. Miller, and L. E. Moses, *AIDS, Sexual Behavior, and Intravenous Drug Use* (Washington, D.C.: National Academy Press, 1989); R. Pillai, B. S. Nair, and R. R. Watson, "AIDS, Drugs of Abuse, and the Immune System: A Complex Immunotoxicological Network," *Arch. Toxico*, 65 (1991): 609–617; H. U. Bryant, K. A. Cunningham, and T. R. Jerrells, "Effects of Cocaine and Other Drugs of Abuse on Immune Responses," in *Cocaine: Pharmacology, Physiology, and Clinical Strategies*, eds. J. M. Lakoski, M. P. Galloway, and F. J. White (Boca Raton, Fla.: CRC Press, 1992), 353–369.
81. Lerner, "Cocaine Abuse," 661–663.
82. Ibid.
83. National Commission on AIDS, *The Twin Epidemics of Substance Use and HIV* (Washington, D.C.: National Commission of AIDS, July 1991).
84. R. T. Michael, J. H. Gagnon, E. O. Laumann, and G. Kolata, *Sex in America: A Definitive Survey* (Boston: Little Brown, 1994).
85. P. E. Larrat and S. Zierler, "Entangled Epidemics: Cocaine Use and HIV Disease," *Journal of Psychoactive Drugs*, 25 (1993): 207–221.
86. Edlin, Irwin, Farouque, McCoy, Word, Serrano, Inciardi, Bowser, Schilling, Holmberg, and Multicenter Crack Cocaine and HIV Infection Study Team, "Intersecting Epidemics," 1422–1427.

87. W. D. Lerner, "Cocaine Abuse and Acquired Immunodeficiency Syndrome: Tale of Two Epidemics," *American Journal of Medicine*, 87 (1989): 1376–1380; F. H. Gawin, "Cocaine Addiction: Psychology and Neurophysiology," *Science*, 251 (1991): 1580–1586.
88. "Cocaine Use Linked to Tuberculosis," *San Francisco Chronicle*, 26 July 1991, A 19; Craffey, "A Killer Returns," *San Francisco Examiner—Image*, 14 June 1992, 6–13.
89. J. H. Braun, C. G. McKerron, R. L. Souhami, D. J. E. Taylor, and H. Andrews, "Severe Systemic Infections Complicating 'Mainline' Heroin Addiction," *Lancet*, ii (1967): 1227–1231; K. Brudney and J. Dobkin, "Resurgent Tuberculosis in New York City," *Am. Rev. Respir. Dis.*, 144 (1991): 744–749.
90. A. S. Evans and H. A. Feldman, eds., *Bacterial Infections of Humans: Epidemiology and Control* (New York/London: Plenum Medical Book Company, 1982).
91. Gawin, "Cocaine Addiction," 1580–1586.
92. Cohen, "'Duesberg Phenomenon,'" 1642–1649.
93. H. S. Fricker and S. Segal, "Narcotic Addiction, Pregnancy, and the Newborn," *Am. J. Dis. Child*, 132 (1978): 360–366; M. H. Lifschitz, G. S. Wilson, E. O. Smith, and M. M. Desmond, "Fetal and Postnatal Growth of Children Born to Narcotic-Dependent Women," *J. Pediatricians*, 102 (1983): 686–691; L. G. Alroomi, J. Davidson, T. J. Evans, P. Galea, and R. Howat, "Maternal Narcotic Abuse and the Newborn," *Arch. Dis. Child*, 63 (1988): 81–83.
94. Layon, Idris, Warzynski, Sherer, Brauner, Patch, McCulley, and Orris, "Altered T-Lymphocyte Subsets in Hospitalized Intravenous Drug Abusers," 1376–1380; C. R. Schuster, foreword, in *Cocaine: Pharmacology, Effects and Treatment of Abuse*, NIDA Research Monograph 50, ed. J. Grabowski (Washington, D.C.: National Institute on Drug Abuse, 1984), VII–VIII; S. Savona, M. A. Nardi, E. T. Lenette, and S. Karpatkin, "Thrombocytopenic Purpura in Narcotics Addicts," *Ann. Intern. Med.*, 102 (1985): 737–741; R. M. Donahoe, C. Bueso-Ramos, F. Donahoe, J. J. Madden, A. Falek, J. K. A. Nicholson, and P. Bokos, "Mechanistic Implications of the Finding that Opiates and Other Drugs of Abuse Moderate T-Cell Surface Receptors and Antigenic Markers," *Ann. N.Y. Acad. Sci.*, 496 (1987): 711–721; P. Espinoza, I. Bouchard, C. Buffet, V. Thiers, J. Pillot, and J. P. Etienne, "High Prevalence of Infection by Hepatitis B Virus and HIV in Incarcerated French Drug Addicts," *Gastroenterologie Clinique et Biologique*, 11 (1987): 288–292; R. Weber, Lederberger, M. Opravil, W. Siegenthaler, and R. Lüthy, "Progression of HIV Infection in Misusers of Injected Drugs Who Stop Injecting or Follow a Programme of Maintenance Treatment with Methadone," *British Medical Journal*, 301 (1990): 1362–1365.
95. Institute of Medicine, *Confronting AIDS—Update 1988* (Washington, D.C.: National Academy Press, 1988).

96. A. Toufexis, "Innocent Victims," *Time*, 19 (1991), 56–60.
97. S. Blanche, C. Rouzioux, M. L. G. Moscato, F. Veber, M. J. Mayaux, C. Jacomet, J. Tricoire, A. Denville, M. Vial, G. Firtion, A. De Crepy, D. Douard, M. Robin, C. Courpotin, N. Ciran-Vineron, F. Le Deist, C. Griscelli, and the HIV Infection in Newborns French Collaborative Study Group, "A Prospective Study of Infants Born to Women Seropositive for Human Immunodeficiency Virus Type 1," *New England Journal of Medicine*, 320 (1989): 1643–1648; Rogers, Rayfield, Thomas, Schoenbaum, Abrams, Krasinski, Selwyn, Moore, Kaul, Grimm, Bamji, Schochetman, the New York City Collaborative Study of Maternal HIV Transmission, and Montefiori Medical Center HIV Perinatal Transmission Study Group, "Use of the Polymerase Chain Reaction," 1649–1654.
98. R. S. Stoneburner, D. C. Des Jarlais, D. Benezra, L. S. Gorelkin, J. L. Sothern, S. R. Friedman, S. Schultz, M. Marmor, D. Mildvan, and R. Maslansky, "A Larger Spectrum of Severe HIV-1–Related Diseases in Intravenous Drug Users in New York City," *Science*, 242 (1988): 916–919; A. Muñoz, D. Vlahov, L. Solomon, J. B. Margolick, J. C. Bareta, S. Cohn, J. Astemborski, and K. E. Nelson, "Prognostic Indicators for Development of AIDS Among Intravenous Drug Users," *J. Aquir. Immune Defic. Syndr.*, 5 (1992): 694–700; Duesberg, "HIV Gap"; Mientjes, van Ameijden, Weigel, van den Hoek, and Countinho, "Clinical Symptoms Associated with Seroconversion," 371–373.
99. D. Des Jarlais, S. Friedman, M. Marmor, H. Cohen, D. Mildvan, S. Yankovitz, U. Mathur, W. El-Sadr, T. J. Spira, and J. Garber, "Development of AIDS, HIV Seroconversion, and Potential Cofactors for T4 Cell Loss in a Cohort of Intravenous Drug Users," *AIDS*, 1 (1987): 105–111; D. Des Jarlais, S. R. Friedman, and W. Hopkins, "Risk Reduction of the Acquired Immunodeficiency Syndrome Among Intravenous Drug Users," in *AIDS and IV Drug Abusers: Current Perspectives*, eds. R. P. Galea, B. F. Lewis, and L. Baker (Owings Mills, Md.: National Health Publishing, 1988), 97–109; P. A. Selwyn, A. R. Feingold, D. Hartel, E. E. Schoenbaum, M. H. Adderman, R. S. Klein, and S. H. Friedland, "Increased Risk of Bacterial Pneumonia in HIV-Infected Intravenous Drug Users Without HIV," *AIDS*, 2 (1988): 267–272; Weber, Lederberger, Opravil, Siegenthaler, and Lüthy, "Progression of HIV Infection," 1362–1365; Duesberg, "AIDS Acquired by Drug Consumption," 201–277; Muñoz, Vlahov, Solomon, Margolick, Bareta, Cohn, Astemborski, and Nelson, "Prognostic Indicators," 694–700; Mientjes, van Ameijden, Weigel, van den Hoek, and Countinho, "Clinical Symptoms Associated with Seroconversion," 371–373.
100. A. Annell, A. Fugelstad, and G. Ågren, "HIV-Prevalence and Mortality in Relation to Type of Drug Abuse Among Drug Addicts in Stockholm 1981–1988," in *Drug Addiction and AIDS*, eds. N. Loimer, R. Schmid, and A. Springer (New York: Springer-Verlag, 1991), 16–22;

F. Bschor, R. Bornemann, C. Borowski, and V. Schneider, "Monitoring of HIV-Spread in Regional Populations of Injecting Drug Users—the Berlin Experience," in *Drug Addiction and AIDS*, eds. N. Loimer, R. Schmid, and A. Springer (New York: Springer-Verlag, 1991), 102–109; K. Püschel and F. Mohsenian, "HIV-1-Prevalence Among Drug Deaths in Germany," in *Drug Addiction and AIDS*, eds. N. Loimer, R. Schmid, and A. Springer (New York: Springer-Verlag, 1991), 89–96; U. Lockemann, F. Wischhusen, K. Püschel et al., "Vergleich der HIV-1-Praevalenz bei Drogentodesfaellen in Deutschland sowie in verschiedenen europaeischen Grosstaedten," *AIDS-Forschung*, 10 (1995): 253–256.
101. Stoneburner, Des Jarlais, Benezra, Gorelkin, Sotheran, Friedman, Schultz, Marmor, Mildvan, and Maslansky, "Larger Spectrum of Severe HIV-1–Related Disease," 916–919.
102. Selwyn, Feingold, Hartel, Schoenbaum, Adderman, Klein, and Friedland, "Increased Risk of Bacterial Pneumonia," 267–272.
103. M. M. Braun, B. I. Truman, B. Maguire, G. T. Di Ferdinando Jr., G. Wormser, R. Broaddus, and D. L. Morse, "Increasing Incidence of Tuberculosis in a Prison Inmate Population, Association With HIV-Infection," *Journal of the American Medical Association*, 261 (1989): 393–397.
104. Donahoe, Bueso-Ramos, Donahoe, Madden, Falek, Nicholson, and Bokos, "Mechanistic Implications," 711–721.
105. Savona, Nardi, Lenette, and Karpatkin, "Thrombocytopenic Purpura," 737–741.
106. Annell, Fugelstad, and Ågren, "HIV-Prevalence and Mortality," 16–22.
107. G. H. C. Mientjes, F. Miedema, E. J. van Ameijden, A. A. van den Hoek, P. T. A. Schellekens, M. T. Roos, and R. A. Coutinho, "Frequent Injecting Impairs Lymphocyte Reactivity in HIV-Positive and HIV-Negative Drug Users, *AIDS*, 5 (1991): 35–41.
108. Des Jarlais, Friedman, and Hopkins, "Risk Reduction," 97–109.
109. Espinoza, Bouchard, Buffet, Thiers, Pillot, and Etienne, "High Prevalence of Infection," 288–292.
110. Brudney and Dobkin, "Resurgent Tuberculosis," 744–749.
111. S. H. Weiss, C. Weston Klein, R. K. Mayur, J. Besra, and T. N. Denny, "Idiopathic CD4+ T-Lymphocytopenia," *Lancet*, 340 (1992): 608–609.
112. Bschor, Bornemann, Borowski, and Schneider, "Monitoring HIV-Spread," 102-109; Püschel and Mohsenian, "HIV-1-Prevalence," 89–96.
113. Lockemann, Wischhusen, Püschel et al., "Vergleich der HIV-1-Praevalenz," 253–256.
114. G. Amendt, *Sucht, Profit, Sucht* (Hamburg, Germany: Rowwohlt Taschenbuch Verlag GmbH), 299–324; Weber, Lederberger, Opravil, Siegenthaler, and Lüthy, "Progression of HIV Infection," 1362–1365.

115. A. Weil and W. Rosen, *Chocolate and Morphine* (Boston: Houghton Mifflin Co., 1983).
116. R. Shilts, *And the Band Played On*, (New York: St. Martin's Press, 1987).
117. Sadownick, "Kneeling at the Crystal Cathedral," 40–45, 86–90.
118. Ibid.
119. Lauritsen, *AIDS War*, 197.
120. R. S. Root-Bernstein, *Rethinking AIDS: The Tragic Cost of Premature Consensus* (New York: The Free Press, 1993), 228.
121. Coulter, *AIDS and Syphilis*, 86.
122. Root-Bernstein, "Do We Know the Cause(s) of AIDS?" 484.
123. Coulter, *AIDS and Syphilis*, 47, 86–87.
124. Lauritsen, *AIDS War*, 199; see also chapter 1.
125. Merck Research Laboratories, *The Merck Manual of Diagnosis and Therapy* (Rahway, N.J.: Merck & Co., 1992).
126. S. Stolberg, "AIDS Tally to Increase Due to New Definition," *Los Angeles Times*, 31 Dec. 1992, A3, A29; see also chapters 9 and 10.
127. Duesberg, "AIDS Acquired by Drug Consumption," 214.
128. P. H. Duesberg, "Foreign-Protein-Mediated Immunodeficiency in Hemophiliacs With and Without HIV," *Genetica*, 95 (1995): 51–70.
129. Ibid.
130. P. H. Duesberg, "Is HIV the Cause of AIDS?" *Lancet*, 346 (1995): 1371–1372.
131. F. I. D. Konotey-Ahulu, "AIDS in Africa: Misinformation and Disinformation," *Lancet* (25 July 1987): 206–207.
132. C. Farber, "Out of Africa, Part 1," *Spin*, March 1993, 61–63, 86–87.
133. N. Hodgkinson, "African AIDS Plague 'a Myth,'" *Sunday Times of London*, 3 Oct. 1993; N. Hodgkinson, "The Plague that Never Was," *Sunday Times of London*, 3 Oct. 1993.
134. Ibid.
135. A. O. Williams, *AIDS: An African Perspective* (Boca Raton, Fla.: CRC Press, 1992), 238.
136. "AIDS and Africa" (Meditel, 24 March 1993).
137. Konotey-Ahulu, *What Is AIDS?* 56–57.
138. Ibid.
139. Ibid., 119.
140. Duesberg, "AIDS Acquired by Drug Consumption."
141. Farber, "Out of Africa, Part 1."
142. "AIDS and Africa."
143. Duesberg, "AIDS Acquired by Drug Consumption."
144. Farber, "Out of Africa, Part 1."

CHAPTER 9

1. *Physician's Desk Reference*, 1993.
2. K. Belani, medical report on Lindsey Nagel, Park Nicollet Medical Center, 15 Feb. 1991.
3. D. Chiu and P. H. Duesberg, "The Toxicity of Azidothymidine (AZT) on Human and Animal Cells in Culture at Concentrations Used for Antiviral Therapy," *Genetica*, 95 (1995): 103–109; P. H. Duesberg, "AIDS Acquired by Drug Consumption and Other Noncontagious Risk Factors," *Pharmacology and Therapeutics*, 55 (1992): 201–277; R. Yarchoan, J. M. Pluda, C.-F. Perno, H. Mitsuya, and S. Broder, "Anti-Retroviral Therapy of Human Immunodeficiency Virus Infection: Current Strategies and Challenges for the Future," *Blood*, 78 (1991): 859–884; G. X. McLeod and S. M. Hammer, "Zidovudine: Five Years Later," *Ann. Intern. Med.*, 117 (1992): 487–501.
4. K. Belani, medical report on Lindsey Nagel, Park Nicollet Medical Center, 14 March 1991.
5. M. Hostetter, letter to Joseph McHugh, 13 Nov. 1991.
6. R. Yogev and E. Conner, "Management of HIV Infection in Infants and Children," *Mosby Year Book* (St. Louis, 1992).
7. *Physician's Desk Reference*; Merck Research Laboratories, *The Merck Manual of Diagnosis and Therapy* (Rahway, N.J.: Merck & Co., Inc., 1992); M. C. Dalakas, I. Illa, G. H. Pezeshkpour, J. P. Laukaitis, B. Cohen, and J. L. Griffin, "Mitochondrial Myopathy Caused by Long-Term Zidovudine Therapy," *New England Journal of Medicine*, 322 (1990): 1098–1105; R. J. Lane, K. A. McLean, J. Moss, and D. F. Woodrow, "Myopathy in HIV Infection: The Role of Zidovudine and the Significance of Tuburoreticular Inclusions," *Neuropathy and Applied Neurobiology*, 19 (1993): 406–413; Duesberg, "AIDS Acquired by Drug Consumption."
8. S. Nagel and C. Nagel, appeal to the Minnesota Board of Medical Practice, 31 March 1993.
9. Ibid.
10. M. Hostetter, letter to Joseph McHugh, 11 Dec. 1992.
11. M. Hostetter, letter to Mr. and Mrs. Steve Nagel, 29 Dec. 1992.
12. Nagel and Nagel, appeal.
13. J. Parsons and K. Chandler, "Girl in Family Stricken with AIDS Virus Dies at Age 5," *Minneapolis Star/Tribune*, 28 June 1993.
14. Ibid.
15. Ibid.
16. M. Samter, D. W. Talmage, M. M. Frank, K. F. Austen, and H. N. Calman, eds., *Immunological Diseases*, I (Boston/Toronto: Little, Brown and Company, 1988); Merck Index; L. S. Young, *Pneumocystis Carinii Pneumonia* (New York/Basel: Marcel Dekker, Inc., 1984).
17. B. Nussbaum, *Good Intentions: How Big Business, Politics, and*

Medicine Are Corrupting the Fight Against AIDS (New York: Atlantic Monthly Press, 1990).
18. S. S. Cohen, "Antiretroviral Therapy for AIDS," *New England Journal of Medicine*, 317 (1987): 629; J. Lauritsen, *Poison by Prescription—The AZT Story* (New York: Asklepios Press, 1990).
19. Nussbaum, *Good Intentions*.
20. Ibid., 23–41.
21. P. A. Furman, J. A. Fyfe, M. St. Clair, K. Weinhold, J. L. Rideout, G. A. Freeman, S. Nusinoff-Lehrman, D. P. Bolognesi, S. Broder, H. Mitsuya, and D. W. Barry, "Phosphorylation of 3'-azido-3'-deoxythymidine and Selective Interaction of the 5'-triphosphate with Human Immunodeficiency Virus Reverse Transcriptase," *Proceedings of the National Academy of Sciences*, 83 (1986): 8333–8337; P. H. Duesberg, "HIV, AIDS, and Zidovudine," *Lancet*, 339 (1992): 805–806.
22. Nussbaum, *Good Intentions*.
23. Duesberg, "AIDS Acquired by Drug Consumption," 201–277; Chiu and Duesberg, "Toxicity of Azidothymidine," 103–109.
24. Nussbaum, *Good Intentions*; E. A. Wyatt, "Rushing to Judgment," *Barron's*, 15 Aug. 1994, 23–27.
25. M. A. Fischl, D. D. Richman, M. H. Grieco, M. S. Gottlieb, P. A. Volberding, O. L. Laskin, J. M. Leedon, J. E. Groopman, D. Mildvan, R. T. Schooley, G. G. Jackson, D. T. Durack, D. King, and the AZT Collaborative Working Group, "The Efficacy of Azidothymidine (AZT) in the Treatment of Patients with AIDS and AIDS-Related Complex," *New England Journal of Medicine*, 317 (1987): 185–191.
26. D. D. Richman, M. A. Fischl, M. H. Grieco, M. S. Gottlieb, P. A. Volberding, O. L. Lasking, J. M.Leedon, J. E. Groopman, D. Mildvan, M. S. Hirsch, G. G. Jackson, D. T. Durack, S. Nusinoff-Lehrman, and the AZT Collaborative Working Group, "The Toxicity of Azidothymidine (AZT) in the Treatment of Patients with AIDS and AIDS-Related Complex," *New England Journal of Medicine*, 317 (1987): 192–197; Duesberg, "AIDS Acquired by Drug Consumption"; Duesberg, "HIV, AIDS, and Zidovudine."
27. E. Burkett, "The Queen of AZT," *Miami Herald Tropic*, 23 Sept. 1990, 8–14.
28. Duesberg, "AIDS Acquired by Drug Consumption."
29. Burkett, "Queen of AZT."
30. Nussbaum, *Good Intentions*.
31. "AZT—Cause for Concern," *Dispatches*, Channel Four Television (London: Meditel, 12 Feb. 1992).
32. Ibid.
33. Duesberg, "AIDS Acquired by Drug Consumption."
34. Burkett, "Queen of AZT."
35. Fischl, Richman, Grieco, Gottlieb, Volberding, Laskin, Leedon, Groopman, Mildvan, Schooley, Jackson, Durack, King, and the AZT Collaborative Working Group, "Efficacy of Azidothymidine," 185–191.

36. Burkett, "Queen of AZT"; J. Lauritsen, *The AIDS War: Propaganda, Profiteering, and Genocide from the Medical-Industrial Complex* (New York: Asklepios, 1993).
37. Lauritsen, *Poison by Prescription*, 34.
38. Ibid.
39. Duesberg, "AIDS Acquired by Drug Consumption."
40. M. Seligmann, D. A. Warrell, J.-P. Aboulker, C. Carbon, J. H. Darbyshire, J. Dormont, E. Eschwege, D. J. Girling, D. R. James, J.-P. Levy, P. T. A. Peto, D. Schwarz, A. B. Stone, I. V. D. Weller, R. Withnall, K. Gelmon, E. Lafon, A. M. Swart, V. R. Aber, A. G. Babiker, S. Ihoro, A. J. Nunn, and M. Vray, "Concorde: MRC/ANRS Randomized Double-Blind Controlled Trial of Immediate and Deferred Zidovudine in Symptom-Free HIV Infection," *Lancet* 343 (1994): 871–881; Duesberg, "AIDS Acquired by Drug Consumption."
41. Cohen, "Antiretroviral Therapy," 629.
42. L. Garrett, "AIDS: The Next Decade," *New York Newsday*, 12 June 1990, 1, 5.
43. G. Kolata, "Imminent Marketing of AZT Raises Problems," *Science*, 235 (1987): 1462–1463.
44. Presidential Commission on the HIV Epidemic, hearings, 19 Feb. 1988, 28.
45. Lauritsen, *Poison by Prescription*, 27.
46. Nussbaum, *Good Intentions*.
47. Nussbaum, *Good Intentions*; Lauritsen, *Poison by Prescription*.
48. Nussbaum, *Good Intentions*, 173.
49. Ibid.
50. Ibid.
51. Sigma Co., Zidovudine product label.
52. Wyatt, "Rushing to Judgment"; Lauritsen, *AIDS War*; M. Chase, "DDI Decision Heralds a New FDA Activism," *Wall Street Journal*, 22 July 1991.
53. Merck Index.
54. Lauritsen, *AIDS War*.
55. Ibid.
56. L. K. Altman, "Experts to Review AZT Role as the Chief Drug for H.I.V.," *New York Times*, 17 Sept. 1995, 38.
57. E. Dournon, S. Matheron, W. Rozenbaum, S. Gharakhanian, C. Michon, P. M. Girard, C. Perrone, D. Salmon, P. DeTruchis, C. Leport, and the Claude Bernard Hospital AZT Study Group, "Effects of Zidovudine in 365 Consecutive Patients with AIDS or AIDS-Related Complex," *Lancet*, ii (1988): 1297–1302.
58. Duesberg, "AIDS Acquired by Drug Consumption."
59. J. M. Pluda, R. Yarchoan, E. S. Jaffe, I. M. Feuerstein, D. Solomon, S. Steinberg, K. M. Wyvill, A. Rabitschek, D. Katz, and S. Broder, "Development of Non-Hodgkin Lymphoma in a Cohort of Patients

with Severe Immunodeficiency Virus (HIV) Infection on Long-Term Antiretroviral Therapy," *Ann. Intern. Med.*, 113 (1990): 276–282.
60. Ibid.
61. "AZT—Cause for Concern."
62. Centers for Disease Control and Prevention, "U.S. HIV and AIDS Cases Reported Through December 1993; Year-End Edition," *HIV/AIDS Surveillance Report*, 5 (1994), 1–33.
63. Cohen, "Antiretroviral Therapy," 629; H. I. Chernov, *Document on New Drug Application 19-655* (Washington, D.C.: Food and Drug Administration, 1986), cited in Lauritsen, *Poison by Prescription*.
64. P. S. Gill, M. Rarick, R. K. Byrnes, D. Causey, C. Loureiro, and A. M. Levine, "Azidothymidine Associated with Bone Marrow Failure in the Acquired Immunodeficiency Syndrome (AIDS)," *Ann. Intern. Med.*, 107 (1987): 502–505.
65. M. Till and K. B. MacDonnell, "Myopathy with Human Immuodeficiency Virus Type 1 (HIV-1) Infection: HIV-1 or Zidovudine?" *Ann. Intern. Med.*, 113 (1990): 492–494.
66. M. Scolaro, R. Durhan, and G. Pieczenik, "Potential Molecular Competitor for HIV," *Lancet*, 337 (1991): 731–732.
67. Duesberg, "AIDS Acquired by Drug Consumption."
68. M. D. Hughes, D. S. Stein, H. M. Gundacker, F. T. Valentine, J. P. Phair, and P. A. Volberding, "Within-Subject Variation in CD4 Lymphocyte Count in Asymptomatic Human Immunodeficiency Virus Infection: Implications for Patient Monitoring," *Journal of Infectious Diseases*, 169 (1994) :28–36.
69. P. A. Volberding, S. W. Lagakos, J. M. Grimes, D. S. Stein, J. Rooney, T.-C. Meng, M. A. Fischl, A. C. Collier, J. P. Phair, M. J. Hirsch, W. D. Hardy, H. H. Balfour, R. C. Reichman, and the AIDS Clinical Trials Group, "A Comparison of Immediate with Deferred Zidovudine Therapy for Asymptomatic HIV-Infected Adults with CD4 Cell Counts of 500 or More per Cubic Millimeter," *New England Journal of Medicine*, 333 (1995): 401–407.
70. N. Ostrom, "Early Intervention: An Idea Whose Time Has Gone?" *New York Native*, 28 Aug. 1995, 35–39.
71. J. D. Hamilton, P. M. Hartigan, M. S. Simberkoff, P. L. Day, G. R. Diamond, G. M. Dickinson, G. L. Drusano, M. J. Egorin, W. L. George, F. M. Gordin, and the Veterans Affairs Cooperative Study Group on AIDS Treatment, "A Controlled Trial of Early Versus Late Treatment with Zidovudine in Symptomatic Human Immunodeficiency Virus Infection, *New England Journal of Medicine*, 326 (1992): 437–443.
72. "AZT—Cause for Concern."
73. Seligmann, Warrell, Aboulker, Carbon, Darbyshire, Dormont, Eschwege, Girling, James, Levy, Peto, Schwarz, Stone, Weller, Withnall, Gelmon, Lafon, Swart, Aber, Babiker, Ihoro, Nunn, and Vray, "Concorde: MRC/ANRS Randomized Double-Blind Controlled Trial."

74. N. Hodgkinson, "The Cure that Failed," *Sunday Times of London*, 4 April 1993.
75. L. K. Altman, "Treatment Guidelines for HIV Amended," *San Francisco Chronicle*, 28 June 1993.
76. H. Bacellar, A. Muñoz, E. N. Miller, E. A. Cohen, D. Besley, O. A. Selnes, J. T. Becker, and J. C. McArthur, "Temporal Trends in the Incidence of HIV-1-Related Neurological Diseases: Multicenter AIDS Cohort Study: 1985-1992," *Neurology*, 44 (1994): 1892-1900.
77. J. J. Goedert, A. R. Cohen, C. M. Kessler, S. Eichinger, S. V. Seremetis, C. S. Rabkin, F. J. Yellin, P. S. Rosenberg, and L. M. Aledort, "Risks of Immunodeficiency, AIDS, and Death Related to Purity of Factor VIII Concentrate," *Lancet*, 344 (1994): 791-792.
78. "Head of National Cancer Institute to Quit," *San Francisco Chronicle*, 23 Dec. 1994, A12.
79. A. J. Saah, D. R. Hoover, Y. Peng, J. P. Phair, B. Visscher, L. A. Kingsley, L. K. Schrager, and the Multicenter AIDS Cohort Study, "Predictors for Failure of Pneumocystis carinii pneumonia prophylaxis," *Journal of the American Medical Association*, 273 (1995): 1197-1202.
80. K. Klinger, "Early AIDS Treatment Questioned," United Press International, 13 July 1995; M. C. Poznansky, R. Coker, C. Skinner, A. Hill, S. Bailey, L. Whitaker, A. Renton, and J. Weber, "HIV Positive Patients First Presenting with an AIDS Defining Illness: Characteristics and Survival," *British Medical Journal*, 311 (1995): 156-158.
81. C. Farber, "AIDS: Words from the Front," *Spin*, 11 Sept. 1995, 103-104.
82. ACT UP, press release, San Francisco, 1995.
83. T. Hand, "Forced H.I.V. Testing Won't Help Newborns; Flaws in AZT Studies" (letter), *New York Times*, 22 July 1995, 18.
84. Duesberg, "AIDS Acquired by Drug Consumption."
85. Ibid.
86. National Institutes of Health, Office of AIDS Research and Division of Safety, "HIV Safety Notice," 23 Feb. 1989.
87. "HIV Seroconversion After Occupational Exposure Despite Early Prophylactic Zidovudine Therapy," *Lancet*, 341 (1993): 1077-1078.
88. J. I. Tokars, R. Marcus, D. H. Culver, C. A. Schable, P. S. McKibben, C. I. Bandea, and D. M. Bell, "Surveillance of HIV Infection and Zidovudine Use Among Health Care Workers After Occupational Exposure to HIV-infected Blood," *Ann. Intern. Med.*, 118 (1993): 913-919.
89. "Zidovudine for Mother, Fetus, and Child: Hope or Poison?" (editorial), *Lancet*, 344 (1994): 207-209.
90. Farber, "AIDS—Words from the Front," 189-193, 214-215.
91. "Zidovudine for Mother."
92. E. M. Connor, R. S. Sperling, R. Gelber, P. Kiselev, G. Scott, M. J. O'Sullivan, R. VanDyke, M. Bey, W. Shearer, R. L. Jacobson, E.

Jimeniz, E. O'Neill, B. Bazin, F.-F. Delfraissy, M. Culnane, R. Coombs, M. Elkins, J. Moye, P. Stratton, J. Balsley, and the Pediatric AIDS Clinical Trials Group Protocol 076 Study Group, "Reduction of Maternal-Infant Transmission of Human Immunodeficiency Virus Type 1 with Zidovudine Treatment," *New England Journal of Medicine*, 331 (1994): 1173–1180; P. Cotton, "Trial Halted After Drug Cuts Maternal HIV Transmission Rate by Two Thirds," *Journal of the American Medical Association*, 271 (1994): 807.

93. Connor, Sperling, Gelber, Kiselev, Scott, O'Sullivan, VanDyke, Bey, Shearer, Jacobson, Jimeniz, O'Neill, Bazin, Delfraissy, Culnane, Coombs, Elkins, Moye, Stratton, Balsley, and the Pediatric AIDS Clinical Trials Group Protocol 076 Study Group, "Reduction of Maternal-Infant Transmission"; Cotton, "Trial Halted"; "Zidovudine for Mother."
94. Cotton, "Trial Halted."
95. Duesberg, "AIDS Acquired by Drug Consumption"; C. A. Thomas Jr., K. B. Mullis, and P. E. Johnson, "What Causes AIDS?" *Reason*, 26 (June 1994): 18–23; P. H. Duesberg, "How Much Longer Can We Afford the AIDS Virus Monopoly?" in *AIDS: Virus or Drug-Induced?* eds. Kluwer and Dordrecht (The Netherlands: Genetica, in press).
96. A. Caplan, "Just as Simple as AZT," *San Diego Union-Tribune*, 8 Nov. 1994, B-7; Cotton, "Trial Halted," 807.
97. Connor, Sperling, Gelber, Kiselev, Scott, O'Sullivan, VanDyke, Bey, Shearer, Jacobson, Jimeniz, O'Neill, Bazin, Delfraissy, Culnane, Coombs, Elkins, Moye, Stratton, Balsley, and the Pediatric AIDS Clinical Trials Group Protocol 076 Study Group, "Reduction of Maternal-Infant Transmission."
98. Cotton, "Trial Halted."
99. Connor, Sperling, Gelber, Kiselev, Scott, O'Sullivan, VanDyke, Bey, Shearer, Jacobson, Jimeniz, O'Neill, Bazin, Delfraissy, Culnane, Coombs, Elkins, Moye, Stratton, Balsley, and the Pediatric AIDS Clinical Trials Group Protocol 076 Study Group, "Reduction of Maternal-Infant Transmission"; Caplan, "Just as Simple as AZT."
100. "Zidovudine for Mother."
101. Farber, "AIDS—Words from the Front."
102. N. Ostrom, "Nightmare on AZT Street," *New York Native*, 4 Sept. 1995, 34–36.
103. R. M. Kumar, P. F. Hughes, and A. Khurranna, "Zidovudine Use in Pregnancy: A Report on 104 Cases and the Occurrence of Birth Defects," *Journal of Acquired Immune Deficiency Syndromes*, 7 (1994): 1034–1039.
104. B. Gavzer, "What We Can Learn from Those Who Survive AIDS," *Parade*, 10 June 1990, 4–7.
105. J. Wells, "We Have to Question the So-Called 'Facts,'" *Capital Gay*, 20 Aug. 1993, 14–15.
106. J. Wells, personal communication, London.

107. J. Nelson, J. Rodack, R. Fitz, and A. B. Smith, "Magic Reeling as Worst Nightmare Comes True—He's Getting Sicker," *National Enquirer*, 10 Dec. 1991, 6.
108. Paul Philpott, personal communication, Tallahassee, Fla.
109. Anderson, *Independent Mail*, 16 Feb. 1986; *Boston Globe* and *Washington Post* (without dates).
110. Duesberg, "AIDS Acquired by Drug Consumption."
111. Lauritsen, *AIDS War*, 324.
112. J. Weisberg, "The Accuser: Kimberly Bergalis, AIDS Martyr," *New Republic*, 21 Oct. 1991, 12–14.
113. M. Rom, "Health-Care Workers and HIV: Policy Choice in a Federal System." *Publius*, Summer 1993, 135–153.
114. B. Hilton, "CDC Accused of Alarming the Public," *San Francisco Examiner*, 10 Feb. 1991, B5.
115. "AIDS Victim's Will Is a Moving 'Good-By,'" *San Francisco Chronicle*, 15 Jan. 1992.
116. Duesberg, "AIDS Acquired by Drug Consumption"; J. Palca, "The Case of the Florida Dentist," *Science*, 255 (1992): 392–394; "No Trial to Come in Florida Dentist Case," *Science*, 255 (14 Feb. 1992), 787.
117. "Study Questions Whether Dentist Spread AIDS," *Orange County Register*, 25 Feb. 1993, A14, describing a study by Ronald DeBry and his colleagues published in the 2-25-93 issue of *Nature*.
118. S. Barr, "The Flawed Case Against Dr. Acer—In defense of AIDS Dentist," *Miami Herald*, 31 March 1994a; S. Barr, "In Defense of the AIDS Dentist," *Lear's*, 2 April 1994b: 68–82; S. Barr, "What If the Dentist Didn't Do It?" *New York Times*, 16 April 1994c, 21.
119. Barr, "What If"; Barr, "In Defense."
120. Barr, "What If."
121. Palca, "Case of the Florida Dentist," 392–394; Barr, "In Defense."
122. Barr, "What If."
123. Barr, "In Defense."
124. K. McMurran and M. Neil, "One Woman's Brave Battle with AIDS," *People*, July 1990, 62–65.
125. Ibid.
126. *MacNeil-Lehrer News Hour*, 10 June 1993.
127. A. Ashe and A. Rampersad, *Days of Grace: A Memoir* (New York: Alfred A. Knopf, 1993), 214.
128. E. Caldwell, *New York Daily News*, 10 Feb. 1993.
129. A. Ashe, "More Than Ever, Magical Things to Learn," *Washington Post*, 11 Oct. 1992.
130. Caldwell, *New York Daily News*.
131. Ashe and Rampersad, *Days of Grace*.
132. Ibid., 213.

CHAPTER 10

1. H. Rubin and G. Stent, National Academy of Sciences members, personal communications with Peter Duesberg.
2. D. Baltimore, letter to Prosper Graf zu Castell-Castell, 9 Sept. 1986.
3. B. Witkop, Ehrlich committee member, personal communication.
4. Institute of Medicine and National Academy of Sciences, *Confronting AIDS: Directions for Public Health* (Washington, D.C.: National Academy Press, 1986), vi.
5. Ibid., 177.
6. Institute of Medicine, *Confronting AIDS—Update 1988* (Washington, D.C.: National Academy Press, 1988)
7. Institute of Medicine and National Academy of Sciences, *Confronting AIDS*, 177.
8. Ibid., 172.
9. Ibid.
10. Ibid.
11. T. McKeown, *The Role of Medicine: Dream, Mirage, or Nemesis?* (Princeton, N.J.: Princeton University Press, 1979).
12. Institute of Medicine and National Academy of Sciences, *Confronting AIDS*, 125.
13. Ibid., 124.
14. Ibid., 130.
15. D. Francis, "Toward a Comprehensive HIV Prevention Program for the CDC and the Nation," *Journal of the American Medical Association*, 268 (1993): 1444–1447.
16. Ibid., 1447.
17. Institute of Medicine and National Academy of Sciences, *Confronting AIDS*, 266.
18. P. Brés, *Public Health Action in Emergencies Caused by Epidemics: A Practical Guide* (Geneva: World Health Organization, 1986), 111–112.
19. "New U.S. AIDS Study Assailed by Activists," *San Francisco Chronicle*, 8 Feb. 1993.
20. Centers for Disease Control, "A Comprehensive HIV Prevention Program," *Public Health Reports* 106 (1991): 699.
21. Ibid., 696.
22. M. Callen, *Surviving AIDS* (New York: Harper Perennial, 1990), 61–62.
23. P. H. Duesberg, "Foreign-Protein-Mediated Immunodeficiency in Hemophiliacs With and Without HIV," *Genetica*, 95 (1995); P. H. Duesberg, "Is HIV the Cause of AIDS?" *Lancet*, 346 (1995): 1371–1372.
24. National Association for People With AIDS, *NAPWA Brochure*, 1993.
25. Centers for Disease Control, "Comprehensive HIV Prevention Program," 676.

26. M. Delaney, P. Goldblum, and J. Brewer, *Strategies for Survival: A Gay Man's Health Manual for the Age of AIDS* (New York: St. Martin's Press, 1987), 32.
27. J. Lauritsen, *The AIDS War: Propaganda, Profiteering, and Genocide from the Medical-Industrial Complex* (New York: Asklepios, 1993).
28. M. Delaney, "Making Regulatory and Research Programs Work," speech at Stanford University School of Medicine (11 Jan. 1990).
29. M. Delaney, "'The Duesberg Phenomenon': Duesberg and Other Voices" (letter), *Science*, 267 (1995): 314.
30. M. Delaney, "Apology," *Science*, 268 (1995): 17.
31. Lauritsen, *AIDS War*, 148.
32. Ibid., 441.
33. J. Lauritsen, "Dissent at the Berlin AIDS Conference," *Rethinking AIDS*, 7 July 1993: 1–2.
34. Lauritsen, *AIDS War*, 441.
35. Ibid., 442.
36. Lauritsen, "Dissent at the Berlin AIDS Conference," 1–2.
37. Centers for Disease Control, "Comprehensive HIV Prevention Program," 673.
38. J. Cohen, "Is a New Virus the Cause of KS?" *Science*, 266 (1994): 1803–1804.
39. P. H. Duesberg, "'The Duesberg Phenomenon': Duesberg and Other Voices" (letter) *Science*, 267 (1995): 313.
40. J. L. Jones, D. L. Hanson, S. Y. Chu, J. E. Ward, and H. W. Jaffe, "AIDS-Associated Kaposi's Sarcoma" (letter), *Science*, 267 (1995): 1077–1078.
41. Ibid.
42. Y. Chang and P. Moore, "AIDS-Associated Kaposi's Sarcoma" (letter), *Science*, 267 (1995): 1078.
43. A. S. Fauci, "CD4+ T-Lymphocytopenia Without HIV Infection—No Lights, No Camera, Just Facts," *New England Journal of Medicine*, 328 (1993): 429–443.
44. J. Cohen, "Doing Science in the Spotlight's Glare," *Science*, 257 (1992): 1033.
45. S. Lang (Yale University) and M. Cochrane (University of California at Berkeley), personal communication.
46. "Doctor Afraid to Speak Out on KS," *Continuum*, 3 (November/December 1995): 3.
47. L. M. Krieger, "An AIDS Reporter's Open Letter to the Community," *San Francisco Chronicle*, 21 March 1993, D2.
48. Ibid.
49. N. Ostrom, "*New York Times* Inserts Propaganda in Letter," *New York Native*, 8 Nov. 1993, 5.
50. "Hunting the Virus Hunter—Q&A: Part 2" (interview with P. H. Duesberg and E. Burkett), *Tony Brown's Journal*, 1410 (June 1991), PBS.
51. W. Booth, "A Rebel Without a Cause of AIDS," *Science*, 239 (1983):

1485–1488; D. Baltimore and M. Feinberg, "Quantification of Human Immunodeficiency Virus in the Blood," *New England Journal of Medicine*, 322 (1990): 1468–1469.
52. Presidential Commission on the HIV Epidemic, hearings (20 Feb. 1988), 85.
53. M. Specter, "Panel Rebuts Biologist's Claims on Cause of AIDS," *Washington Post*, 10 April 1988.
54. A. S. Fauci, "Writing for My Sister Denise," *AAAS Observer*, 1 Sept. 1989, 4.
55. L. Chieco-Bianchi and G. B. Rossi, letter to the editor, *Nature*, 364 (1993): 96.
56. Health and Human Services, memorandum of the Office of the Secretary, 28 April 1987 (C. Kline, Media Alert).
57. F. S. Karlsberg, memorandum to Peter Fischinger, Howard Streicher, William Blattner, and Robert Gallo, National Institutes of Health, 28 April 1987.
58. W. A. Blattner, memorandum to Deputy Director, NCI (National Cancer Institute, 30 April 1987). See also F. Wong-Staal, memorandum to Peter Fischinger, Deputy Director, NCI (National Institutes of Health, 1 May 1987); and National Cancer Institute, Office of Cancer Communications, statement, June 1987.
59. F. S. Karlsberg, memorandum to Linda re: NCI response to Peter Duesberg, National Cancer Institute, 30 Dec. 1987. On the bottom is a handwritten note to Fischinger, dated January 7, 1988, and another handwritten note from "PVN" (Paul van Nevel) to "Eleanor" and "Pat."
60. L. Chieco-Bianci and G. B. Rossi, "Duesberg Rights and Wrongs," *Nature* (London), 364 (1993): 96.
61. B. Ellison, personal communication.
62. W. Hunter, letter to Dr. David Schryer, 20 Nov. 1992.
63. M. Singer, letter to Peter Duesberg, 30 June 1988.
64. I. Dawid, letter to Peter Duesberg, 20 July 1988.
65. P. H. Duesberg, "Human Immunodeficiency Virus and Acquired Immunodeficiency Syndrome: Correlation but Not Causation," *Proceedings of the National Academy of Sciences*, 86 (1989): 755–764.
66. I. Dawid, letter to Peter Duesberg, 12 Feb. 1991.
67. L. Bogorad, letter to Peter Duesberg, 13 May 1991.
68. J. Cole, letter to Peter Duesberg, 22 Oct. 1990.
69. M. S. Ascher, H. W. Sheppard, W. Winkelstein Jr., and E. Vittinghoff, "Does Drug Use Cause AIDS?" *Nature* (London), 362 (1993): 103–104.
70. B. J. Ellison, A. B. Downey, and P. H. Duesberg, "HIV as a Surrogate Marker for Drug Use: A Re-analysis of the San Francisco Men's Health Study," *Genetica*, 95 (1995): 165–171.
71. M. S. Ascher, H. W. Sheppard, and W. Winkelstein Jr., "AIDS-Associated Kaposi's Sarcoma" (letter), *Science*, 267 (1995): 1080.

72. Ellison, Downey, and Duesberg, "HIV as a Surrogate Marker."
73. Ascher, Sheppard, Winkelstein, and Vittinghoff, "Does Drug Use Cause AIDS?"; Ascher, Sheppard, and Winkelstein, "AIDS-Associated Kaposi's Sarcoma."
74. P. H. Duesberg, "HIV and the Aetiology of AIDS," *Lancet* 341 (1993): 957–958; P. H. Duesberg, "Can Epidemiology Determine Whether Drugs or HIV Cause AIDS?" *AIDS-Forschung*, 12 (1993): 627–635.
75. J. Maddox, "Has Duesberg a Right of Reply?" *Nature*, 363 (1993): 109.
76. Duesberg, "HIV and the Aetiology of AIDS."
77. Ellison, Downey, and Duesberg, "HIV as a Surrogate Marker."
78. P. H. Duesberg, "AIDS Data," *Science*, 268 (1995): 350–351.
79. G. Pantello, C. Graziosi, J. F. Demarest, L. Butini, M. Montroni, C. H. Fox, J. M. Orenstein, D. P. Kotler, and A. S. Fauci, "HIV Infection Is Active and Progressive in Lymphoid Tissue During the Clinically Latent Stage of Disease," *Nature* (London), 362 (1993) 355–358.
80. M. Craddock, "A Critical Appraisal of the Vancouver Men's Study," in *AIDS: Virus or Drug-Induced?* eds. Kluwer and Dordrecht (The Netherlands: Genetica, in press).
81. J. Maddox, "Where the AIDS Virus Hides Away," *Nature* (London), 362 (1993): 287.
82. H. Temin and D. Bolognesi, "Where Has HIV Been Hiding?" (1993).
83. H. W. Sheppard, M. S. Ascher, and J. F. Krowka, "Viral Burden and HIV Disease," *Nature* (London), 364 (1993): 291–292.
84. H. W. Haverkos, and J.A. Dougherty, eds., "Health Hazards of Nitrite Inhalants," *NIDA Research Monograph 83* (Washington, D.C.: U.S. Department of Health and Human Services, 1988).
85. Maddox, "Has Duesberg a Right of Reply?" 109.
86. J. Maddox, "Duesberg and the New View of HIV," *Nature* (London), 373 (1995a): 189.
87. J. Maddox, "More Conviction on HIV and AIDS," *Nature* (London), 377 (1995b): 1.
88. P. H. Duesberg, "Antecedents of a Nobel Prize," *Nature* (London), 343 (1989): 302–303; P. H. Duesberg, "Oncogenes and Cancer" (letter) *Science*, 267 (1995): 1407–1408.
89. J. D. Dingell, "Shattuck Lecture—Misconduct in Medical Research," *New England Journal of Medicine*, 328 (1993): 1610–1615; S. Lang, "Questions of Scientific Responsibility: The Baltimore Case," *Ethics and Behavior*, 3 (1993): 3–72.
90. S. Hall, "David Baltimore's Final Days," *Science*, 254 (1991): 1576–1579.
91. Ibid.
92. Institute of Medicine and National Academy of Sciences, *Confronting AIDS: Directions for Public Health* (Washington, D.C.: National Academy Press, 1986).

CHAPTER 11

1. Drug Strategies, *Keeping Score—What We Are Getting for Our Federal Drug Control Dollars* (1995).
2. Ibid.
3. M. S. Ascher, H. W. Sheppard, W. Winkelstein Jr., and E. Vittinghoff, "Does Drug Use Cause AIDS?" *Nature* (London), 362 (1993): 103–104; J. Maddox, "Has Duesberg a Right of Reply?" *Nature* (London), 363 (1993): 109; J. Cohen, "Is a New Virus the Cause of KS?" *Science*, 266 (1994b): 1803–1804.
4. J. Cohen, "'The Duesberg Phenomenon': Duesberg and Other Voices," *Science*, 266 (1994a): 1642–1649.
5. H. W. Haverkos and D. P. Drotman, "Measuring Inhalant Nitrite Exposure in Gay Men: Implications for Elucidating the Etiology of AIDS-related Kaposi's Sarcoma," *Genetica*, 95 (1995a): 157–164.
6. M. Craddock, "A Critical Appraisal of the Vancouver Men's Study," in *AIDS: Virus or Drug-Induced?* eds. Kluwer and Dordrecht (The Netherlands: Genetica, in press).
7. P. H. Duesberg, "Aetiology of AIDS," *Lancet*, 341 (1993a): 1544; P. H. Duesberg, "HIV and the Aetiology of AIDS," *Lancet*, 341 (1993b): 957–958; P. H. Duesberg, "Can Epidemiology Determine Whether Drugs or HIV Cause AIDS?" *AIDS-Forschung*, 12 (1993c): 627–635; B. J. Ellison, A. B. Downey, and P. H. Duesberg, "HIV as a Surrogate Marker for Drug Use: A Re-analysis of the San Francisco Men's Health Study," *Genetica*, 95 (1995): 165–171.
8. R. A. Kaslow, W. C. Blackwelder, D. G. Ostrow, D. Yerg, J. Palenicek, A. H. Coulson, and R. O. Valdiserri, "No Evidence for a Role of Alcohol or Other Psychoactive Drugs in Accelerating Immunodeficiency in HIV-1–Positive Individuals," *Journal of the American Medical Association*, 261 (1989): 3424–3429; J. P. Vandenbroucke and V. P. A. M. Pardoel, "An Autopsy of Epidemiologic Methods: The Case of 'Poppers' in the Early Epidemic of the Acquired Immunodeficiency Syndrome (AIDS)," *J. Epidemiol.*, 129 (1989): 455–457; D. Baltimore and M. B. Feinberg, "Quantification of Human Immunodeficiency Virus in the Blood, *New England Journal of Medicine*, 322 (1990): 1468–1469; R. Weiss and H. Jaffe, "Duesberg, HIV, and AIDS," *Nature* (London), 345 (1990): 659–660; Ascher, Sheppard, Winkelstein, and Vittinghoff, "Does Drug Use Cause AIDS?" 103–104.
9. M. Marmor, A. E. Friedman-Kien, L. Laubenstein, R. D. Byrum, D. C. William, S. D'Onofrio, and N. Dubin, "Risk Factors for Kaposi's Sarcoma in Homosexual Men," *Lancet*, i (1982): 1083–1087; H. W. Jaffe, K. Choi, P. A. Thomas, H. W. Haverkos, D. M. Auerbach, M. E. Guinan, M. F. Rogers, T. J. Spira, W. W. Darrow, M. A. Kramer, S. M., Friedman, J. M. Monroe, A. E. Friedman-Kien, L. J. Laubenstein, M. Marmor, B. Safai, S. K. Dritz, S. J. Crispi, S. L.

Fannin, J. P. Orkwis, A. Kelter, W. R. Rushing, S. B. Thacker, and J. W. Curran, "National Case-Control Study of Kaposi's Sarcoma and *Pneumocystis Carinii* Pneumonia in Homosexual Men: Part 1, Epidemiologic Results," *Ann. Intern. Med.*, 99 (1983): 145–151; U. Mathur-Wagh, R. W. Enlow, I. Spigland, R. J. Winchester, H. S. Sacks, E. Rorat, S. R. Yancovitz, M. J. Klein, D. C. William, and D. Mildwan, "Longitudinal Study of Persistent Generalized Lymphadenopathy in Homosexual Men: Relation to Acquired Immunodeficiency Syndrome," *Lancet*, i (1984): 1033–1038; H. W. Haverkos, P. F. Pinsky, D. P. Drotman, and D. J. Bregman, "Disease Manifestation Among Homosexual Men with Acquired Immunodeficiency Syndrome: A Possible Role of Nitrites in Kaposi's Sarcoma," *J. Sex. Trans. Dis.*, 12 (1985): 203–208; G. R. Newell, P. W. A. Mansell, M. R. Spitz, J. M. Reuben, and E. M. Hersh, "Volatile Nitrites: Use and Adverse Effects Related to the Current Epidemic of the Acquired Immune Deficiency Syndrome," *Am. J. Med.*, 78 (1985a): 811–816; G. R. Newell, P. W. A. Mansell, M. B. Wilson, H. K. Lynch, M. R. Spitz, and E. M. Hersh, "Risk Factor Analysis Among Men Referred for Possible Acquired Immune Deficiency Syndrome," *Preventive Med.*, 14 (1985b): 81–91; J. Lauritsen and H. Wilson, *Death Rush, Poppers, and AIDS* (New York: Pagan Press, 1986); H. W. Haverkos and J. A. Dougherty, eds., *Health Hazards of Nitrite Inhalants*, NIDA Research Monograph 83 (Washington, D.C.: U.S. Department of Health and Human Services, 1988b).

10. G. M. Oppenheimer, "Causes, Cases, and Cohorts: The Role of Epidemiology in the Historical Construction of AIDS," in *AIDS: The Making of a Chronic Disease*, eds. E. Fee and D. M. Fox (Berkeley, Calif.: University of California Press, 1992), 49–83.

11. P. H. Duesberg, "AIDS Epidemiology: Inconsistencies with Human Immunodeficiency Virus and with Infectious Disease," *Proc. Natl. Acad. Sci. USA*, 88 (1991): 1575–1579; P. H. Duesberg, "AIDS Acquired by Drug Consumption and Other Noncontagious Risk Factors," *Pharmacology and Therapeutics*, 55 (1992): 201–277; P. H. Duesberg, "Infectious AIDS—Stretching the Germ Theory Beyond its Limits," *Int. Arch. Allergy Immunol.*, 103 (1994): 131–142.

12. P. H. Duesberg, "How Much Longer Can We Afford the AIDS Virus Monopoly?" in *AIDS: Virus or Drug-Induced?*, eds. Kluwer and Dordrecht (The Netherlands: Genetica, 1996).

13. Ascher, Sheppard, Winkelstein, and Vittinghoff, "Does Drug Use Cause AIDS?" 103–104; M. T. Schechter, K. J. P. Craib, K. A. Gelmon, J. S. G. Montaner, T. N. Le, and M. V. O'Shaughnessy, "HIV-1 and the Aetiology of AIDS," *Lancet*, 341 (1993a): 658–659; Cohen, "'The Duesberg Phenomenon,'" 1642–1649; Cohen, "Is a New Virus the Cause of KS?" 1803–1804.

14. Ascher, Sheppard, Winkelstein, and Vittinghoff, "Does Drug Use Cause AIDS?"

15. Baltimore and Feinberg, "Quantification of Human Immunodeficiency Virus," 1468–1469; Weiss and Jaffe, "Duesberg, HIV, and AIDS," 659–660; J. Maddox, "Humbling of World's AIDS Researchers," *Nature* (London), 358 (1992): 367; Maddox, "Has Duesberg a Right of Reply?" 109; Cohen, "'The Duesberg Phenomenon,'" 1642–1649; J. Maddox, "Duesberg and the New View of HIV," *Nature* (London), 373 (1995a): 189; J. Maddox, "More Conviction on HIV and AIDS," *Nature* (London), 377 (1995b): 1.
16. Drug Strategies, *Keeping Score*.
17. Duesberg, "AIDS Acquired by Drug Consumption," 201–277; Drug Strategies, *Keeping Score*; Duesberg, "How Much Longer."
18. Haverkos and Dougherty, *Health Hazards of Nitrite Inhalants*; Duesberg, "AIDS Acquired by Drug Consumption," 201–277; Haverkos and Drotman, "Measuring Inhalant Nitrite Exposure," 157–164; H. W. Haverkos and D. P. Drotman, "NIDA Technical Review: Nitrite Inhalants" (National Institute on Drug Abuse, Washington, D.C., and Centers for Disease Control, Atlanta, 1995b); B. Mirken, "Everything You Always Wanted to Know About Poppers; The "Gay Drug" Is Still Here Despite a Ban, and So Is the Controversy," *San Francisco Frontiers Newsmagazine*, 14, 20 July 1995, 16–19; Duesberg, "How Much Longer."
19. Duesberg, "AIDS Acquired by Drug Consumption"; Duesberg, "How Much Longer."
20. S. Lang, "To Fund or Not to Fund, That Is the Question: Proposed Experiments on the Drug–AIDS Hypothesis. To Inform or Not to Inform, That Is Another Question," in *AIDS: Infectious or Not?* eds. Kluwer and Dordrecht (The Netherlands: Genetica, in press).
21. J. Lauritsen, "NIH Reconsiders Nitrites' Link to AIDS," *Bio/Technology*, 12 (1994): 762–763; D. P. Drotman, T. A. Peterman, and A. E. Friedman-Kien, "Kaposi's Sarcoma. How Can Epidemiology Help Find the Cause?" *Dermatoepidemiology*, 13 (1995): 575–582.
22. Haverkos and Drotman, "Measuring Inhalant Nitrite Exposure," 157–164.
23. Cohen, "'The Duesberg Phenomenon,'" 1642–1649.
24. Duesberg, "AIDS Acquired by Drug Consumption," 201–277; Duesberg, "How Much Longer."
25. Duesberg, "How Much Longer."
26. Duesberg, "AIDS Acquired by Drug Consumption," 201–277; Drug Strategies, *Keeping Score*; Duesberg, "How Much Longer."
27. Duesberg, "AIDS Acquired by Drug Consumption," 201–277; P. H. Duesberg, "Infectious AIDS—Stretching the Germ Theory Beyond Its Limits," *Int. Arch. Allergy Immunol.*, 103 (1994): 131–142; Duesberg, "How Much Longer."
28. R. A. Weiss, "How Does HIV Cause AIDS?" *Science*, 260 (1993): 1273–1279; J. Cohen, "Researchers Air Alternative Views on How HIV Kills Cells," *Science*, 269 (1995): 1044–1045.

29. Centers for Disease Control and Prevention, "U.S. HIV and AIDS Cases Reported Through June 1994; Mid-Year Edition," *HIV/AIDS Surveillance Report*, 6 (1994c): 1–27; Centers for Disease Control and Prevention, "U.S. HIV and AIDS Cases Reported Through December 1993; Year-End Edition," *HIV/AIDS Surveillance Report*, 5 (1994a), 1–33.
30. Duesberg, "AIDS Acquired by Drug Consumption," 201–277; Ascher, Sheppard, Winkelstein, and Vittinghoff, "Does Drug Use Cause AIDS?" 103–104; Duesberg, "HIV and the Aetiology of AIDS," 957–958; Duesberg, "Aetiology of AIDS," 1544; Duesberg, "Can Epidemiology Determine," 627–635; D. Parke, "Key Factor" (letter) *Sunday Times*, 19 Dec. 1993; Schechter, Craib, Gelmon, Montaner, Le, and O'Shaughnessy, "HIV-1 and the Aetiology of AIDS," 658–659; Mirken, "Everything You Always Wanted to Know About Poppers," 16–19; Duesberg, "How Much Longer."
31. Duesberg, "AIDS Acquired by Drug Consumption," 201–277.
32. Centers for Disease Control, "Heterosexually Acquired AIDS—United States, 1993," *Morb. Mortal. Weekly Reports*, 43 (1994): 155–160; Centers for Disease Control and Prevention, "U.S. HIV and AIDS Cases Reported Through December 1993; Year-End Edition," *HIV/AIDS Surveillance Report*, 5 (1994a): 1–33.
33. Drug Strategies, *Keeping Score*.
34. A. R. Lifson, W. W. Darrow, N. A. Hessol, P. M. O'Malley, J. L. Barnhart, H. W. Jaffe, and G. W. Rutherford, "Kaposi's Sarcoma in a Cohort of Homosexual and Bisexual Men: Epidemiology and Analysis of Cofactors," *Am. J. Epidemiol.*, 131 (1990): 221–231; Duesberg, "AIDS Acquired by Drug Consumption," 201–277; Ascher, Sheppard, Winkelstein, and Vittinghoff, "Does Drug Use Cause AIDS?" 103–104; Duesberg, "Can Epidemiology Determine," 627–635; Schechter, Craib, Gelmon, Montaner, Le, and O'Shaughnessy, "HIV-1 and the Aetiology of AIDS," *Lancet*, 341 (1993a): 658–659; M. T. Schechter, K. J. P. Craib, J. S. G. Montaner, T. N. Le, M. V. O'Shaughnessy, and K. A. Gelmon, "Aetiology of AIDS," *Lancet* 341 (1993c): 1222–1223.
35. D. G. Ostrow, "Substance Abuse and HIV Infection," *Psychiatric Manifestations of HIV Disease*, 17 (1994): 69–89.
36. H. W. Jaffe, K. Choi, P. A. Thomas, H. W. Haverkos, D. M. Auerbach, M. E. Guinan, M. F. Rogers, T. J. Spira, W. W. Darrow, M. A. Kramer, S. M. Friedman, J. M. Monroe, A. E. Friedman-Kien, L. J. Laubenstein, M. Marmor, B. Safai, S. K. Dritz, S. J. Crispi, S. L. Fannin, J. P. Orkwis, A. Kelter, W. R. Rushing, S. B. Thacker, and J. W. Curran, "National Case-Control Study of Kaposi's Sarcoma and *Pneumocystis Carinii* Pneumonia in Homosexual Men: Part 1, Epidemiologic Results," *Ann. Intern. Med.*, 99 (1983): 145–151; W. W. Darrow, D. F. Echenberg, H. W. Jaffe, P. M. O'Malley, R. H. Byers, J. P. Getchell, and J. W. Curran, "Risk Factors for Human Immunodeficiency Virus

(HIV) Infections in Homosexual Men," *Am. J. Publ. Health*, 77 (1987): 479–483; Lifson, Darrow, Hessol, O'Malley, Barnhart, Jaffe, and Rutherford, "Kaposi's Sarcoma in a Cohort of Homosexual and Bisexual Men," 221–231; D. G. Ostrow, M. J. Van Raden, R. Fox, L. A. Kingsley, J. Dudley, R. A. Kaslow, and the Multicenter AIDS Cohort Study (MACS), "Recreational Drug Use and Sexual Behavior Change in a Cohort of Homosexual Men," *AIDS*, 4 (1990): 759–765; Ascher, Sheppard, Winkelstein, and Vittinghoff, "Does Drug Use Cause AIDS?" 103–104; Ellison, Downey, and Duesberg, "HIV as a Surrogate Marker," 165–171; Schechter, Craib, Gelmon, Montaner, Le, and O'Shaughnessy, "HIV-1 and the Aetiology of AIDS," 658–659; Craddock, "Critical Appraisal"; D. G. Ostrow, W. J. DiFranceisco, J. S. Chmiel, D. A. Wagstaff, and J. Wesch, "A Case-Control Study of Human Immunodeficiency Virus Type 1 Seroconversion and Risk-Related Behaviors in the Chicago MACS/CCS Cohort, 1984–1992," *American Journal of Epidemiology*, 142 (1995) 875–883.

37. Duesberg, "AIDS Acquired by Drug Consumption," 201–277; Duesberg, "Can Epidemiology Determine"; Ellison, Downey, and Duesberg, "HIV as a Surrogate Marker for Drug Use," 165–171; M. S. Ascher, H. W. Sheppard, and W. Winkelstein Jr., "AIDS-Associated Kaposi's Sarcoma," *Science*, 267 (1995): 1080; Duesberg, "How Much Longer."
38. Duesberg, "AIDS Acquired by Drug Consumption," 201–277.
39. Ibid.
40. Duesberg, "Foreign-Protein–Mediated Immunodeficiency in Hemophiliacs With and Without HIV," *Genetica*, 95 (1995b): 51–70; Duesberg, "How Much Longer."
41. Duesberg, "AIDS Acquired by Drug Consumption," 201–277; Duesberg, "How Much Longer."
42. J. Q. Mok, A. De Rossi, A. E. Ades, C. Giaquinto, I. Grosch-Woerner, and C. S. Peckham, "Infants Born to Mothers Seropositive for Human Immunodeficiency Virus," *Lancet*, i (1987): 1164–1168; European Collaborative Study, "Children Born to Women With HIV-1 Infection: Natural History and Risk of Transmission," *Lancet*, 337 (1991): 253–260; Duesberg, "AIDS Acquired by Drug Consumption," 201–277; Duesberg, "How Much Longer."
43. Duesberg, "AIDS Acquired by Drug Consumption," 201–277.
44. Ibid.
45. Ibid.
46. Ibid; S. V. Meddis, "Heroin Use Said to Near Crisis Level," *USA Today*, 25 May 1994, 1, 3A; Drug Strategies, *Keeping Score*.
47. J. Gettman, "Heroin Returning to Center Stage," *High Times*, Dec. 1994, 23; Meddis, "Heroin Use," 1, 3A; Drug Strategies, *Keeping Score*.
48. Drug Strategies, *Keeping Score*.

49. Newell, Mansell, Spitz, Reuben, and Hersh, "Volatile Nitrites," 811–816; Haverkos and Dougherty, "Health Hazards of Nitrite Inhalants"; Haverkos and Drotman, "Measuring Inhalant Nitrite Exposure in Gay Men," 157–164; Mirken, "Everything You Always Wanted to Know About Poppers," 16–19.
50. Haverkos and Drotman, "NIDA Technical Review: Nitrite Inhalants."
51. T. J. Flanagan and K. Maguire, *Sourcebook of Criminal Justice Statistics (1989)—Bureau of Justice Statistics NCJ-124224* (Washington, D.C.: U.S. Department of Justice, U.S. Government Printing Office, 1989).
52. Bureau of Justice Statistics, *Catalog of Federal Publications on Illegal Drug and Alcohol Abuse* (Washington, D.C.: U.S. Department of Justice, 1991).
53. Drug Strategies, *Keeping Score*.
54. Rauschgiftbilanz 1994, "Starke Nachfrage nach synthetischen Drogen," *Deutsches Aerzteblatt*, 92 (1995): C-422.
55. Centers for Disease Control and Prevention, "U.S. HIV and AIDS Cases Reported Through June 1994," 1–27.
56. Duesberg, "AIDS Acquired by Drug Consumption," 201–277; D. Thomas, "Risky Business: Taking Stock of AZT's Future," *Men's Style*, May/June 1995, 54–56, 102–106.
57. Duesberg, "AIDS Acquired by Drug Consumption," 201–277; Drug Strategies, *Keeping Score*.
58. Duesberg, "AIDS Acquired by Drug Consumption," 201–277.
59. Centers for Disease Control and Prevention, "U.S. HIV and AIDS Cases Reported Through June 1994," 1–27.
60. Drug Strategies, *Keeping Score*.
61. Centers for Disease Control and Prevention, U.S. HIV and AIDS Cases Reported Through December 1994," 1–39.
62. Duesberg, "AIDS Acquired by Drug Consumption," 201–277; Thomas, "Risky Business."
63. Duesberg, "AIDS Acquired by Drug Consumption," 201–277.
64. J. Layon, A. Idris, M. Warzynski, R. Sherer, D. Brauner, O. Patch, D. McCulley, and P. Orris, "Altered T Lymphocyte Subsets in Hospitalized Intravenous Drug Abusers," *Arch. Intern. Med.* 144 (1984): 1376–1380; C. R. Schuster, foreword, in "Cocaine: Pharmacology, Effects, and Treatment of Abuse," *NIDA Research Monograph 50*, ed. J. Grabowski (Washington, D.C.: National Institute on Drug Abuse, 1984), VII–VIII; S. Savona, M. A. Nardi, E. T. Lenette, and S. Karpatkin, "Thrombocytopenic Purpura in Narcotics Addicts," *Ann. Intern. Med.*, 102 (1985): 737–741; R. M. Donahoe, C. Bueso-Ramos, F. Donahoe, J. J. Madden, A. Falek, J. K. A. Nicholson, and P. Bokos, "Mechanistic Implications of the Findings that Opiates and Other Drugs of Abuse Moderate T-Cell Surface Receptors and Antigenic Markers," *Ann. N.Y. Acad. Sci.*, 496 (1987): 711–721; P. Espinoza, I. Bouchard, C. Buffet, V. Thiers, J. Pillot, and J. P. Etienne,

"High Prevalence of Infection by Hepatitis B Virus and HIV in Incarcerated French Drug Addicts," *Gastroenterologie Clinique et Biologique*, 11 (1987): 288–292; R. Weber, W. Ledergerber, M. Opravil, W. Siegenthaler, and R. Lüthy, "Progression of HIV Infection in Misusers of Injected Drugs Who Stop Injecting or Follow a Programme of Maintenance Treatment with Methadone," *British Medical Journal*, 301 (1990): 1362–1365.

65. Newell, Mansell, Spitz, Reuben, and Hersh, "Volatile Nitrites," 811–816; V. Beral, T. A. Peterman, R. L. Berkelman, and H. W. Jaffe, "Kaposi's Sarcoma Among Persons With AIDS: A Sexually Transmitted Infection?" *Lancet*, 335 (1990): 123–128; Lifson, Darrow, Hessol, O'Malley, Barnhart, Jaffe, and Rutherford, "Kaposi's Sarcoma in a Cohort of Homosexual and Bisexual Men," 221–231; Duesberg, "AIDS Acquired by Drug Consumption," 201–277.

66. Duesberg, "AIDS Acquired by Drug Consumption," 201–277; R. Lewis-Thorton, "Facing AIDS," *Essence (New York)*, Dec. 1994, 63, 64, 124, 126, 130.

67. Haverkos and Dougherty, "Health Hazards of Nitrite Inhalants"; Beral, Peterman, Berkelman, and Jaffe, "Kaposi's Sarcoma Among Persons With AIDS," 123–128.

68. Marmor, Friedman-Kien, Laubenstein, Byrum, William, D'Onofrio, and Dubin, "Risk Factors for Kaposi's Sarcoma," 1083–1087; Haverkos, Pinsky, Drotman, and Bregman, "Disease Manifestation Among Homosexual Men," 203–208.

69. E. Sloand, P. N. Kumar, and P. F. Pierce, "Chemotherapy for Patients With Pulmonary Kaposi's Sarcoma: Benefit of Filgrastim (G–CSF) in Supporting Dose Administration," *Southern Medical Journal*, 86 (1993): 1219–1224; G. U. Meduri, D. E. Stover, M. Lee, P. L. Myskowski, J. F. Caravelli, and M. B. Zama, "Pulmonary Kaposi's Sarcoma in the Acquired Immune Deficiency Syndrome: Clinical, Radiographic, and Pathologic Manifestations," *Am. J. Med.*, 81 (1986): 11–18; S. M. Garay, M. Belenko, E. Fazzini, and R. Schinella, "Pulmonary Manifestations of Kaposi's Sarcoma," *Chest*, 91 (1987): 39–43; P. S. Gill, B. Akli, P. Coletti, M. Rarick, C. Louriero, M. Bernstein-Singer, M. Krailo, and L. A. M., "Pulmonary Kaposi's Sarcoma: Clinical Findings and Results of Therapy," *Am. J. Med.*, 87 (1989): 57–61.

70. Gill, Akli, Coletti, Rarick, Louriero, Bernstein-Singer, Krailo, and L. A. M., "Pulmonary Kaposi's Sarcoma," 57–61; D. H. Irwin and L. D. Kaplan, "Pulmonary Manifestations of Acquired Immunodeficiency Syndrome–Associated Malignancies," *Seminars in Respiratory Infections*, 8 (1993): 139–148.

71. M. Kaposi, "Idiopathisches Multiples Pigmentsarkom der Haut," *Archiv für Dermatologie und Syphilis*, 2 (1872): 265–273.

72. Sloand, Kumar, and Pierce, "Chemotherapy for Patients With Pulmonary Kaposi's Sarcoma," 1219–1224.

73. Meduri, Stover, Lee, Myskowski, Caravelli, and Zama, "Pulmonary

Kaposi's Sarcoma in the Acquired Immune Deficiency Syndrome," 11–18; Garay, Belenko, Fazzini, and Schinella, "Pulmonary Manifestations of Kaposi's Sarcoma," 39–43; Gill, Akli, Coletti, Rarick, Louriero, Bernstein-Singer, Krailo, and L. A. M., "Pulmonary Kaposi's Sarcoma," 57–61; Irwin and Kaplan, "Pulmonary Manifestations," 139–148.

74. Meduri, Stover, Lee, Myskowski, Caravelli, and Zama, "Pulmonary Kaposi's Sarcoma in the Acquired Immune Deficiency Syndrome," 11–18; D. P. Drotman and H. W. Haverkos, "What Causes Kaposi's Sarcoma? Inquiring Epidemiologists Want to Know," *Epidemiology*, 3 (1992): 191–193; Cohen, "'The Duesberg Phenomenon,'" 1642–1649.

75. Meduri, Stover, Lee, Myskowski, Caravelli, and Zama, "Pulmonary Kaposi's Sarcoma in the Acquired Immune Deficiency Syndrome," 11–18.

76. Garay, Belenko, Fazzini, and Schinella, "Pulmonary Manifestations," 39–43.

77. Haverkos and Dougherty, "Health Hazards of Nitrite Inhalants"; Duesberg, "AIDS Acquired by Drug Consumption," 201–277.

78. R. B. Nieman, J. Fleming, R. J. Coker, J. R. Harris, and D. M. Mitchell, "The Effect of Cigarette Smoking on the Development of AIDS in HIV-1–Seropositive Individuals," *AIDS*, 7 (1993): 705–710.

79. M. Seligmann, L. Chess, J. L. Fahey, A. S. Fauci, P. J. Lachmann, J. L'Age-Stehr, J. Ngu, A. J. Pinching, F. S. Rosen, T. J. Spira, and J. Wybran, "AIDS—An Immunologic Reevaluation," *New England Journal of Medicine*, 311 (1984): 1286–1292.

80. Layon, Idris, Warzynski, Sherer, Brauner, Patch, McCulley, and Orris. "Altered T-Lymphocyte Subsets," 1376–1380; R. L. Stoneburner, D. C. Des Jarlais, D. Benezra, L. Gorelkin, J. L. Sotheran, S. R. Friedman, S. Schultz, M. Marmor, D. Mildvan, and R. Maslansky, "A Larger Spectrum of Severe HIV-I–Related Disease in Intravenous Drug Users in New York City," *Science*, 242 (1988): 916–919; R. Pillai, B. S. Nair, and R. R. Watson, "AIDS, Drugs of Abuse, and the Immune System: A Complex Immunotoxicological Network," *Arch. Toxicol.*, 65 (1991): 609–617; Duesberg, "AIDS Acquired by Drug Consumption," 201–277; G. H. C. Mientjes, E. J. C. van Ameijden, H. M. Weigel, J. A. R. van den Hoek, and R. A. Countinho, "Clinical Symptoms Associated With Seroconversion for HIV-1 Among Misusers of Intravenous Drugs: Comparison with Homosexual Seroconverters and Infected and Noninfected Intravenous Drug Misusers," *British Medical Journal*, 306 (1993): 371–373.

81. Duesberg, "AIDS Acquired by Drug Consumption," 201–277.

82. U. Lockemann, F. Wischhusen, K. Püschel et al., "Vergleich der HIV-1-Praevalenz bei Drogentodesfaellen in Deutschland sowie in verschiedenen europaeischen Grosstaedten (Stand: 31.12.1993)," *AIDS-Forschung*, 10 (1995): 253–256.

83. Stoneburner, Des Jarlais, Benezra, Gorelkin, Sotheran, Friedman, Schultz, Marmor, Mildvan, and Maslansky, "A Larger Spectrum of Severe HIV-I-Related Disease," 916–919.
84. Duesberg, "AIDS Acquired by Drug Consumption," 201–277; Drug Strategies, *Keeping Score*.
85. Centers for Disease Control, "Revision of the CDC Surveillance Case Definition for Acquired Immunodeficiency Syndrome," *Journal of the American Medical Association*, 258 (1987): 1143–1154; Centers for Disease Control and Prevention, "1993 Revised Classification System for HIV Infection and Expanded Surveillance Case Definition for AIDS Among Adolescents and Adults," *Morb Mort Weekly Rep*, 41, no. RR17 (1992): 1–19; Duesberg, "AIDS Acquired by Drug Consumption," 201–277.
86. Duesberg, "AIDS Acquired by Drug Consumption," 201–277.
87. J. P. Freiman, K. E. Helfert, M. R. Hamrell, and D. S. Stein, "Hepatomegaly with Severe Steatosis in HIV-Seropositive Patients," *AIDS*, 7 (1993): 379–385.
88. A. J. Saah, D. R. Hoover, Y. Peng, J. P. Phair, B. Visscher, L. A. Kingsley, L. K. Schrager, and the Multicenter AIDS Cohort Study, "Predictors for Failure of *Pneumocystis Carinii* Pneumonia Prophylaxis," *Journal of the American Medical Association*, 273 (1995): 1197–1202.
89. J. M. Pluda, R. Yarchoan, E. S. Jaffe, I. M. Feuerstein, D. Solomon, S. Steinberg, K. M. Wyvill, A. Raubitschek, D. Katz, and S. Broder, "Development of Non Hodgkin Lymphoma in a Cohort of Patients With Severe Human Immunodeficiency Virus (HIV) Infection on Long-Term Antiretroviral Therapy," *Ann. Intern. Med.*, 113 (1990): 276–282; Duesberg, "AIDS Acquired by Drug Consumption," 201–277; G. X. McLeod and S. M. Hammer, "Zidovudine: Five Years Later," *Ann. Intern. Med*, 117 (1992): 487–501; Freiman, Helfert, Hamrell, and Stein, "Hepatomegaly with Severe Steatosis," 379–385; H. Bacellar, A. Muñoz, E. N., Miller, B. A. Cohen, D. Besley, O. A. Selnes, J. T. Becker, and J. C. McArthur, "Temporal Trends in the Incidence of HIV-1–Related Neurologic Diseases: Multicenter AIDS Cohort Study, 1985–1992," *Neurology*, 44 (1994): 1892–1900; W. B. Parker and Y. C. Cheng, "Mitochondrial Toxicity of Antiviral Nucleoside Analogs," *Journal of NIH Research*, 6 (1994): 57–61; *Physicians' Desk Reference*, "Retrovir" (Orandell, N.J.: Medical Economics Co., 1994).
90. Pluda, Yarchoan, Jaffe, Feuerstein, Solomon, Steinberg, Wyvill, Raubitschek, Katz, and Broder, "Development of Non Hodgkin Lymphoma," 276–282.
91. J. J. Goedert, A. R. Cohen, C. M. Kessler, S. Eichinger, S. V. Seremetis, C. S. Rabkin, F. J. Yellin, P. S. Rosenberg, and L. M. Aledort, "Risks of Immunodeficiency, AIDS, and Death Related to Purity of Factor VIII Concentrate," *Lancet*, 344 (1994): 791–792.
92. Seligmann, Chess, Fahey, Fauci, Lachmann, L'Age-Stehr, Ngu,

Pinching, Rosen, Spira, and Wybran, "AIDS—An Immunologic Reevaluation," 1286–1292.
93. M. C. Poznansky, R. Coker, C. Skinner, A. Hill, S. Bailey, L. Whitaker, A. Renton, and J. Weber, "HIV Positive Patients First Presenting With an AIDS Defining Illness: Characteristics and Survival," *British Medical Journal*, 311 (1995): 156–158.
94. M. Scolaro, R. Durham, and G. Pieczenik, "Potential Molecular Competitor for HIV," *Lancet*, 337 (1991): 731–732; J. Learmont, B. Tindall, L. Evans, A. Cunningham, P. Cunningham, J. Wells, R. Penny, J. Kaldor, and D. A. Cooper, "Long-Term Symptomless HIV-1 Infection in Recipients of Blood Products from a Single Donor," *Lancet*, 340 (1992): 863–867; Y. Cao, L. Quin, L. Zhang, J. Safrit, and D. D. Ho, "Virologic and Immunologic Characterization of Long-Term Survivors of Human Immunodeficiency Virus Type 1 Infection," *New England Journal of Medicine*, 332 (1995): 201–208.
95. M. H. Merson, "Slowing the Spread of HIV: Agenda for the 1990s," *Science*, 260 (1993): 1266–1268; World Health Organization, *The Current Global Situation of the HIV/AIDS Pandemic*, 1995.
96. Duesberg, "AIDS Acquired by Drug Consumption," 201–277.
97. World Health Organization, *Current Global Situation*.
98. Duesberg, "'The Duesberg-Phenomenon,'" 313.
99. Cao, Quin, Zhang, Safrit, and Ho, "Virologic and Immunologic Characterization," 201–208.
100. A. Muñoz, "Disease Progression 15 Percent of HIV-Infected Men Will Be Long-Term Survivors," *AIDS Weekly (News Report)*, 15 and 29 May 1995: 5–6, 3–4.
101. B. Gavzer, "What We Can Learn from Those Who Survive AIDS." *Parade*, 10 June 1990, 4–7; J. Wells, "We Have to Question the So-Called 'Facts'," *Capital Gay*, 20 Aug. 1993, 14–15; B. Gavzer, "Love Has Helped Keep Me Alive," *Parade*, 16 April 1995, 4–6; R. S. Root-Bernstein, "Five Myths About AIDS that Have Misdirected Research and Treatment," *Genetica*, 95 (1995): 111–132.
102. W. Lang, H. Perkins, R. E. Anderson, R. Royce, N. Jewell, and W. Winkelstein Jr., "Patterns of T Lymphocyte Changes with Human Immunodeficiency Virus Infection: From Seroconversion to the Development of AIDS," *J. Acquir. Immune Defic. Syndr.*, 2 (1989): 63–69.
103. W. Lang, R. E. Anderson, H. Perkins, R. M. Grant, D. Lyman, W. Winkelstein Jr., R. Royce, and J. A. Levy, "Clinical, Immunologic, and Serologic Findings in Men at Risk for Acquired Immunodeficiency Syndrome," *J. Am. Med. Assoc.*, 257 (1987): 326–330.
104. W. W. Darrow, D. F. Echenberg, H. W. Jaffe, P. M. O'Malley, R. H. Byers, J. P. Getchell, and J. W. Curran, "Risk Factors for Human Immunodeficiency Virus (HIV) Infections in Homosexual Men," *Am. J. Publ. Health*, 77 (1987): 479–483; A. R. Moss, "AIDS and Intravenous Drug Use: The Real Heterosexual Epidemic," *British Medical Journal*, 294 (1987): 389–390; Ascher, Sheppard, Winkelstein, and

Vittinghoff, "Does Drug Use Cause AIDS?" 103–104; Duesberg, "Can Epidemiology Determine," 627–635; Ellison, Downey, and Duesberg. "HIV as a Surrogate Marker," 165–171.
105. S. A. Marion, M. T. Schechter, M. S. Weaver, W. A. McLeod, W. J. Boyko, B. Willoughby, B. Douglas, K. J. P. Craib, and M. V. O'Shaughnessy, "Evidence that Prior Immune Dysfunction Predisposes to Human Immunodeficiency Virus Infection in Homosexual Men," *J. Acquir. Immune Defic. Syndr.*, 2 (1989): 178–186.
106. C. P. Archibald, M. T. Schechter, T. N. Le, K. J. P. Craib, J. S. G. Montaner, and M. V. O'Shaughnessy, "Evidence for a Sexually Transmitted Cofactor for AIDS-Related Kaposi's Sarcoma in a Cohort of Homosexual Men," *Epidemiology*, 3 (1992): 203–209; P. H. Duesberg, "The HIV Gap in National AIDS Statistics," *Bio/Technology*, 11 (1993d): 955–956; Schechter, Craib, Montaner, Le, O'Shaughnessy, and Gelmon, "Aetiology of AIDS," 1222–1223.
107. Kaslow, Blackwelder, Ostrow, Yerg, Palenicek, Coulson, and Valdiserri, "No Evidence for a Role of Alcohol," 3424–3429.
108. R. A. Kaslow, J. P. Phair, H. B. Freidman, R. E. Lyter, R. E. Solomon, J. Dudley, F. Polk, and W. Blackwelder, "Infection with the Human Immunodeficiency Virus: Clinical Manifestations and Their Relationship to Immunodeficiency," *Ann. Intern. Med.*, 107(1987): 474–480.
109. Ibid.
110. Ostrow, Van Raden, Fox, Kingsley, Dudley, Kaslow, and the Multicenter AIDS Cohort Study (MACS), "Recreational Drug Use," 759–765.
111. J. Phair, L. Jacobson, R. Detels, C., Rinaldo, A. Saah, L. Schrager, and A.D. Muñoz, "Acquired Immune Deficiency Syndrome Occurring within Five Years of Infection with Human Immunodeficiency Virus Type-1: The Multicenter AIDS Cohort Study," *Journal of Acquired Immune Deficiency Syndromes*, 5 (1992): 490–496.
112. Haverkos and Dougherty, "Health Hazards of Nitrite Inhalants"; Ostrow, Van Raden, Fox, Kingsley, Dudley, Kaslow, and the Multicenter AIDS Cohort Study (MACS), "Recreational Drug Use and Sexual Behavior Change," 759–765; Duesberg, "AIDS Acquired by Drug Consumption," 201–277; D. Parke, "Key Factor."
113. Duesberg, "AIDS Acquired by Drug Consumption"; S. N. Seidman and R. O. Rieder, "A Review of Sexual Behaviour in the United States," *American Journal of Psychiatry*, 151 (1994): 330–341.
114. D. C. Des Jarlais, S. R. Friedman, M. Marmor, D. Mildvan, S. Yancovitz, J. L. Sotheran, J. Wenston, and S. Beatrice. "CD4 Lymphocytopenia Among Injecting Drug Users in New York City," *Journal of Acquired Immune Deficiency Syndromes*, 6 (1993): 820–822.
115. A. Nicolosi, M. Musico, A. Saracco, S. Molinari, N. Ziliani, and A. Lazzarin, "Incidence and Risk Factors of HIV Infection: A Prospective Study of Seronegative Drug Users from Milan and Northern Italy, 1987–1989," *Epidemiology*, 1 (1990): 453–459.

116. Duesberg, "HIV Gap," 955–956.
117. Kaslow, Phair, Freidman, Lyter, Solomon, Dudley, Polk, and Blackwelder, "Infection with the Human Immunodeficiency Virus," 474–480; Lang, Anderson, Perkins, Grant, Lyman, Winkelstein, Royce, and Levy, "Clinical, Immunologic, and Serologic Findings," 326–330; European Collaborative Study, "Children Born to Women With HIV-1 Infection," 253–260; S. H. Weiss, C. Weston Klein, R. K. Mayur, J. Besra, and T. N. Denny, "Idiopathic CD4+ T-Lymphocytopenia." *Lancet*, 340 (1992): 608–609; Ellison, Downey, and Duesberg, "HIV as a Surrogate Marker," 165–171; Moore, and Chang, "Detection of Herpesvirus-like DNA Sequences," 1181–1185.
118. Duesberg, "HIV Gap," 955–956; Duesberg, "Foreign-Protein–Mediated Immunodeficiency," 51–70.
119. Duesberg, "AIDS Acquired by Drug Consumption," 201–277.
120. Scolaro, Durham, and Pieczenik, "Potential Molecular Competitor for HIV," 731–732.
121. M. Till and K. B. MacDonnell, "Myopathy with Human Immunodeficiency Virus Type 1 (HIV-1) Infection: HIV-1 or Zidovudine?" *Ann. Intern. Med.*, 113 (1990): 492–494.
122. P. S. Gill, M. Rarick, R. K. Byrnes, D. Causey, C. Loureiro, and A. M. Levine, "Azidothymidine Associated with Bone Marrow Failure in the Acquired Immunodeficiency Syndrome (AIDS)," *Ann. Intern. Med.*, 107 (1987): 502–505.
123. Weber, Ledergerber, Opravil, Siegenthaler, and Lüthy, "Progression of HIV Infection," 1362–1365.
124. D. C. Des Jarlais, S. R. Friedman, M. Marmor, H. Cohen, D. Mildvan, S. Yancovitz, U. Mathur, W. El-Sadr, T. J. Spira, and J. Garber, "Development of AIDS, HIV Seroconversion, and Potential Cofactors for T4 Cell Loss in a Cohort of Intravenous Drug Users," *AIDS*, 1 (1987): 105–111.
125. Wells, "We Have to Question," 14–15.
126. J. Wells, personal communication, London.
127. M. D. Hughes, D. S. Stein, H. M. Gundacker, F. T. Valentine, J. P. Phair, and P. A. Volberding, "Within-Subject Variation in CD4 Lymphocyte Count in Asymptomatic Human Immunodeficiency Virus Infection: Implications for Patient Monitoring," *Journal of Infectious Diseases*, 169 (1994): 28–36.
128. S. Blanche, M.-J. Mayaux, C. Rouzioux, J.-P. Teglas, G. Firtion, F. Monpoux, N. Ciraru-Vigneron, F. Meier, J. Tricoire, C. Courpotin, E. Vilmer, C. Griscelli, J.-F. Delfraissy, and the French Pediatric HIV Infection Study Group, "Relation of the Course of HIV Infection in Children to the Severity of the Disease in their Mothers at Delivery," *New England Journal of Medicine*, 330 (1994): 308–312.
129. European Collaborative Study, "Natural History of Vertically Acquired Human Immunodeficiency Virus-1 Infection," *Pediatrics*, 94 (1994): 815-819.

702 ■ *Notes*

130. Ibid.
131. Mok, De Rossi, Ades, Giaquinto, Grosch-Woerner, and Peckham, "Infants Born to Mothers Seropositive," 1164–1168: Duesberg, "AIDS Acquired by Drug Consumption," 201–277.
132. European Collaborative Study, "Children Born to Women With HIV-1 Infection," 253–260.
133. Ibid.
134. Blanche, Mayaux, Rouzioux, Teglas, Firtion, Monpoux, Ciraru-Vigneron, Meier, Tricoire, Courpotin, Vilmer, Griscelli, Delfraissy, and the French Pediatric HIV Infection Study Group, "Relation of the Course of HIV Infection in Children," 308–312.
135. Ibid. The Blanche et al. (1994) study did include mothers with AIDS who were not intravenous drug users.
136. P. Nair, L. Alger, S. Hines, S. Seiden, R. Hebel, and J. P. Johnson, "Maternal and Neonatal Characteristics Associated with HIV Infection in Infants of Seropositive Women," *Journal of Acquired Immune Deficiency Syndromes*, 6 (1993): 298–302.
137. "Zidovudine Use Can Reduce Numbers of HIV Infected Babies Being Born," *AIDS Weekly*, 2 Oct. 1995, 11.
138. Duesberg, "Foreign-Protein–Mediated Immunodeficiency," 51–70.
139. D. J. Martin, J. G. Sim, G. J. Sole, L. Rymer, S. Shalekoff, A. B. N. van Niekerk, P. Becker, C. N. Weilbach, J. Iwanik, K. Keddy, G. B. Miller, B. Ozbay, A. Ryan, T. Viscovic, and M. Woolf, "CD4+ Lymphocyte Count in African Patients Co-Infected With HIV and Tuberculosis," *Journal of Acquired Immune Deficiency Syndromes and Human Retrovirology*, 8 (1995): 386–391.
140. Duesberg, "Infectious AIDS," 131–142.
141. Cohen, "Is a New Virus the Cause of KS?" 1803–1804; D. J. DeNoon, "Duesberg Redux (Commentary)." *AIDS Weekly*, 9 Jan. 1995, 1–2.
142. Duesberg, "'The Duesberg Phenomenon,'" 313.
143. Haverkos and Dougherty, "Health Hazards of Nitrite Inhalants"; Duesberg, "AIDS Acquired by Drug Consumption," 201–277.
144. "Government Congressman Questions Funding for AIDS," *AIDS Weekly*, 22 May 1995.
145. Drug Strategies, *Keeping Score*.
146. "Studies Find Ex-Smokers Still Risk Lung Cancer," *San Francisco Chronicle*, 23 May 1995, A5.

CHAPTER 12

1. J. Cohen, "AIDS: The Unanswered Questions," *Science*, 260 (1993): 1262.
2. Ibid., 1254.
3. Ibid., 1219, 1253–1293.

4. World Health Organization, *The Current Global Situation of the HIV/AIDS Pandemic*, 1995.
5. J. Cohen, "'The Duesberg Phenomenon': Duesberg and Other Voices," *Science*, 266 (1994a): 1642–1649; N. Birkett, "'The Duesberg Phenomenon'" (letters) *Science*, 267 (1995): 315; M. Seligmann, D. A. Warrel, J.-P. Aboulker, C. Carbon, J. H. Darbyshire, J. Dormont, E. Eschwege, D. J. Girling, D. R. James, J.-P. Levy, P. T. A. Peto, D. Schwarz, A. B. Stone, I.V.D. Weller, R. Withnall, K. Gelmon, E. Lafon, A. M. Swart, V. R. Aber, A. G. Babiker, S. Lhoro, A. J. Nunn, and M. Vray, "Concorde: MRC/ANRS Randomised Double-Blind Controlled Trial of Immediate and Deferred Zidovudine in Symptom-Free HIV Infection," *Lancet*, 343 (1994): 871–878; P. H. Duesberg, "How Much Longer Can We Afford the AIDS Virus Monopoly?" in *AIDS: Virus or Drug-Induced?* eds. Kluwer and Dordrecht (The Netherlands: Genetica, in press).
6. *Physicians' Desk Reference*, "Retrovir" (Orandell, N.J.: Medical Economics Co., 1994).
7. D. Swinbanks, "AIDS Chief Promises a Shift Towards Basic Research," *Nature* (London), 370 (1994): 494.
8. Ibid.
9. S. Rey, "The Good News Is, the Bad News Is the Same," *Spy*, Feb. 1993, 19.
10. P. H. Duesberg, "Infectious AIDS—Stretching the Germ Theory Beyond Its Limits," *International Archives of Allergy and Immunology*, 103 (1994): 131–142.
11. J. McDonald, "Foreword," *Genetica*, 95 (1995): 1.
12. Cohen, "'The Duesberg Phenomenon': Duesburg and Other Voices" *Science*, 266 (1994a): 1642–1649.
13. D. Thomas, "Risky Business: Taking Stock of AZT's Future," *Men's Style*, May/June 1995, 54–56, 102–106.
14. P. H. Duesberg, "Duesberg on AIDS Causation: The Culprit Is Non-Contagious Risk Factors," *Scientist*, 9 (March 20, 1995a): 12.
15. J. Lauritsen, "NIH Reconsiders Nitrites' Link to AIDS," *Bio/Technology*, 12 (1994): 762–763.
16. J. Le Fanu, "Telescope," *London Sunday Telegraph*, 29 Nov. 1992.
17. S. Threakall, letter to Peter Duesberg, 31 May 1992; N. Hodgkinson, "Court Battles Launched Over Anti-AIDS Drug," *Sunday Times*, 30 Jan. 1994.
18. N. Hodgkinson, "Court Battles Launched."
19. Thomas, "Risky Business," 54–56, 102–106.
20. J. W. Ward, T. J. Bush, H. A. Perkins, L. E. Lieb, J. R. Allen, D. Goldfinger, S. M. Samson, S. H. Pepkowitz, L. P. Fernando, P. V. Holland, and the Study Group from the AIDS Program, "The Natural History of Transfusion-Associated Infection with Human Immunodeficiency Virus," *New England Journal of Medicine*, 321 (1989): 947–952.

21. Dr. John Ward, CDC, personal communication, October 1989.
22. D. P. Drotman, T. A. Peterman, and A. E. Friedman-Kien, "Kaposi's Sarcoma. How Can Epidemiology Help Find the Cause?" *Dermatoepidemiology*, 13 (1995): 575–582.
23. N. Ostrom, "Shalala Defends HIV Against Congressman's Challenge," *New York Native*, 15, 31 July 1995, 30.
24. E. Marshall, "Committee Treats Healy Gently," *Science*, 251 (1991): 1423.
25. J. Crewdson, "The Great AIDS Quest," *Chicago Tribune*, 19 Nov. 1989, 1–16, Section 5; J. Crewdson, "Burden of Proof: Gallo Case Spotlights a Key Question: Can U.S. Science Be Believed?" *Chicago Tribune*, 6 Dec. 1992.
26. "GEN's Fifty Molecular Millionaires," *Genetic Engineering News*, Feb. 1987, 15.
27. P. H. Duesberg and J. R. Schwartz, "Latent Viruses and Mutated Oncogenes: No Evidence for Pathogenicity," *Prog. Nucleic Acid Res. Mol. Biol.*, 43 (1992): 135–204; P. H. Duesberg, "Oncogenes and Cancer" (letter) *Science*, 267 (1995b): 1407–1408.
28. G. Condon, "The Assault on Germs," *San Francisco Chronicle*, 28 July 1993, C3–C4.
29. J. Cairns, *Cancer: Science and Society* (San Francisco: W. H. Freeman and Company, 1978).
30. L. Siegel, "AIDS Fear Manifesting Itself in Irrational Ways," *Daily Recorder*, 23 March 1987.
31. T. Schmitz, "A Personal Experience (An Open Letter)," 8 Dec. 1993.
32. K. Day, "After AIDS, Superbugs Give Medicine the Jitters," *Int. Herald Trib.*, 28 June 1995, 2.
33. C. Ortleb, "The Chantarelle Syndrome," *New York Native*, 17 Jan. 1994, 4–5.
34. World Health Organization, *Current Global Situation*.
35. J. Cohen, "Is a New Virus the Cause of KS?" *Science*, 266 (1994b): 1803–1804.

Index

Aaron Diamond AIDS Research Center, 340
Abelson, Philip, 66
Acer, David, 348-353
Acquired immune deficiency syndrome. *See also* Drug-AIDS hypothesis; War on AIDS: adoption of retrovirus-AIDS hypothesis, 155-156; AIDS-defining diseases in the U.S. in 1992-93 (table), 187; AIDS statistics for U.S., Europe and Africa (table), 257; autoimmunity hypothesis of AIDS, 247-248; CDC search for epidemics, 137-145; convergence of the virus hunters, 131-168; cytomegalovirus hypothesis, 153; as a fabricated epidemic, 169-217; HIV-AIDS correlation in the U.S. since 1981 (fig.), 194; inventing AIDS, 145-159; lymphadenopathy-associated virus as cause, 157; naming of, 150
ACS. *See* American Cancer Society
ACT UP. *See* AIDS Coalition to Unleash Power
Adult T-cell Leukemia, 125-126
Advocate, 261
Africa: AIDS and malnutrition correlation, 295-296; AIDS epidemic, 170; AIDS in the heterosexual population, 289-296; AIDS infection rates, 214-215; AIDS statistics, 5, 258-259; AIDS statistics for U.S., Europe and Africa (table), 257; dangerous AIDS experiments, 163; HIV infection in tuberculosis patients, 293; Kaposi's sarcoma in, 259; malignant lymphoma in children, 99-100, 101; manifestations of AIDS, 215; misdiagnosis of diabetes as AIDS, 294-295; myth of African AIDS epidemic, 291-292; population growth, 290; slim disease, 215, 216, 293; typical AIDS patient, 255-256; urban prostitutes, 295
Age factors: AIDS, 258, 261; liver cancer, 114; papilloma virus, 111
Agency for International Development, 371
AID. *See* Agency for International Development
AIDS: Virus or Drug-Induced?, 441
AIDS Coalition to Unleash Power, 171, 229, 266, 333-334, 379-381
The AIDS Cover-up: The Real and Alarming Facts About AIDS, 172
The AIDS Debate, 240
AIDS-Forschung, 245
AIDS funding, 5-6, 7, 174, 364, 367
AIDS Prevention Act of 1990, 173
AIDS Research, 221, 223, 225
AIDS Research and Human Retroviruses, 225

706 ■ Index

The AIDS War: Propaganda, Profiteering, and Genocide from the Medical-Industrial Complex, 231
Alcoholics: hepatitis and, 83
Alkylated nitrites, 222, 271
Alkylnitrites, 261, 270. *See also* Nitrite inhalants
Altman, Lawrence, 136, 144, 158, 386
American Association for the Advancement of Science, 64-65
American Association of Blood Banks, 86, 127
American Cancer Society, 127-128
American Foundation for AIDS Research, 249, 377, 378, 389
American Laboratory, 251
American Medical Foundation, 224
American Navaho Indians, 138
American Pediatric Association, 40
American press. *See also individual publications by name*: favorable attention to HIV dissidents, 442
American Red Cross, 125, 146, 151, 375
Americans for a Sound AIDS Policy, 376-377
AMF. *See* American Medical Foundation
AmFAR. *See* American Foundation for AIDS Research
Amphetamines, 261, 281, 415, 420, 421
Anal intercourse, 223, 228, 270, 417, 427
And the Band Played On, 147
Animal studies: AIDS, 182; anthrax, 33; AZT, 323; cholera, 35-36; hepatitis, 84-85; hepatitis B virus, 115; kuru, 77-78; *Legionella pneumophila,* 55; mycoplasmas, 233-234; nitrite inhalants, 275; pellagra, 47, 50-51; polio, 94; scurvy, 40; SMON syndrome, 16-17; tuberculosis, 35
Anthrax, 33

Antibiotic use among homosexuals, 282-283. *See also specific antibiotics by name*
Antonio, Gene, 172
Argentina, 21
Arizona State University, 79
Armed Forces Institute of Pathology, 233
Arsenic compounds: as syphilis treatment, 53
ASAP. *See* Americans for a Sound AIDS Policy
Ascher, Michael, 400-401
Ashe, Arthur, 356-358
Asia: hepatitis B, 113
Asian flu epidemic, 140
ATL. *See* Adult T-cell Leukemia
ATLV. *See* Japanese virus
Australia: AZT study, 326-327; reports of SMON symptoms, 21
Autoimmunity hypothesis of AIDS, 247-248
Azidothymidine, 301, 309
AZT. *See* Azidothymidine; Zidovudine

Babick, Christopher, 318
Bacteria hunting. *See also* Virus hunting: beriberi, 42-44; Legionnaires' disease, 54-59; pellagra plague, 44-52; plagues of malnutrition, 37-44; scurvy, 37-41; syphilis, 52-54; tuberculosis, 34-35
Bactrim, 300
Baltimore, David, 8, 73, 165, 372; censorship of Duesberg, 388-389; as director of the Whitehead Institute, 363; as high wattage panelist, 438; as leader of the war on AIDS, 362, 364; opposition to Duesberg's honors, 363; as virus researcher, 118; vulnerability of, 406-408
Barry, David, 311-314, 315, 322-323

Beecham, Jim, 143-144
Beijerinck, Martinus Willem, 69
Beppu, H., 18
Bergalis, Kimberly, 348-354
Beriberi, 42-44
Berkelman, Ruth, 139
Bialy, Harvey, 249-250
Bio/Technology, 197, 249, 250, 252
Bishop, Michael, 363
Blattner, William, 197; arguments for HIV based on antibody correlations, 207; arguments for HIV based on evasion, 204-206; standard defense of HIV hypothesis, 390-392
Blood transfusion recipients, 4, 83, 150, 183, 259
Blood transfusions: emergency transfusions, 284-286; long-term transfusion as prophylaxis for hemophilia, 286-288
"Bob Club," 164, 225, 311, 312, 402
Bolognesi, Dani, 225, 311-314, 399, 402
Brewer, J., 378
Bristol-Myers Squibb, 325, 380
Bristol-Myers Squibb Foundation, 378
British Medical Journal, 249, 331
Broder, Sam, 225, 311-314, 322, 325, 331
Burkitt, Dennis, 99
Burkitt, Elinor, 388
Burkitt's lymphoma, 17, 100-102, 153
Burroughs Wellcome: advertising for AZT, 377; AZT as preventive drug, 328; AZT as treating drug, 301, 310-312, 314, 315-316, 321, 322-323; financing AIDS activist groups, 377, 378; stock of, 330; support for ACT UP, 380
Burton, Phillip, 151
Bush, President George, 171

Cable News Network, 392
Caceres, Cesar, 266
California Institute of Technology, 116
Callen, Michael, 339, 374-375
Cambridge Bioscience Corporation, 123
Cancer. *See* War on cancer; *specific cancers by name*
Cancer Research, 196, 197, 230, 237, 390-391
Candida, 193, 255, 256
Caribbean virus, 125, 160
CDC. *See* Centers for Disease Control and Prevention; Communicable Disease Center
Censorship: in the media, 388-396; in professional literature, 396-406
Centers for Disease Control and Prevention, 8, 134. *See also* War on AIDS; AIDS fills CDC's need for a new epidemic, 145-159; Asian flu epidemic, 140; consideration of nitrite as an explanation of AIDS epidemic, 272; Ebola Virus pandemic, 138; fast-track homosexuals study, 267; first genuine success of, 139; Hepatitis Laboratories Division, 155; Kimberly Bergalis case, 349-353; Legionnaires' epidemic, 54-58; *Morbidity and Mortality Weekly Report*, 148; Navajo flu, 138; official list of AIDS indicator diseases (table), 210-211; premature panics, 138-139; steps to expand CDC authority, 369-371; "superbugs" crisis, 138-139; swine flu epidemic, 141-143, 145
Centocor, 166
Cervical cancer: as AIDS indicator disease, 186; as first AIDS disease to affect only women, 209; viruses to cause, 106-113
Challenges against the HIV hypothesis, 219-253

Chemotherapy, 309
Chicago Tribune, 161-162, 163, 167, 394
Chicken sarcoma virus. *See* Rous sarcoma virus
Chirac, Prime Minister Jacques, 165
Chiron Corporation, 84-86
Chlamydia: as cervical cancer cause, 109
Christopher Street, 245
Ciba-Geigy: lawsuit against, 21, 26
Civil rights organizations: as recipients of CDC funds, 376
Clean needle program, 262
Clioquinol. *See also* Emaform; Entero-vioform: SMON syndrome and, 18-23
CMV. *See* Cytomegalovirus
"Cocaine Abuse and Acquired Immunodeficiency Syndrome: A Tale of Two Epidemics," 276
Cocaine use. *See also* Crack cocaine use: AIDS cured by withdrawal from, 428; cocaine/heroin epidemics in the U.S. since 1980 (fig.), 194; DEA confiscation of cocaine, 261; drug-AIDS hypothesis and, 412, 415, 419, 421; effect on unborn children, 278; escalation of, 260
Cofactor hypothesis of AIDS, 242-243
Cohen, Seymour, 321
Cold Spring Harbor research labs, 238, 239
Cold Spring Harbor Symposium, 105, 116
Columbia University, 225
The Coming Plague, 362
Command science: powers of, 452-463
Committee to Monitor Poppers, 228
Communicable Disease Center, 134. *See also* Centers for Disease Control and Prevention; creation of Epidemic Intelligence Service, 135-137; powers to deal with potential emergencies, 135
Communications media: funding of, 381-382
Community Research Initiative, 225
Community Research Initiative on AIDS, 227
Compound S. *See* Zidovudine
Conant, Marcus, 273-274
Concorde study, 436, 437
Confronting AIDS: Directions for Public Health, Health Care, and Research: report recommendations, 365-372
Connor, Steve, 160, 161
Continuum magazine, 386
Cook, James, 38
Coronavirus: multiple sclerosis and, 83
Cowley, Geoffrey, 385
Coxsackie viruses: diabetes and, 80; poliovirus and, 74; SMON syndrome and, 15
Coyle, William Bryan, 264, 265
Crack cocaine use, 260; "Intersecting Epidemics—Crack Cocaine Use and HIV Infection," 277
Creutzfeld-Jakob disease, 78-79
Crewdson, John, 161-162, 163, 394
CRIA. *See* Community Research Initiative on AIDS
Crime Control Act of 1990, 271
Cromwell Hospital (London), 290
"Crystal." *See* Methamphetamine
Cuba, 173
Cumming, Paul, 146
Curran, James, 148-149, 156, 224, 385
Cytomegalovirus, 152-153, 176, 193, 219

Dan, Bruce, 136
Dannemeyer, Rep. William, 172, 173
Davaine, Casimir Joseph, 33

Day One, 442
Days of Grace: A Memoir, 357-358
ddC. *See* Dideoxycytidine
ddI. *See* Dideoxyinosine
Death Rush: Poppers and AIDS, 229
Delaney, Martin, 322, 333, 378-379
Dementia: AIDS dementia, 4, 54, 186, 187; AZT use for AIDS dementia, 331; syphilis and, 53-54
Deoxyribonucleic acid, 321; AZT and, 309; human DNA sequence (fig.), 310; liver cancer and, 114-115; reverse transcriptase and, 307-308
Department of Agriculture: pellagra research, 49
Department of Health, Education, and Welfare, 129; secretary of, 71, 72
Department of Health and Human Services, 166; premature endorsement of AIDS hypothesis, 10; secretary of, 7-8, 10, 158, 390, 399-400, 413-414, 446
Department of Veterans Affairs, 330
Detroit Cancer Foundation, 309
Diabetes: Coxsackie virus and, 80; misdiagnosis as AIDS, 294-295
Dideoxycytidine, 303, 326, 353-354
Dideoxyinosine, 4, 325, 353-354, 355, 358; FDA approval, 380-381
Diet *See also* Malnutrition: beriberi and, 42-44; pellagra and, 44-52; scurvy and, 37-41
Digene Diagnostics, 113
Diiodohydroxyquin, 283
Dingell, Rep. John, 167, 173, 406-407
Diphtheria, 35
Discovering, 246
DNA. *See* Deoxyribonucleic acid
Do we know the Cause(s) of AIDS?", 246
Dormant viruses, 97-98

Drotman, P., 450-451
Drug-AIDS hypothesis: AIDS cured by recreational drug withdrawal, 428-431; drugs as plausible causes of AIDS, 414-416; examples of how drug hypothesis predicts AIDS, 416-417, 419-431; gender factor, 416-417; long-term drug consumption, 411-412; non-correlation between HIV and AIDS, 425-428; proving the hypothesis, 409-433; risk behavior and, 416; risk-group-specific AIDS diseases, 422-425; savings to American taxpayers, 433; testing the hypothesis, 432
Drug Enforcement Administration, 261
Drug overdoses: deaths from, 262
Drug use, 214, 222. *See also* Intravenous drug use; Recreational drug use; *specific drugs by name*; and AIDS as the same epidemic, 260-269; AIDS through chemistry, 270-284; medical reports of, 278-279
Duesberg, Peter, 69, 86, 103; Baltimore's opposition to, 363; as book editor, 441; censorship in professional literature, 396-406; censorship in the media, 388-396; challenge against the HIV hypothesis, 196-198, 226, 227, 229-230, 235, 239-242; Delaney's attacks, 379; dissident papers, 439-440; excerpt from Ashe's book, 358; HIV-negative AIDS patients, 386; letter from Raphael Lombardo, 341-348; letter to *Science*, 383; Lindsey Nagel case, 303-304; Outstanding Investigator Grant, 398-400; retrovirus research, 118-119; *Science* editorial on, 441; speaking invitations, 441-442; support from Harry Rubin, 248-249

Duke University, 225, 311
Dutch Colonial Administration, 42-43
Dysplasias, 112

E. coli bacteria, 36-37, 283
The Early Homosexual Rights Movement (1864-1935), 227
"Early Intervention: An Idea Whose Time Has Gone?," 329
Ebola Fever, 154
Ebola Virus, 138, 362
EBV. See Epstein-Barr virus
Echovirus. See Enteric cytopathogenic human orphan virus
Edwards, Charles, 129
Ehrlich, Paul, 53
Eigen, Manfred, 240
Eijkman, Christiaan, 42-43
EIS. See Epidemic Intelligence Service
Eisenhower, President Dwight, 64
Ellermann, Vilhelm, 91
Ellner, Michael, 339
Emaform, 14-15. See also Clioquinol
Enders, John, 71
England. See also Great Britain: AZT study, 326; crack in wall of silence on the HIV debate, 442-443; London AIDS survivor group, 340
"Entangled Epidemics: Cocaine Use and HIV Disease," 277
Enteric cytopathogenic human orphan virus, 14, 74
Entero-vioform, 14-15. See also Clioquinol
Epidemic Intelligence Service, 135-137, 458
Epstein, M. Anthony, 100
Epstein-Barr virus, 100-103, 153, 176, 219. See also Burkitt's lymphoma
Essex, Max, 155, 156, 164, 225
Etheridge, Elizabeth, 149, 154

The Etiology of AIDS, 221
Europe. See also specific countries in Europe: AIDS statistics for U.S., Europe and Africa (table), 257; European Collaborative Study, 430-431; gender and AIDS, 416-417; low birth weight of AIDS babies, 424; mental retardation of AIDS babies, 424; pediatric AIDS, 419; publication of dissident views, 442; speaking invitations to dissidents, 441-442
Evatt, Bruce, 151

Factor VIII, 287
FAIDS. See Feline AIDS
Farber, Celia, 291, 333, 337, 338
Farr's law, 191-192, 201-202; hypothetical flu epidemic compared to long-established microbes (fig.), 192
Fast-track homosexuals, 262-263, 265, 266, 267, 282-283
Fauci, Anthony, 8, 152, 225, 340, 385; advisor to Martin Delaney, 378; attendance at ACT UP meetings, 380; as director of the National Institute of Allergy and Infectious Diseases, 189-190, 325-326; paper on finding HIV in lymph nodes of infected people, 400, 401-402; as stand-in for Duesberg-AIDS television programs, 392-393; trials of AZT on pregnant mothers, 336-339
FDA. See Food and Drug Administration
Feline AIDS, 167-168
Feline Leukemia Virus, 155, 167-168
Fischl, Margaret, 315-317, 319-320, 328
Flagyl, 283
Florida State University, 352
Food and Drug Administration, 8; approval for AZT, 311, 314, 315, 316, 319, 320, 321, 323, 324,

Index ■ 711

329; approval for ddI, 325, 380-381; approval of hepatitis C test, 85-86; *Treatment IND* permit, 322-323
Ford, President Gerald, 142, 143
France: Aids foundation report on methamphetamine, 281-282; AZT study, 326; Concorde trial, 330-331; dispute with U.S. over discovery of virus, 164-165; dissident papers, 439-440; trials of AZT on pregnant mothers, 336-337
Francis, Donald, 153-156, 369-371
Fraser, David, 142-143, 144
Freedom of Information Act, 318-319, 320
Fumento, Michael, 213

Gajdusek, Carleton, 16-17, 76-80, 82, 96-97, 126, 459
Gallo, Robert, 7-8, 10, 225; anger at Montagnier for breaking ranks, 241; arguments for HIV based on antibody correlations, 207; arguments for HIV based on evasion, 204-206; arguments for HIV by ignoring the facts, 199-203; exchanges with Peter Duesberg, 196-199; human leukemia virus search, 123-127; as panel member at Ninth Annual Congress of Immunology, 438-439; patent for virus antibody test, 158-169; refusal to attend conferences with Duesberg, 396-397; research papers, 82, 83; scandal in the establishment, 159-168; search for AIDS retrovirus, 156-159; support for Duesberg, 441
Gann, Paul, 284-286
Garrett, Laurie, 355, 362
Gay Men's Health Crisis, 379
Gender factors: AIDS, 5, 27, 262; genital warts, 112; infectious diseases, 191; SMON syndrome, 12, 27

General Motors Award, 363
Genetica, 249, 401, 440
Genital warts, 112. *See also* Human papilloma virus
Georgia: pellagra outbreak, 47-48
Germ theory of disease, 32
Germany: censorship of Duesberg, 395; dissident papers, 439; Ninth International AIDS Conference, 435-436; recreational drug use, 420
Gertz, Alison, 354-356
Gilbert, Walter, 236-237
GMHC. *see* Gay Men's Health Crisis
Goettingen University, 32
Goldberger, Joseph, 29, 51-52, 70
Goldblum, Peter, 378
Gonorrhea: as cervical cancer cause, 109
Good Intentions, 314, 317
Good Morning, America, 392
Gottlieb, Michael, 27-28, 146-148, 152-153, 155, 219, 225, 255, 377
Great Britain: Concorde trial, 330-331; dissident papers, 440; reports of SMON symptoms, 21; study of AZT effects on HIV-positive babies, 336-337
Greenberg, Daniel S., 104, 128-129, 166
Griffin, Beverly, 251
Groopman, Jerome, 322
Gross, Ludwik, 93, 95, 98
Group for the Scientific Reappraisal of the HIV/AIDS Hypothesis, 245, 252, 439
Gutknecht, Gil, 413-414, 445-449

Haiti: AIDS victims, 215, 296; HIV population, 5; malnutrition in, 150, 296; tuberculosis in, 296
HAM. See HTLV-Associated Myelopathy
Hameau, Jean-Marie, 45-46
Hand, Timothy H., 334

Hantavirus epidemic, 138
Harper's magazine, 395
Harvard University, 227, 236, 244
Haseltine, William, 164
Haverkos, Harry, 402, 403
Hawkins, Sir Richard, 37-38
HEAL. *See* Health-Education-AIDS Liaison
Health care workers: AIDS and, 183
The Health Century, 131
Health-Education-AIDS Liaison, 339
"Health Hazards of Nitrite Inhalants," 275, 402
Healy, Bernadine, 453-454
Heckler, Margaret, 158
Helicon Foundation, 244
Hemophilia, 239; long-term transfusion as prophylaxis of, 286-288
Hemophiliacs, 4, 5, 27, 150, 183-184, 215, 243, 259
Henle, Jakob, 32
Hepatitis: as profitable virus-hunting opportunity, 83-87
Hepatitis A: transmission of, 83
Hepatitis B: homosexuality and, 214; liver cancer hypothesis and, 113-116; transmission of, 83; virus particles and, 175
Hepatitis C, 83-86
Heroin use, 5, 27, 83, 150, 260; AIDS cured by withdrawal from, 428; cocaine/heroin epidemics in the U.S. since 1980 (fig.), 194; drug-AIDS hypothesis and, 412, 415, 419, 421; effect on unborn children, 278; health problems from, 275-276
Herpes simplex virus-1, 109
Herpes simplex virus-2, 109, 112-113, 153
Herpes simplex viruses. *See also specific forms*: cervical cancer and, 109-110; "hit-and-run" hypothesis, 110; as a model for HIV, 203
Herpes virus, 176, 192-193

Herpesvirus-like DNA sequences, 383-384
Hirohito, Emperor, 85-86
HIV. *See* Human immunodeficiency virus
HIV-AIDS correlation in the U.S. since 1981 (fig.), 194, 211, 212
HIV testing, 178, 369
Ho, David, 340
Hodgkinson, Neville, 442-443
Hoffmann-LaRoche, 326
Holland: AZT study, 327
Homosexual community, 5, 27. *See also* Fast-track homosexuals; drug use, 148, 152, 214; sexual activity, 149-150, 214
Homosexual press, 228-229, 245. *See also individual publications by name*
Hopkins, Donald R., 372
Horwitz, Jerome, 309, 326
Hostetter, Margaret, 302-306
Hot Zone, 362
HPV. *See* Human papilloma virus
HTLV. *See* Human T-cell leukemia virus
HTLV-Associated Myelopathy, 126
HTLV-I. *See* Human T-cell Leukemia Virus-I
HTLV-II. *See* Human T-cell Leukemia Virus-II
HTLV-III. *See* Human T-cell Leukemia Virus-III
Huebner, Robert J., 94, 105
Human immunodeficiency virus: arguments for HIV based on antibody correlations, 207-209, 211-216; arguments for HIV based on evasion, 204-206; arguments for HIV based on inappropriate models, 203-204; arguments for HIV that ignore facts, 199-203; challenges against the HIV hypothesis, 219-253; examples of how HIV hypothesis fails to predict AIDS, 416-417, 419-431; HIV-AIDS

correlation in the U.S. since 1981 (fig.), 194; Koch's postulates and, 174-186; latent periods, 181; mosquito transmission of, 169; multiple sclerosis and, 83; naming of, 165; as new cause of old diseases, 3-4; nonprogressors, 425-426; patent for antibody test, 159; scientists' rejection of Koch's postulates, 188-190; subacute sclerosing panencephalitis and, 82; survival of the HIV hypothesis, 167-168; transmission of, 179-180

Human papilloma virus, 107; as cervical cancer cause, 110-111; herpes simplex virus-2 as cofactor, 112-113; incidence, 176; papilloma virus test, 113

Human T-cell Leukemia Virus, 125-127. *See also* Human T-cell Lymphotropic Virus

Human T-cell Leukemia Virus-I, 28, 127; as cause of AIDS, 156, 157, 219

Human T-cell Leukemia Virus-II, 127, 157

Human T-cell Leukemia Virus-III, 157, 223

Human T-cell Lymphotropic Virus, 156

Hygienic Laboratory, 70. *See also* National Institutes of Health

Hyperplasias: cervical cancer and, 112

Hypothetical flu epidemic compared to long-established microbes (fig.), 192

"Ice." *See* Amphetamines

ICL. *See* Idiopathic CD4-lymphocytopenia

Idiopathic CD4-lymphocytopenia, 189-190, 274, 385-386

Immigration and tourism restrictions, 173

Indiana Hemophilia Foundation, 288

Infants. *See also* Perinatal transmission: born to drug addicted mothers, 268-269, 278, 419; latency period, 215; low birth weight of AIDS babies, 424; mental retardation of AIDS babies, 424; psychomotor indices, 268-269; recovery of HIV-positive babies born to drug-addicted mothers, 429-430

Infectious diseases. *See specific diseases by name*

Inoue, Shigeyuki, 23-25

Institute of Medicine, 364

Interferon, 220, 224, 226-227

International AIDS Conferences, 236, 245, 274, 355, 381, 435-437

International Archives of Allergy and Immunology, 440

International Cancer Conferences, 118

Intravenous drug use, 262, 279-280 416, 423-424, 426-427

Italy: censorship of Duesberg, 392, 389

Iwanowski, Dmitri, 69

Jackson, Frederick, 39-40

Jaffe, Harold, 200, 205, 206, 207, 383

JAMA. *See Journal of the American Medical Association*

Japan. *See also* Hirohito, Emperor; SMON syndrome: Human T-cell leukemia virus on island of Kyushu, 125; International AIDS Conference, 436-438; Ministry of Health and Welfare, 20-21

Japanese Ministry of Health and Welfare, 14, 15

Japanese Society of Internal Medicine, 13

Japanese virus, 160

Jenner, Edward, 68

Johns Hopkins University, 134

Johnson, Earvin "Magic," 340-341

Joswig, Christian, 381
Journal of AIDS Research, 198
Journal of NIH Research, 67
Journal of Psychoactive Drugs, 277
Journal of the American Medical Association, 128, 136, 213, 223

Kaposi, Moritz, 273
Kaposi's sarcoma, 3, 148, 175; African patient, 256; claim of "new virus," 274; diagnoses of, 215; direct link to nitrites, 272-273; herpes virus as cause, 190; news media and, 382-384; normal immune systems and, 186-187; removal from list of HIV diseases, 463; as risk-group-specific AIDS disease, 422-423
Kaposi's Sarcoma and Opportunistic Infections task force, 148, 224, 286
Karlsberg, Florence, 390-391
Karpas, Abraham, 159
Kessler, David, 325
Klebs, Edwin, 33
Koch, Robert, 9, 33-37, 39, 44, 89
Koch-Pasteur model, 9
Koch's postulates: Burkitt's lymphoma and, 101; definition of, 35; Gallo's rationale for bypassing, 199-201; hepatitis C and, 84-85; HIV and, 174-186; Legionnaires' disease and, 55, 58; mycoplasma, 233-234; pellagra and, 46; syphilis and, 53
Koehnlein, Claus, 331-332
Kono, Reisaku: SMON nonvirological research, 11-14, 16-18, 22-23, 24, 28, 29
Konotey-Ahulu, Felix, 290, 293-294, 295, 296
Koop, C. Everett, 100, 169
Koprowski, Hilary, 80-83, 165-166, 363
Koshland, Dan, Jr., 85
Kramer, Larry, 266, 379

Krieger, Lisa, 387
Krim, Mathilde, 224-225
Krynen, Evelyne, 291-292
Krynen, Philippe, 291-292
KSOI. *See* Kaposi's Sarcoma and Opportunistic Infections task force
Kuru disease, 16, 76-79
Kurume University, 14
Kusui, Kenzo, 12
Kyoto University, 14, 23

Labor unions: as recipients of CDC funds, 376
Lancet, 25, 42, 245, 246, 249, 290, 336, 338, 382, 401
Lange, Michael, 318
Langmuir, Alexander, 134-136, 458
Lapp, Ralph E., 62
Larry King Live program, 392-393
Lasker Prize, 160, 164
Latent-period concept. *See* Slow viruses
Lauritsen, John, 245, 267; challenge against the HIV hypothesis, 226, 227-231; observations on ACT UP, 380, 381; treatment of venereal disease with antibiotics, 282
LAV. *See* Lymphadenopathy-Associated Virus
Lavinder, Claude, 48
Lawyers. *See* Legal profession
"Leaders of People With Aids," 267
Lederle: development of polio vaccine, 80-81
Lee, Philip R., 339
Legal profession: interest in the HIV debate, 444
Legionella pneumophila, 54-57
Legionnaires' disease, 54-59, 143, 144-145
Leonard, Mike, 339-340
Lerner, W.D., 276-277
Leukemia. *See also* Adult T-cell Leukemia; Human T-cell Leukemia Virus: EIS investigation of clusters of leukemia cases, 140-

141; Gallo as presumed discoverer of leukemia virus, 160; hunt for human viruses, 99; mouse leukemia virus, 91-95, 98, 168; retroviruses in leukemia patients, 124
Levy, Jay, 160, 322
Lifestyle hypothesis. *See* Drug-AIDS hypothesis
Lilly, Frank, 389
Lind, James, 38
Lister, Joseph, 32, 40
Liver cancer: and hepatitis B virus hypothesis, 113-116
Liver disease. *See* Hepatitis
Lo, Shyh-Ching, 233-234, 243
Lombardo, Raphael, 341-348
London University, 39
Long, John D., 48-49
Long-term survivors of HIV. *See* Nonprogressors
Los Angeles Times, 106-107, 395
Lwoff, André, 97-98, 165, 415
Lymphadenopathy-associated virus, 157, 160, 161
Lymphoma, 4, 99, 327

MacArthur Prize, 245-246
MacNeil-Lehrer News Hour, 355-356, 392
Maddox, John, 7, 188, 332-333, 401, 402, 405, 406
Maekawa, Magojiro, 14
MAIDS. *See* Mouse AIDS
Malaria, 293
Malaria Control in War Areas, 133, 135
Malnutrition, 37-44, 283-284. *See also* Diet
Manhattan Project, 62, 63, 73
Mann, Jonathan, 135
Marine Hospital Service, 69-70
Massachusetts Institute of Technology, 118, 362, 406
Max Planck Institute, 240
McClintock, Barbara, 237-239

McCormick, Douglas, 250
McDonald, John, 440-441
McKenna, Joan, 266-267, 283
MCWA. *See* Malaria Control in War Areas
Measles virus: multiple sclerosis and, 82; subacute sclerosing panencephalitis and, 82
Media censorship, 382, 388-396. *See also* American press; Communications media
Medical establishment: consensus through commercialization, 454-456; enforced consensus through peer review, 452-454; fear of disease, 456-463
Men's Style magazine, 444-445
Mercury: as syphilis treatment, 53
Merson, Michael, 135
Methamphetamine, 281-282
Miami Herald, 388
Middlesex Hospital (London), 17, 100
Mie University, 11
Milstein, Cesar, 165-166
Minneapolis Children's Hospital, 306-307
Minority organizations: as recipients of CDC funds, 376
MIT. *See* Massachusetts Institute of Technology
Modern Medicine, 40
Monoclonal antibodies, 165
Mononucleosis, 100. *See also* Epstein-Barr virus
Montagnier, Luc: cofactor hypothesis of AIDS, 241-243; LAV hypothesis of AIDS, 156-157, 160-161, 163-164, 201-203
Morbidity and Mortality Weekly Report, 148
Mortality rates: AIDS, 261-262; HIV, 213-214
Mosquito transmission of HIV, 169
Mouse AIDS, 168
Mouse leukemia virus, 91-95, 98, 168

Mullis, Kary, 180, 245, 251-252, 393
Multifactorial hypothesis of AIDS, 220-225, 249
Multiple sclerosis: virus hunters and, 82-83
Mumps virus: multiple sclerosis and, 82-83
Murphy, Michael, 444-445
Murphy, Rep. John, 56-57, 144
Mycoplasma, 17, 109, 233-234, 241, 242-243
Mycoplasma incognitus, 233, 243
The Myth of Heterosexual AIDS, 213

Nagel, Lindsey, 299-306
NAPWA. *See* National Association for People With AIDS
Nasopharyngeal carcinoma: Epstein-Barr virus and, 102
National Academy of Engineering, 364
National Academy of Sciences, 65, 165, 229, 235, 365, 372
National Association for People With AIDS, 376
National Cancer Act, 104
National Cancer Conferences, 95
National Cancer Institute: Acute Leukemia Task Force establishment, 103; AIDS task force establishment, 152; budget, 103, 104; EIS investigation support, 141; Laboratory of Tumor Cell Biology, 124; lymphoma as complication of AZT, 327; revival of tumor virology, 92-93; War on Cancer and, 127-128
National Commission on AIDS, 171, 277, 372
National Conference on Pellagra, 48, 50
National Foundation for Infantile Paralysis, 70
National Hemophilia Foundation, 375

National Institute of Allergy and Infectious Diseases, 105, 152; Fauci as director, 189-190, 225, 325-326; project to study AZT as a preventive drug, 328; trials of AZT on pregnant mothers, 336-339
National Institute on Drug Abuse, 8, 261, 275, 402, 403-404, 441
National Institutes of Health, 8, 16, 28; budget, 67, 72-73, 104; conference on slow and unconventional viruses, 81; dominance of cancer virology, 103; mobilization of the scientific community, 132, 133; new guidelines for AZT use, 331; Office of Research Integrity, 161, 166, 167; opportunity for a new epidemic in 1980, 146; Outstanding Investigator Grant, 398-400; posting of "HIV Safety Notice," 336; promotion of Gajdusek to head Laboratory of Central Nervous System Studies, 79; renamed from Hygienic Laboratory, 70
National Panel of Consultants on the Conquest of Cancer, 104
National Science Foundation, 63
Nature, 7, 17, 83, 110, 118, 161, 165, 198, 240, 245, 331-333, 400, 401-402, 443
Naturwissenschaften, 240
Navajo flu, 138
NCI. *See* National Cancer Institute
Neoplasia, 112
Neurosyphilis, 53-54, 232
New England Journal of Medicine, 147, 182, 198, 230-231, 245, 277, 315-316, 385-386
New Guinea, 16, 76, 79
New Scientist magazine, 160
New York Native, 229, 230, 231, 329-330, 339
New York Safe Sex Committee, 228
New York Times, 51, 52, 136, 144,

Index ■ 717

158, 325-326, 334-335, 353, 387, 394
Newsday, 322
Newsweek, 152, 385, 394
NFIP. *See* National Foundation for Infantile Paralysis
NIAID. *See* National Institute of Allergy and Infectious Diseases
NIDA. *See* National Institute on Drug Abuse
Nightline, 393, 442
NIH. *See* National Institutes of Health
Niigata University, 19
Ninth Annual Congress of Immunology, 438
Nitrite inhalants, 402. *See also* Alkylated nitrites; failure to study hazards of long-term use, 412-413; increased use of, 419-420, 421; nitrate inhalant-AIDS link conference, 441; studies of, 267-268; toxicity of, 270-272, 274-275
Nixon, President Richard, 104-106
Nobel Prize, 69, 71, 78-79, 95, 118, 165, 180, 239, 262
Nonprogressors, 425-426
Null, Gary, 357
Nussbaum, Bruce, 312, 314, 317

Occupational Safety and Health Administration, 351
OIG. *See* Outstanding Investigator Grant
Oncogenic retroviruses, 119-120
Opportunistic infections, 139. *See also specific infections by name*
Oral contraceptives, 112
Orphan viruses, 14, 74
Ortleb, Charles, 229, 230, 245, 462
OSHA. *See* Occupational Safety and Health Administration
Outstanding Investigator Grant, 398-400

Pan Data Systems Inc., 162
Pantox, 244

Papadopulos-Eleopolus, Elini, 252
Papilloma virus. *See* Human papilloma virus
Parade magazine, 339
Paris Academy of Medicine, 39
Parkinson's disease, 16
Parkman, Paul, 323-324
Passenger viruses, 188-190
Pasteur, Louis, 32, 40, 68-69
Pasteur Institute, 97, 123, 157, 161, 164, 242
Paul, William, 437
Paul Ehrlich Award, 363
PCR. *See* Polymerase Chain Reaction
Pediatric AIDS, 419, 421. *See also* Infants
Pekelharing, Dr., 42-43
Pellagra plague: comparison to AIDS epidemic, 49-50; as most devastating vitamin deficiency epidemic, 44-52
Penhoet, Edward, 85
Penicillin: as syphilis treatment, 53
Penile cancer, 112
Pennsylvania: Legionnaires' disease incident, 54-55
People magazine, 354-355
People With AIDS Coalition, 318
Perinatal transmission: HIV and, 179, 185-186, 215
Petit, Charles, 438
Phantom viruses, 83-87
Philadelphia Gay News, 228
PHS. *See* Public Health Service
Physician's Desk Reference, 300, 313
Pinching, Anthony, 355-356
Pneumocystis carinii, 147, 193, 221, 255, 256, 283, 309, 331
Pneumonia, 4, 35, 274-275. *See also Legionella pneumophila*
Poison by Prescription: The AZT Story, 231
Policy Review, 234
Poliomyelitis: epidemic, 4; as model

718 ■ Index

for failed wars on cancer and AIDS, 70-73
Poliovirus research, 11, 23, 94
Polymerase Chain Reaction, 180, 251-252
Polyoma virus, 93, 98, 105
Popovic, Mikulas, 161, 162
"Poppers." *See* Nitrite inhalants
Pregnant women: delay of pregnancy recommendations, 368-369; trials of AZT on pregnant mothers, 336-339
Prescription drugs. *See specific drugs by name*
Presidential Commission on the HIV Epidemic, 372
Press. *See* American press; Homosexual press; Media censorship; *individual publications by name*
Preston, Richard, 362
Proceedings of the National Academy of Sciences, 198, 239, 385, 397-398
Professional literature. *See also specific publications*: censorship in, 396-406
Project AIDS International, 444
Project Inform, 318-319, 322, 333, 378, 379
Protein deficiency, 283
Pseudomonas bacteria, 19
Ptomaine poisoning, 35, 40
Public Health Action in Emergencies Caused by Epidemics, 371
Public Health Service, 70, 368; creation of Communicable Disease Center, 134; establishment of Malaria Control in War Areas, 133-134; pellagra research, 48, 50-51; role in blaming AIDS on a virus, 132-137
Purtilo, David, 223-224
Pyrard, François, 37-38

Raabe, Otto, 402, 403, 413
Rabies virus, 81

Rauscher, Frank, 106
Reagan, President Ronald, 165
Reappraising AIDS, 439
Recreational drug use, 148, 149, 283, 414, 415. *See also* Drug use; Intravenous drug use; Risk behavior; *specific drugs by name*; AZT and, 428-429; types of drugs used (table), 418
Reed, Walter, 9
Research and development funding in the United States, 63-64, 67
Research in Immunology, 198, 242
Research in Virology, 242
Rethinking AIDS, 246-247
Retrovirus-AIDS hypothesis, 155-156
Retroviruses, 116-127; oncogenic retroviruses, 119-120; reverse transcriptase, 118, 308, 312-313
Reverse transcriptase, 118, 308, 312-313
Review of Medical Microbiology, 25
Ribonucleic acid: hepatitis C and, 84; reverse transcriptase and, 308
Risk behavior, 5, 189, 191, 217, 222, 416. *See also* Drug use; Fast-track homosexuals
RNA. *See* Ribonucleic acid
RNA tumor viruses. *See* Retroviruses
Rockefeller, David, 407-408
Rockefeller Institute, 80, 91, 92
Rockefeller University, 407
Rogers, David, 372
Rolling Stone, 395
Roosevelt, President Franklin, 70
Root-Bernstein, Robert, 245-248
Rous, Peyton, 91, 92, 105-106, 112
Rous sarcoma virus, 91, 93, 117, 119, 363. *See also* Retroviruses
Royal Postgraduate Medical School, 251
Royal Society of London, 40
RSV. *See* Rous sarcoma virus

Rubin, Harry, 69, 116-117, 248-249, 398

Sabin, Albert, 4, 81, 95, 234-236
Salahuddin, Syed Zaki, 162
Salivary gland tumors, 93
Salk, Jonas, 4, 71, 73, 81, 165, 310, 436
Salk vaccine, 72, 73, 139
Sambon, Louis, 49
San Francisco Chronicle, 138, 395, 401, 438
San Francisco Examiner, 273-274, 387
Sarin, Prem, 162-163
Schmidt, Peter, 381
Schmitz, Teresa, 460-462
Science, 65, 66, 79, 85, 156, 190, 197-198, 241, 245, 249-250, 271, 382-383, 436, 441
Science & Government Report, 104
Scientific journals. *See specific journals by name*
The Scientist, 441
Scott, Frederick, 251
Scurvy, 37-41; syphilis and, 40-41
Semen, 223
Semmelweis, Ignaz, 32
Septra, 300-301, 302
Sex in America: A Definitive Survey, 277
Sexual transmission, 179-180, 185, 189. *See also* Fast track homosexuals; Risk behavior
Shalala, Donna, 413-414, 446, 449, 451
Shandera, Wayne, 147
Shannon, James: as director of NIH, 72-73, 133, 458-459; launching of Virus-Cancer Program, 308-309, 459; leadership of National Cancer Institute, 103; retirement of, 104
Sharrar, Robert, 143-144
Shilts, Randy, 147, 154, 155, 358-359
Shingu, Masahisa, 14

Shope, Robert, 139
Shorter, Edward, 131-132
Simian immunodeficiency virus, 168, 203
Simian Virus 40, 94, 98
Simon, Candice, 306-307
SIV. *See* Simian immunodeficiency virus
Slim disease, 215, 216, 293
Slow viruses, 96-103. *See also* Subacute sclerosing panencephalitis; cervical cancer and, 111; concept, 74-75; connection between HIV and AIDS, 75-76; definition, 16; hypothesis for HIV, 4-5, 178, 203, 214; kuru syndrome, 76-79
SMON syndrome, 11-29, 80; AIDS as a possible encore to SMON disaster, 27-29; drug connection and, 18-23; formal name, 13-14; symptoms, 12-13; "Toda disease," 13; virus hunt revived, 23-27
Sodium nitrite, 270-271
Sonnabend, Joseph: challenges against the HIV hypothesis, 220-227; questions about AZT, 316
South Carolina: pellagra research, 48, 49
Spencer, David, 141-142, 145
Spiegelman, Sol, 123-124
Spin, 291, 333
SSPE. *See* Subacute sclerosing panencephalitis
Stanford University, 274
Stanley, Wendell M., 69, 71, 95-96, 118
Steadman, Lyle, 79
Stetten, Hans Dewitt, 89-90
Stewart, C.P., 41
Stewart, Gordon, 249
Stewart, Sarah, 93
Stewart, William, 135
Strategies for Survival, 378
Subacute Myelo-Optico-Neuropathy. *See* SMON syndrome

Subacute sclerosing panencephalitis, 81-83
Sulfa drugs, 283. *See also specific drugs*
Sullivan, Louis, 399-400
Sunday Telegraph (London), 443
Sunday Times of London, 250, 442-443
"Superbugs" crisis, 138-139
Surgeon General, 133, 135, 390. *See also specific Surgeons General*
Surviving AIDS, 339
SV40. *See* Simian Virus 40
Sweden, 21
Swine flu epidemic, 141-143, 145
Switzerland: problems in trying to control AIDS in Zurich, 280-281
Syphilis, 35, 52-54. *See also* Neurosyphilis; as alternative hypothesis for HIV, 232-233; as cervical cancer cause, 109

TAG. *See* Treatment Action Group
Takaki, Kanehiro, 42
Takasaki, Hiroshi, 11
Teachers' unions: as recipients of CDC funds, 376
Temin, Howard, 165, 197, 402; arguments for HIV based on antibody correlations, 207; arguments for HIV based on evasion, 204-206; enzyme isolation, 117-118
Terry Higgins Trust, 393-394
Tetanus, 35
Tetracycline, 282-283
Thailand: AIDS in the heterosexual population, 289; urban prostitutes, 295
Third World countries. *See also specific countries by name*: AIDS in the heterosexual population, 289-297; contagious diseases, 31; diseases confused with AIDS, 293; Ebola Fever, 154; hepatitis, 83; liver cancer, 114; polio, 70; protein-and vitamin-deficient diets, 283-284
Thomas, Charles, Jr., 243-245
Thomas, Jr., Charles, 246, 250
Thompson-McFadden Commission, 50, 51-52
Thrombocytopenia, 283
Tien, Chang-Lin, 404
Time magazine, 138, 152, 395
Tobacco mosaic virus, 69
Toda disease, 13
Tony Brown's Journal, 442
Totsuka, Etsuro, 13, 20, 24, 26
Treatment Action Group, 380
Trichomonas protozoa: as cervical cancer cause, 109
Tsubaki, Tadao, 19-20
Tuberculosis, 9, 34-35, 39; in Africa, 293; among cocaine and crack addicts, 277-278; CDC definition, 212; HIV-antibody and, 4; intravenous drug addicts and, 215
Tufts University, 406
Turck, Marvin, 136
The Twin Epidemics of Substance Abuse and HIV, 277

United States Conference of Mayors, 374
University of Alabama, 276
University of California, Los Angeles, 146
University of California at Berkeley, 85, 103, 249, 399, 404; Department of Molecular and Cell Biology, 86; Virus Lab, 69, 116, 118
University of California at Davis, 402, 413
University of California at Los Angeles, 27
University of California at San Francisco, 268, 322
University of Edinburgh, 41
University of Miami, 249, 315, 349
University of Minnesota, 302

University of Prague, 33
University of San Francisco, 273-274
University of Vienna, 32
University of Washington, 136
University of Wisconsin, 117
U.S. Army, 185
U.S. House of Representatives: Subcommittee on Oversight and Investigations, 167
USCM. *See* United States Conference of Mayors

Vaccination, 71
van Leeuwenhoek, Antony, 31-32, 37
Vancouver study, 267-268
Varmus, Harold, 166-167, 363
Veterans Administration Hospital, Bronx, New York, 93
Villemin, Jean-Antoine, 39
Virions, 175
Virus-Cancer Program, 103, 105, 106, 124, 140, 308, 361, 459
Virus hunting, 61-87; from early virology to polio, 68-73; profitable virus-hunting opportunities, 83-87; slow viruses, 74-83; tobacco mosaic virus, 69
Virus Hunting—AIDS, Cancer, and the Human Retrovirus: A Story of Scientific Discovery, 199
Viruses. *See specific viruses by name*
Vitamin deficiency epidemic, 37-52
Vogt, Peter, 363
Volberding, Paul, 327, 328-329
Voyage des Krynen en Tanzanie, 291

Wall Street Journal, 223, 266
War on AIDS: AIDS as an encore of the SMON disaster, 27-29; blueprint for, 364-372; committee members, 364-365; committee recommendations, 365-372; *Confronting AIDS: Directions for Public Health, Health Care, and Research* as blueprint for war, 365; failure of, 3-29; funding for, 364; media censorship, 388-396; parallel efforts abroad, 371-372; project sponsors, 364; public financing, 367; public health measures, 367-371; question of infectious AIDS as the right choice, 7-10; research agenda, 365-367; as rigged debate, 372-387; SMON fiasco, 11-27; strategy, 364
War on cancer: chemotherapy treatments for cancer, 103; creation of, 73; effort to find cancer-causing chemicals in the environment, 103; evaluation of, 127-129; fading of, 153; first known tumor virus, 91; hepatitis B virus and liver cancer hypothesis, 113-116; President Nixon and, 104-106; retroviruses, 116-127; slow viruses and, 96-103; virologists and, 89-129; Virus-Cancer Program, 103; viruses to cause cervical cancer, 106-113
Warner, Jim, 394
Washington Post, 357, 394-395
Watson, James, 105, 238, 239-240
Waxman, Rep. Henry, 151
Wecht, Cyril H., 145
Weinberg, Robert, 363
Weiss, Robin, 164, 200, 205, 206, 207
Weller, Ian, 330
White, Ryan, 287-288
Whitehead, Edwin C. "Jack," 362
Whitehead Institute, 362
WHO. *See* World Health Organization
Williams, Robert, 44
Wilson, Hank, 228, 229, 267
Wistar Institute, 81, 165
Witkin, Steven, 223, 224
Wong-Staal, Flossie, 399

World Health Organization, 81, 115, 135, 154, 177, 214, 371

Yellow fever virus, 9

Zagury, Daniel, 163
Zidovudine, 4, 28; AIDS cured by discontinuation of AZT, 428-431; Alison Gertz case, 354-356; Arthur Ashe case, 356-358; articles on, 231; AZT cover-up, 314-324; AZT product label (fig.), 324; breakdown of trust in medical authority, 339-348; Candice Simon case, 306-307; criticism of, 226; death and resurrection of, 307-314; dideoxyinosine, 325-326; French study, 326-327; Kimberly Bergalis case, 348-354; Lindsey Nagel case, 301-307; Phase I trials, 307-314; Phase II trials, 314-324; prevention of HIV infection as last stand of AZT lobby, 335-339; side effects from, 4, 424; stories from believers in AZT, 348-359
Zinc deficiency, 283
zur Hausen, Harald, 110-111